Continued on back

Understanding Robust
and Exploratory
Data Analysis

Understanding Robust and Exploratory Data Analysis

Edited by

DAVID C. HOAGLIN
Harvard University and Abt Associates Inc.

FREDERICK MOSTELLER
Harvard University

JOHN W. TUKEY
Princeton University and Bell Laboratories

John Wiley & Sons, Inc.
New York · Chichester · Brisbane · Toronto · Singapore

Library of Congress Cataloging in Publication Data:
Main entry under title:

Understanding robust and exploratory data anlaysis.

(Wiley series in probability and mathematical
statistics. Applied probability and statistics, ISSN
0271-6356)
   Bibliography: p.
   Includes index.
   1. Mathematical statistics.   I. Hoaglin, David Caster, 1944–   .
II. Mosteller, Frederick, 1916–   .   III. Tukey,
John Wilder, 1915–   .   IV. Series.

QA276.U5 1982   519.5   82-8528
ISBN 0-471-09777-2   AACR2

Printed in the United States of America

10 9 8 7 6 5 4

To the memory of

William G. Cochran

Preface

In current statistical practice, both exploratory data analysis and robust and resistant methods have gained important roles. To apply such methods most effectively, the user needs to understand why they are needed and how they work—and can be helped by some insight into how they were devised. This book provides conceptual, logical, and, sometimes, mathematical support for the simpler of these new techniques.

The techniques of exploratory data analysis, particularly as embodied in the book of that title by Tukey (Addison–Wesley, 1977), may seem to have sprung from nowhere and to be supported only by anecdote. The attitudes underlying exploration, though long used by skilled data analysts, have been little exposed to public view. Many of the purposes parallel those of more conventional techniques. Indeed, we can express some of the justifications for particular techniques by using the concepts of classical statistical theory. This book explains and illustrates such connections.

The robust and resistant techniques that we discuss have considerable support in the statistical research literature, both at a highly abstract mathematical level and in extensive Monte Carlo studies. The book provides the basis for an adequate understanding of these techniques using examples and a much reduced level of mathematical sophistication.

By studying this book the user will become more effective in handling robust and exploratory techniques, the student better able to understand them, and the teacher better able to explain them.

Robust and resistant techniques and those of exploratory data analysis have arisen mainly under the guidance of experience, skilled insight, empirical studies of performance, and even analogy with classical techniques. Today these techniques, in part because of this diversity of guidance, do not seem to follow naturally from any unifying structure. Nevertheless, the connections with classical theory do enable us to explain many of the grounds for choosing techniques.

Classical theory emphasizes large-sample notions of consistency, asymptotic variance, and asymptotic relative efficiency. Knowing the behavior of a statistic as the sample becomes large has some utility when, as so frequently happens, large-sample behavior is simpler than small-sample behavior. Above all, we need to recognize that data sets are usually small and that their behavior often lacks the simplicity of large samples. Useful examination of a technique will, therefore, often require new small-sample studies of its performance. Several of these studies, whose results we include, have formed part of the research leading to this volume.

Large-sample considerations can provide a unifying structure for some robust and resistant techniques. The strongest unifying theme underlying exploratory data analysis is expressed in "Look at the data and think about what you are doing." Proceeding largely from these bases, we provide a broad overview of the simpler aspects of data analysis, emphasizing exploratory and robust techniques.

Our intent is that each chapter be reasonably self-contained except for a few generally applicable techniques from the early chapters. Thus, we explain each method in conjunction with an example or two. We continue to use these examples when we describe variations on the technique, explain connections with other techniques (both classical and exploratory), and present results on performance.

Examples are generally small, and almost all rely on real data. They help introduce the reader to techniques and illustrate why and when one method is preferable to another. Through them we give some empirical evidence for the efficacy of each technique in a concrete application.

A brief collection of exercises at the end of each chapter enables the reader to participate more directly in applying the techniques to other sets of data, establishing their properties, or extending them to new situations.

Although our presentation does not generally involve a great deal of mathematics or theoretical argument, we do use them where they seem appropriate. Sometimes, when more formal approaches are unrewarding, we have recourse to numerical simulation. The mathematical arguments we employ are of three types: (a) a proof that a technique meets some desirable objective, (b) an argument that a certain property of a given technique is valuable, and (c) mathematical analogy to extend a technique.

Books in the mathematical sciences may be pitched at various levels and written using different plans. One plan assumes that readers have a particular level of mathematical background and then consistently maintains this level. The advantages are clear—reader and author have a definite contract. Against it, note that some not so well prepared readers who might benefit from parts of the work can be frozen out because they cannot meet the level chosen. Some almost accessible ideas may then have to be held back for reasons only of level. Another plan uses different levels in different parts of a work, with the intention of keeping as many readers as possible in touch with each part.

This latter plan, which we have adopted, requires tolerance on the part of the readers. The well prepared need to appreciate that they are not being talked down to, whereas the less well prepared must be willing to skip along when too much background is required. Such a plan can produce seemingly inconsistent writing, such as defining a factorial and yet seeming to assume that the same reader finds the gamma function an old friend. The approach we take also means that rigor, when available, may sometimes be deliberately sacrificed in order to communicate the main idea to more readers.

The mathematical prerequisite for the reader is not high for most chapters. At the same time, mathematical sophistication is matched to the requirement for explaining each technique, and so the level rises especially in Chapters 8 and 11. These we have marked by a star (*) before the chapter number in the table of contents. The reader may omit the starred chapters at first reading with no loss of continuity.

Many of the techniques we discuss also appear in *Exploratory Data Analysis* by John W. Tukey and *Data Analysis and Regression* by Frederick Mosteller and John W. Tukey (Addison–Wesley, 1977). In the present book, the emphasis is more on the rationale and development of the methods, and less on illustrating their use. Our exposition is self-contained, but a reader wishing to see more examples and different applications may find profit in referring to one or both of the books just mentioned or to *Applications, Basics, and Computing of Exploratory Data Analysis* by Paul F. Velleman and David C. Hoaglin (Duxbury Press, 1981).

We are preparing a further volume to provide a similar rationale for additional techniques of exploratory data analysis and other robust and resistant methods.

DAVID C. HOAGLIN
FREDERICK MOSTELLER
JOHN W. TUKEY

Saconesset Hills, Massachusetts
July 1982

Acknowledgments

This book grew out of a working group on exploratory data analysis in the Department of Statistics at Harvard University that began in the spring of 1977 and has involved students, faculty, academic visitors, and others. Those who have participated (at one time or another) are Nancy Romanowicz Cook, John D. Emerson, Miriam Gasko, John P. Gilbert (deceased), Katherine Godfrey, Colin Goodall, David C. Hoaglin, Boris Iglewicz, Lois Kellerman, Guoying Li, Lillian Lin, Frederick Mosteller, Anita Parunak, James L. Rosenberger, Andrew Siegel, Keith A. Soper, Michael A. Stoto, Judith Strenio, John W. Tukey, Paul F. Velleman, George Wong, and Cleo Youtz—ten of whom have spent time at Princeton University. Through their sharing of ideas and friendly criticism, all contributed to the development of the material.

In addition to funding from the National Science Foundation (through grants SES 75-15702 and SES 8023644), these activities received partial support from the Middlebury College Faculty Leave Program, the National Institutes of Health (grant CA-23415), the National Cancer Institute (grant T32 CA09337-01), and for John W. Tukey's activities at Princeton, the U.S. Army Research Office (Durham).

Persi Diaconis, Peter J. Huber, David A. Lax, Lincoln E. Moses, Paul F. Velleman, and especially Erich L. Lehmann generously provided comments on various draft chapters.

Katherine Bell Krystinik and Stephan Morgenthaler assembled numerical results at Princeton University related to Chapter 11. Karen Kafadar made available unpublished results from her work on biweight estimators. Virginia C. Klema and Susan Vinal provided information for checking Table 11-3. Jorge Martinez carried out the simulation work for Chapter 12.

Holly Grano, Marjorie Olson, and Cleo Youtz prepared the manuscript with great care and efficiency.

D. C. H.
F. M.
J. W. T.

Contents

Introduction

The classical statistical techniques are designed to be the best possible when stringent assumptions apply. However, experience and further research have forced us to recognize that classical techniques can behave badly when the practical situation departs from the ideal described by such assumptions. The more recently developed robust and exploratory methods are broadening the effectiveness of statistical analyses.

The techniques of exploratory data analysis help us to cope with a set of data in a fairly informal way, guiding us toward structure relatively quickly and easily. Good statistical practitioners have always looked in detail at the data before producing summary statistics and tests of hypotheses. Exploratory data analysis provides us with an extensive repertoire of methods for the detailed study of a set of data. The emphasis is on flexible probing of the data, often before comparing them to any probabilistic model.

Robust and resistant methods, instead of being the best possible in a narrowly defined situation, are "best" compromises for a broad range of situations and, surprisingly often, are close to "best" for each situation alone. Whereas distribution-free methods treat all distributions equally, robust and resistant methods discriminate between those that are more and less plausible.

Broad Phases of Data Analysis

One description of the general steps and operations that make up practical data analysis identifies two broad phases: exploratory and confirmatory. Exploratory data analysis isolates patterns and features of the data and reveals these forcefully to the analyst. It often provides the first contact with the data, preceding any firm choice of models for either structural or stochastic components, and it also serves to uncover unexpected departures from familiar models. An important element of the exploratory approach is flexibility, both in tailoring the analysis to the structure of the data and in responding to patterns that successive steps of analysis uncover.

1

Confirmatory data analysis assesses the reproducibility of the observed patterns or effects. Its role is closer to that of traditional statistical inference in providing statements of significance and confidence; but the confirmatory phase often includes steps such as (a) incorporating information from an analysis of another, closely related body of data and (b) validating a result by collecting and analyzing new data.

In brief, exploratory data analysis emphasizes flexible searching for clues and evidence, whereas confirmatory data analysis stresses evaluating the available evidence.

A cycle of alternating uses of exploratory and confirmatory techniques, either on successive smaller bodies of data or on a single substantial one, is not uncommon and is often very desirable.

Four Themes

Throughout exploratory data analysis, four main themes appear and often combine. These are *resistance*, *residuals*, *re-expression*, and *revelation*.

Resistance provides insensitivity to localized misbehavior in data. A resistant method produces results that change only slightly when a small part of the data is replaced by new numbers, possibly very different from the original ones. Resistant methods pay much attention to the main body of the data and little to outliers. The median is a resistant statistic, whereas the sample mean is not. Attention to resistance reflects the understanding that "good" data seldom contain less than a few percent of gross errors or blunders, so that protection against the adverse effects of such errors should always be available.

In theoretical discussions, one seeks to limit the effect of any "small" change in the sample. In this sense, small changes include minor perturbations in all the data, drastic shifts in a small fraction of the data, and numerous possibilities between these two extremes. In a particular instance, we may need to be concerned with only some of the possible small changes. Thus we might speak of "resistance to wild values" or "resistance to rounding and grouping." Most commonly, because their presence can so easily produce serious distortion, we have wild values in mind when we discuss resistance. Also, it is generally easier to overcome the lesser difficulties that arise when an estimator, such as the sample median, is not resistant to rounding and grouping.

We distinguish between resistance and the related notion of robustness. Robustness generally implies insensitivity to departures from assumptions surrounding an underlying probabilistic model. (Some discussions regard resistance as one aspect of "qualitative robustness.")

In summarizing the location of a sample, the median is highly resistant. A number of exploratory techniques for more structured forms of data provide resistance because they are based on the median. In terms of efficiency, the median, for all its resistance, is not highly robust because other estimators achieve appreciably greater efficiency across a broader range of distributions. By contrast, the mean is both badly nonresistant and badly nonrobust

Residuals are what remain after a summary or fitted model has been subtracted out of the data according to the schematic equation

$$\text{residual} = \text{data} - \text{fit}.$$

For example, if the data are the pairs (x_i, y_i) and the fit is the line $\hat{y}_i = a + bx_i$, then the residuals are $r_i = y_i - \hat{y}_i$.

A key attitude of exploratory data analysis asserts that an analysis of a set of data is not complete without a careful examination of the residuals. This analysis can and should take advantage of the tendency of resistant analyses to provide a clear separation between dominant behavior and unusual behavior in the data. When the bulk of the data follows a consistent pattern, that pattern determines a resistant fit. The resistant residuals then contain any drastic departures from the pattern, as well as chance fluctuations. Unusual residuals call for a check on the details of how the corresponding observations were made and handled. As in more traditional practice, the residuals—properly analyzed and displayed—can warn of important systematic aspects of data behavior that may need attention, such as curvature, nonadditivity, and nonconstancy of variability.

Re-expression involves finding what scale (e.g., logarithmic or square root) would simplify the analysis of the data. Exploratory data analysis emphasizes the benefits of considering, at an early stage, whether the original scale of measurement for the data is satisfactory. If not, a re-expression into another scale may help to promote symmetry, constancy of variability, straightness of relationship, or additivity of effect, depending on the structure of the data. A view that the original scale of measurement has a preferred status may cause reluctance to consider re-expression. That view will often not stand examination. True, the physicist sometimes has a cogent theoretical basis for deciding whether to work with volts or (volts)^2. However, in circumstances where cogent theory does not guide the choice, the original scale of measurement does not have a similar claim to preferred status. Thus the response of an animal's liver to some treatment may be no more naturally reflected in w, the weight, than in $\log w$ or $\sqrt{w}$, at least until quantitative understanding has advanced.

Revelation through displays meets the analyst's need to see behavior—of data, of fits, of diagnostic measures, and of residuals—and thus to grasp the unexpected features as well as the familiar regularities. Emphasis on visual displays, including many new graphical techniques, has been a major contribution of exploratory data analysis.

Terminology

Readers who have become acquainted with exploratory data analysis (EDA) after studying traditional statistical methods may wonder why relatively few of the traditional technical terms carry over. Because this book often makes connections between EDA and existing background in statistics, the new words and the familiar ones frequently appear almost side by side. Although such passages may help to clarify how EDA terms relate to traditional ones, they do not always attempt to explain the need for new words. We now offer a few of the more general reasons for the EDA terminology.

First, EDA is more concerned with earlier stages in the overall process of working on data and with different operations and emphases. New techniques often require new technical terms. Stem-and-leaf displays, letter values, and boxplots all illustrate this fact.

Second, some EDA terms are related to, but not equivalent to, more traditional notions. Here the preferred approach is to avoid disturbing the definitions already embedded in the literature. For example, the "hinge" or "fourth" is not exactly a "quartile," and "batch" does not include the assumptions of independence and identical distribution usually associated with "sample."

Third, gains may come from avoiding misleading words. A primary example is "normal," as used in "normal distribution" and in "normal equations" (in least-squares regression), which conflict with this word's usage in normal parlance. Thus we frequently refer to the "Gaussian distribution," rather than the "normal distribution," to avoid any suggestion that this shape customarily underlies actual data.

Finally, some terms, although perhaps unfamiliar, are standard in a particular area of statistics whose results are related to EDA. Some examples in Chapters 9 through 12 come from the field of robustness, where a spurt of research in recent years has established new concepts and many valuable theoretical results.

Estimation

When we engage in a careful discussion of estimation, as in Chapters 9 through 12, we often need to distinguish between the procedure (which we would apply to any sample) and the numerical value (which we obtain by applying the procedure to a particular sample). We use "estimator" for the procedure and "estimate" for the value, and we have tried to maintain this usage throughout the book. On occasion we use "estimand" for the value that an estimator would produce if it were applied, conceptually, to an entire theoretical distribution. Often the estimand is a familiar parameter of the distribution, but we would also use the term for other quantitative descriptions of what an estimator seems to be estimating.

Sampling Situations

Some of the data sources assembled to challenge proposed robust estimators are technically not distributions. For example, sets of 20 observations in which exactly 19 always come from the standard Gaussian distribution and exactly one always comes from a Gaussian distribution with a larger variance are not samples from either of these two distributions or from a mixture distribution. When we must be careful about this distinction (as in Chapters 10 and 11), we refer to such a data source as a "situation."

Iteration

Resistant and robust techniques more often involve iteration than do classical ones. Thus instead of finding a solution in a single step, we often take an initial value and successively refine it, bringing it closer and closer to the final answer.

In this book the main examples are the three-group resistant line (Chapter 5), median polish (Chapter 6), and the M-estimators of location (Chapter 11). The resistant line procedure defines the slope of the fitted line in terms of the residuals in a way that requires calculating the residuals from a preliminary fit and then adjusting the preliminary slope if necessary. Similarly, in two-way tables, the process of adjusting a preliminary row effect may leave some column medians nonzero. Here, median polish makes alternate adjustments of row and column effects. In doing this, it seeks a two-way table of residuals whose rows and columns all have their medians equal to zero, although we conventionally stop after taking only a few steps toward this goal. In general, the calculations for an M-estimator involve solving, again usually approximately, a nonlinear equation in which the previous estimate enters through the definition of the corresponding residuals.

Often, a certain amount of iteration is to be expected as part of the price of resistance or robustness; the procedures that yield a fit or an estimate without iteration may not be adequately resistant or robust. Happily, the iterative adjustment procedures in this book are simple, and they seldom require many steps.

ADDITIONAL LITERATURE

Besag, J. (1981). "On resistant techniques and statistical analysis," *Biometrika*, **68**, 463–469.

Hampel, F. R. (1971). "A general qualitative definition of robustness," *Annals of Mathematical Statistics*, **42**, 1887–1896.

Tukey, J. W. (1972). "Data analysis, computation, and mathematics," *Quarterly of Applied Mathematics*, **30**, 51–65.

——— (1979). "Robust techniques for the user." In R. L. Launer and G. N. Wilkinson (Eds.), *Robustness in Statistics*. New York: Academic, pp. 103–106.

——— (1980). "We need both exploratory and confirmatory," *The American Statistician*, **34**, 23–25.

CHAPTER 1

Stem-and-Leaf Displays

John D. Emerson
Middlebury College

David C. Hoaglin
Harvard University and Abt Associates Inc.

The most common data structure is a batch of numbers. Even this simple structure of data may have characteristics not easily discerned by scanning or studying the numbers. The stem-and-leaf display enables us to organize the numbers graphically in a way that directs our attention to various features of the data. This basic but versatile exploratory technique is the most widely used; we call upon it in later chapters, for example, to compare batches and to examine residuals.

The stem-and-leaf display enables us to see the batch as a whole and to notice such features as:

How nearly symmetric the batch is.
How spread out the numbers are.
Whether a few values are far removed from the rest.
Whether there are concentrations of data.
Whether there are gaps in the data.

In exposing the analyst to these features of data, the stem-and-leaf display has much in common with its close relative, the histogram. By using the digits of the data values themselves instead of merely enclosing area, this display offers advantages in some situations. When we work by hand, it is easier to construct, and it takes a major step in sorting the data. Thus we can readily find the median and other summaries based on the ordered batch. It can help us to see the distribution of data values within each interval, as well as patterns in the data values. For example, we might discover that all values are multiples of 3 or that the person recording motor vehicle speeds reported them as multiples of 5 miles per hour. By preserving more early digits of the data values, the stem-and-leaf display also shortens the link back to the individual observation and any identifying information that accompanies it.

The stem-and-leaf display does not involve any elaborate theory. Rather, because the interaction between analyst and display is a personal one, we face considerations of taste and esthetics, particularly in choosing the number of intervals or the interval width. As often happens in exploratory data analysis, we aim to provide flexibility through a number of variations. We begin with the basic stem-and-leaf display and the steps in its construction, and then we add variations. For historical interest, we reproduce a close predecessor of the stem-and-leaf display. We also briefly pursue the connections with sorting to put the data in order. Choosing the number of lines (or the interval width) provides a point of contact with work (some of it more theoretical) on these choices for histograms.

1A. THE BASIC DISPLAY

The stem-and-leaf display (Tukey, 1970, 1972) provides a flexible and effective technique for starting to look at a batch or sample of data. The most significant digits of the data values themselves do most of the work of sorting the batch into numerical order and displaying it.

To explain the display and how one constructs it, we begin with an example. Royston and Abrams (1980) give the mean menstrual cycle length and the mean preovulatory basal body temperature for 21 healthy women who were using natural family planning. We reproduce these data in Table 1-1 and show one stem-and-leaf display for the cycle lengths in Figure 1-1.

If we follow the first data value, 22.9, we see that it appears in the display as 22 | 9. To construct this simplest form of stem-and-leaf display, we proceed as follows. First we choose a suitable pair of adjacent digits in the data—in the example, the ones digit and the tenths digit. Next we split each data value between these two digits:

data value		split		stem	and	leaf
		↓				
22.9	→	22	9	→	22	and	9

TABLE 1-1. **Mean menstrual cycle length and mean preovulatory basal body temperature (BBT) for 21 women.**

i	Cycle Length (days)	BBT (°C)
1	22.9	36.44
2	26.3	36.21
3	26.6	36.71
4	26.8	36.13
5	26.9	36.25
6	26.9	36.53
7	27.5	36.41
8	27.6	36.45
9	27.6	36.53
10	28.0	36.31
11	28.4	36.63
12	28.4	36.54
13	28.5	36.52
14	28.8	36.62
15	28.8	36.40
16	29.4	36.48
17	29.9	36.39
18	30.0	36.37
19	30.3	36.77
20	31.2	36.76
21	31.8	36.50

Source: J. P. Royston and R. M. Abrams (1980). "An objective method for detecting the shift in basal body temperature in women," *Biometrics*, **36**, 217–224 (data from Table 1, p. 221, used by permission).

Then we allocate a separate line in the display for each possible string of leading digits (the *stem*)—for Figure 1-1, the necessary lines (10 in all) run from 22 to 31. Finally we write down the first trailing digit (the *leaf*) of each data value on the line corresponding to its leading digits.

The finished display, Figure 1-2, includes a reminder that all data values are in units of .1 day, as well as a column of depths (which we will define shortly) to the left of the stems. When the raw data values have not been sorted, the initial display will not usually have its leaves in increasing order. As an option, the final display can then sort the leaves. This can happen

(unit = .1 day)

```
22 | 9
23 |
24 |
25 |
26 | 3 6 8 9 9
27 | 5 6 6
28 | 0 4 4 5 8 8
29 | 4 9
30 | 0 3
31 | 2 8
```

Figure 1-1. One stem-and-leaf display for mean menstrual cycle length.

automatically when a computer produces the display (Velleman and Hoaglin, 1981). In overall appearance the display resembles a histogram with an interval width of 1 day; the leaves add numerical detail, and in this instance they preserve all the information in the data.

Figures 1-1 and 1-2 show that two-thirds of the women have mean cycle length between 26.3 and 28.8 days. All but one of the other third have longer mean cycle lengths. One woman, who appears unusual, has a mean cycle length of only 22.9 days. We regard such an unusual data point as an outlier, and we might well treat it separately from the other data points. If we do, then a stem-and-leaf display for the remaining cycle lengths can give more detail by choosing the stems differently. We return to this example in Section 1B.

Depths of Data Values

A data value can be assigned a *rank* by counting in from each end of the ordered batch. For example, in Figure 1-2, 26.3 has rank 2 when counting up from 22.9 and rank 20 when counting down from 31.8. The *depth* of the data value is the smaller of these two ranks, 2 in the example. Because a number of summary values (such as the median and the quartiles or fourths, see Chapter 2) can easily be defined in terms of their depths, it is helpful to present a set of depths with the display. Except for one middle line, the number in the depth column is the maximum depth associated with data values on that line. Thus the depth of 29.4 is 6.

The "middle line" includes the median, and the depth column shows in parentheses the number of leaves on this line. In the example, this number is 6. (When the batch size is even *and* the median falls between lines, we do not need this special feature.) If the display has been prepared by hand,

```
Depths    (unit = .1 day)
   1      22 | 9
          23 |
          24 |
          25 |
   6      26 | 3 6 8 9 9
   9      27 | 5 6 6
  (6)     28 | 0 4 4 5 8 8
   6      29 | 4 9
   4      30 | 0 3
   2      31 | 2 8
```

Figure 1-2. A stem-and-leaf display with sorted leaves for mean menstrual cycle length ($n = 21$).

adding the count on the middle line and the depths on the two adjacent lines provides a simple check that no data values have been omitted. In the example, $9 + 6 + 6 = 21$, the total number of women in the study. (Incidentally, care is needed when determining the depths of numbers above the median: in the example one might tend, erroneously, to give 4, not 3, as the depth of 30.3.) Chapters 2 and 3 use the depths further in forming other summaries of data.

Organizing the Display

An effective choice of the number of lines in a stem-and-leaf display involves the number of data values in the batch and the range to be covered, as well as some judgment. To get started, we set a maximum number of lines:

$$L = [10 \times \log_{10} n], \qquad (1)$$

where n is the number of data values and $[x]$ is the largest integer not exceeding x. This rule seems to give values of L that produce effective displays over the range $20 \leq n \leq 300$, where most applications fall. For the example, which has $n = 21$, it gives

$$L = [10 \times \log_{10} 21]$$

$$= [10 \times 1.32]$$

$$= 13.$$

Thus unless we decide to treat the value 22.9 as special (as we often would), the 10-line stem-and-leaf display of Figures 1-1 and 1-2 seems satisfactory.

Values of n smaller than 20 may need special treatment. These are also more likely to arise when comparing several batches in parallel stem-and-leaf displays, a situation we would want to handle differently anyway (see Section 1B and Chapter 3). Batches of 300 or so are usually cumbersome in a stem-and-leaf display, but the rule should still cope with them reasonably well.

Using L as a rough limit on the number of lines in the display, we must now determine the interval of values corresponding to each line. The simple way to do this uses a power of 10 as the interval width. (We discuss other interval widths in Section 1B.) Thus we divide R, the range of the batch, by L and round the quotient up, if necessary, to the nearest power of 10. In the example, the range $R = 31.8 - 22.9 = 8.9$ and $L = 13$, so that $R/L = .68$. Rounding up to the nearest power of 10 gives the value 1 as the interval width. This is the value used in Figures 1-1 and 1-2, which actually require only 10 lines.

1B. SOME VARIATIONS

A segment of stems for the basic stem-and-leaf display might look like this:

$$
\begin{array}{c|}
0 \\
1 \\
2 \\
3 \\
\end{array}
$$

and each line may receive leaves 0 through 9. But sometimes this format is too crowded, having too many leaves per line. One effective response is to split lines and repeat each stem:

$$
\begin{array}{c|}
0* \\
0\cdot \\
1* \\
1\cdot \\
2* \\
2\cdot \\
\end{array}
$$

putting leaves 0 through 4 on the $*$ line and 5 through 9 on the $\cdot$ line. In such a display, the interval width is 5 times a power of 10.

EXAMPLE: HARDNESS OF ALUMINUM DIE CASTINGS

Shewhart (1931, p. 42) gives the hardness of 60 aluminum die castings. For the first 30 of these, the data values and two stem-and-leaf displays appear in Figure 1-3. We note that $L = [10 \times \log_{10} 30] = [14.77] = 14$, $R = 95.4 - 50.7 = 44.7$, and $R/L = 44.7/14 = 3.19$. If we rounded 3.19 up to the nearest power of 10, we would obtain 10 as the indicated interval width. This width is used for the basic stem-and-leaf display on the left in Figure 1-3. Because this display has relatively few lines, we split the lines and repeat each stem. This, of course, corresponds to rounding 3.19 up to 5. The result, which we feel is an improvement, appears at the right in Figure 1-3.

The stem-and-leaf displays of Figure 1-3 also illustrate how we handle digits that come after those that serve as the leaves in the display. Such low-order digits of data values are preferably truncated rather than rounded. This practice makes it easier to recover the original data value corresponding to a leaf in the display. Thus the values 55.3, 55.7, and 55.7 are all truncated at the decimal point and appear as 5s on the 5 · line in the display; the values at 55.7 are not rounded up to 56. To recover the three data values, we simply locate the three numbers in the raw data set whose first two digits are 55.

Sometimes the display is still too crowded with two lines per stem and too straggly with one line per stem at the next lower power of 10. To cure

Data: Hardness of aluminum die castings

53.0	70.2	84.3	55.3	78.5	63.5	71.4	53.4
82.5	67.3	69.5	73.0	55.7	85.8	95.4	51.1
74.4	54.1	77.8	52.4	69.1	53.5	64.3	82.7
55.7	70.5	87.5	50.7	72.3	59.5		

Displays:

($n = 30$) Depths	(unit = 1)		($n = 30$) Depths	(unit = 1)	
11	5	0 1 2 3 3 3 4 5 5 5 9	7	5 *	0 1 2 3 3 3 4
(5)	6	3 4 7 9 9	11	5 ·	5 5 5 9
14	7	0 0 1 2 3 4 7 8	13	6 *	3 4
6	8	2 2 4 5 7	(3)	6 ·	7 9 9
1	9	5	14	7 *	0 0 1 2 3 4
			8	7 ·	7 8
			6	8 *	2 2 4
			3	8 ·	5 7
				9 *	
			1	9 ·	5

Figure 1-3.   Splitting to get two lines per stem. *Source*:   W. A. Shewhart (1931). *Economic Control of Quality of Manufactured Product.* Princeton, NJ: D. Van Nostrand, Inc. [data from Table 3 (specimens 1 through 30), p. 42].

these troubles, we have a third form, five lines per stem:

$$
\begin{array}{r|}
0* \\
t \\
f \\
s \\
0\cdot
\end{array}
$$

with leaves 0 and 1 on the $*$ line, 2 (two) and 3 (three) on the **t** line, 4 (four) and 5 (five) on the **f** line, 6 (six) and 7 (seven) on the **s** line, and 8 and 9 on the $\cdot$ line. As a reminder in starting to place leaves, the three lettered lines contain leaves whose words begin with that letter. Here the interval width is 2 times a power of 10.

EXAMPLE: TUMOR PROGRESSION IN PATIENTS WITH GLIOBLASTOMA

Dinse (1982) gives the times until tumor progression for 172 patients with glioblastoma, a brain tumor. The times for 83 of the patients are *censored* times in that the tumors for these patients had not progressed at the time the study was concluded. These times are indicated by " $+$ " in the data portion of Figure 1-4, which also shows them in a stem-and-leaf display with 5 lines per stem. The display has a total of 19 lines, and each line has width 2 months. Note that for this example, $L = [10 \log_{10} 83] = 19$ lines, $R = 37 - 1 = 36$ months, and $R/L = 36/19 = 1.89$. When this quotient is rounded up to 1, 2, or 5 times the nearest power of 10, we obtain 2×10^0, or 2, as the interval width. This is consistent with the width adopted for the display.

In this display we immediately see the asymmetry of the main part of the data, the clump of values from 30 months to 37 months, and the single value at 25 months. The large number of small censoring times most likely indicates that these patients have not been participating in the study long enough to suffer a progression of their tumors. The practice of recording these times in whole months, so that times within the first month are recorded as 1, explains why there are no values at 0, and it may mean that progression times and censoring times are both rounded up. The times at 25 months and beyond do not fit into a regular pattern with the rest of the censoring times; they may represent an initial bulge in the rate of entry of patients into the study.

Another variation in the stem-and-leaf display accommodates data that include both positive and negative values. Generally, residuals (to which we

Data: Time to tumor progression (in months) for patients with glioblastoma
9 + , 3 + , 6 + , 6, 5, 34 + , 10, 22, 9, 2, 14 | , 3, 9, 6 | , 0, 0 | , 3, 3, 11,
4 + , 9, 5, 17 + , 9 + , 17, 13 + , 3, 5, 3 + , 14, 11 + , 3 + , 9 + , 13, 15 + , 3,
3 + , 4 + , 11, 3, 1 + , 9, 16, 14 + , 6, 2 + , 24, 22, 10, 34 + , 10, 4, 1, 3, 15 + ,
6 + , 28, 3 + , 4, 31 + , 6, 2, 9 + , 4 + , 13 + , 21, 8 + , 11 + , 37 + , 6, 1 + , 4 + , 15 + ,
7, 4, 3, 19, 2 + , 18 + , 9 + , 6, 9 + , 9, 10 + , 35 + , 23, 33 + , 16 + , 5 + , 34 + ,
13 + , 2, 12 + , 3 + , 10, 8, 3 + , 4 + , 1 + , 7, 3, 8, 9 + , 10 + , 10, 6 + , 10, 3, 1 + , 5, 4, 2,
1, 5, 4 + , 5 + , 1 + , 2, 6, 3, 7, 1 + , 7 + , 10 + , 6, 2 + , 11 + , 5, 10 + , 9 + , 18, 3 + , 6, 4 + ,
2, 7, 25 + , 2, 30 + , 2 + , 4, 13 + , 5, 19, 9, 5 + , 4, 32 + , 23, 19, 10 + , 5 + , 6, 9, 13 + ,
5, 13, 1, 15, 4 + , 8, 9 + , 20 + , 16 + , 19, 8 + , 4, 7 + , 5, 5, 7 + , 6

Stem-and-leaf: Times at which patients' observations were censored ($n = 83$)

Depths		Censoring time (unit = 1 month)
6	0 ∗	1 1 1 1 1 1
18	t	2 2 2 2 3 3 3 3 3 3 3 3
30	f	4 4 4 4 4 4 4 4 5 5 5 5
37	s	6 6 6 6 7 7 7
(12)	0 ·	8 8 8 9 9 9 9 9 9 9 9 9
34	1 ∗	0 0 0 0 0 1 1 1
26	t	2 3 3 3 3 3
20	f	4 4 5 5 5
15	s	6 6 7
12	1 ·	8
11	2 ∗	0
	t	
10	f	5
	s	
	2 ·	
9	3 ∗	0 1
7	t	2 3
5	f	4 4 4 5
1	s	7

Figure 1-4. A stem-and-leaf display with five lines per stem. *Source*: G. E. Dinse (1982). "Nonparametric estimation for partially-complete time and type of failure data," *Biometrics*, **38**, 417–431 (data from Table 1, p. 426, used by permission).

devote Chapter 7) are centered at 0, and some other types of data take both signs. When we start to display such data, we see that they require a −0 stem. The only tricky detail is that values exactly equal to 0 belong to either or both the −0 stem and the +0 stem, so that we usually share them roughly equally between the two.

To illustrate this variation, we use the residuals from fitting a straight line (by the "resistant line" method discussed in Chapter 5) to the basal temperature data of Table 1-1. Here

$$y = \text{basal body temperature,}$$
$$x = \text{cycle length,}$$

and the fitted line is

$$\hat{y} = .02813x + 35.68.$$

Figure 1-5 shows the stem-and-leaf display, with the residuals in units of .01°C. We notice some tendency for the values to pile up around zero, and the value at $-.30$ attracts some attention. Together with the one at $+.28$, it probably deserves a closer look.

In Figure 1-5 and in other stem-and-leaf displays that involve both positive and negative values, we have chosen, perhaps arbitrarily, to have the data values increase from the top of the display toward the bottom. We could equally well handle plus and minus by having the entries increase from bottom to top, and others may prefer this direction. In any one book or paper, however, we usually adopt one of these directions.

Resistance

Resistant methods are little affected by a small fraction of unusual data values and are an important part of exploratory data analysis. Thus it is unwise for the scale of a stem-and-leaf display to depend on the largest and smallest data values. (The stem-and-leaf display of the mean cycle lengths in Figure 1-2 is clearly influenced by the unusually small value at 22.9 days.) Instead, we often begin by setting aside any unusual data values, and we then base the choice of scale for the display on the rest of the data. (One rule of thumb for setting aside low and high values appears in Section 2C.) We list those outlying values on the lines labeled "low" and "high," beyond the set of stems. To emphasize further the separate treatment of these values, we may place parentheses around the lists. A comma after each

```
Depths
(n = 21)        unit = .01°C
    1       -3 | 0
    2       -2 | 0
    5       -1 | 3 5 8
   (6)      -0 | 0 2 4 7 8 9
   10        0 | 3 6 7 9
    6        1 | 1 2 5
    3        2 | 0 3 8
```

Figure 1-5. Stem-and-leaf display of residuals from a line for mean basal body temperature against mean cycle length.

value in the list serves as a reminder that these entries are data values (represented as multiples of the unit in the display) and not strings of leaves.

EXAMPLE: MENSTRUAL CYCLE LENGTHS

The clear separation between the rest of the data and the value at 22.9 days, which we saw in Figure 1-2, suggests that we place that one value on a line labeled "low." To avoid confusion, we recommend putting the entries on such lines in parentheses, as well as separating individual entries by commas. The first display in Figure 1-6 does this. By avoiding the three empty lines in Figure 1-2, we focus more attention on the bulk of the data at the expense of no longer showing just how far the value at 22.9 stands off from

One Line per Stem with "Low" Line		
Depths (n = 21)	(unit = .1 day)	
1	low	(229,)
6	26	3 6 8 9 9
9	27	5 6 6
(6)	28	0 4 4 5 8 8
6	29	4 9
4	30	0 3
2	31	2 8

Two Lines per Stem with "Low" Line		
Depths (n = 21)	(unit = .1 day)	
1	low	(229,)
2	26 *	3
6	26 ·	6 8 9 9
	27 *	
9	27 ·	5 6 6
(3)	28 *	0 4 4
9	28 ·	5 8 8
6	29 *	4
5	29 ·	9
4	30 *	0 3
	30 ·	
2	31 *	2
1	31 ·	8

Figure 1-6. Two stem-and-leaf displays for mean cycle length.

the rest. We can now calculate the scale that formula (1) suggests for the remaining 20 data values:

$$L = [10 \times \log_{10} 20] = 13$$

$$R = 31.8 - 26.3 = 5.5$$

$$\frac{R}{L} = .42$$

and we round R/L up to .5—a display with two lines per stem. The second part of Figure 1-6 shows the result, which probably has too many lines for some viewers. Both displays list the 22.9 separately, and the choice between them is a matter of taste.

With its variations, the stem-and-leaf display has proved to be a versatile technique for the analyst's first look at a batch of numbers. The three ways of factoring 10 (1×10, 2×5, and 5×2) provide adequate control over scaling (1, 2, or 5 lines per stem) especially when combined with "low" and "high" lines for unusual data values. Setting aside potential outliers in this way often leads to a more detailed and more effective display; it also focuses attention on the unusual data values so that we do what we can to probe their surrounding circumstances for clues.

1C. AN HISTORICAL NOTE

The histogram, the better known relative of the stem-and-leaf display, has been in use for many years. Beniger and Robyn (1978) trace the origin of the term to Karl Pearson in 1895, and they mention earlier examples going back to the bar chart published by William Playfair in 1786 in his *Commercial and Political Atlas*.

Of course, the primary difference between the histogram and the stem-and-leaf display is the use of a digit from each data value to form the display. Without any systematic search for possible predecessors of the stem-and-leaf display, we report finding a digit-based display (Figure 1-7) in the text by Dudley (1946, p. 22). In the way it groups matching "leaves" together and spreads the groups out over the line, this technique serves more as a sorting and tallying device than as a semigraphic display. Still, its similarity to the stem-and-leaf display emphasizes the convenience of working with digits from the data values.

First
Digits | Remaining Digit (tenths)

```
163 | 0
162 | 0   1 22    3333        6   7        9
161 | 00001 2222223333 44 555      77777 118 99999
160 | 000011 22   333 4444 55 666666 777 8888 999
159 | 0   11 222   3   4   555 6        999
158 |             3          6          9
157 |     1       3       5            9
```

Note: When necessary, allow two or more lines for each item of left-hand column.

Figure 1-7. Dudley's transcription of data in order of magnitude. From *Examination of Industrial Measurements* by John W. Dudley, Jr. Copyright © 1946 by McGraw-Hill Book Company, Inc. Used with permission of McGraw-Hill Book Company.

The role and development of graphical methods in statistics have received considerable attention in recent years. Exploratory data analysis has contributed several novel displays. Fienberg (1979) discusses the history and use of graphical methods and includes examples of several recent innovations. [See also Wainer and Thissen (1981).]

1D. SORTING

Because many exploratory techniques work with the ordered observations—the simplest way of gaining resistance—we devote this section to a discussion of sorting, the mechanical process of putting a set of numbers in order (usually, from smallest to largest). Readers who care little about these details may skip this section without loss.

In hand work, as we mentioned in Section 1A, constructing a stem-and-leaf display accomplishes much of the task of sorting the data. All that remains is to rearrange the leaves on each line. Technically, it is a form of "bucket sort" or "radix sort," with the lines playing the role of the "buckets." The skeleton structure of stems is easy to set up after a glance at the data establishes the range, and then one quickly places each leaf on its proper line. Because each line usually has relatively few leaves, completing the process by putting the leaves in order on the lines goes rapidly.

Until electronic digital computers became widespread, most sorting of substantial data sets worked with punch cards and relied on electromechanical card-sorting equipment. Again, the sorting operation was a form of radix sort, with one pass (initiated by hand) for each digit of the data values. The operator set the machine for the right-most digit of the data and fed in the cards, and the machine routed each card to the bin corresponding to its digit, 0 through 9. The operator then reset the machine for the next digit to the left and fed in the cards again, this time carefully placing the 0s first, then the 1s, and so on, to preserve the order from the previous step. The process continued until the leftmost digit had been processed, leaving the deck or file of data cards in order from smallest to largest value.

Relative to the number of data values, such use of a card sorter was not especially time-consuming. Sorting n numbers, each having at most k digits, required kn machine operations. The need for operator intervention, with the possibility of scrambling or even dropping the cards, made this method less effective than purely electronic ones.

Once digital computers were able to retain all the data in memory simultaneously, interest shifted to sorting procedures that use as little additional storage as possible. Now, instead of routing a card into a bin, the basic operations are comparisons and interchanges between numbers. These operations can be combined in numerous ways, and a great many general-purpose and special-purpose sorting algorithms have been developed. We cannot begin to study them here. Our primary need is to appreciate the relationship between n and the time that a good sorting algorithm takes to sort a list of n numbers.

It is easy to see that, with $n(n-1)/2$ comparisons, we can put n numbers into increasing order. We compare the first number in the list to each of the other $n-1$, making whatever interchanges are necessary to select the smallest observation. We now repeat the process with the remaining $n-1$ observations, making $n-2$ comparisons, and so on, until we have finally selected the largest observation. The total number of comparisons is then

$$(n-1) + (n-2) + \cdots + 1 = \frac{n(n-1)}{2}.$$

As n becomes large, this total increases like n^2, and so the time required for this simpleminded sorting algorithm grows as a constant times n^2.

Fortunately, more efficient methods are readily available. These organize the process to take advantage of the results of earlier comparisons, and the time that they require is proportional to $n \log n$. To explain the reason for this form of dependence on n, we begin by noting that a comparison between two numbers, x and y, has two possible outcomes, "$x < y$" and "$x \geq y$." More generally, k decisions, each involving two alternatives, can combine to yield at most 2^k outcomes. We say "at most" because some sequences of comparisons may reach a definitive outcome with fewer than k comparisons. As a partial example, suppose that we are sorting x, y, and z and that we have already established $x < y$. If we next compare x and z and discover $x \geq z$, we can conclude that $z \leq x < y$ without further effort. If, however, we find $x < z$, we do not yet know whether $z < y$, and we must make this comparison to find out.

Formally, sorting n numbers is equivalent to determining which of the $n!$ permutations connects their original order to their sorted order. Thus we must make at least enough comparisons to identify the permutation, for any list of n data values. Because 2^k is the number of outcomes for a set of k comparisons, k must satisfy (at least approximately)

$$2^k = n!,$$

so that

$$k = \log_2(n!)$$

is the minimum number of comparisons. Now, by Stirling's approximation (see, e.g., Feller, 1957, or Johnson and Kotz, 1969)

$$n! \sim \left(\frac{n}{e}\right)^n \sqrt{2\pi n},$$

so that for large n,

$$\log(n!) \approx n \log n - n + \tfrac{1}{2} \log(2\pi n).$$

Because the time for sorting already involves a proportionality constant, the base for the logarithm does not matter, and so the dominant term in the minimum number of comparisons is $n \log n$.

The advantage of good sorting algorithms is that they attain this minimum behavior and thus are able to take far less time for moderate n (say, $n \geq 11$) than algorithms whose time is proportional to n^2. For certain limited objectives, such as finding the median, even more speed is possible through "partial sorting," which does not attempt to order all the data values.

For readers who would like to pursue sorting more seriously, we mention three references. Martin (1971) gives a tutorial introduction, including an extensive decision diagram that suggests when each sorting technique should be used. Aho, Hopcroft, and Ullman (1974) discuss the analysis of several important techniques, including partial sorting. Knuth (1973) provides a more technical treatment of sorting algorithms and the methods for analyzing them.

1E. BACKGROUND ON NUMBER OF LINES

Choosing the number of lines for a stem-and-leaf display appears to have much in common with determining the number of intervals or the interval width for a histogram. One important difference is that, at least where theoretical work is involved, a desirable interval width for a histogram will bring it close to an assumed density function for the data. For the stem-and-leaf display, on the other hand, the major objectives are to produce an interval width that is 2, 5, or 10 times a power of 10, to allow for various unexpected patterns of data, and to provide convenient access to the individual data values. In this section we briefly compare five rules: three giving the number of lines and two giving an interval width.

Rules for the Number of Lines

In Section 1A we introduced $L = [10 \times \log_{10} n]$ as a reasonable upper limit on the number of lines in a stem-and-leaf display. We found this rule in a paper by Dixon and Kronmal (1965), who use it for histograms. Considerable experience enables us to report that it is generally quite effective in practice.

As a matter of taste, however, one sometimes wants fewer lines when n is small (say, 50 or less), and Velleman (1976) has suggested $L = [2\sqrt{n}\,]$.

Taking a different approach, Sturges (1926) argued that, when n is a power of 2, the proper frequency distribution follows the appropriate sequence of binomial coefficients. "For example, 16 items would be divided normally into 5 classes, with class frequencies 1, 4, 6, 4, 1." This is equivalent to taking $2^{L-1} = n$ or $L = 1 + \log_2 n$. Actually, one would use $[1 + \log_2 n]$.

Sturges' rule is an interesting argument by analogy. It is difficult to see why anyone would feel compelled to accept this result, but it may deserve comparison with the other suggestions. It would seem Sturges intended his rule as one providing an aesthetic picture, rather than as something effective in transferring information. When we are concerned with information transfer or ease of manipulation, his rule has little support.

Table 1-2 shows the values of L under these three rules for selected n between 10 and 300. (For convenience, we have not rounded the entries back to integer values.) The 10-log-n rule and the $2\sqrt{n}$ rule cross at $n = 100$, so we might use $[2\sqrt{n}]$ below 100 and $[10 \log_{10} n]$ above 100. Sturges' rule calls for substantially fewer lines or intervals—roughly 40 percent as many as the 10-log-n rule. These relationships are clearer in Figure 1-8, which uses a logarithmic scale for n.

For sample sizes up to about 30 or 40, Sturges' rule is probably satisfactory, especially if one uses the next *smaller* interval width among those of the form 1, 2, or 5 times a power of 10 (and thus somewhat more lines) as Sturges suggests. For larger sample sizes, this rule will give too few lines in a stem-and-leaf display, or, equivalently, it will tend to pile the data up too much on some lines. For example, the ideal frequencies for $n = 64$ would be 1, 6, 15, 20, 15, 6, 1, and the 20 leaves on the center line will often be tedious to work with in subsequent steps of analysis.

Another criticism of Sturges' rule is that it makes no allowances for outliers, skewed data, and multiple clumps separated by gaps. To accom-

TABLE 1-2. **Number of lines for a stem-and-leaf display or number of intervals for a histogram, as suggested by three rules.**

	Rule (Integer Part of)		
n	$10 \log_{10} n$	$2\sqrt{n}$	$1 + \log_2 n$
10	10.0	6.3	4.3
20	13.0	8.9	5.3
30	14.7	10.9	5.9
40	16.0	12.6	6.3
50	16.9	14.1	6.6
75	18.7	17.3	7.2
100	20.0	20.0	7.6
150	21.7	24.4	8.2
200	23.0	28.2	8.6
300	24.7	34.6	9.2
16	12.0	8.0	5
32	15.1	11.3	6
64	18.1	16.0	7
128	21.1	22.6	8
256	24.1	32.0	9
512	27.1	45.3	10

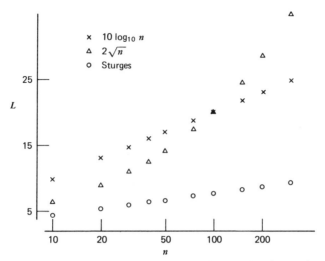

Figure 1-8.   Number of lines or intervals given by three rules.

modate skewness, Doane (1976) proposes adding extra intervals according to

$$\log_2\left[1 + \frac{\sqrt{b_1}}{\sigma\left(\sqrt{b_1}\right)}\right], \tag{2}$$

where $\sqrt{b_1}$ is the sample measure of skewness,

$$\sqrt{b_1} = \frac{\Sigma(x - \bar{x})^3 \sqrt{n}}{\left[\Sigma(x - \bar{x})^2\right]^{3/2}}, \tag{3}$$

whose standard deviation for Gaussian data is

$$\sigma\left(\sqrt{b_1}\right) = \sqrt{\frac{6(n - 2)}{(n + 1)(n + 3)}}$$

(we presume that, when $\sqrt{b_1}$ is negative, its absolute value would be used).
For simple exploratory work, we feel that this modification involves too
much calculation, and the nonresistance of the sample third moment is a
serious drawback.

Rules for Interval Width

Theoretical considerations can provide a way of defining an optimum interval width for histograms that are serving as estimators of an assumed underlying density function, f. In this setting, one constructs the sample histogram (on an equally spaced mesh) with area equal to 1 and then, for a given point x, estimates the height of the density function, $f(x)$, by the height of the histogram in the interval containing x. We denote this estimate by $\hat{f}_n(x)$.

Scott (1979) begins with x fixed and defines the *mean squared error* at x as

$$\text{MSE}(x) = E\left[\hat{f}_n(x) - f(x)\right]^2. \qquad (4)$$

To move from this pointwise measure of accuracy to an overall measure, he uses the *integrated mean squared error*:

$$\text{IMSE} = \int E\left[\hat{f}_n(x) - f(x)\right]^2 dx. \qquad (5)$$

The objective then is to choose the interval width, h_n, in a way that minimizes IMSE. For densities, f, that have two bounded and continuous derivatives, Scott shows that as n becomes large, taking

$$h_n = \left\{\frac{6}{\int_{-\infty}^{+\infty} [f'(x)]^2 \, dx}\right\}^{1/3} n^{-1/3} \qquad (6)$$

minimizes IMSE. Using simulation for the Gaussian distribution, he also finds that this formula works well for sample sizes as small as 25.

To apply this rule to data, we note that a change of scale from x to $y = \sigma x$ simply multiplies h_n by σ. Because knowledge of the true f is rare, Scott adopts the Gaussian density as a standard and proposes

$$h_n = 3.49 s n^{-1/3} \qquad (7)$$

as a data-based choice of interval width. Here s is an estimate related to the standard deviation, but not necessarily the usual sample standard deviation.

One alternative to integrated mean squared error uses the absolute value of the difference between $\hat{f}_n(x)$ and $f(x)$ instead of the square. The resulting interval-width formula is similar and again depends on $n^{-1/3}$. We do not pursue the details.

A second alternative is the maximum absolute deviation of the sample probability histogram $\hat{f}_n$ from the underlying density f:

$$\max_x |\hat{f}_n(x) - f(x)|.$$ (8)

Freedman and Diaconis (1981a) investigate the behavior of this difference and suggest (1981b) an interval width that works well for a variety of criteria. Roughly, their approach trades off pointwise bias, $E\hat{f}(x) - f(x)$, against the typical size of the maximum absolute deviation. These considerations lead to

$$h_n = c(f)\left(\frac{\log_e n}{n}\right)^{1/3},$$ (9)

where $c(f)$ is a constant that depends on the assumed density, f. Specifically, assuming that f has a unique maximum value f_0 at $x = x_0$ and is locally quadratic at x_0, and calculating

$$f_1 = \max_x |f'(x)|,$$

the expression for $c(f)$ is

$$c(f) = \left(\frac{2f_0}{3f_1^2}\right)^{1/3}.$$ (10)

A change of scale has the same effect as in Scott's formula, and the Gaussian value of $c(f)$ is 1.66, so that the corresponding formula based on Freedman and Diaconis (1981a) is

$$h_n = 1.66s\left(\frac{\log_e n}{n}\right)^{1/3}.$$ (11)

Again, one would generally want to use a resistant estimate as s.

Working with a different theoretical criterion, Freedman and Diaconis (1981b) arrive at

$$h_n = \frac{2(\text{IQR})}{n^{1/3}},$$ (12)

where IQR is the interquartile range of the sample. This rule is simpler than the one in equation (11) and nearly equivalent for a broad range of sample sizes.

To compare the rules of equations (7) and (12), we consider the same values of n that we used in Table 1-2, and we assume that the data are actually standard Gaussian, so that $s = 1$ and $IQR = 1.349$ are appropriate. Table 1-3 and Figure 1-9 show the resulting values of h_n. In this situation, the Freedman–Diaconis rule calls for narrower intervals. At all sample sizes, the ratio of the two values of h_n is $2.698/3.49$ or about .77. In practical applications, however, the two rules will often lead to the same choice of interval width because we would use a nearby value of the form 1, 2, or 5 times a power of 10.

Perhaps the most interesting feature of these two rules for h_n is that they both depend primarily on $n^{-1/3}$. If we turned such an interval width into a suggested number of intervals, it would behave like $n^{1/3}$, a functional form between $\log(n)$ and $\sqrt{n}$.

One way to accomplish this conversion between interval width and number of intervals, for rough comparison with the three rules described earlier, continues to assume Gaussian data and divides h_n into EW_n, the average value of the range in Gaussian samples of n. A table in Harter (1970) makes EW_n easy to calculate, and Table 1-4 shows the results (for illustration and without trying to substitute a nearby round value for h_n). When we compare the values of EW_n/h_n to Table 1-2, we see that they are

TABLE 1-3. Interval widths h_n for histograms, as suggested by two rules applied to Gaussian data (widths are in units of the estimated standard deviation).

n	Scott	Freedman–Diaconis
10	1.620	1.252
20	1.286	0.994
30	1.123	0.868
40	1.020	0.789
50	0.947	0.732
75	0.828	0.640
100	0.752	0.581
150	0.657	0.508
200	0.597	0.461
300	0.521	0.403

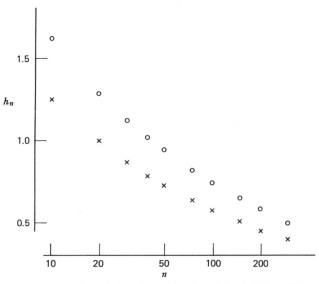

Figure 1-9. Interval widths calculated by the rules of Scott (○) and Freedman and Diaconis (×).

TABLE 1-4. Expected value of the Gaussian range and the corresponding numbers of intervals for the two interval-width rules.

n	EW_n	EW_n/h_n Scott	EW_n/h_n Freedman–Diaconis
10	3.078	1.90	2.46
20	3.735	2.91	3.76
30	4.086	3.64	4.71
40	4.322	4.23	5.48
50	4.498	4.95	6.14
75	4.806	5.81	7.51
100	5.015	6.67	8.63
150	5.298	8.07	10.43
200	5.492	9.20	11.90
300	5.756	11.04	14.28

substantially closer to Sturges' rule than to the other two. Thus for exploration we would still prefer to begin a stem-and-leaf display or a histogram with roughly $[10 \log_{10} n]$ lines or intervals.

We conclude this brief discussion of rules for interval width by emphasizing that such theoretical results for histograms have little connection with stem-and-leaf displays. The usual measures of the deviation of a histogram from the underlying density are strongly affected by the *corners* of the histogram bars, and we do not even see corners in stem-and-leaf displays.

1F. SUMMARY

The stem-and-leaf display, like the histogram, summarizes the shape of a batch of data. The digits themselves do most of the work of sorting the batch and are the elements used for the construction of the display. Several variants of the stem-and-leaf display provide flexibility for summarizing a wide variety of data sets. In particular, these different versions of the display give flexibility in determining both the number of lines in the display and the method of handling unusual data values.

We prefer the stem-and-leaf display to the histogram because it retains the most significant digits of the data. This feature enables us

To see patterns in the data.

To see the distribution of data values within an interval.

To go more easily from a value in the display to the datum that produced it.

Of primary importance is the use of the stem-and-leaf display for examining the residuals when a model is fit to a data set.

The equation

$$L = [10 \log_{10} n]$$

provides a useful upper bound on the number of lines to be used in a display, and alternative rules are available. Theoretical work by Scott and by Freedman and Diaconis on interval widths for histograms serves as a valuable point of comparison.

Many of the chapters in this volume use the stem-and-leaf display at several stages in understanding the structure of a batch of data or a batch of residuals. As we encounter these applications of the stem-and-leaf display, its utility and its ability to illuminate will become more evident.

REFERENCES

Aho, A. V., Hopcroft, J. E., and Ullman, J. D. (1974). *The Design and Analysis of Computer Algorithms*. Reading, MA: Addison-Wesley.

Beniger, J. R. and Robyn, D. L. (1978). "Quantitative graphics in statistics: A brief history," *The American Statistician*, **32**, 1–11.

Dinse, G. (1982). "Nonparametric estimation for partially-complete time and type of failure data," *Biometrics*, **38**, 417–431.

Dixon, W. J. and Kronmal, R. A. (1965). "The choice of origin and scale for graphs," *Journal of the Association for Computing Machinery*, **12**, 259–261.

Doane, D. P. (1976). "Aesthetic frequency classifications," *The American Statistician*, **30**, 181–183.

Dudley, J. W., Jr. (1946). *Examination of Industrial Measurements*. New York: McGraw–Hill.

Feller, W. (1957). *An Introduction to Probability Theory and Its Applications*, Vol. 1, second edition. New York: Wiley.

Fienberg, S. E. (1979). "Graphical methods in statistics," *The American Statistician*, **33**, 165–178.

Freedman, D. and Diaconis, P. (1981a). "On the maximum deviation between the histogram and the underlying density," *Zeitschrift für Wahrscheinlichkeitstheorie und verwandte Gebiete*, **58**, 139–167.

Freedman, D. and Diaconis, P. (1981b). "On the histogram as a density estimator: L_2 theory," *Zeitschrift für Wahrscheinlichkeitstheorie und verwandte Gebiete*, **57**, 453–476.

Harter, H. L. (1970). *Order Statistics and Their Use in Testing and Estimation, Vol. 2: Estimates Based on Order Statistics of Samples from Various Populations*. Washington, DC: U.S. Government Printing Office.

Johnson, N. L. and Kotz, S. (1969). *Discrete Distributions*. Boston, MA: Houghton Mifflin.

Knuth, D. E. (1973). *Sorting and Searching (The Art of Computer Programming*, Vol. 3). Reading, MA: Addison–Wesley.

Martin, W. A. (1971). "Sorting," *Computing Surveys*, **3**, 147–174.

Royston, J. P. and Abrams, R. M. (1980). "An objective method for detecting the shift in basal body temperature in women," *Biometrics*, **36**, 217–224.

Scott, D. W. (1979). "On optimal and data-based histograms," *Biometrika*, **66**, 605–610.

Shewhart, W. A. (1931). *Economic Control of Quality of Manufactured Product*. Princeton, NJ: D. Van Nostrand, Inc.

Sturges, H. A. (1926). "The choice of a class interval," *Journal of the American Statistical Association*, **21**, 65–66.

Tukey, J. W. (1970). *Exploratory Data Analysis* (Limited Preliminary Edition), Vol. 1. Reading MA: Addison–Wesley.

Tukey, J. W. (1972). "Some graphic and semigraphic displays." In T. A. Bancroft (Ed.), *Statistical Papers in Honor of George W. Snedecor*. Ames, IA: Iowa State University Press.

Velleman, P. F. (1976). "Interactive computing for exploratory data analysis I: display algorithms," *1975 Proceedings of the Statistical Computing Section*. Washington, DC: American Statistical Association.

Velleman, P. F. and Hoaglin, D. C. (1981). *Applications, Basics, and Computing of Exploratory Data Analysis*. Boston, MA: Duxbury Press.

Wainer, H. and Thissen, D. (1981). "Graphical data analysis," *Annual Review of Psychology*, **32**, 191–241.

EXERCISES

1. Suppose that the residuals from fitting a model to a 20-point data set are:

    $$-1.4, 1.2, 0.5, -0.3, -0.8, 0.4, -1.3, 0.5, 0.7, -0.2$$
    $$0.1, 2.3, 0.1, -0.8, -2.6, 0.7, -0.9, -0.4, 0.2, 1.3.$$

    Determine whether this batch has any outside values and provide a suitable stem-and-leaf display.

2. For the 21 women whose data appear in Table 1-1, additional data were gathered that pertain to the preovulatory basal body temperatures of the women. These data, in the order corresponding to that used in Table 1-1, are listed below.

    (a) Mean temperature in degrees Celsius (listed in Table 1-1).

    (b) Standard deviation of temperature.

    $$0.12, 0.15, 0.15, 0.10, 0.13, 0.15, 0.14,$$
    $$0.08, 0.11, 0.09, 0.14, 0.09, 0.13, 0.14,$$
    $$0.12, 0.16, 0.15, 0.13, 0.12, 0.12, 0.09.$$

    (c) Number of temperature readings.

    $$88, 66, 71, 63, 101, 55, 76,$$
    $$49, 63, 38, 91, 79, 41, 36,$$
    $$73, 55, 42, 49, 50, 90, 51.$$

    Give stem-and-leaf displays for each of these batches of data.

3. Dinse (1982) gives the survival times in weeks for 79 patients with lymphocytic nonHodgkins lymphoma. About 35% of the times are censored; these are indicated by a " $+$ " sign. (The others ended at the patient's death.)

    124, 188, 225, 398 +, 64, 24, 365 +, 294, 189, 355 +, 381 +,
    362 +, 354 +, 35, 135, 21, 198, 369 +, 58, 353 +, 357 +, 359,
    281, 360 +, 371 +, 342 +, 189, 76, 358, 79, 400 +, 242, 262,
    378 +, 388 +, 75, 55, 347, 50, 48, 292, 250, 301, 281, 157,
    139, 159, 349 +, 346 +, 49, 151, 58, 316 +, 112, 141, 349 +,
    356 +, 96, 306 +, 152, 339 +, 329 +, 327 +, 31, 174, 212, 37,
    110, 276, 239, 327 +, 132, 102, 68, 300 +, 299 +, 257, 291 + , 222.

    (a) Give stem-and-leaf displays for this data set, using (i) two, and (ii) five lines per stem. Which display seems more effective and why?

    (b) Give parallel stem-and-leaf displays for the censored and the completed survival times. What do you think accounts for the differences between the two batches?

4. For each of the stem-and-leaf displays suggested in Exercises 3a and 3b, what interval widths are indicated by the rules of Scott and of Freedman and Diaconis?

5. (Mathematical). Using the Gaussian density with mean μ and variance σ^2 as f and Scott's asymptotic formula (6), justify the suggested estimate for h_n given in formula (7).

6. Continuation. Apply expression (6) to a uniform distribution on the interval $(-A, A)$ and obtain the formula corresponding to (7).

CHAPTER 2

Letter Values:
A Set of Selected
Order Statistics

David C. Hoaglin
Harvard University and Abt Associates Inc.

The classical summary statistics for a batch of data consisting of n observations, $x_1, x_2, \ldots, x_n$, involve only simple arithmetic operations on all the data, such as addition, multiplication, division, and perhaps square root. Most familiar are the *sample mean* $\bar{x}$, given by

$$\bar{x} = \frac{x_1 + \cdots + x_n}{n},$$

and the *sample variance*, most often given by

$$s^2 = \frac{1}{n-1} \sum_{i=1}^{n} (x_i - \bar{x})^2.$$

Especially for exploratory purposes, it is often advantageous to use simpler summaries based on sorting and counting. Among other merits, such summaries can be resistant; that is, an arbitrary change in a small part of the batch can have only a small effect on the summary. The sample mean and sample variance cannot behave in this way, and a single wild data value can have a substantial adverse effect on both of them.

This chapter derives and discusses the "letter values," a set of summary values that have a variety of uses. Letter values are essentially a collection of observations drawn systematically from the batch, more densely from the tails than from the middle. Sorting and ranking operations can extract these

33

summary values from data, and the letter values can, in turn, be used to define resistant measures of location and of the amount of spread in the batch. They also help in the search for outliers. They can economically summarize the data in a batch in such a way that most observations can be approximately recovered. Background information on letter values comes from theory and from examining their behavior in some idealized situations.

2A. SORTING AND RANKING

Either by hand (e.g., using a stem-and-leaf display) or with the aid of a computer, we can sort the data $x_1, \ldots, x_n$ into ascending order. The result of this operation is the ordered sample, which we denote by

$$x_{(1)}, x_{(2)}, \ldots, x_{(n)};$$

that is, $x_{(i)}$ is the ith smallest observation. More formally, $x_{(1)}, \ldots, x_{(n)}$ are called the *order statistics* of the sample $x_1, \ldots, x_n$, and $x_{(i)}$ is the ith order statistic. (The notation $x_{i|n}$ can be used to show explicitly that there are n observations in the sample. Chapter 10 gives some further discussion of order statistics.)

On the basis of the ordering, we can define the *rank* of an observation in either of two ways: we may count up from the smallest value, or we may count down from the largest. The first of these yields the observation's *upward rank*; that is, $x_{(2)}$ has upward rank 2 and, in general, $x_{(i)}$ has upward rank i. Counting down from the largest yields an observation's *downward rank*; $x_{(n-1)}$ has downward rank 2 and $x_{(i)}$ has downward rank $n + 1 - i$. Considering both of these rankings together, we see that for any data value

upward rank PLUS downward rank $= n + 1$.

Sometimes it is useful to think in terms of the original observations. For example, if, through the sorting process, the raw observation x_i becomes the order statistic $x_{(j)}$, then the upward rank of x_i is j.

Often we want to give equal attention to both ends of a batch. A convenient way of handling this is to use the two ranks, upward and downward, in defining depth.

DEFINITION: The *depth* of a data value in a sample is the smaller of its upward rank and its downward rank.

By using the notion of depth, we can specify how to extract various exploratory summary values from a sample. Probably the most familiar of these summaries is the *median*, which gives the center of the sample in terms of counting. Its depth is $(n + 1)/2$. If n is even, we find that $(n + 1)/2$ involves a fraction, $\frac{1}{2}$; by convention we interpolate whenever a depth is not an integer. If n is even, say $n = 2k$, then the median falls halfway between $x_{(k)}$ and $x_{(k+1)}$:

$$\text{median} = \tfrac{1}{2}(x_{(k)} + x_{(k+1)}), \qquad n = 2k.$$

Two examples illustrate the calculation of the depth of the median. When $n = 3$, the depth of the median is $(3 + 1)/2 = 2$, so that the median is $x_{(2)}$. And when $n = 4$, we get $(4 + 1)/2 = 2\frac{1}{2}$ so that the median is $\frac{1}{2}(x_{(2)} + x_{(3)})$.

The simplest data values to extract are the *extremes*, the two data values with depth 1, namely, the largest value and smallest value in the batch. Here, as for all summary values other than the median, a given depth identifies two (data) values, one below the median and the other above the median.

To the median and the extremes we add another pair of summary values, the hinges (a form of quartile) or, as we call them, the *fourths*, by defining

$$\text{depth of fourth} = \frac{[\text{depth of median}] + 1}{2}; \qquad (1)$$

the brackets in $[x]$ stand, as before, for the operation of finding the largest integer not exceeding x. So the rule for finding the depth of a fourth says, "Drop any fraction from the depth of the median, add 1, and halve." This simplifies interpolation because the depth of a fourth can be only an integer or an integer plus $\frac{1}{2}$. In terms of counting, each fourth comes halfway between the median and the corresponding extreme, and so the two fourths bracket the middle half of the batch.

Taken together, the median, fourths, and extremes are the ingredients of the five-number summary, which we will discuss in Section 2D as a letter-value display. We now turn to an example.

EXAMPLE: 5-NUMBER SUMMARY

Bendixen (1977, p. 374) gives the percentage of survivors for each of 11 categories of hospital patients who required respiratory support (a form of intensive care) for more than 24 hours: 37, 52, 90, 56, 58, 45, 66, 75, 36, 68, 100. What percentage seems to be typical, and how much variation is there among categories? We approach this by finding the median, the fourths, and the extremes for these data.

The ordered observations are

i:	1	2	3	4	5	6	7	8	9	10	11
$x_{(i)}$:	36	37	45	52	56	58	66	68	75	90	100

Because $n = 11$, the depth of the median is $(11 + 1)/2 = 6$, and so the median is 58. The depth of a fourth is $(6 + 1)/2 = 3.5$, so that the fourths are $(45 + 52)/2 = 48.5$ and $(75 + 68)/2 = 71.5$. Finally, the extremes are 36 and 100. Thus for all 11 categories, we can say that typically 58 percent of the patients survive and, although the survival rate can be as high as 100 percent or as low as 36 percent, half the categories had rates from 48.5 to 71.5 percent.

2B. LETTER VALUES

In large batches it is often useful to summarize with somewhat more detail. The five-number summary can easily be enlarged to accommodate two more summary values, the *eighths*, which are determined by

$$\text{depth of eighth} = \frac{[\text{depth of fourth}] + 1}{2}. \qquad (2)$$

This simply follows the pattern which began in defining the fourths; each eighth comes halfway (in counting) between the corresponding fourth and extreme.

With the inclusion of the eighths, the 5-number summary has become a *7-number summary*.

EXAMPLE: EIGHTHS

In order to complete the summary values for a 7-number summary, we find the eighths of the patient survival data. Because the depth of a fourth is 3.5 for this batch, the depth of an eighth is $([3.5] + 1)/2 = (3 + 1)/2 = 2$, and the eighths are 37 and 90. We show the 7-number summary in schematic form in Section 2D.

Underlying the name "eighth" is the fact that roughly one-eighth of the data in the batch lies below the lower eighth and one-eighth lies above the upper eighth. For larger and larger batches we can continue to add pairs of summary values by halving the fraction of the data remaining beyond the previous (nonextreme) summary value at each end of the batch. Thus we would move from eighths to *sixteenths* to *thirty-seconds* and so on, calculating each depth in the sequence according to

$$\frac{[\text{previous depth}] + 1}{2} \tag{3}$$

and stopping when this new depth reaches 1 (bringing us to the extremes).

The letters as tags.   For convenience of notation and display, we often use 1-letter tags for the summary values that we have been extracting from a batch. Thus we begin with M for median and F for fourth. The extremes have no tag other than 1, their depth. If we tried to invent, for these new summary values (eighths and sixteenths, etc.), tags that directly reflect the fraction of data remaining beyond each, we would soon find it difficult to use only one letter per tag. An alternative is to notice that in the alphabet, E (for eighth) immediately precedes F and then to use as our tags the letters of the alphabet, working backward from F and wrapping around so that Z follows A. Through this rememberable association with 1-letter tags, these summary values have come to be called *letter values*.

As a standard summary for general use, the 5-number summary provides about the right amount of detail. More information is available in larger batches, however, and we may want to use a fuller set of seven or more letter values to summarize such batches further. Section 2G gives some background related to this. Learning more about the shape of the data requires large batches and more detail than a 5-number summary can provide, especially toward the ends of the batch. The additional letter values are well suited to this purpose, and we use them to study distribution shape in a companion volume.

If we were working with a continuous distribution instead of with finite batches, there would be no need to interpolate, and the fraction remaining outside a letter value would be known as a *tail area*. These tail areas would be powers of $\frac{1}{2}$. Table 2-1 shows the relationship between the usual letters and such tail areas. For finite n it is possible to define tail areas in several ways. We discuss these when we come to ideal letter values in Section 2E.

We have mentioned using letter values to provide resistant measures of location. For almost all exploratory purposes we can use the median to summarize the location, center, or typical value of a batch. Another location summary, which uses more order statistics but still involves only hand arithmetic, is the *trimean*, defined by

$$\text{trimean} = \tfrac{1}{4}(\text{lower fourth}) + \tfrac{1}{2}(\text{median}) + \tfrac{1}{4}(\text{upper fourth}).$$

TABLE 2-1. **Relationship between letter values and tail areas for continuous distributions.**

Tag	Tail Area
M	$\frac{1}{2} = .5$
F	$\frac{1}{4} = .25$
E	$\frac{1}{8} = .125$
D	$\frac{1}{16} = .0625$
C	$\frac{1}{32} = .03125$
B	$\frac{1}{64} = .015625$
A	$\frac{1}{128} = .0078125$
Z	$\frac{1}{256} = .00390625$
Y	$\frac{1}{512} = .001953125$
X	$\frac{1}{1024} = .0009765625$

We discuss this and more sophisticated measures or estimators of location in Chapters 10 and 11.

2C. SPREADS

So far we have talked of summarizing a batch or sample, and we have derived the letter values as a convenient set of summary values. We can combine different letter values to describe several aspects of the data. In the present chapter we emphasize two aspects: location and spread. The two most common summaries of location based on letter values, the median and the trimean, have appeared in the previous sections.

In summarizing spread, we seek an indication of how concentrated the data values are. A simple resistant measure is the *fourth-spread*, defined by

$$\text{fourth-spread} = (\text{upper fourth}) - (\text{lower fourth}), \qquad (4)$$

which gives the width of the middle half of the batch. More briefly, we may call it the F-spread, and we write $d_F = F_U - F_L$. Of course, the *range*, the difference between the extremes, also reflects spread, but stray values influence the range so strongly that it has negligible resistance.

Some readers may be familiar with the *interquartile range*, which is very close to the fourth-spread because quartiles are nearly the same as fourths.

Choice of scale. Just as we use the median to compare location of batches, we can use the fourth-spread to compare spread among batches. For analytical purposes, we also use it, when we have several batches, in choosing an appropriate scale such as the original raw value, the logarithm, or the square root of the observations. As discussed in Chapter 3, the relation between fourth-spread and median across batches helps us choose a scale of measurement that makes spread comparable from batch to batch.

Identifying outlying values. To examine the data for outlying values, we need a measure of spread that is insensitive to them. The fourth-spread provides such a measure, whereas the range and sample standard deviation do not. More generally, we need a measure of spread that emphasizes the behavior of the central portion of the data rather than the extremes, and the fourth-spread does this.

It would be nice if we could clearly identify "outliers," corresponding to different underlying behavior for certain values as compared with that for the bulk of the data. The explanation for the difference may lie in how the quantity we are studying truly behaves, how we measured it, or how we mishandled the measurements. Identifying "outliers" with near certainty is, unfortunately, not possible for the sorts of sample sizes we often encounter. It seems to require, say, 500 or more to do well. The best we can do in small or moderate-sized samples is to fence off some values for special attention. Values that lie outside the *outside cutoffs* we will call *outside* values. Many of these will not be "outliers"—they will have the same underlying behavior as the bulk of the data. We shall get an idea about how many below. Some will be "outliers" in truth and deserve some special attention. Other discussions in the statistical literature do not always make this distinction so clearly. Often an observation that is sufficiently unlikely under the assumed probability model is classified as an outlier. Some other chapters of this book follow that usage and speak of "outlier cutoffs."

We now give an approach for identifying outlying values based on the fourth-spread. The rule of thumb that we use designates certain data values as outside. We use the fourth-spread, d_F, as a measure of distance and lay off a multiple, specifically $1\frac{1}{2}$, of it upward from F_U and downward from F_L. We regard observations outside these cutoffs as outliers, deserving close scrutiny. Figure 2-1 shows the fourths and these cutoffs.

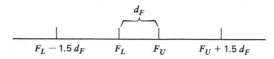

Figure 2-1. Position of the fourths and outside cutoffs.

As an aid to interpreting our designation of certain observations as "outside," we can study the chance that a data value will fall beyond the "outside" cutoffs, $F_L - 1.5d_F$ and $F_U + 1.5d_F$, when the observations come from a Gaussian (normal) distribution. Because it is simplest, we begin with the distribution itself (corresponding to a very very large sample) instead of with a finite sample. The fourth corresponds to a tail area of .25, so in a Gaussian distribution the fourths are $\mu - 0.6745\sigma$ and $\mu + 0.6745\sigma$, yielding an F-spread of 1.349σ. This places the "outside" cutoffs at $\mu - 2.698\sigma$ and $\mu + 2.698\sigma$, corresponding to a tail area of .00349 in each tail. Thus the fraction of the distribution that lies below $F_L - 1.5d_F$ or above $F_U + 1.5d_F$ is .00698. In finite samples, the average fraction of observations beyond the (sample) "outside" cutoffs is substantially larger than the population value (still double for samples of 30). From a simulation study, Hoaglin, Iglewicz, and Tukey (1981) develop the approximation

$$0.00698 + \frac{0.4}{n}, \tag{5}$$

which gives a somewhat conservative upper bound on the average fraction of "outside values" in Gaussian samples with $n \geq 5$.

If we ask for the average number "outside," we have to multiply expression (5) by n, finding

$$0.4 + 0.007n$$

for a single batch and hence

$$0.4(\text{number of batches}) + 0.007(\text{number of observations})$$

for several batches. The results can be of considerable help to us, provided we remember that they apply to Gaussian distributions and that heavy-tailed distributions will have more values "outside" by right.

Replacements for standard deviation or variance. When we would otherwise think in terms of a standard deviation or a variance, we can obtain a resistant analogue from the F-spread. We simply ask what standard deviation a Gaussian distribution would need to have in order to yield the same F-spread as our data. The fourths of a Gaussian distribution are $\mu - 0.6745\sigma$ and $\mu + 0.6745\sigma$, so its F-spread is 1.349σ and σ would have to be

$$\frac{\text{data F-spread}}{1.349}.$$

This ratio is known as the *F-pseudosigma*, and its square (analogous to σ^2) is the *F-pseudovariance*.

When the data are actually Gaussian, the F-pseudosigma yields an estimate of σ, and its value will usually be close to that of s, the sample standard deviation. Even when the data are not Gaussian, we can use the F-pseudosigma as a measure of spread; Chapter 12 discusses this and other estimators of scale. Using the resistant F-pseudosigma in addition to, or instead of, the nonresistant s has advantages, particularly when the data may be close to Gaussian except for a few wild observations. If the two estimates differ substantially, we should prefer the F-pseudosigma and look for the observations that have inflated s.

We can also use other pairs of letter values to define a pseudosigma and the corresponding pseudovariance. All that is required is the value of the spread for that pair of letter values in a standard Gaussian population so that we can calculate

$$\frac{\text{data letter-spread}}{\text{standard Gaussian value}}. \tag{6}$$

We pursue the details in a companion volume, where we use letter values to study distribution shape.

2D. LETTER-VALUE DISPLAYS

Defining and obtaining easily interpretable summary values is an important step, but if we are to make effective use of this information, we must present it in a format that reveals the important numerical features of a batch and invites simple calculations related to location and spread. A variety of horizontal and vertical styles have been tried, but the most useful and flexible seems to be a rectangular format. In the simplest form, also known as a *5-number summary*, the skeleton of the letter-value display looks like this:

```
#   n

M   depth of median   |              median               |
F   depth of fourth    | lower fourth      upper fourth    |
    1                  | lower extreme     upper extreme   |
```

At the top left of this arrangement, # n reminds the reader that the batch has n values. The letters on the next two lines are the tags, M for median and F for fourth. The extremes have no tag other than 1, their depth. Some of the advantages of this form of layout will become clearer later, when we have included more letter values.

EXAMPLE: 5-NUMBER SUMMARY

As an example, we again use the patient survival data, for which we repeat the 11 ordered observations:

i:	1	2	3	4	5	6	7	8	9	10	11
$x_{(i)}$:	36	37	45	52	56	58	66	68	75	90	100.

The display for the 5-number summary is

#11	percent surviving	
M 6	58	
F 3.5	48.5	71.5
1	36	100

We have used the space atop the box as a further reminder that the data are the percent of patients surviving.

Expanding the 5-number summary into a 7-number summary involves simply adding a line for the eighths: the tag E, the depth of the eighth, and the lower and upper eighths in their respective columns. Continuing with the same set of data, we have

#11	percent surviving	
M 6	58	
F 3.5	48.5	71.5
E 2	37	90
1	36	100

To add further pairs of letter values in a larger data set, we would follow this same pattern.

2E.   IDEAL LETTER VALUES

The definition of the median, as splitting a batch of data or a distribution into two halves, is familiar. The fourths, except for the specific way of computing them by using depths, are essentially the quartiles and also may be familiar. The other letter values are likely to be new to most readers. We offer an explanation of why their depths are found according to equation (3), namely by calculating ([previous depth] + 1)/2, instead of using $n/4$ for fourths, $n/8$ for eighths, and so on. One reason is that we are keeping the interpolation simple for hand calculation by always using either a data value or the average of two consecutive data values. The definition of depth does this, and expressions such as $n/4$ and $n/32$ do not. This is not the whole story, however, as we can see for $n = 11$, where $n/4 = 2\frac{3}{4}$, which we might round to 3, but the depth of a fourth is $3\frac{1}{2}$. The explanation lies in the way we have chosen to define the fraction of the data to the left of any specified point x.

Although we want ultimately to go from a specified fraction in the tail of a batch to the corresponding value in the data, it is simplest to approach this by determining the fraction of the data at or beyond each of the order statistics. We begin by reviewing three of the definitions in common use: simple fractions, fractions that give intervals of equal probability, and fractions that correspond (approximately) to the median of the distribution of each order statistic. To keep the algebra tidy, we concentrate on the left tail.

Simple fraction. In calculating fractions from data, one usually has in mind some underlying distribution, so that the fraction found from the data might be used as an estimate of the corresponding fraction for that distribution. For this reason, the simplest definition of the fraction to the left of $x_{(i)}$, namely i/n, is not likely to be satisfactory because it does not treat the two tails symmetrically and because for $x_{(n)}$, the upper extreme, it yields the fraction $n/n = 1$, a result that we surely do not want for any continuous distribution. The standard remedy for this difficulty splits each data value in half; that is, we count only half of it as lying at or below $x_{(i)}$. Thus the fraction corresponding to $x_{(i)}$ is $(i - \frac{1}{2})/n$.

*Intervals of equal probability.** A rather different approach recognizes that the n order statistics divide the range of the distribution into $n + 1$ intervals, each of which contains probability $1/(n + 1)$ on the average. More specifically, if we work with the order statistics $X_{(1)} \leq \cdots \leq X_{(n)}$ of a random sample from the distribution whose cumulative distribution function is F, then the probability in the random interval $X_{(i-1)} \leq x \leq X_{(i)}$ is $F(X_{(i)}) - F(X_{(i-1)})$. The average value of this probability is not difficult to compute, because transforming a random variable X by its own cumulative distribution function F yields a uniform random variable, $U = F(X)$. Because F is nondecreasing, the same transformation takes the order statistics $X_{(i)}$ ($i = 1, \ldots, n$) into the order statistics of a uniform sample, $U_{(i)}$ ($i = 1, \ldots, n$). In terms of the $U_{(i)}$, the probability in the random interval

---

*In this discussion we distinguish between a random variable and a value of the variable by using capital letters for random variables [X, $X_{(i)}$] and small letters for values [x, $x_{(i)}$].

$X_{(i-1)} \leqslant x \leqslant X_{(i)}$ is $U_{(i)} - U_{(i-1)}$. The expected value of $U_{(i)}$ is $i/(n + 1)$, so that each interval contains probability $1/(n + 1)$ on the average.

Median of the distribution of each order statistic. The two preceding definitions fit into a more general framework, suggested by Blom (1958), for the relationship between order statistics and fractions. Blom describes a family of definitions, parametrized by α:

$$\left(\text{fraction} \leqslant x_{(i)}\right) = \frac{i - \alpha}{n + 1 - 2\alpha}. \tag{7}$$

We see that taking $\alpha = \frac{1}{2}$ gives $(i - \frac{1}{2})/n$ and $\alpha = 0$ gives $i/(n + 1)$. The definition of "fraction $\leqslant x_{(i)}$" that underlies our choice of depths for letter values is another member of this family, namely $\alpha = \frac{1}{3}$, which gives

$$\left(\text{fraction} \leqslant x_{(i)}\right) = \frac{i - \frac{1}{3}}{n + \frac{1}{3}}. \tag{8}$$

We can motivate this particular choice in part by writing the fraction for any value x (not just an $x_{(i)}$) as

$$\frac{(\text{count below } x) + \frac{1}{2}(\text{count at } x) + \frac{1}{6}}{n + \frac{1}{3}} \tag{9}$$

The constant $\frac{1}{6}$ that has been added can be interpreted as reflecting the possibility of data values beyond the largest and smallest values that occurred in the sample. If the data have no ties, the numerator in equation (9) simplifies to $(i - 1) + \frac{1}{2} + \frac{1}{6} = i - \frac{1}{3}$ for $x = x_{(i)}$. Also, when d is a depth that requires averaging two adjacent order statistics, the formula $(d - \frac{1}{3})/(n + \frac{1}{3})$ gives the correct fraction. Thus we can use it for any letter-value depth.*

Another motivation, more important for our uses of letter values, is the fact that the median of the distribution of $X_{(i)}$ in a sample of n is, very closely, at the point where the value of the cumulative distribution function equals $(i - \frac{1}{3})/(n + \frac{1}{3})$. This is true for any continuous distribution, so we need make no assumptions about a particular distribution underlying the data.

Because one seldom works with the median of the distribution of $X_{(i)}$, we examine it in more detail. The definition is straightforward: the median of

---

*The definition in equation (9) is the same as that for "ss-fraction below," which is discussed in Tukey (1977, Section 15A).

$X_{(i)}$ is that value x such that $\text{Prob}[X_{(i)} \leqslant x] = \frac{1}{2}$. We do, however, want to be careful to distinguish it from the median of the sample. Each order statistic $X_{(i)}$ has a distribution, and each of these distributions has a (population) median. We must usually obtain the median of $X_{(i)}$ by numerical calculations with the cumulative distribution function F and the corresponding density function f. The density function of $X_{(i)}$ is

$$\frac{n!}{(i-1)!(n-i)!}[F(x)]^{i-1}[1-F(x)]^{n-i}f(x), \qquad -\infty < x < +\infty,$$

so that, denoting the desired median by $x_{i,.5}$, we must determine the value of $x_{i,.5}$ such that

$$\frac{n!}{(i-1)!(n-i)!}\int_{-\infty}^{x_{i,.5}}[F(x)]^{i-1}[1-F(x)]^{n-i}f(x)\,dx = \frac{1}{2}.$$

When we recall that the differential $dF(x) = f(x)\,dx$, we see that it suffices to solve this equation for the uniform distribution:

$$\frac{n!}{(i-1)!(n-i)!}\int_{0}^{F_{i,.5}}F^{i-1}(1-F)^{n-i}\,dF,$$

where $F_{i,.5} = F(x_{i,.5})$. Interpolation based on tables of the incomplete beta function now yields a reasonable approximation to $F_{i,.5}$.

Some useful results can be obtained with less effort. For example, if n is odd, $n = 2k + 1$, then the density function of the sample median is

$$\frac{(2k+1)!}{k!k!}[F(x)]^{k}[1-F(x)]^{k}f(x).$$

The symmetry produced by equal powers of $F(x)$ and $1 - F(x)$ means that the median of the distribution of the sample median equals the population median because

$$\frac{(2k+1)!}{k!k!}\int_{0}^{1/2}F^{k}(1-F)^{k}\,dF = \frac{1}{2}.$$

Interestingly, this result holds even when the population F is not symmetric. For another example, we consider the lower extreme, $X_{(1)}$. The density is

$$n[1-F(x)]^{n-1}f(x),$$

and we obtain the median of the distribution of $X_{(1)}$ by solving

$$n \int_0^{F_{1,.5}} (1 - F)^{n-1} \, dF = \frac{1}{2}$$

$$-(1 - F)^n \Big|_0^{F_{1,.5}} = \frac{1}{2}$$

$$(1 - F_{1,.5})^n = \frac{1}{2}$$

$$F_{1,.5} = 1 - \left(\frac{1}{2}\right)^{1/n}.$$

Thus when $n = 10$, $F_{1,.5} = .0670$. For comparison, $(1 - \frac{1}{3})/(10 + \frac{1}{3})$ $= \frac{2}{3}/\frac{31}{3} = .0645$, roughly 3.7% low.

The approximation $(i - \frac{1}{3})/(n + \frac{1}{3})$ faces its toughest test for small n and extreme i. Using the uniform distribution on $[0, 1]$, Table 2-2 compares the median of $X_{(i)}$ and the approximation for $n = 2, 3, 4,$ and 5. Except for $i = 1$, the approximation is quite good, and even at $i = 1$ it is adequate for most purposes.

Using the third definition, we can now determine the ideal depth corresponding to a particular fraction or tail area by rewriting equation (8):

$$(\text{depth}) = \left(n + \tfrac{1}{3}\right) \times (\text{tail area}) + \tfrac{1}{3}. \tag{10}$$

When we substitute into this the ideal tail areas for letter values, namely powers of $\frac{1}{2}$, we find the results displayed in Table 2-3. To see how our conventional depths, found from equation (3), compare with the ideal

TABLE 2-2.   Accuracy of $(i - \frac{1}{3})/(n + \frac{1}{3})$ in approximating the median of $X_{(i)}$ in small samples from the uniform distribution.

n	i	Median of $X_{(i)}$	$(i - \frac{1}{3})/(n + \frac{1}{3})$
2	1	.293	.286
3	1	.206	.200
4	1	.159	.154
4	2	.386	.385
5	1	.129	.125
5	2	.314	.312

TABLE 2-3.   Ideal depths for letter values.

Tag	Tail Area	Ideal Depth
M	$\frac{1}{2}$	$\frac{1}{2}(n + 1)$
F	$\frac{1}{4}$	$\frac{1}{4}(n + \frac{5}{3})$
E	$\frac{1}{8}$	$\frac{1}{8}(n + 3)$
D	$\frac{1}{16}$	$\frac{1}{16}(n + \frac{17}{3})$
C	$\frac{1}{32}$	$\frac{1}{32}(n + 11)$
B	$\frac{1}{64}$	$\frac{1}{64}(n + \frac{65}{3})$
A	$\frac{1}{128}$	$\frac{1}{128}(n + 43)$
Z	$\frac{1}{256}$	$\frac{1}{256}(n + \frac{257}{3})$
Y	$\frac{1}{512}$	$\frac{1}{512}(n + 171)$

depths, we must go into more detail. The fourths are the easiest place to begin, and we represent n as $n = 4j + i$ with $i = 1, 2, 3,$ or 4 to allow for the four possible cases. For $i = 1$, the median is at depth $2j + 1$, and our conventional choice for the depth of the fourth is $j + 1$, which is within $\frac{1}{3}$ of the ideal value $j + \frac{2}{3}$. For $i = 2$, the median is at depth $2j + \frac{3}{2}$, and the conventional depth for the fourth is again $j + 1$, which is now to be compared with $j + \frac{11}{12}$, the ideal choice, revealing a departure of only $\frac{1}{12}$. Similar calculations for $i = 3$ yield a conventional depth of $j + \frac{3}{2}$ and an ideal depth of $j + \frac{7}{6}$. Finally, for $i = 4$, the conventional and ideal values are $j + \frac{3}{2}$ and $j + \frac{17}{12}$, respectively. On the whole, these differences between the conventional depth and the ideal depth for the fourths are not large: half are $\frac{1}{3}$ and half are $\frac{1}{12}$.

We can carry out a similar detailed analysis for eighths, sixteenths, and so on, representing n in the form $8j + i$, $16j + i$, and so on. Table 2-4 summarizes the results for fourths and also gives a summary of the analysis for sixteenths. Table 2-5 examines, for letters as far as Y, the cases in which the conventional depth exceeds the ideal depth by $\frac{1}{2}$ or more. The pattern that emerges is this:

1. The conventional depth is always deeper into the batch than the ideal depth.
2. Most of the time the difference is less than $\frac{1}{2}$ unit (always through F, $\frac{3}{4}$ of the time for E and D, and about $\frac{2}{3}$ of the time thereafter).
3. The conventional depth is never as much as 1 unit deeper into the batch than the ideal depth.

TABLE 2-4. **Comparison of conventional and ideal depths for fourths and sixteenths.**

		Depth MINUS j		
		Conventional	Ideal	Difference
Fourths	$n - 4j =$			
	1	1	$\frac{2}{3}$	$+\frac{1}{3}$
	2	1	$\frac{11}{12}$	$+\frac{1}{12}$
	3	$\frac{3}{2}$	$\frac{7}{6}$	$+\frac{1}{3}$
	4	$\frac{3}{2}$	$\frac{17}{12}$	$+\frac{1}{12}$
Sixteenths	$n - 16j =$			
	1	1	$\frac{5}{12}$	$\frac{7}{12}$
	2	1	$\frac{23}{48}$	$\frac{25}{48}$
	3	1	$\frac{13}{24}$	$\frac{11}{24}$
	4	1	$\frac{29}{48}$	$\frac{19}{48}$
	5	1	$\frac{2}{3}$	$\frac{1}{3}$
	6	1	$\frac{35}{48}$	$\frac{13}{48}$
	7	1	$\frac{19}{24}$	$\frac{5}{24}$
	8	1	$\frac{41}{48}$	$\frac{7}{48}$
	9	$\frac{3}{2}$	$\frac{11}{12}$	$\frac{7}{12}$
	10	$\frac{3}{2}$	$\frac{47}{48}$	$\frac{25}{48}$
	11	$\frac{3}{2}$	$\frac{25}{24}$	$\frac{11}{24}$
	12	$\frac{3}{2}$	$\frac{53}{48}$	$\frac{19}{48}$
	13	$\frac{3}{2}$	$\frac{7}{6}$	$\frac{1}{3}$
	14	$\frac{3}{2}$	$\frac{59}{48}$	$\frac{13}{48}$
	15	$\frac{3}{2}$	$\frac{31}{24}$	$\frac{5}{24}$
	16	$\frac{3}{2}$	$\frac{65}{48}$	$\frac{7}{48}$

Because the fractions involved become increasingly complex, we would not want to use the ideal depths in hand calculation. And when we are mixing computer work with hand calculation, a need for uniformity will generally dissuade us from using the ideal definition in the computer programs.

We would *not* want to use the ideal depth when n is small because this loses resistance. For example, when $n = 5$, the ideal depth of a fourth is $1\frac{2}{3}$, and, as a result, the value of the fourth would be affected by the presence of a single wild value. The conventional depth of a fourth, on the other hand, is 2, and the fourth calculated in this way would be unaffected by the single wild value.

TABLE 2-5. Cases in which the conventional depth exceeds the ideal depth by $\frac{1}{2}$ or more.

Tag	Tail Area	Conventional MINUS Ideal $\geqslant \frac{1}{2}$ For What n	Fraction of Cases
F	$\frac{1}{4}$	none	0
E	$\frac{1}{8}$	$n = 4j + 1$	$\frac{1}{4} = .25$
D	$\frac{1}{16}$	$n = 8j + 1, 8j + 2$	$\frac{1}{4} = .25$
C	$\frac{1}{32}$	$n = 16j + 1$ to $16j + 5$	$\frac{5}{16} = .31$
B	$\frac{1}{64}$	$n = 32j + 1$ to $32j + 10$	$\frac{5}{16} = .31$
A	$\frac{1}{128}$	$n = 64j + 1$ to $64j + 21$	$\frac{21}{64} = .33$
Z	$\frac{1}{256}$	$n = 128j + 1$ to $128j + 42$	$\frac{21}{64} = .33$
Y	$\frac{1}{512}$	$n = 256j + 1$ to $256j + 85$	$\frac{85}{256} = .33$

Some further analysis of bias and variance for conventional and ideal letter values indicates that little would be gained in these terms by using the ideal definition.

2F. WHEN THE LETTER VALUES ARE EQUALLY SPACED

In working toward a better understanding of the letter values and what they tell us about a batch or a distribution, we may find it helpful to ask whether there is a (theoretical) distribution whose letter values are equally spaced; that is, C is as far beyond D as D is beyond E, and so on. Exercise 8 involves constructing such a distribution. Also, it turns out that a well-known distribution has this property to a reasonably good approximation from the eighths on out.

Table 2-6 shows the upper letter values of the logistic distribution, whose cumulative distribution function is

$$F(x) = \frac{e^x}{1 + e^x}. \tag{11}$$

The differences from M to F and from F to E are substantially larger than those further out, but beyond D the difference changes by only a few percent. For most purposes, then, we can think of the logistic distribution as having equally spaced letter values. As the tail area becomes smaller, the difference between successive letter values approaches the limiting value of

$\log_e 2 = 0.69315$. This is easy to see because the inverse function F^{-1} can be written in the closed form [by straightforward algebra from equation (11)]:

$$F^{-1}(u) = \log_e\left(\frac{u}{1-u}\right). \qquad (12)$$

For an upper tail area of 2^{-k}, $u = 1 - 2^{-k}$ and $F^{-1}(u) = \log_e(2^k - 1)$, so that the difference between this letter value and the one just beyond it is

$$\log_e\left(\frac{2^{k+1} - 1}{2^k - 1}\right) = \log_e\left(\frac{2 - 1/2^k}{1 - 1/2^k}\right), \qquad (13)$$

and as k becomes large, this goes to $\log_e 2$.

For many purposes, the Gaussian distribution serves as the ideal model for symmetric data, so we have included its letter values and their differences for comparison in Table 2-6. The Gaussian and logistic distributions are not so very different in shape, and, as a result, the Gaussian letter

TABLE 2-6.   Letter values and differences for the standard logistic and standard Gaussian distributions.

Tag	Tail Area	Logistic		Gaussian	
		Letter Value	Diff	Letter Value	Diff
M	$\frac{1}{2}$	0		0	
			1.099		.674
F	$\frac{1}{4}$	1.099		0.674	
			.847		.476
E	$\frac{1}{8}$	1.946		1.150	
			.762		.384
D	$\frac{1}{16}$	2.708		1.534	
			.726		.329
C	$\frac{1}{32}$	3.434		1.863	
			.709		.291
B	$\frac{1}{64}$	4.143		2.154	
			.701		.264
A	$\frac{1}{128}$	4.844		2.418	
			.697		.242
Z	$\frac{1}{256}$	5.541		2.660	
			.695		.226
Y	$\frac{1}{512}$	6.236		2.886	
			.694		.212
X	$\frac{1}{1024}$	6.930		3.097	

values do not depart violently from equal spacing, but they do trend steadily closer together as the tail area decreases. (Of course, this must happen because the Gaussian probability density function goes to zero more rapidly.) The letter values and differences are larger for the logistic distribution, primarily because we have simply taken two "standard" distributions and made no attempt to match them with respect to scale or spread. For example, the standard logistic distribution has variance $\pi^2/3 = 3.29$ and hence standard deviation 1.81. Another indication of the difference in spread is the pseudosigmas [defined in equation (6)], which range from 1.63 at F to 2.24 at X. Even without matching, however, the pattern of the differences is clear, and that is what matters.

2G. LETTER VALUES AS SELECTED ORDER STATISTICS

By defining the letter-value depths as we did in equation (3), we have selected the letter values from the full set of order statistics for a sample of n, interpolating as necessary halfway between adjacent order statistics. To conclude this chapter, we make a rough assessment of how accurately this particular set of selected order statistics summarizes the remaining, unselected, order statistics.

If we were to approach this task quite systematically, we could identify at least two steps in selecting a set of order statistics: first decide how many to select and then choose which ones. This is sometimes done for a particular purpose in estimation, such as estimating the location parameter or the scale parameter of a distribution by using a linear combination of order statistics. In those situations, it is possible to guide the choice by minimizing the variance of the resulting estimator, but we simply want to summarize the batch without emphasizing any single attribute of a possible underlying distribution. Thus we ask how well we can predict the value of an unselected order statistic by using the nearest selected order statistic. We measure this in terms of the squared correlation coefficient, just as is often done in regression, as a way of giving the percentage of variance explained and therefore the closeness of prediction.

As random variables, the full set of order statistics is

$$X_{(1)}, X_{(2)}, \ldots, X_{(n)}.$$

If we let $p = i/n$ and $q = j/n$, then when the data come from a continuous distribution satisfying rather mild regularity conditions, it follows from a theorem of Mosteller (1946; see also David, 1970, Section 9.2) that as n

becomes large,

$$\text{corr}^2\left[X_{(i)}, X_{(j)}\right] \approx \frac{p/(1-p)}{q/(1-q)},\tag{14}$$

as long as $p \leqslant q$. We can use this result to find the approximate correlation between successive letter values in large samples. For example, to calculate the correlation between the median and the upper fourth, we take $p = \frac{1}{2}$ and $q = \frac{3}{4}$. Evaluating equation (14) gives $\frac{1}{3}$ as the value of corr^2, and the value of the correlation is .577. Table 2-7 shows the results for various letter values. The squared correlation is lowest for M and F (.333), but it reaches nearly .500 fairly rapidly.

Let us look at the worst situation. The closer an order statistic is to one of the selected ones, the better its value will be predicted by the value of the selected order statistic. Thus between two consecutive selected order statistics, $X_{(i)}$ and $X_{(j)}$, the order statistic least accurately predicted by either $X_{(i)}$ or $X_{(j)}$ lies essentially halfway between them. Its squared correlation with

TABLE 2-7.   Correlation between successive letter values in large samples.

Tag	Tail Area	corr^2	corr
M	$\frac{1}{2}$		
		.333	.577
F	$\frac{1}{4}$		
		.429	.655
E	$\frac{1}{8}$		
		.467	.683
D	$\frac{1}{16}$		
		.484	.696
C	$\frac{1}{32}$		
		.492	.701
B	$\frac{1}{64}$		
		.496	.704
A	$\frac{1}{128}$		
		.498	.706
Z	$\frac{1}{256}$		
		.499	.706
Y	$\frac{1}{512}$		
		.500	.707
X	$\frac{1}{1024}$		

each of $X_{(i)}$ and $X_{(j)}$ is approximately equal to the square root of $\text{corr}^2[X_{(i)}, X_{(j)}]$. For the more extreme letter values, this is roughly $\sqrt{.500}$ = .707, so that the correlation with each of $X_{(i)}$ and $X_{(j)}$ is roughly $\sqrt{.707}$ = .841. For unselected order statistics between the median and the fourth, the correlation (with one or the other) is at least $\sqrt{.577}$ = .760. These correlations, which come about because the order statistics are in order and hence subject to rather strong restrictions on how much their values can vary when adjacent order statistics are taken as fixed, are already quite substantial; they involve only an unselected order statistic and the one nearest selected order statistic.

To get an indication of the situation in moderate-sized samples, we can use the available tables of variances and covariances of order statistics in Gaussian samples (Tietjen, Kahaner, and Beckman, 1977) and calculate the squared correlation exactly. Table 2-8 shows the results for $n = 33$, chosen so that all the letter values are order statistics and so that there are more than a few unselected order statistics. As expected, the squared correlation is smallest for the order statistics $[X_{(13)}, X_{(7)}, \text{and } X_{(4)}]$ that lie halfway

TABLE 2-8. Prediction of order statistics by nearest letter values in Gaussian samples, $n = 33$.

Tag	Depth	corr^2 with		Large-Sample corr^2
		Less Extreme	More Extreme	
M	17	↓		
	16	.889		
	15	.789		
	14	.699		
	13	.618	.578	.577
	12		.657	
	11		.750	
	10		.863	
F	9	↓	↑	
	8	.853		
	7	.716	.656	.655
	6		.798	
E	5	↓	↑	
	4	.764	.709	.683
D	3		↑	
C	2			
1	1			

between letter values. In this instance, each of these order statistics has two nearest letter values, and the squared correlation is lower with the more extreme of the two. Between M and F and between F and E, this smaller value is very nearly equal to the theoretical value for large samples. Thus the large-sample result seems quite useful in moderate-sized samples, and prediction from only the nearest letter value would be satisfactory.

If we analyzed the accuracy of prediction by using the squared correlation between an unselected order statistic and all possible linear combinations of the two adjacent selected order statistics, the picture would be even more encouraging. In summary, then, the letter values are a very effective set of selected order statistics. Heuristically speaking, little of the information in the ordered sample is lost when we use the letter values to summarize it.

2H. SUMMARY

In comparison with classical summary measures such as the sample mean and standard deviation, exploratory techniques gain resistance by using summaries based on the ordered observations. One set of selected order statistics, the letter values, begins with the median and the fourths and reaches to the extremes by repeatedly halving the tail area. To simplify hand computation, the letter values involve no more than averaging adjacent order statistics, but their depths remain close to the ideal values that result when we adopt $(i - \frac{1}{3})/(n + \frac{1}{3})$ as the fraction to the left of $x_{(i)}$.

A two-column display format makes it easy to scan the letter values and calculate such spread measures as the fourth-spread, $d_F = F_U - F_L$. A convenient rule of thumb uses the fourths and the fourth-spread to set up cutoffs at $F_L - 1.5d_F$ and $F_U + 1.5d_F$ and flags any observations beyond these as "outside," deserving further investigation as possible outliers.

Some study of the theoretical spacing of the letter values reveals that the more extreme letter values of the Gaussian distribution become progressively closer together, whereas those of the logistic distribution become equally spaced.

As a basis for predicting other order statistics in large samples, the nearest letter value yields a squared correlation equal to at least .577 between the median and either fourth. Between more extreme pairs of letter values, the minimum squared correlation with the nearest letter value is higher, rising quickly to nearly .707. These and other encouraging results indicate that the letter values capture much of the information in large and small samples.

REFERENCES

Bendixen, H. H. (1977). "The cost of intensive care." In J. P. Bunker, B. A. Barnes, and F. Mosteller (Eds.), *Costs, Risks, and Benefits of Surgery*. New York: Oxford University Press, pp. 372–384.

Blom, G. (1958). *Statistical Estimates and Transformed Beta-Variables*. New York: Wiley.

David, H. A. (1970). *Order Statistics*. New York: Wiley.

Hoaglin, D. C., Iglewicz, B., and Tukey, J. W. (1981). "Small-sample performance of a resistant rule for outlier detection," *1980 Proceedings of the Statistical Computing Section*. Washington, DC: American Statistical Association, pp. 148–152.

Mosteller, F. (1946). "On some useful 'inefficient' statistics," *Annals of Mathematical Statistics*, **17**, 377–408.

Tietjen, G. L., Kahaner, D. K., and Beckman, R. J. (1977). "Variances and covariances of the normal order statistics for sample sizes 2 to 50." In D. B. Owen and R. E. Odeh (Eds.), *Selected Tables in Mathematical Statistics, Vol. 5*. Providence, RI: American Mathematical Society, pp. 1–73.

Additional Literature

Barnett, V. (1978). "The study of outliers: purpose and model," *Applied Statistics*, **27**, 242–250.

Barnett, V. and Lewis, T. (1978). *Outliers in Statistical Data*. Chichester: Wiley.

Brown, B. M. (1981). "Symmetric quantile averages and related estimators," *Biometrika*, **68**, 235–242.

Lin, P.-E., Wu, K.-T., and Ahmad, I. A. (1980). "Asymptotic joint distribution of sample quantiles and sample mean with applications," *Communications in Statistics*, **A9**, 51–60.

Parzen, E. (1979a). "Nonparametric statistical data modeling (with discussion)," *Journal of the American Statistical Association*, **74**, 105–131.

———— (1979b). "A density-quantile function perspective on robust estimation." In R. L. Launer and G. N. Wilkinson (Eds.), *Robustness in Statistics*. New York: Academic, pp. 237–258.

Rosner, B. (1975). "On the detection of many outliers," *Technometrics*, **17**, 221–227.

EXERCISES

1. Show that the expected value of the trimean for a symmetric distribution is the same as the expected values for the mean and median of the batch.

2. Calculate the trimean for the patient survival data. Compared to the median, what does the trimean tell us about the symmetry of the data?

3. For the mean menstrual cycle lengths in Table 1-1, calculate the
   (a) Mean
   (b) Median
   (c) Trimean
   (d) 5-Number summary.

4. Using the data on aluminum die castings in Figure 1-3, calculate the values as in Exercise 3.

5. Compare results for Exercises 3 and 4. How well do the three measures of location agree among themselves for each of the three data sets? What does this suggest about the underlying distributions? What additional information can we get from the 5-number summaries for these three data sets?

6. We can use the 5-number summary to learn about the shape of a distribution. For example, if the difference between the upper fourth and the median is greater than that between the median and the lower fourth, what kind of shape does the underlying distribution have? Why might we expect this behavior in counted data such as population figures?

7. Find the letter values M, F, E, and D, and their differences (using published tables where helpful) for the following distributions:
   (a) t-distribution on 4 degrees of freedom
   (b) t-distribution on 30 degrees of freedom
   (c) Chi-squared distribution on 5 degrees of freedom
   (d) Uniform distribution on [0, 1].
   How do these results compare with those in Table 2-6?

8. Find the probability density function of a continuous distribution centered at zero that has its letter values equally spaced.

9. About how many outside values should we expect (on Gaussian theory)
   (a) In a single batch of 120 observations?
   (b) In total for two batches of 60?
   (c) In total for batches of 40, 30, 20, 10, 5, 5, 5, and 5?

10. If you had a distribution that you knew was heavy-tailed and found j values outside in a batch of 120, how would you feel if j were 5, 10, 20, 40, 58, or 62?

11. For other values of n near 33 (so that not all the letter values have integer depths), use a table of the covariances of Gaussian order

statistics to construct tables of squared correlations like Table 2-8. Compare the minimum squared correlations among these tables (and Table 2-8).

12. From available tables of order-statistic covariances for nonGaussian distributions (e.g., logistic, Cauchy) construct tables of squared correlations parallel to Table 2-8 (all for $n = 33$). Examine how the minimum squared correlations vary with distribution.

CHAPTER 3

Boxplots and Batch Comparison

John D. Emerson
Middlebury College

Judith Strenio
Westat Inc.

A graphical display of the five-number summary of a batch of numbers—the boxplot—shows much of the structure of the batch. From a boxplot we can pick out the following features of a batch:

Location
Spread
Skewness
Tail length
Outlying data points.

Thus the boxplot provides a visual impression of several important aspects of the empirical distribution of a batch of data.

This compact visual display is especially useful for comparing several batches of data. By drawing a boxplot for each batch and arranging them in parallel, we can compare the batches with respect to location and spread, and perhaps also skewness and tail heaviness. In this comparison, we may find that the data from the different batches do not all fit well into the same scale. In particular, batches located far from the origin may be much more spread out than batches located near the origin. Thus if the batches are plotted on a common scale, the details of batches close to the origin will be harder to see.

An appropriate transformation can often alleviate this difficulty by making the variability of the batches more nearly comparable. In drawing some guidance from the data as to what transformations may achieve these objectives, a spread-versus-level plot may suggest a power transformation that tends to equalize spread across different levels, or locations, of the batches.

Throughout this chapter, we restrict our examples to batches of measured or counted data. We assume that observations are nonnegative and possibly quite large. The origin then provides a lower bound, but there is no upper bound. Thus we do not consider the special features of such types of data as fractions bounded above by 1, and percents bounded above by 100%. (Such assumptions are not needed in Sections 3A and 3B.)

3A. THE BOXPLOT FOR A SINGLE BATCH

We introduce the boxplot for a single batch of data, using an example from the 1960 Census.

EXAMPLE:

The World Almanac (1967) reported the populations of United States cities; Table 3-1 gives the populations (to the nearest 10,000) of the 15 largest cities. We form the boxplot in order to pick out major features of the batch. In particular, we ask whether the batch of 15 cities is skewed and whether it has outlying data points.

As a first step in analysis, we construct the 5-number summary (see Section 2D), and we calculate the fourth-spread and the cutoffs for outliers based on the fourth-spread. The 5-number summary displays the median, fourths, and extremes of the batch:

#	15	U.S. Cities	
M	8	88	
F	4.5	74	184
	1	63	778

The *fourth-spread* is the range of the data defined by the upper fourth and lower fourth. It is closely related to the *interquartile range*, although technical differences between quartiles and fourths distinguish the two concepts.

Data values that are far enough beyond the fourths are considered as potential *outliers*. We use the fourth-spread to make this vague concept precise and give technical meaning to the term "outlier." Specifically, we

TABLE 3-1. Populations of the 15 largest U.S. cities in 1960.

City	Population (10,000s)
New York	778
Chicago	355
Los Angeles	248
Philadelphia	200
Detroit	167
Baltimore	94
Houston	94
Cleveland	88
Washington, DC	76
St. Louis	75
Milwaukee	74
San Francisco	74
Boston	70
Dallas	68
New Orleans	63

Source: *The World Almanac*, 1967 edition. New York: Newspaper Enterprise Association, Inc. (Data from p. 323).

define $F_L - \frac{3}{2}d_F$ and $F_U + \frac{3}{2}d_F$ as the *outlier cutoffs*, where F_L and F_U denote the fourths and d_F is $F_U - F_L$, the fourth-spread. Data values that are smaller than $F_L - \frac{3}{2}d_F$ or larger than $F_U + \frac{3}{2}d_F$ are called outliers and will receive special attention.

For the 15 cities,

$$d_F = 184 - 74 = 110,$$

$$F_L - \tfrac{3}{2}d_F = 74 - \tfrac{3}{2} \times 110 = -91,$$

and

$$F_U + \tfrac{3}{2}d_F = 184 + \tfrac{3}{2} \times 110 = 349.$$

Thus the outlier cutoffs are -91 and 349, and so cities with populations over 3,490,000 (namely, New York and Chicago) are classified as outliers.

To construct the boxplot, we first draw a box with ends at the lower fourth and upper fourth and a crossbar at the median. Next, we draw a line from each end of the box to the most remote point that is not an outlier.

The resulting figure schematically represents the body of data minus the outliers. The outliers are represented individually by xs situated beyond the outlier cutoffs.

Figure 3-1 shows the boxplot for the U.S. cities. To conserve space, this plot is drawn horizontally, but vertical plots may be more appropriate in some settings.

The boxplot shows at a glance the location, spread, skewness, tail length, and outlying data points. The *location* of the batch is summarized by the median, the crossbar in the interior of the box. The length of the box shows the *spread*, using the fourth-spread. From the relative positions of the median, the upper fourth, and the lower fourth, we also see some of the *skewness*; the median is much closer to the lower fourth than to the upper fourth, indicating that the batch is positively skewed—a common situation with unbounded positive data. The plot indicates *tail length* by the lines extended to New Orleans and to Los Angeles, and by the *outliers* (Chicago and New York).

The message of the boxplot for the 15 largest U.S. cities is strong: the batch is heavily skewed, and there are two outlying data points. But the greatest value of the boxplot is its ability to convey visually some important information about the shape of this batch of data.

Resistance of the Boxplot

We have seen that the construction of the rectangular box in the boxplot needs the median and the fourths of a data set. Because the median and the fourths are resistant to the impact of a few wild data values, the boxplot is also resistant to gross influence by these values. More specifically, up to 25% of the data values can be made arbitrarily large ("wild") without greatly disturbing the median, the fourths, or the rectangular box in the boxplot.

The "tails" of the boxplot are determined primarily by the most extreme data values that are within the outlier cutoffs. Thus they are relatively undisturbed by gross changes in the values of any outliers, and they can be only modestly affected by gross changes of values originally within the outlier cutoffs. Of course, the outlier cutoffs themselves are defined using

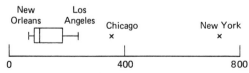

Figure 3-1. Boxplot for the 15 largest U.S. cities in 1960.

only the fourths of a batch. Thus they can resist gross disturbances in up to 25% of the data.

The resistance properties of the boxplot make it attractive for use in exploratory data analysis. Although an analogous plot could be based on the sample mean and sample standard deviation, such a plot would necessarily lack resistance to the influence of even a single wild data value.

Our definition of outliers as data values that are smaller than $F_L - \frac{3}{2}d_F$ or larger than $F_U + \frac{3}{2}d_F$ is somewhat arbitrary, but experience with many data sets indicates that this definition serves well in identifying values that may require special attention. Section 2C describes the relationship between fourth-spread and standard deviation for a Gaussian distribution. We show below that if the cutoffs are applied to a Gaussian distribution, then .7% of the population is outside the outlier cutoffs; this figure provides a standard of comparison for judging the placement of the outlier cutoffs (cp. p. 40).

Comparison with Classical Methods

The boxplot shows characteristics that derive from the actual data, not from an assumed distributional form. It is helpful to contrast the boxplot with the familiar visual display of the sample mean, $\bar{x}$, plus and minus 1.96 times the sample standard deviation, s. The latter sketch is often used when a batch resembles a random sample from a population believed to be single-humped, perhaps vaguely like a Gaussian distribution:

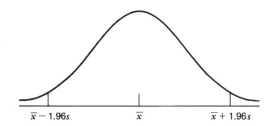

$$\overline{x} - 1.96s \qquad \overline{x} \qquad \overline{x} + 1.96s$$

The interval mentioned would contain roughly 95% of the population if the true mean and standard deviation of the population were $\bar{x}$ and s. When we cannot or do not assume a distributional form for our data, we can use the boxplot to show analogous features of location and width.

We know what a boxplot shows about the data in hand, regardless of their origin. So in the distribution-free setting, we can more easily interpret a boxplot. But we also may wonder what the boxplot represents for a random sample from a Gaussian population. We next explore this question for very large samples.

EXAMPLE: APPLICATION TO A GAUSSIAN POPULATION

Consider the standard Gaussian distribution, with mean 0 and variance 1. We look for population values of this distribution that are analogous to the sample values used in the boxplot.

For a symmetric distribution, the median equals the mean, so the population median of the standard Gaussian distribution is 0. The population fourths are -0.6745 and 0.6745, so the population fourth-spread is 1.349, or about $\frac{4}{3}$. Thus $\frac{3}{2}$ times the fourth-spread is 2.0235 (about 2). The population outlier cutoffs are ± 2.698 (about $2\frac{2}{3}$), and they contain 99.3% of the distribution.

We can gain some further understanding of the values chosen to define the boxplot by considering the population values of the median, fourths, and outlier cutoffs for several familiar distributions. Table 3-2 shows these values, together with the probabilities beyond the outlier cutoffs.

In this table we see that for large samples from extremely short-tailed symmetric distributions, exemplified here by the uniform distribution, all the data values tend to fall within the outlier cutoffs. For the standard Gaussian distribution just discussed for very large samples, we expect only seven-tenths of one percent of the values to be outliers. We choose the t-distributions with various numbers of degrees of freedom to represent symmetric, heavy-tailed distributions. As the tails become heavier, we have a greater probability of observing outliers. Thus we can judge whether our data seem heavier-tailed than Gaussian by how many points fall beyond the outlier cutoffs.

Table 3-2 also includes the chi-squared distributions as examples of asymmetric distributions. These range from the extremely skewed χ_1^2 to the more nearly symmetric χ_5^2 and χ_{20}^2. We find one trait that often occurs in skewed situations (it happened in our example of U.S. cities): the lower outlier cutoff is below the smallest possible data value. Thus the probability of an outlier on this side is 0, and so we get an indication of skewness not only from the relative position of median and fourths, but also from whether all outliers are on one side of the box.

The preceding discussion describes what happens for boxplots of very large samples from some familiar distributions. What happens for smaller samples? The smallest sample size that allows comparison is five, for which the fourths are given by the second smallest and second largest data points. In a sample of size four, the largest data value cannot be outside the outlier cutoffs. This follows from the fact that the upper fourth is the average of the two largest data values and thus involves the extreme value in a direct way.

To examine the question of sampling behavior for small samples, we carried out a simulation study for samples of size five from a standard Gaussian distribution. The results of this experiment suggest that 67% of the

TABLE 3-2. **Population values of median, fourths, and outlier cutoffs, and percent outliers for various distributions.**

Distribution	M^a	Upper[b] Fourth	Outlier[c] Cutoffs	Total[d] % Out	Value[e] of 1.96σ	% Outside $\mu \pm 1.96\sigma$
			Symmetric			
$U(-1,1)$	0	0.500	±2.000	none	1.132	none
$N(0,1)$	0	0.674	±2.698	0.70	1.960	5.00
t_{20}	0	0.687	±2.748	1.24	2.066	5.20
t_{10}	0	0.700	±2.800	1.88	2.191	5.32
t_5	0	0.727	±2.908	3.35	2.530	5.25
t_1	0	1.000	±4.000	15.59	—	—
			Nonsymmetric			
		0.102	-1.730^f		-1.772	
χ_1^2	0.45			7.58		5.22
		1.323	3.155		3.772	
		2.675	-3.252^f		-1.198	
χ_5^2	4.35			2.80		4.78
		6.626	12.552		11.198	
		15.452	2.888		7.604	
χ_{20}^2	19.34			1.39		4.53
		23.828	36.392		32.396	

[a]M = median of distribution. Defined so that $F(M) = .5$, where F is the cumulative distribution function.

[b]Upper fourth is the value above which .25 of the probability lies. (Lower fourth has .25 of probability below it.) For the nonsymmetric distributions, the entries in this column are the lower fourth and the upper fourth.

[c]Upper outlier cutoff = upper fourth $+ \frac{3}{2} \times$ (upper fourth $-$ lower fourth). (Lower outlier cutoff = lower fourth $-$ same quantity.)

[d]% Out = percent of probability below the lower outlier cutoff or above the upper outlier cutoff.

[e]For $U(-1, 1)$, $\sigma = \sqrt{1/3} \approx .58$ and $\mu = 0$. For t_ν, $\sigma = \sqrt{\nu/(\nu - 2)}$ for $\nu > 2$ and $\mu = 0$. For χ_ν^2, $\mu = \nu$ and $\sigma = \sqrt{2\nu}$.

[f]For skewed distributions, one of the pair of cutoffs often falls beyond the range of possible values.

samples had no values beyond the outlier cutoffs. For the remaining samples with values beyond the cutoffs, 24% had one value outside the outlier cutoffs and 9% had two values that were outliers, a total of 33%. Thus we can expect

About a tenth of Gaussian samples of five to have both extremes outside the outlier cutoffs,

About a quarter of them to have just *one* extreme outside the outlier cutoffs, and

About two-thirds of them to have no outliers.

How should we compare these results with the large-sample results? We must examine the details carefully because the two sets of results do not have parallel structure. For samples of size five, 0% (none) of the data values are outliers with probability about .67, 20% (one value in five) are outliers with probability about .24, and 40% (two values in five) are outliers with probability .09. For "infinitely large samples," the analogous result (Table 3-2) is that .7% of the data lie beyond the outlier cutoffs *all* of the time. (The approximate formulas on p. 40 help to fill the gap.)

The directions and, probably, something of the amounts of difference in Table 3-2 do apply in a qualitative way to small samples. Table 3-2 can occasionally be helpful, but it is far from the whole story about such samples.

With Table 3-2 and the sampling experiment, we conclude our introduction to the boxplot. We are now ready to consider its use in the comparison of batches.

3B. COMPARING BATCHES USING BOXPLOTS

A display of parallel boxplots can facilitate the comparison of several batches of data. From the display we can see similarities and differences among the batches with respect to each of the five features already discussed.

EXAMPLE: LARGEST CITIES IN 16 COUNTRIES

The 1967 *World Almanac* lists 16 countries that have 10 or more large cities; among these, we selected the 10 most populous cities. Table 3-3 gives the populations of these cities, in 100,000s. Table 3-4 provides the 16 5-number summaries, the outlier cutoffs, and the outliers for these batches.

TABLE 3-3. Population (in 100,000s) of 10 largest cities[a] for 16 countries[b] as reported by the 1967 *World Almanac*, using latest available census figures or official estimates. The populations are reported in 100,000s (not rounded).

(1) Sweden		(2) Netherlands		(3) Canada	
Stockholm	7.87	Amsterdam	8.68	Montreal	11.91
Goteborg	4.22	Rotterdam	7.31	Toronto	6.72
Malmo	2.49	The Hague	6.02	Vancouver	3.84
Norrkoping	0.94	Utrecht	2.64	Edmonton	2.81
Vasteras	0.89	Eindhoven	1.75	Hamilton	2.73
Uppsala	0.87	Haarlem	1.72	Ottawa	2.68
Orebro	0.81	Groningen	1.51	Winnipeg	2.65
Halsingborg	0.78	Tilburg	1.42	Calgary	2.49
Linkoping	0.71	Enschede	1.31	Quebec	1.71
Boras	0.69	Arnhem	1.29	London	1.69

(4) France		(5) Mexico		(6) Argentina	
Paris	28.11	Mexico City	31.18	Buenos Aires	29.66
Marseilles	7.83	Guadalajara	10.12	Rosario	7.61
Lyon	5.35	Monterrey	8.06	Cordoba	6.35
Toulouse	3.30	Juarcz	3.79	La Plata	4.10
Nice	2.94	Puebla	3.46	Avellaneda	3.80
Bordeaux	2.54	Mexicali	2.91	Santa Fe	2.75
Nantes	2.46	Leon	2.71	Mar del Plata	2.70
Strasbourg	2.33	Torreon	2.17	General San Martin	2.69
St. Etienne	2.03	Chihuahua	2.06	Tucuman	2.51
Lille	1.99	San Luis Potosi	1.86	Lanus	2.44

(7) Spain		(8) England		(9) Italy	
Madrid	25.99	London	79.86	Rome	23.59
Barcelona	16.96	Birmingham	11.02	Milan	15.80
Valencia	5.01	Liverpool	7.22	Naples	11.82
Seville	4.74	Manchester	6.38	Turin	11.14
Zaragoza	3.57	Leeds	5.09	Genoa	7.84
Bilboa	3.34	Sheffield	4.88	Palermo	5.90
Malaga	3.12	Bristol	4.30	Florence	4.54
Murcia	2.64	Coventry	3.30	Bologna	4.44
Cordoba	2.14	Nottingham	3.10	Catania	3.61
Palma	1.69	Kingston-upon-Hull	2.99	Venice	3.36

TABLE 3-3. (*Continued*)

(10) West Germany		(11) Brazil		(12) Soviet Union	
West Berlin	21.92	Sao Paulo	49.81	Moscow	63.34
Hamburg	18.56	Rio de Janeiro	38.57	Leningrad	36.36
Munich	11.42	Recite	9.68	Kiev	13.32
Cologne	8.27	Belo Horizonte	9.52	Baku	11.37
Essen	7.28	Salvador	8.08	Tashkent	10.90
Dusseldorf	7.02	Porto Alegre	8.03	Gorky	10.84
Frankfurt	6.94	Fortaleza	6.99	Kharkov	10.70
Dortmund	6.53	Curitiba	5.02	Novosibirsk	10.27
Bremen	5.84	Belem	4.95	Kuibyshev	9.50
Hannover	5.66	Niterol	2.78	Sverdlovsk	9.17

(13) Japan		(14) United States		(15) India	
Tokyo	110.21	New York City	77.81	Bombay	45.37
Osaka	32.14	Chicago	35.50	Calcutta	30.03
Nagoya	18.88	Los Angeles	24.79	Delhi	22.98
Yokohama	16.39	Philadelphia	20.02	Hyderabad	20.62
Kyoto	13.37	Detroit	16.70	Madras	17.25
Kobe	11.95	Baltimore	9.39	Howrah	16.11
Kita Kyushu	10.70	Houston	9.38	Ahmedabad	11.49
Kawasaki	7.89	Cleveland	8.76	Kanpur	9.47
Fukuoka	7.71	Washington, DC	7.63	Bangalore	9.07
Sapporo	7.04	St. Louis	7.50	Poona	7.21

(16) China[c]	
Shanghai	69.00
Beijing	40.10
Hong Kong	36.92
Tianjin	32.20
Shenyang	24.11
Wuhan	21.46
Chongqing	21.21
Canton	16.50
Xian	15.00
Nanjing	11.13

Source: *The World Almanac*, 1967 edition. New York: Newspaper Enterprise Association, Inc. (data from pp. 323, 672-673).

[a] If greater metropolitan area population and city limits population both given, the city limits population was recorded.

[b] These are the 16 countries with 10 or more cities listed. The countries are ordered by the median populations of the 10 cities.

[c] English spellings follow the Pinyin system introduced in 1979.

TABLE 3-4. 5-Number summaries, outlier cutoffs, and outliers for the populations of 10 largest cities from 16 countries. Populations in 100,000s.

(1) # 10   Sweden

M	5.5	0.9		d_F
F	3	0.8	2.5	1.7
	1	0.7	7.9	

1.5 d_F = 2.5
Outlier cutoffs: $(-1.7, 5.0)$
Outliers: 7.9

(2) # 10   Netherlands

M	5.5	1.7		d_F
F	3	1.4	6.0	4.6
	1	1.3	8.7	

1.5 d_F = 6.9
Outlier cutoffs: $(-5.5, 12.9)$
Outliers: none

(3) # 10   Canada

M	5.5	2.7		d_F
F	3	2.5	3.8	1.3
	1	1.7	11.9	

1.5 d_F = 1.9
Outlier cutoffs: $(0.6, 5.7)$
Outliers: 6.7, 11.9

(4) # 10   France

M	5.5	2.7		d_F
F	3	2.3	5.4	3.1
	1	2.0	28.1	

1.5 d_F = 4.6
Outlier cutoffs: $(-2.3, 10.0)$
Outliers: 28.1

(5) # 10   Mexico

M	5.5	3.2		d_F
F	3	2.2	8.1	5.9
	1	1.9	31.2	

1.5 d_F = 8.9
Outlier cutoffs: $(-6.7, 17.0)$
Outliers: 31.2

(6) # 10   Argentina

M	5.5	3.3		d_F
F	3	2.7	6.3	3.6
	1	2.4	29.7	

1.5 d_F = 5.4
Outlier cutoffs: $(-2.7, 11.7)$
Outliers: 29.7

(7) # 10   Spain

M	5.5	3.5		d_F
F	3	2.6	5.0	2.4
	1	1.7	26.0	

1.5 d_F = 3.6
Outlier cutoffs: $(-1.0, 8.6)$
Outliers: 17.0, 26.0

(8) # 10   England

M	5.5	5.0		d_F
F	3	3.3	7.2	3.9
	1	3.0	79.9	

1.5 d_F = 5.9
Outlier cutoffs: $(-2.6, 13.1)$
Outliers: 79.9

TABLE 3-4. (*Continued*)

(9)	# 10	Italy

M	5.5	6.9	d_F
F	3	4.4 11.8	7.4
	1	3.4 23.6	

$1.5\, d_F = 11.1$

Outlier cutoffs: $(-6.7, 22.9)$

Outliers: 23.6

(10)	# 10	West Germany

M	5.5	7.2	d_F
F	3	6.5 11.4	4.9
	1	5.7 21.9	

$1.5\, d_F = 7.3$

Outlier cutoffs: $(-0.8, 18.7)$

Outliers: 21.9

(11)	# 10	Brazil

M	5.5	8.1	d_F
F	3	5.0 9.7	4.7
	1	2.8 49.8	

$1.5\, d_F = 7.0$

Outlier cutoffs: $(-2.0, 16.7)$

Outliers: 38.6, 49.8

(12)	# 10	USSR

M	5.5	10.9	d_F
F	3	10.3 13.3	3.0
	1	9.2 63.3	

$1.5\, d_F = 4.5$

Outlier cutoffs: $(5.8, 17.8)$

Outliers: 36.4, 63.3

(13)	# 10	Japan

M	5.5	12.7	d_F
F	3	7.9 18.9	11.0
	1	7.0 110.2	

$1.5\, d_F = 16.5$

Outlier cutoffs: $(-8.6, 35.4)$

Outliers: 110.2

(14)	# 10	United States

M	5.5	13.0	d_F
F	3	8.8 24.8	16.0
	1	7.5 77.8	

$1.5\, d_F = 24.0$

Outlier cutoffs: $(-15.2, 48.0)$

Outliers: 77.8

(15)	# 10	India

M	5.5	16.7	d_F
F	3	9.5 23.0	13.5
	1	7.2 45.4	

$1.5\, d_F = 20.2$

Outlier cutoffs: $(-10.7, 43.2)$

Outliers: 45.4

(16)	# 10	China

M	5.5	22.8	d_F
F	3	16.5 36.9	20.4
	1	11.1 69.0	

$1.5\, d_F = 30.6$

Outlier cutoffs: $(-14.1, 67.5)$

Outliers: 69.0

The data set gives rise to many questions about the cities in these 16 nations. How do the median populations of major cities compare across nations? Are the smallest major cities in China larger than the largest cities of some other countries? Do the countries having larger cities tend to show more variability in city populations? Which cities are outliers, relative to other cities in their own countries? How much asymmetry is present in the various batches? These questions can, in principle, be answered by using the 5-number summaries from Table 3-4. But, in practice, a parallel display of

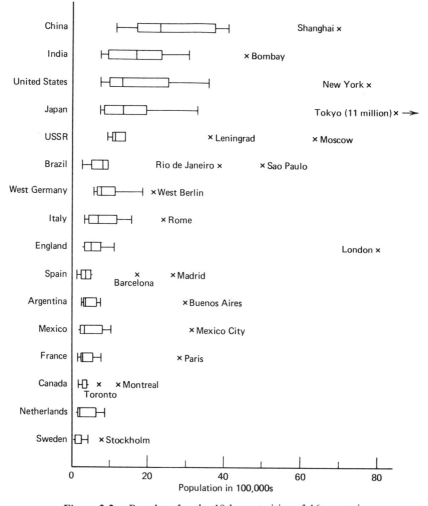

Figure 3-2. Boxplots for the 10 largest cities of 16 countries.

the boxplots for the 16 batches makes the answers to these and similar questions more readily apparent. Figure 3-2 gives this display, with the boxplots parallel to each other and to the population axis.

By ordering the countries in the display according to median city population for the 10 largest cities, the display enables easy comparison of location or level across countries. For example, we see that the largest cities in China tend to be larger than those for any other nation. All of China's largest cities are larger than all cities in the Netherlands and in Sweden. We can also easily compare spread for these 16 batches by using the lengths of the boxes. The data values for Canada are the closest together among the countries, and the values for China are the most dispersed.

We can use the parallel display to compare the *symmetry*, *tail length*, and *outliers* of the batches as well. Nearly all the countries give some indications of asymmetry in the direction of larger cities; only India and Brazil have boxes that are skewed to the left, but both of these countries have a few cities whose populations are substantially greater than the cities represented by the boxes. The largest city for every country except the Netherlands is a designated outlier, although some countries have more than one outlier among their 10 most populous cities. Thus we have noted two irregular features of these data: asymmetry and many outliers.

Because we ordered the countries by median value, we can pick out another comparative feature: a strong tendency for spread to increase as level does. This tendency is not compatible with a belief in equal variability across batches. Such an equality, when true or nearly true, often simplifies further analysis. To promote equality of spread and to reduce the dependence of spread on level, we can try re-expressing, or transforming, the data. We let the data point toward an appropriate power transformation by means of a spread-versus-level plot, introduced in Section 3C.

EXAMPLE: FUEL ECONOMY ESTIMATES FOR 1979 CARS AND LIGHT TRUCKS

For several years, the U.S. Environmental Protection Agency (EPA) has required that each model of automobile and light truck sold in the United States be tested for its fuel economy. Each make and type of car or truck is often represented by several different models that vary in engine size and number of cylinders, carburetion, transmission type, and other mechanical factors. Station wagons are considered distinct from sedans and are tested separately. Thus the following models are distinguished by the EPA and are separately certified for their fuel economy:

Chevrolet Impala sedan with 305 cubic inch V-8 engine and automatic transmission
Same car with 350 cubic inch engine and 4-barrel carburetor

Chevrolet Malibu sedan with V-6 engine and manual 3-speed transmission

The same model station wagon and with a 305 cubic inch V-8 engine and automatic transmission

Because of the distinctions made, over 500 different 1979 model automobiles and light trucks were tested and reported upon just before these models became available for sale (EPA, September 1978). Because of the heterogeneity among these vehicles, the EPA classifies them, by type and by carrying capacity, into 13 categories. Table 3-5 enumerates and explains the classifications, and Table 3-6 shows the number of models tested in each category.

As a result of the classification scheme, some heavier vehicles are classified as being smaller than other lighter vehicles. For example, a 1979 Lincoln Continental Mark V is a mid-size car, whereas the much lighter Chevrolet Impala is a large car. Many potential buyers of a new car or truck would presumably agree with the EPA that the carrying capacity of a vehicle is the most useful index of its size. The relationship between this index of size and fuel economy is probably of greater interest to the

TABLE 3-5.   EPA vehicle classes.

Two-Seater: Cars designed primarily to seat only two adults.
Sedans
  Minicompact: Less than 85 cubic feet of passenger and luggage volume.
  Subcompact: Between 85 and 100 cubic feet of passenger and luggage volume.
  Compact: Between 100 and 110 cubic feet of passenger and luggage volume.
  Mid-size: Between 110 and 120 cubic feet of passenger and luggage volume.
  Large: More than 120 cubic feet of passenger and luggage volume.
Station wagons
  Small: Less than 130 cubic feet of passenger and cargo volume.
  Mid-size: Between 130 and 160 cubic feet of passenger and cargo volume.
  Large: 160 or more cubic feet of passenger and cargo volume.
Truck classes
  Small pickup: Trucks having GVWR[a] under 4500 pounds.
  Standard pickup: Trucks having GVWR from 4500 to 6000 pounds.
  Van
Other special-purpose vehicles: All other light vehicles not
  included in another car or truck class.

[a]Gross Vehicle Weight Rating (GVWR) equals truck weight plus carrying capacity.

TABLE 3-6.   Number of models per category.

Cars	
Two-seater	15
Minicompact	25
Subcompact	109
Compact	36
Mid-size	91
Large	35
Station wagons	
Small	50
Mid-size	38
Large	13
Trucks	
Small pickup	23
Standard pickup	28
Van	25
Special-purpose vehicles	17
Total	505

consumer than is (say) the relationship between vehicle weight and fuel economy.

Twice each year, the EPA publishes a booklet that contains all the information mentioned above (and more); it also contains the fuel mileage ratings in miles per gallon for each of the approximately 500 vehicles tested. We do not report the findings here because they are extensive and because the booklet must by law be made available to consumers at all automobile and light truck dealerships.

The booklet published by the EPA confronts the consumer with a mass of numerical information about the 505 vehicles tested. The volume of information may discourage some consumers from finding the answers to their questions. Must increased interior space exact a penalty in fuel economy? Are station wagons less economical than sedans of comparable size? Are some mid-size cars as economical, or more economical, than most compact cars? For a given vehicle type, do certain vehicles get unusually good (or unusually bad) mileage? Figure 3-3 provides a visual summary of the fuel mileage data for vehicles in 13 classes; it enhances comparisons among classes, and it points to unusual data values in each vehicle group.

Remarks on Fuel Economy of 1979 Vehicles

At the risk of diverting attention from the statistical issues at hand, we yield to the temptation to add some remarks concerning the mileage figures for the 1979 vehicles. However, our remarks are limited to a few of the more visible aspects of Figure 3-3.

The parallel boxplots show that the smallest vehicles do tend to achieve higher mileage than the largest vehicles. However, this correspondence is far from perfect; note, for example, that mid-size cars achieve better mileage as

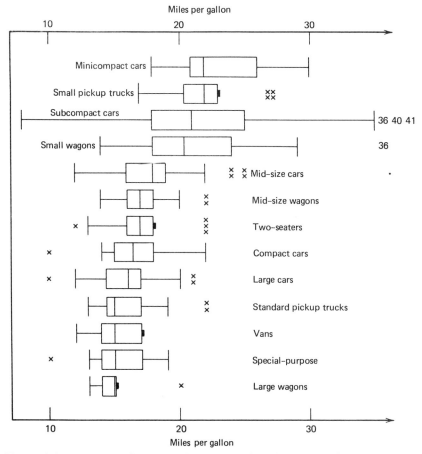

Figure 3-3. Boxplots of EPA fuel economy data for 1979 vehicles. (Numbers between 36 and 41 identify individual models and are not plotted to scale.) Data from *1979 Gas Mileage Guide*, Washington, DC: U.S. Department of Energy.

a group than do compact cars. And a few subcompact cars achieve the best mileage of all—better, even, than any of the minicompact cars. Although station wagons generally achieve mileage that is roughly comparable to their sedan counterparts, the largest wagons have the lowest mileage figures of all 13 groups. The variability in mileage for the largest wagons is also the smallest of all groups.

A striking feature of Figure 3-3 is the great variation in mileage figures for subcompact cars. The fact that this group contains the largest number of vehicles accounts for only a part of this great variation. A look at the raw data reveals that this class contains some strange bedfellows. It includes an exotic sports car (Aston Martin V8, at 8 mpg), an exotic high-performance sedan (Jaguar XJS, 12 cylinders, at 10 mpg), and a high-performance American GT coupe (Pontiac Firebird Trans-Am 400, at 12 mpg). At the opposite end of the scale, it includes many economical four-cylinder domestic and imported sedans, including the Chevettes, Datsuns, Toyotas, and Volkswagens. A more detailed analysis than the EPA's might categorize still further the 109 subcompact vehicles.

It is often interesting to take a closer look at the outliers. There are 26 outliers, representing 5% of the data. The four outliers that are below the lower outlier cutoff are displayed below.

Vehicle	MPG
Porsche 930	12
Jaguar XJ, 12 cylinder	10
Cadillac Limousine	10
Cadillac Commercial Chassis	10

The reader may be startled that the Jaguar is included among the compact vehicles. The anomaly lies in the classification system as we discussed above.

Table 3-7 gives the outliers above the upper outlier cutoff. A clear message is that diesel-engine vehicles get good mileage. Nearly all such vehicles are high-mileage outliers; the remaining diesels had mileage figures just inside the upper outlier cutoffs. (As we write this, diesel fuel is usually a few cents per gallon cheaper than gasoline. Diesel engines do not need routine tuneups. Perhaps the advantages of the diesel engine outweigh the sometimes-mentioned disadvantages: hard starting in cold weather and a characteristic diesel odor!)

Variations of Boxplots

The boxplots of Figure 3-3 do not take into account important differences in the numbers of models included in the various vehicle classes. Although

TABLE 3-7. High-mileage outliers.

Vehicle	MPG
Mazda Pickup (4 cylinders, 5-speed)	27
Mazda Pickup (4 cylinders, 4-speed)	27
Ford Courier Pickup (4 cylinders, 5-speed)	27
Ford Courier Pickup (4 cylinders, 4-speed)	27
Volkswagen Rabbit *Diesel* (5-speed)	41
Volkswagen Rabbit *Diesel* (4-speed)	40
Volkswagen Dasher *Diesel* Sedan	36
Volkswagen Dasher *Diesel* Wagon	36
Oldsmobile Cutlass Supreme *Diesel* (automatic)	24
Oldsmobile Cutlass Supreme *Diesel* (5-speed)	25
Oldsmobile Cutlass Salon *Diesel* (automatic)	24
Oldsmobile Cutlass Salon *Diesel* (5-speed)	25
Oldsmobile Cutlass *Diesel* Wagon	22
Chevrolet Malibu Wagon (V-6)	22
Triumph Spitfire Sedan (with overdrive)	22
Triumph Spitfire (4 cylinders)	22
MG Midget (4 cylinders)	22
Oldsmobile 98 *Diesel*	21
Oldsmobile Delta 88 *Diesel*	21
GMC Caballero Pickup (V-6)	22
Chevrolet El Camino Pickup (V-6)	22
Oldsmobile *Diesel* Wagon	20

the plots for mid-size wagons and for two-seaters are very similar, the former category contains 38 models, whereas the latter contains 15. Table 3-6 indicates that class sizes range from 13 to 109. For situations like this, McGill, Tukey, and Larsen (1978) introduce variable-width boxplots, in which the width of the boxes is made proportional to the square root of the batch size. Their idea is to allow large batches to have boxes of greater area; these boxes then have appropriately greater visual impact.

Other and more complicated variations of boxplots than the variable-width plot are possible. McGill, Tukey, and Larsen also propose several of these. One such modification employs notches of varying length in the edges of the boxes. These provide a rough measure of the significance of differences among the medians.

Further modifications of boxplots are likely to be proposed, and they may prove useful in visually conveying information about batches. We encourage the user to adopt such modifications when they seem worthwhile.

However, it is essential that these new techniques be clearly identified and explained when they are used.

Computer Implementation of Boxplots

Automated boxplots are available in Minitab, an interactive statistical system; see Ryan, Joiner, and Ryan (1976 and 1981). Velleman and Hoaglin (1981) provide programs for boxplots in both Fortran and Basic computer languages.

3C. THE SPREAD-VERSUS-LEVEL PLOT

When a comparison of batches shows a systematic relationship between spread and level, we search for a re-expression, or transformation, of the raw data that reduces or eliminates this dependency. If such a transformation can be found, the re-expressed data will be better suited both for visual exploration and for common analytic techniques of batch comparison. For example, one-way analysis of variance is simpler and more effective when there is, exactly or approximately, equal variance across groups.

Power Transformations

We define the power transformation with power (or exponent) p as the transformation that replaces x by x^p. For $p = 0$ we use $\log x$ rather than x^0. (Explanations for this convention appear in Chapter 4.)

The spread-versus-level plot guides us to an appropriate power transformation if one exists for the data in hand. Power transformations are not the only way to re-express data, but together with the logarithm they form a convenient family of transformations useful in a wide variety of situations. For a fuller discussion of data re-expression through power transformations, we refer the reader to Chapters 4 and 8.

Construction of Spread-versus-Level Plots

Heuristically, we want to remove a relationship between spread and level. To understand this relationship, it seems reasonable to plot a measure of spread against a measure of level.

We can suppose that the fourth-spread is proportional to a power of the median:

$$d_F = cM^b, \qquad (1)$$

or

$$\log d_F = \log c + b \log M,$$

or, letting $k = \log c$,

$$\log d_F = k + b \log M. \tag{2}$$

Thus the logarithms of the fourth-spreads and the logarithms of the medians are linearly related. The spread-versus-level plot springs from this leading case—we introduce it through an illustration using the data on largest cities in 16 countries.

EXAMPLE: SPREAD-VERSUS-LEVEL PLOT FOR LARGEST CITIES DATA

Table 3-8 uses results of Table 3-4 and gives the logarithm (base 10) of the median and fourth-spread for each of the 16 countries. A plot of $\log d_F$ against $\log M$ for all batches is a spread-versus-level plot; Figure 3-4 shows this plot for the 16 countries.

The points in the spread-versus-level plot of Figure 3-4 show a tendency for $\log d_F$ to increase as $\log M$ increases; furthermore, to a first approxima-

TABLE 3-8.   **Logs of medians and fourth-spreads for largest cities in 16 countries.**

Country	$\log M$	$\log d_F$
Sweden	$-.06$	.23
Netherlands	.24	.66
Canada	.43	.13
France	.44	.48
Mexico	.50	.77
Argentina	.51	.56
Spain	.54	.38
England	.70	.59
Italy	.84	.87
West Germany	.85	.69
Brazil	.91	.67
USSR	1.04	.48
Japan	1.10	1.04
United States	1.12	1.20
India	1.22	1.13
China	1.36	1.31

tion, the relationship appears almost linear. Our goal is to use the plot to determine the value of b in equation (2). The transformation $z = x^{1-b}$ of the data x gives re-expressed values z whose fourth-spread does not depend, at least approximately, on the level. Under our assumptions, we have the following:

Procedure for diagnosing power to stabilize spread. If b is the slope of the spread-versus-level plot, then $p = 1 - b$ is the approximate value of the exponent for a power transformation of x to stabilize spread. When $p = 0$, we use logarithms.

We now apply this procedure to the largest-cities data.

EXAMPLE: (CONTINUED)

We eye-fit a line to the points in Figure 3-4. Although two people using this fitting method will not arrive at the same slope, they will almost certainly draw a line whose slope is between $\frac{1}{2}$ and 1, and probably closer to 1. If we add a line of slope 1 to Figure 3-4, the line will be only slightly too steep.

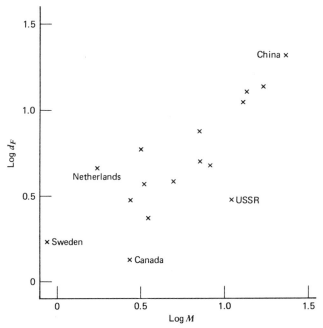

Figure 3-4. Spread-versus-level plot for largest cities in 16 countries.

The procedure for diagnosing a value of p to stabilize spread uses the relationship $p = 1 - b$. For $b = 1$, p is 0, and the logarithm transformation emerges. Similarly, $b = \frac{1}{2}$ leads to $p = \frac{1}{2}$, the square-root transformation. Although a power between 0 and $\frac{1}{2}$ might do better than either of these at stabilizing spread, simplicity and interpretability argue for considering first the log transformation and the square-root transformation.

To help fine-tune our decision, we apply each of these transformations. The medians and fourth-spreads of the re-expressed data are displayed in Table 3-9. The fourth-spreads of these new batches are plotted against the medians in Figures 3-5 and 3-6. The relative magnitudes of the slopes of these plots enable us to choose between the two transformations. The fourth-spreads of log population decrease slightly with level, whereas the fourth-spreads of square-root population increase with level. Thus although either transformation would probably serve our needs, the square-root transformation is somewhat weak for stabilizing fourth-spreads, whereas the log transformation is slightly strong. We would choose the logarithm.

TABLE 3-9.   Medians and fourth-spreads for log values and for square-root values of populations of 10 largest cities in 16 countries. (Logs are expressed to base 10.)

Country	Log Population		Square-Root Population	
	Median (Plus 5)	F-Spread	Median (Times 10^2)	F-Spread
(1) Sweden	−.06	.51	2.97	2.19
(2) Netherlands	.22	.62	4.17	3.98
(3) Canada	.43	.18	5.21	1.20
(4) France	.44	.36	5.24	2.49
(5) Mexico	.50	.57	5.64	4.32
(6) Argentina	.52	.37	5.72	2.78
(7) Spain	.54	.28	5.88	2.12
(8) England	.70	.34	7.07	2.75
(9) Italy	.84	.42	8.29	4.21
(10) West Germany	.85	.24	8.46	2.61
(11) Brazil	.91	.29	8.98	2.75
(12) USSR	1.04	.11	10.43	1.41
(13) Japan	1.10	.39	11.25	4.86
(14) United States	1.12	.45	11.42	6.38
(15) India	1.22	.38	12.92	5.42
(16) China	1.36	.35	15.09	6.37

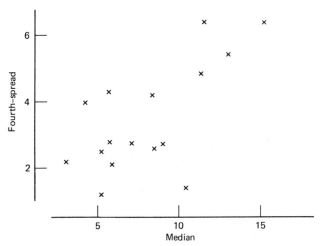

Figure 3-5. Fourth-spread versus median for square root of populations of 10 largest cities in 16 countries.

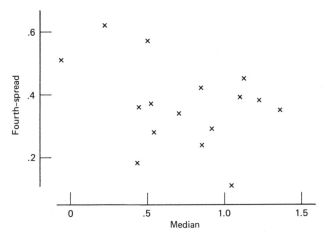

Figure 3-6. Fourth-spread versus median for log of populations of 10 largest cities in 16 countries.

Making the Choice

How can we decide among the log transformation, the square-root transformation, and some other power transformation with p between 0 and 1? Many factors can influence this decision and they do not always lead to a common choice.

Ideally, a transformation not only equalizes spreads, but also has a subject-matter explanation. For example, in demography, one widely-used model assumes that populations tend to grow exponentially. If so, the logarithm of population will grow approximately in a linear way. Advantages of linear growth such as interpretability, ease of detecting departure from fit, and convenience in interpolation suggest the logarithm as a sensible transformation for human populations.

When we do not have a subject-matter basis for adopting a particular transformation, we will usually have a clearer sense about what the transformation does if we choose a power that is an integer multiple of $\frac{1}{2}$. (Admittedly, the cube-root transformation, with $p = \frac{1}{3}$, sometimes does have physical meaning.) Table 3-10 shows some frequently used power transformations, together with the slopes that indicate them.

In some fields of science, the choice of transformation is so deeply set that failing to use it creates objections, even though the transformation may be inappropriate for a class of data. About the only way around this situation is to analyze the data both ways and explain one's preference while still giving the standard transformation to those who want it.

For the largest-cities data set, we decide to use the logarithmic transformation because we are dealing with population data. The simple theoretical

TABLE 3-10.   **Power transformations most often used.**[a]

Transformation	Power	Slope of Spread-versus-Level Plot
Cube	3	-2
Square	2	-1
No change	1	0 (flat plot)
Square root	$\frac{1}{2}$	$\frac{1}{2}$
Logarithm	0	1
Reciprocal root	$-\frac{1}{2}$	$1\frac{1}{2}$
Reciprocal	-1	2

[a]The transformations in this list are the main members of Tukey's "ladder of powers."

model favoring the logarithm, even more than strong evidence in the data, provides much of the basis for this decision.

Reanalysis in the Log Scale

We now apply the logarithmic transformation to the data and evaluate its effect on the 16 batches. Because power transformations are monotonic for positive data values, the order statistics of the transformed data will equal the transformed original order statistics (except for the effects of rounding and interpolation). Thus to draw the new boxplots, we need only apply the transformation to the 5-number summary and to the outliers. Of course, after we recalculate the fourth-spread and the outlier cutoffs, the cities that were originally high-side outliers may no longer be, whereas some that were not outliers may now be low-side outliers, though this is unlikely to happen when we deal with the ten highest. The final example for this section looks at these new boxplots of the log of population in order to assess the impact of the transformation on the individual data sets.

EXAMPLE:

Table 3-11 gives the logs of the 5-number summaries and new calculations of outliers. The new boxplots of the log data appear in Figure 3-7.

To see how well the log has done, we compare Figure 3-7 with Figure 3-2. The log has done a good job of equalizing spreads—the boxes are now more similar in length, and the remaining inequality does not seem to be much related to level.

The new scale has pulled in all of the outliers as well. Of the 19 outliers in the raw scale, eight cities are no longer outliers and the others have moved relatively closer to the upper outlier cutoffs.

Another improvement from using logarithms is both aesthetically and practically important. The new boxplots are much easier to look at, and the countries are now displayed at much the same level of detail. In the raw scale of Figure 3-2, the values for Sweden, the Netherlands, and Canada are much harder to read from the plot than those for India and China. In the plots for the log scale in Figure 3-7, the details show up equally well for all countries.

In sum, we have looked carefully at the 16 batches once again, now that they are in the logarithmic scale. This examination convinces us that the transformation does improve the data in several important ways. The log scale does seem to be an appropriate scale for working with the population data for these largest cities.

The spread-versus-level plot is one of many similar plots used in various situations to let the data guide us toward an appropriate transformation.

TABLE 3-11. 5-Number summaries of log values of populations of 10 largest cities in 16 countries, with outlier cutoffs and outliers noted.
[Log(population in 100,000s) is shown. Add 5 to get log(population).]

(1) # 10   Sweden

M	5.5	-0.06		d_F
F	3	-0.11	0.40	0.51
	1	-0.16	0.89	

1.5 $d_F = 0.83$
Outlier cutoffs: $(-0.94, 1.23)$
Outliers: none

(2) # 10   Netherlands

M	5.5	0.24		d_F
F	3	0.16	0.78	0.62
	1	0.11	0.94	

1.5 $d_F = 0.93$
Outlier cutoffs: $(-0.82, 1.71)$
Outliers: none

(3) # 10   Canada

M	5.5	0.43		d_F
F	3	0.40	0.58	0.18
	1	0.23	1.08	

1.5 $d_F = 0.27$
Outlier cutoffs: $(0.13, 0.85)$
Outliers: 1.08

(4) # 10   France

M	5.5	0.44		d_F
F	3	0.37	0.73	0.36
	1	0.30	1.45	

1.5 $d_F = 0.54$
Outlier cutoffs: $(-0.17, 1.27)$
Outliers: 1.45

(5) # 10   Mexico

M	5.5	0.50		d_F
F	3	0.34	0.91	0.57
	1	0.27	1.49	

1.5 $d_F = 0.86$
Outlier cutoffs: $(-0.52, 1.77)$
Outliers: none

(6) # 10   Argentina

M	5.5	0.52		d_F
F	3	0.43	0.80	0.37
	1	0.39	1.47	

1.5 $d_F = 0.56$
Outlier cutoffs: $(-0.13, 1.36)$
Outliers: 1.47

(7) # 10   Spain

M	5.5	0.54		d_F
F	3	0.42	0.70	0.28
	1	0.23	1.41	

1.5 $d_F = 0.42$
Outlier cutoffs: $(0.00, 1.12)$
Outliers: 1.23, 1.41

(8) # 10   England

M	5.5	0.70		d_F
F	3	0.52	0.86	0.34
	1	0.48	1.90	

1.5 $d_F = 0.51$
Outlier cutoffs: $(0.01, 1.37)$
Outliers: 1.90

TABLE 3-11. (*Continued*)

(9) # 10   Italy

M	5.5	0.84		d_F
F	3	0.65	1.07	0.42
	1	0.53	1.37	

1.5 d_F = 0.63
Outlier cutoffs: (0.02, 1.70)
Outliers: none

(10) # 10   West Germany

M	5.5	0.85		d_F
F	3	0.82	1.06	0 24
	1	0.75	1.34	

1.5 d_F = 0.36
Outlier cutoffs: (0.46, 1.42)
Outliers: none

(11) # 10   Brazil

M	5.5	0.91		d_F
F	3	0.70	0.99	0.29
	1	0.44	1.70	

1.5 d_F = 0.44
Outlier cutoffs: (0.26, 1.43)
Outliers: 1.59, 1.70

(12) # 10   USSR

M	5.5	1.04		d_F
F	3	1.01	1.12	0.11
	1	0.96	1.80	

1.5 d_F = 0.17
Outlier cutoffs: (0.84, 1.29)
Outliers: 1.56, 1.80

(13) # 10   Japan

M	5.5	1.10		d_F
F	3	0.89	1.28	0.39
	1	0.85	2.04	

1.5 d_F = 0.59
Outlier cutoffs: (0.30, 1.87)
Outliers: 2.04

(14) # 10   United States

M	5.5	1.12		d_F
F	3	0.94	1.39	0.45
	1	0.88	1.89	

1.5 d_F = 0.68
Outlier cutoffs: (0.06, 2.07)
Outliers: none

(15) # 10   India

M	5.5	1.22		d_F
F	3	0.98	1.36	0.38
	1	0.86	1.66	

1.5 d_F = 0.57
Outlier cutoffs: (0.41, 1.93)
Outliers: none

(16) # 10   China

M	5.5	1.36		d_F
F	3	1.22	1.57	0.35
	1	1.06	1.84	

1.5 d_F = 0.53
Outlier cutoffs: (0.87, 1.92)
Outliers: none

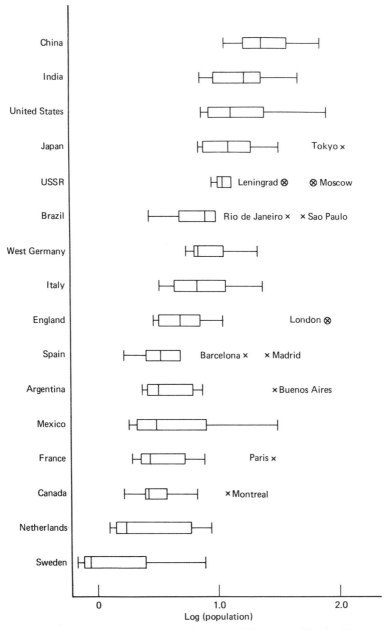

Figure 3-7. Boxplots for log of populations of 10 largest cities in 16 countries.

See, for example, the transformation plot for symmetry (Chapter 4) and the diagnostic plot for two-way tables (Chapter 6). In this section, we presented the spread-versus-level plot, using the 16-country example to illustrate how it works in practice. We give some motivation for the plot in Section 3D and a more detailed mathematical discussion in Chapter 8.

3D. BACKGROUND FOR SPREAD-VERSUS-LEVEL PLOTS

We have used the spread-versus-level plot to suggest a power transformation. The transformation enabled us to equalize the spreads of several batches of data for the largest-cities example. We now sketch a brief development of the spread-versus-level plot. We aim for directness and simplicity of presentation by omitting mathematical details that are implicit throughout this algebraic development. The underlying mathematical issues, along with some of the history of their development, are presented in much greater detail in Chapter 8.

We assume that our batches have underlying distributions with similar shapes, although different batches are at different levels and have different spreads. If X_i is a random variable representing an observation from batch i, the level of X_i depends on the particular batch represented. For simplicity, we suppress in the notation this dependence of X_i on level, and we use X to denote a random variable corresponding to any batch. We keep in mind that X really represents several variables having different levels and spreads but otherwise similar shapes.

Consider the power transformations X^p, where p is any fixed real number. Suppose that for some p, the fourth-spread of X^p is constant over all batches at the various levels of X. We establish assumptions and notation as follows:

For X^p

$$
\begin{aligned}
\text{median} &= m & (m > 0) \\
\text{upper fourth} &= m + d & (d > 0) \\
\text{lower fourth} &= m - c & (c > 0) \\
\text{fourth-spread} &= d + c & (\text{constant, independent of } m)
\end{aligned}
\tag{3}
$$

For X

$$
\begin{aligned}
\text{median} &= m^{1/p} \\
\text{upper fourth} &= (m + d)^{1/p} \\
\text{lower fourth} &= (m - c)^{1/p} \\
\text{fourth-spread} &= (m + d)^{1/p} - (m - c)^{1/p}
\end{aligned}
\tag{4}
$$

Because the spread-versus-level plot involves the log of the fourth-spread of X and the log of the median, we ask how these quantities are related.

Consideration of this question requires that c be less than m. Otherwise, the lower fourth would be negative in the raw data, which are all positive. We actually expect c and d to be no larger than, say, $m/2$. Thus it may be reasonable to expand the expression for the fourth-spread of X in terms of d/m and c/m. For convenience, we define q as $1/p$, and we obtain for the fourth-spread in display (4):

$$(m + d)^q - (m - c)^q$$

$$= m^q \left[\left(1 + \frac{d}{m} \right)^q - \left(1 - \frac{c}{m} \right)^q \right]$$

$$= m^q \left[1 + q\frac{d}{m} + \frac{q(q-1)}{2} \left(\frac{d}{m} \right)^2 + \frac{q(q-1)(q-2)}{6} \left(\frac{d}{m} \right)^3 + \cdots \right.$$

$$\left. -1 + q\frac{c}{m} - \frac{q(q-1)}{2} \left(\frac{c}{m} \right)^2 + \frac{q(q-1)(q-2)}{6} \left(\frac{c}{m} \right)^3 + \cdots \right]$$

$$= m^{q-1} \left[q(d + c) + \frac{q(q-1)}{2} \frac{d^2 - c^2}{m} \right.$$

$$\left. + \frac{q(q-1)(q-2)}{6} \frac{d^3 + c^3}{m^2} + \cdots \right].$$

Thus

$$(m + d)^q - (m - c)^q = q(d + c)m^{q-1} \left[1 + \frac{(q-1)}{2} \frac{(d-c)}{m} \right.$$

$$\left. + \frac{(q-1)(q-2)}{6} \frac{d^2 - dc + c^2}{m^2} + \cdots \right] \quad (5)$$

The leading term is that outside the square brackets, namely $q(d + c)m^{q-1}$. We have

$$\log(\text{leading term}) = \log q + \log(d + c) + (q - 1)\log m.$$

Because

$$(q - 1)\log m = [(q - 1)/(1/p)]\log(m^{1/p}),$$

we see that log(leading term) is linear in log(median of X) = log($m^{1/p}$), with a slope of

$$\frac{(q - 1)}{1/p} = p\left(\frac{1}{p} - 1\right) = 1 - p.$$

Thus we should take

$$p \approx 1 - \text{slope} \tag{6}$$

as the power derived from the spread-versus-level plot.

It is evident that the preceding development is an approximate one and does not apply for $p = 0$. We next consider two special cases, including the important case with $p = 0$. In equation (5), note that for $q = 2$ ($p = \frac{1}{2}$),

$$(m + d)^2 - (m - c)^2 = 2(d + c)m + (d^2 - c^2).$$

Thus when $c = d$ we obtain

$$(m + d)^2 - (m - d)^2 = 4\,dm,$$

so that the leading term is exact in this special case.

For $p = 0$ (the logarithm), separate algebra is necessary. In this case, the analogs of (3) and (4) are:

For log X

$$\text{median} = m$$

$$\text{upper fourth} = m + d$$

$$\text{lower fourth} = m - c \tag{7}$$

$$\text{fourth-spread} = d + c$$

For X

$$\text{median} = 10^m$$

$$\text{upper fourth} = 10^{m+d} \tag{8}$$

$$\text{lower fourth} = 10^{m-c}$$

$$\text{fourth-spread} = 10^m(10^d - 10^{-c})$$

Thus for this case,

$$\log(\text{fourth-spread of } X) = m + \log(10^d - 10^{-c})$$

and, because $\log(\text{median of } X) = m$, the slope is exactly 1. No approximation from an expansion is assumed.

It is natural for us to wonder how accurate the approximation is when we base our transformation only on the leading term of expansion (5). To pursue this question, we can ask how close the expression

$$\frac{\text{fourth-spread of } X}{\text{leading term of } (5)} = \frac{(m + d)^q - (m - c)^q}{q(d + c)m^{q-1}}$$

is to 1. The above expression, by factoring out m^q in the numerator and then canceling m^q, reduces to

$$\frac{(1 + d/m)^q - (1 - c/m)^q}{q(d + c)/m}.$$

Because d is the distance from the median of X^p to the upper fourth of X^p, the fraction d/m is the ratio of a measure of spread to the median for X^p. Similarly, c is the distance from the lower fourth of X^p to the median of X^p, so c/m has a similar interpretation. For distributions with the fourths relatively near the median in the transformed scale (i.e., d/m and c/m not

TABLE 3-12.   Values of the ratio: [fourth-spread of X]/[leading term of expansion (5)].

		\multicolumn{6}{c}{$(d/m, c/m)$}				
q	p	$(.2, .2)$	$(.2, .1)$	$(.4, .4)$	$(.4, .2)$	$(.5, .5)$	$(.8, .5)$
1	1	1.00	1.00	1.00	1.00	1.00	1.00
2	$\frac{1}{2}$	1.00	.95	1.00	1.10	1.00	1.15
3	$\frac{1}{3}$	1.01	1.11	1.05	1.24	1.08	1.46
4	$\frac{1}{4}$	1.04	1.23	1.16	1.43	1.25	2.01
—	0	1.00	1.00	1.00	1.00	1.00	1.00
-2	$-\frac{1}{2}$	1.04	.92	1.19	.89	1.78	1.42
-1	-1	1.08	.90	1.42	.88	1.33	1.11
$-\frac{1}{2}$	-2	1.03	.94	1.11	.91	1.20	1.03

too large), we expect the leading term of expansion (5) to offer a good approximation. Next, we examine this belief numerically.

Table 3-12 displays values of this expression for various $(d/m, c/m)$ combinations. Clearly, for $d/m = .2$, the leading term is very close to the fourth-spread of X. In a few combinations for $d/m - .4$, the ratio is large enough to give us some pause. For larger values of d/m, the approximation often appears rather inadequate.

In sum, we would probably feel comfortable with the applicability of the spread-versus-level plot whenever, in the transformed scale, the fourths are relatively close to the median in that they are a small fraction of the median away from the median. Judging p from the slope of log(fourth-spread of X) against log(median of X) is an approximate process, but except for sampling fluctuations, the approximation is often a very good one.

The fact that it is approximate, however, leads us to recommend replotting the [log(median), log(fourth-spread)] pairs for the transformed scale, just as a check. For the largest cities example Table 3-13 provides the necessary data, and the new spread-versus-level plot for log(population) appears as Figure 3-8. The result shows little systematic variation of log(fourth-spread) with log(median); the values of log(fourth-spread) show much greater variability than do the values of log(median). Close scrutiny of this plot indicates that there may be a slight downward trend, suggesting

TABLE 3-13. Calculations for the spread-versus-level plot for log(population) of 10 largest cities in 16 countries. (Base-10 logarithms are used.)

Country	Median	d_F	Log(M)	Log(d_F)
(1) Sweden	4.94	.51	.69	$-.29$
(2) Netherlands	5.22	.62	.72	$-.21$
(3) Canada	5.43	.18	.73	$-.74$
(4) France	5.44	.36	.74	$-.44$
(5) Mexico	5.50	.57	.74	$-.24$
(6) Argentina	5.52	.37	.74	$-.43$
(7) Spain	5.54	.28	.74	$-.55$
(8) England	5.70	.34	.76	$-.47$
(9) Italy	5.84	.42	.77	$-.38$
(10) West Germany	5.85	.24	.77	$-.62$
(11) Brazil	5.91	.29	.77	$-.54$
(12) USSR	6.04	.11	.78	$-.96$
(13) Japan	6.10	.39	.79	$-.41$
(14) United States	6.12	.45	.79	$-.35$
(15) India	6.22	.38	.79	$-.42$
(16) China	6.36	.35	.80	$-.46$

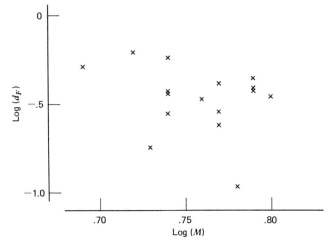

Figure 3-8. Spread-versus-level plot for largest-cities data re-expressed in the log scale. Base-10 logarithms are used throughout.

that a slightly less vigorous re-expression than the logarithm might be worthy of consideration, as we mentioned earlier.

3E. SUMMARY

The boxplot is a simple visual display of a batch of data; it shows the level, spread, skewness, tail length, and outliers. When several batches of similar data need comparison, parallel boxplots make the comparison easy. Boxplots are suitable display forms for exploratory data analysis because they summarize the body of data without being influenced by outliers and then allow the outliers to stand out as points needing special attention. In particular, boxplots provide one basis for defining cutoffs for data outliers.

The spread-versus-level plot can point to a power transformation to stabilize spread when multiple batches show different spreads of batches at different levels. Transformation of the batches sometimes allows the details of the batches to stand out. Furthermore, stable spread can be essential for subsequent analysis of multiple batches; it is, for example, an assumption of classical analysis of variance. Parallel boxplots in the transformed scale enable us to view the effects of transformation and to compare the shapes of batches in the new scale to those in the raw scale.

Chapter 4 begins a more thorough study of transformation—its purposes, implementation, effects, and interpretation for data in several common forms.

REFERENCES

McGill, R., Tukey, J. W., and Larsen, W. A. (1978). "Variations of box plots," *The American Statistician*, **32**, 12–16.

Ryan, T. A., Joiner, B. L., and Ryan, B. F. (1976). *Minitab Student Handbook*. North Scituate, MA: Duxbury Press.

Ryan, T. A., Joiner, B. L., and Ryan, B. F. (1981). *Minitab Reference Manual*. University Park, PA: Minitab Project, The Pennsylvania State University.

U.S. Environmental Protection Agency. *1979 Gas Mileage Guide*, First Edition, September 1978. Washington, DC: U.S. Department of Energy.

Velleman, P. F. and Hoaglin, D. C. (1981). *Applications, Basics, and Computing of Exploratory Data Analysis*. Boston: Duxbury Press.

Additional Literature

Tukey, J. W. (1972). "Some graphic and semigraphic displays." In T. A. Bancroft (Ed.), *Statistical Papers in Honor of George W. Snedecor*. Ames, IA: The Iowa State University Press, pp. 293–316.

Wainer, H. and Thissen, D. (1981). "Graphical Data Analysis," *Annual Review of Psychology*, **32**, 191–241.

EXERCISES

1. The mean menstrual cycle lengths for 21 women are recorded in Table 1-1.
   (a) Find the median, lower fourth, upper fourth, fourth-spread, and outlier cutoffs for this data set.
   (b) Construct the boxplot for this batch and report any outliers that you discover.

2. Repeat Exercise 1 for the basal body temperatures recorded in Table 1-1 for 21 women.

3. Show that in a sample of size 4, no data value can lie outside the outlier cutoffs, and thus verify the remark made in Section 3A.

4. In Section 3C, transformation of the largest-cities data led us to conclude that the square-root transformation was too weak and that the log transformation was slightly too strong (although satisfactory for our purposes). These observations suggest that an intermediate transformation such as the fourth-root transformation might also do well at stabilizing spread for the 16 batches.
   (a) What line (i.e., what slope), if fit to the spread-versus-level plot of Figure 3-4, corresponds to $p = \frac{1}{4}$? With a pencil, lightly

sketch a line with this slope in Figure 3-4 and assess its suitability in representing the 16 points.

(b) Apply the fourth-root transformation to the data and assess its effect on the batches by plotting fourth-spreads against medians in the new scale.

(c) Construct parallel boxplots for the data re-expressed in the fourth-root scale. Compare the results with those we presented for the log scale. Of these two transformations, which seems to be better for stabilizing spread?

5. Consider the parallel boxplot display in Figure 3-3 for the EPA mileage data. Although you lack the raw data, some qualitative judgments can be made from the boxplots.

(a) Does spread depend strongly on level? Would you expect the spread-versus-level plot to have positive slope or negative slope? Large slope or small slope?

(b) Give arguments both for and against transforming these data, bearing in mind the approximate power that you suspect might emerge from the spread-versus-level plot.

6. A spread-versus-level plot can be constructed by using sample means for the levels of batches and sample standard deviations for their spreads. The relationship between slope and power is unchanged by this modification.

(a) Using a scientific calculator or computer software, find the means and standard deviations needed to provide this modified spread-versus-level plot for the largest-cities data of Table 3-3.

(b) Construct the modified spread-versus-level plot and compare it to that shown in Figure 3-4. What power p is indicated?

(c) Comment on any advantages and/or disadvantages in using this modification of the spread-versus-level plot.

Exercises 7, 8, and 9 pertain to the following.

Certain key "marker" words have been used in analyzing the writings of Alexander Hamilton. These words can be grouped according to their number of letters, ranging from 2 to 6. Mosteller and Wallace recorded the number of occurrences of each of 51 words in a specified "block" of Hamilton's writing; this number ranges from 0 to 14. In all, 247 blocks of Hamilton's writing were examined in this way. Table 3-14 records the numbers of blocks containing words of a given length a specified number of times.

TABLE 3-14. Numbers of blocks of text by Hamilton that exhibit marker words of fixed length and frequency.

Number of Letters	Frequencies of Words											
	0	1	2	3	4	5	6	7	8	9	⋯	14
2	1592	533	229	71	26	11	7	1				
3	2398	685	242	82	24	13	9	1	2	1	⋯	1
4	3696	958	328	138	39	20	6	2				
5	532	234	100	53	33	17	6	8	4	1		
6	397	69	26	2								

Source: Frederick Mosteller and David L. Wallace (1964), *Inference and Disputed Authorship: The Federalist*. Reading, MA: Addison–Wesley. [Data based on Table 2.3-3, p. 29.]

7. Consider the batch of word frequencies for words having 2 letters; note that these frequencies occur with multiplicities—for example, there are 71 occurrences of the frequency 3 (meaning that 71 blocks contained words of length 2 that appear exactly 3 times in the block.)

    (a) Find the median frequency for words two letters long and give the associated fourth-spread.

    (b) Repeat part (a) for the four other batches.

    (c) Give the 5-number summaries, the outlier cutoffs, and the outliers for each of the five batches of word frequencies.

    (d) Construct parallel boxplots for the five batches.

8. Use the results from Exercise 7 to construct the spread-versus-level plot for the five batches of data.

    (a) Construct the spread-versus-level plot and fit a line to it.

    (b) Estimate the slope of the line and determine the power of a transformation that you believe will stabilize spread for the five batches.

    (c) Compare plots of fourth-spread against median for the raw batch and for your re-expressed batch. Has most of the dependence of spread on level been removed by your transformation?

9. Reanalyze the word-frequency data in the transformed scale adopted in Exercise 8. Construct a new set of parallel boxplots and compare the results to those of Exercise 7.

*10. Consider the exponential distribution with mean 1. The density function is $f(x) = e^{-x}$ for positive x, and the standard deviation is also 1.

(a) Use integral calculus to verify that the median is $1/\log_e 2$.

(b) Find the fourths and the fourth-spread for this distribution. Use your answers to determine the outlier cutoffs for this exponential distribution.

(c) Again use integration to determine the percent of this exponential distribution that lies beyond the outlier cutoffs.

(d) Determine the fraction of the distribution that is outside $\mu \pm 1.96\sigma$.

(Note that you have now added an entry to the nonsymmetric distributions in Table 3-2.)

*11. Use series expansions to verify that expansion (5) follows from the assumptions in display (4).

CHAPTER 4

Transforming Data

John D. Emerson
Middlebury College

Michael A. Stoto
Harvard University

Raw data may require a change of expression to produce an informative display, effective summaries, or an uncomplicated analysis. That is, we may need to change not only the units in which the data values are stated, but also the basic scale of measurement. Difficulties may arise because the raw data have

Strong asymmetry,

Many outliers in one tail,

Batches at different levels with widely differing spreads,

Large and systematic residuals from fitting a simple model to the data.

By altering the shape of the batch or batches we may alleviate these problems. To accomplish this change, we transform the data by applying a single mathematical function to all raw data values.

To change shape, we must do more than change the origin and/or unit of measurement. Changes of origin and scale are linear transformations, and they leave shape alone. Stronger transformations such as the logarithm and square root change shape. One way to approach shape is to think of what a histogram of the data looks like when you erase the numbers on horizontal and vertical scales.

After a brief description of the uses of transformations in data analysis, this chapter begins a systematic study of power transformations. We discuss some reasons for transforming batches. We tell how these relate to each other; usually a transformation that achieves one particular objective is also simultaneously helpful in achieving other objectives. We try to suggest why such lucky effects occur, and we briefly explore the alternative possibility that objectives in transforming could be at odds with one another. Section 4E discusses matching, a procedure that makes the transformed data more nearly comparable to the original data. We conclude this chapter with a few suggestions about worthwhile opportunities for transforming.

4A. POWER TRANSFORMATIONS

A *transformation* of the batch $x_1, x_2, \ldots, x_n$, is a function T that replaces each x_i by a new value $T(x_i)$ so that the transformed values of the batch are $T(x_1), \ldots, T(x_n)$. In illustrating the spread-versus-level plot, Section 3C introduced an important family of transformations, the power transformations. We used one of these, the log transformation, to re-express the populations of largest cities in 16 countries. We also introduced, in Table 3-10, a *ladder of re-expression* for use when we wish to choose a few power transformations with integer or half-integer exponents. We use the terms *transformation* and *re-expression* on a roughly interchangeable basis.

Our definition of transformation is quite general. It includes the possibility of mapping all raw data values to a constant, though such a procedure is clearly not helpful in any way. It includes the categorization of data. For example, cities could be categorized as small, medium, or large, according to their populations. Categorization transformations are often useful, but we do not mention them further in this chapter. The definition includes complicated functions that may lack the appealing properties of elementary and familiar functions such as $y = x^2$ and $y = \log x$.

In this chapter we discuss simple transformations that are known to be useful for sets of data consisting of amounts or counts. These transformations have the following properties:

1. They preserve the order of data in a batch; that is, they are strictly increasing functions. Data values that are larger in the original scale will be larger in the re-expressed scale, but the spacing may change.
2. They preserve letter values of a batch except for small differences that may result from interpolating between data points. In particular, because letter values rely on order, medians are transformed to medians, and fourths to fourths.

3. They are continuous functions; this guarantees that points that are very close together in the raw batch will also be very close together in the re-expressed batch, at least relative to the scale being used.

4. They are smooth functions in that the functions we use have derivatives of all orders. This requirement guarantees that the functions do not have sharp corners. We exploit this property in Chapter 8, where we also provide a mathematical discussion of transformation.

5. They are specified by elementary functions, so that re-expression with the aid of all but the simplest hand-held calculators is quick and easy.

Only the first of these properties depends at all on the choice of additive and multiplicative constants, and it will hold if the constant multiplying a positive power of x is positive, while that multiplying a negative power is negative.

Most transformations discussed in this book belong to the family of power transformations; these are among the simplest and most widely used. We now give a more formal definition than that introduced in Chapter 3.

DEFINITION: Power transformations have the form

$$T_p(x) = \begin{cases} ax^p + b & (p \neq 0) \\ c \log x + d & (p = 0), \end{cases} \tag{1}$$

where a, b, c, d, and p are real numbers. We shall require $a > 0$ for $p > 0$ and $a < 0$ for $p < 0$, so that conditions 1 to 5 hold. The values of a, b, c, and d are otherwise rather arbitrary, usually chosen for convenience, whereas p matters and is chosen to help us analyze the data.

The context in which we work with transformations helps to determine the choices for the constants a, b, c, and d. Three common situations arise:

1. When we want to re-express a batch in a particularly easy way, we specialize the transformations (1) to

$$T_p(x) = \begin{cases} x^p & (p > 0) \\ \log x & (p = 0) \\ -x^p & (p < 0). \end{cases} \tag{2}$$

2. When we want to compare transformations to each other and examine

their mathematical and geometric properties, we specialize to

$$T_p^*(x) = \begin{cases} \dfrac{x^p - 1}{p} & p \neq 0 \\ \ln x & p = 0, \end{cases} \tag{3}$$

where $\ln x$ denotes the log to the base e. These transformations also offer the five properties described above. We will see that they satisfy some additional mathematical properties as well.

3. When we want to re-express a batch so that the new data resemble the raw data in location and spread, we choose the constants by a process called *matching*. Section 4E discusses this topic.

When p is fixed, any choice of the constants a and b (or c and d, if $p = 0$) represents a linear transformation of any other choice of constants. For example, when a is nonzero, $ax^p + b$ is transformed to $Ax^p + B$ using the multiplier A/a and the additive constant $B - (A/a)b$:

$$\frac{A}{a}(ax^p + b) + \left(B - \frac{A}{a}b\right) = Ax^p + B.$$

Because any nonconstant linear transformation merely moves the origin and gives a uniform change of scale (e.g., feet to inches), we emphasize that the choices of the constants are made for convenience and ease of interpretation, and not for necessity or essential change of behavior.

The family of power transformations is convenient in many respects. When the constants are specified in one of the three ways discussed above, the power transformations satisfy the five characteristics already described. Other helpful properties of these transformations include the following:

Concavity. Any power transformation is concave up ($\smile$) or concave down ($\frown$) throughout its domain of positive numbers. There is no "inflection point" at which concavity changes. A consequence is that a particular power transformation either compresses the scale for larger data values more than it does for smaller ones [e.g., $T(x) = \log x$] or does the reverse [e.g., $T(x) = x^2$]. It cannot do something more complicated, such as expanding the scale for large *and* small values of x while compressing it for intermediate values of x. [An example of a transformation that does behave this way is

$$T(x) = (x - 10)^3,$$

where 10 is "intermediate" in the range of values of x.]

Flexibility. In spite of our restriction to a limited family of functions with convenient mathematical properties, the power transformations offer adequate flexibility for practical work. In Section 4B we see that a power transformation can achieve each of several objectives. For example, a member of this family can often stabilize the spread for several batches (see Section 3C). In practice, the flexibility offered by the power family is often greater than is needed, and one may wish to restrict attention to members of the ladder of powers.

Geometric unity. In certain other ways, the power transformations (including the log transformation) belong together. We next explore these ways using some geometric arguments and a little calculus. In particular, we explain why we include the log function as a member of this family of transformations.

Consider the curves specified in expression (1) simultaneously for all p. Because translation and change of scale do not alter subsequent analysis, we lose no generality by making particular choices for the arbitrary constants (unless a or c is 0). We choose the constants so that the curves share a common point and have a common slope at that point; we choose them so the curves have value 0 and slope 1 when $x = 1$. The constants in equation (3) meet this objective. Figure 4-1 gives the graphs of these functions. The graphs of power functions with coefficient 1 and constant term 0 appear in Figure 4-2 for comparison. The "matching" that is evident in Figure 4-1 is absent in Figure 4-2.

Figure 4-1 shows that the representation of the power transformations in the form (3) gives some nice geometric properties. In addition to sharing a common point, $(1, 0)$, the curves also pass through that point in the same direction. Thus the curves have the same slope at $x = 1$, and they nearly coincide at nearby points. The curves are ordered by the values of p, in that larger p give curves that are above those with smaller p. For example, if the curve corresponding to $p = \frac{3}{4}$ were added to Figure 4-1, it would lie about halfway between those shown for $p = \frac{1}{2}$ and $p = 1$.

What are the implications, when re-expressing a batch of data, if we use the transformations shown in Figure 4-1 instead of those in Figure 4-2? In isolation, for one transformation applied to a set of data, how we choose to write the power transformation does not matter greatly. When we study several alternative transformations at once (as we do in this chapter) and when we need especially to compare the raw and transformed data, we find advantages in removing as much confusion as possible.

The transformations $y = x^p$ pose several possible difficulties. For negative values of p, $y = x^p$ is a decreasing function of x when x is positive.

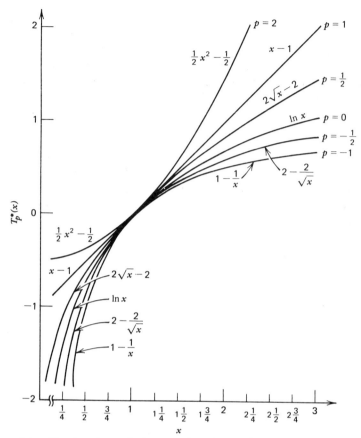

Figure 4-1. Graphs of matched power functions, $T_p^*(x)$, for selected values of p.

Thus the transformation reverses the order of a data set when p is negative; this minor difficulty does not arise with the form shown in equation (2) for the power transformations.

Although the curves pictured in Figure 4-2 share the common point $(1, 1)$, they represent transformations with very different behavior even for values arbitrarily close to 1. The differences in slopes at the point $(1, 1)$ produce this phenomenon. As an illustration, consider the curves $y = x$ and $y = x^2$ at the point $(1, 1)$. Since the tangent lines to the curves at that point have slopes of 1 and 2, respectively, the value $x = 1.001$ is mapped nearly twice as far away from $y = 1$ by $y = x^2$ as by $y = x$. This "stretching" by $y = x^2$ is even greater at all values of x greater than 1. In this sense, the members of the family $y = x^p$ are not comparable in scale. Sometimes, this

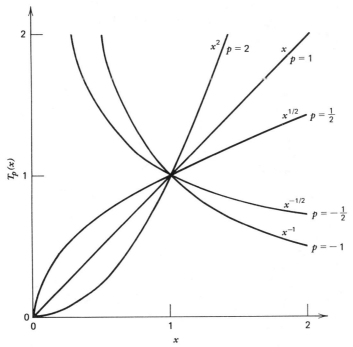

Figure 4-2. Graphs of basic power functions $T_p(x) = x^p$ for selected values of p.

fact does not trouble us; if not, for simplicity we feel free to use the functions given in equation (2). These transformations ensure monotonicity. At other times we want comparability of scale at some point that is shared by the transformations; if so, we use the functions in equation (3) or else a similar set of "matched" power transformations.

The graphs in Figure 4-1 suggest correctly that T_p^* changes continuously as a function of the parameter p. By this we mean roughly that if p_1 and p_2 are close together, then $T_{p_1}^*$ and $T_{p_2}^*$ will be close together as functions. Closeness of functions is a sophisticated mathematical topic. Of course, for any one x, the closeness of $T_{p_1}^*(x)$ and $T_{p_2}^*(x)$ is simply expressed by the magnitude of their difference.

Figure 4-1 also suggests geometrically that the log transformation fits between the power transformations $T_{1/2}^*(x) = 2\sqrt{x} - 2$ and $T_{-1/2}^*(x) = 2 - 2/\sqrt{x}$. By plotting the graphs of additional transformations, we can verify that for any small positive number λ, the graph of $T_0^*(x) = \ln x$ lies between the graphs of $T_\lambda^*(x)$ and $T_{-\lambda}^*(x)$ for all x. This observation gives geometric justification to our use of the logarithmic transformation as the appropriate power transformation when p is 0. In practice, we almost always use logs to the base 10 for re-expression; these differ from logs to the base e only by a multiplicative constant:

$$\ln x \approx 2.303 \log_{10} x.$$

[Notice that $\log_{10} x$ cannot simply be substituted for $\ln x$ in equation (3).] One reason for emphasizing base 10 is clear: data are almost always given as *decimal* numbers. (If we match, the base chosen will not matter.)

4B. REASONS FOR TRANSFORMING

The reasons for transforming a batch include the following:

Facilitate interpretation in a natural way.
Promote symmetry in a batch.
Promote stable spread in several batches.
Promote a straightline relationship between two variables.
Simplify the structure of a two-way or higher-dimensional table so that a simple additive model can help us understand the characteristics of the data.

We next consider each of these reasons, and we explore the first two in some detail.

Transforming to Enhance Interpretability

Sometimes, changing the scale of measurement is natural because it gives an alternative way to report information. For example, when scientists report temperature in Celsius degrees (C) instead of Fahrenheit degrees (F), the implied transformation is to the more convenient scale. One merit is that the zero of the Celsius scale corresponds to a natural and widely understood phenomenon, the freezing point of water. (The choice may be more familiar now that home thermometers, radio reports, and displays on buildings give the temperature in both scales. These practices do facilitate the transformation process!) This transformation is linear:

$$C = \tfrac{5}{9}(F - 32).$$

A nonlinear transformation may also be quite natural. Consider again the Environmental Protection Agency (EPA) estimates of gas mileage for automobiles (as used in examples in Chapter 3). An indirect method provides these figures. That method measures the composition of the exhaust of the automobiles. After positioning cars in an isolation chamber, the researchers "drive them on a moving driveway" while making exhaust measurements. A formula converts these measurements into estimates of the mileage figures. The EPA reports these estimates in miles per gallon, a scale that is familiar to the public. In this instance, the transformation, which is unknown to us, is probably rather complicated. But all who use the EPA data understand the transformed scale.

Sometimes, by considering a data set in a new scale, we enhance our understanding, even when the re-expressed scale might seem less natural than the original scale. For instance, suppose that the population of a developing city grows at a rate that is nearly proportional to the current population; that is, the population is an exponential function of time. If so, the log of the population, when plotted against time, will lie near a straight line. Although it isn't as easy to visualize the numbers on the log scale as on the raw scale, we do gain additional understanding of the growth pattern of the city if we view it in the log scale. We can see whether and when the growth is proportional to size, and we can explore reasons for strong deviations from this pattern when they occur. Part of the reason is that the linear function is the easiest to interpolate in and extrapolate from, both by eye and by arithmetic.

Of course, re-expression may not always bring a convenient interpretation. Even when the new scale has no ready physical interpretation, transforming a batch may teach us something. We can consider the negative reciprocal square root of a measurement even though we may be unable to give it a convenient interpretation. We explore below some reasons for our willingness to consider transformations that may lack physical motivation.

4C.   TRANSFORMING FOR SYMMETRY

Symmetry of a batch is often a desirable property; many estimates of location work best and are best understood when the data come from a symmetric distribution. A simple way to check on symmetry is to define a set of *midsummaries*, one for each pair of letter values (as defined in Section 2B). Each *midsummary* is the average of the two corresponding letter values. The abbreviation is "mid" followed by the letter used to tag the letter value —for example, midF. Another name, *midrange*, is commonly used for the midextreme.

Once we have calculated midsummaries for all the pairs of letter values, we can readily examine them for evidence of systematic skewness. (If

apparent skewness were due to one or two stray values, only the most
extreme letter values and their midsummaries would be affected. Thus using
the full set of midsummaries provides more resistance.) In a perfectly
symmetric batch, all midsummaries would be equal to the median. If the
data were skewed to the right, the midsummaries would increase as they
came from letter values further into the tails. For data skewed to the left,
they would decrease.

EXAMPLE: HOUSEHOLD INCOME

A sample of 994 households reported their annual income in dollars, and
the letter-value display is in Table 4-1. The calculations for the midsummaries are straightforward:

$$midF = \tfrac{1}{2}(2412 + 4944) = 3678,$$

and so on. The values of the mids increase steadily as we work down the
column, and midY is about 1.6 times as large as the median. This pattern is
solid evidence of very substantial skewness. Only when we move from midY
to midextreme do we fail to see an increase, and such fluctuation is not
surprising in a batch of this size.

We may recall that economists often prefer to work with the logarithm of
income rather than income itself. Re-expressing these data as logarithms

TABLE 4-1.   **Letter-value displays with midsummaries for a sample of reported
household incomes; original scale and log scale (base 10).**

# 994		Original Scale Household Income in Dollars			Log Scale Household Income		
M	497.5		3480			3.54	
F	249	2412	3678	4944	3.38	3.535	3.69
E	125	1788	4115.5	6443	3.25	3.53	3.81
D	63	1517	4400.5	7284	3.18	3.52	3.86
C	32	1248	4799	8350	3.10	3.51	3.92
B	16.5	963.5	4978.75	8994	2.98	3.465	3.95
A	8.5	727.5	5241	9754.5	2.86	3.425	3.99
Z	4.5	579	5394.5	10210	2.76	3.385	4.01
Y	2.5	345	5510.25	10675.5	2.50	3.265	4.03
	1	114	5494	10874	2.06	3.05	4.04

Source: Unpublished data from the Housing Allowance Demand Experiment.
Used by permission of Abt Associates Inc.

leads to the following sequence of midsummaries, from median to midextreme: 3.54, 3.535, 3.53, 3.52, 3.51, 3.465, 3.425, 3.385, 3.265, 3.05. There is now some tendency toward skewness to the left, as indicated by the monotone decreasing trend in the midsummaries. We now attempt to find another transformation that more nearly achieves symmetry for this batch.

In search of a re-expression that promotes symmetry in the household income data, we introduce a graphical technique analogous to the spread-versus-level plot (see Section 3C). We call the graph a *transformation plot for symmetry*.

DEFINITION: TRANSFORMATION PLOT FOR SYMMETRY: Let M be the median of a batch and let x_L and x_U represent lower and upper letter values for various letters. The transformation plot for symmetry places

$$\frac{(x_U - M)^2 + (M - x_L)^2}{4M} \tag{4}$$

on the horizontal axis and

$$\frac{x_U + x_L}{2} - M \tag{5}$$

on the vertical axis. If the resulting graph is nearly linear, then one-minus-slope is the indicated power of a transformation for symmetry having the form

$$T(x) = kx^p. \tag{6}$$

Like the other diagnostic plots, the purpose of the transformation plot for symmetry is to give us a first approximation to a good choice of transformation. It tells us what to try next, but does not guarantee that what we try will be best. (It is usually a lot better.)

We examine the origins and interpretation of this transformation plot in detail in Chapter 8. In this section, we illustrate the use of the plot with the household income data. First, we note that expression (5) measures the distance between the median and a midsummary; this applies to each letter value. These differences would all be 0 in a batch with perfect symmetry. For p nonnegative, the constant k is often chosen to be 1, and for p negative, the value of k is often -1. When $p = 0$, the logarithmic transformation is used.

The coordinates for the transformation plot for the household income data are shown in Table 4-2. We construct a graph which has the approxi-

TABLE 4-2. Computations for the transformation plot for symmetry in the household income example.[a]

Letter Value	x_L	x_U	$\dfrac{x_L + x_U}{2} - M$	$\dfrac{(x_U - M)^2 + (M - x_L)^2}{4M}$	Estimate of p
F	2412	4944	198	235.9	0.16
E	1788	6443	635.5	836.4	0.24
D	1517	7284	920.5	1316.4	0.30
C	1248	8350	1319	2061.7	0.36
B	963.5	8994	1498.75	2639.2	0.43
A	727.5	9754.5	1761	3372.5	0.48
Z	579	10210	1914.5	3858.4	0.50
Y	345	10675.5	2030.25	4425.5	0.54

[a] The median, M, is 3480.
The values in the first three columns are exact; those in column 4 are rounded to the nearest tenth, and those in the last column are to the nearest hundredth.

mate form

$$\frac{x_U + x_L}{2} - M = (1 - p)\frac{(x_U - M)^2 + (M - x_L)^2}{4M}. \tag{7}$$

Solving this equation yields an estimate of p for each letter value; these estimates form the last column in Table 4-2. We note that these estimates vary as we change letter values; the pattern is consistent with the observation that the data show a departure from apparent symmetry beyond what a log transformation can adjust for.

The transformation plot for symmetry appears in Figure 4-3. The eight points shown exhibit some systematic departure from linearity. Points corresponding to the inner letter values indicate greater asymmetry (in terms of re-expression exponent) than do the points for letter values farther into the tails of the data set. This phenomenon has interesting implications when we choose a transformation.

There are many ways to fit a line to the eight points shown in Figure 4-3. We desire to fit a line through the origin by a resistant method (i.e., by a method that is not heavily influenced by a few points that show considerable deviation from the pattern of the other points). One quick way to do this is to consider the eight lines that pass through each of the eight points and the origin, and to choose a line whose slope is the median of the eight slopes. This approach gives a median slope of .60, and the resulting line is

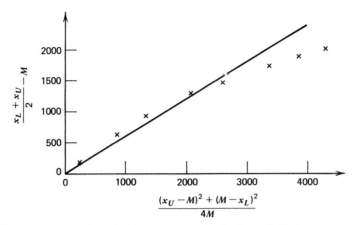

Figure 4-3. Transformation plot for symmetry in the household income example.

shown in Figure 4-3. The relationship

$$\text{power} = 1 - \text{slope}$$

indicates a power of .40 for re-expression with a power transformation.

In practice, we do not often use a transformation with a power such as .40. Instead we round the exponent to the nearest half-integer; here $p = \frac{1}{2}$ is nearest, so we try the square-root transformation. If the resulting batch were not sufficiently symmetric, we would try the log transformation (corresponding to $p = 0$) or perhaps the fourth-root transformation (corresponding to $p = \frac{1}{4}$). We show the new letter-value displays and midsummaries for the square-root and fourth-root transformations in Table 4-3; the results for the log transformation were given in Table 4-1.

We compare asymmetry in the original batch and the three transformed batches, whose letter-value displays appear in Tables 4-1 and 4-3, by examining trends in the midsummaries. The raw data have a strong monotonically increasing trend in the mids through the next-to-last letter value. This trend indicates the right-skewness of the income distribution. The log of income shows a monotonically decreasing trend in the midsummaries, indicating that the log of income is asymmetrically distributed with left-skewness. Although this pattern is consistent, it is slight for the early letter values, but it becomes stronger as we look further into the tails. If our primary interest is in symmetry in the main part of the batch and not in the extremes, we might be satisfied with the log transformation. Furthermore, experience in economics has shown that income distributions can often be

TABLE 4-3. **Letter-value displays with midsummaries for a sample of reported household incomes; square-root and fourth-root scales.**

#	994	Square-Root Scale Household Income			Fourth-Root Scale Household Income		
M	497.5		58.99			7.68	
F	249	49.11	59.71	70.31	7.01	7.70	8.39
E	125	42.28	61.28	80.27	6.50	7.73	8.96
D	63	38.95	62.15	85.35	6.24	7.74	9.24
C	32	35.33	63.36	91.38	5.94	7.75	9.56
B	16.5	31.04	62.94	94.84	5.57	7.66	9.74
A	8.5	26.97	62.87	98.77	5.19	7.57	9.94
Z	4.5	24.06	62.55	101.04	4.91	7.48	10.05
Y	2.5	18.57	60.95	103.32	4.31	7.24	10.16
	1	10.68	57.48	104.27	3.28	6.75	10.21

modeled with the lognormal distribution. If this were true here, application of the log transformation would give Gaussianity and thus symmetry.

The square-root transformation produces midsummaries that are monotonically increasing to letter *C* and monotonically decreasing from there to the extremes. All midsummaries but the last are larger than the median, so we might consider that this batch retains some of the right-skewness present in the raw data. The fourth-root transformation gives midsummaries that are monotonically increasing to letter *C* and monotonically decreasing thereafter. Unlike the square-root transformation, the fourth-root deviations are well balanced; four midsummaries are larger than the median and five are smaller. Because of the stronger asymmetry near the center of the raw household income data and the weaker symmetry in the extremes of that batch, a power transformation cannot symmetrize the batch completely.

The choice of transformation is often partly a matter of judgment, especially when symmetry is the objective. Several factors can help us in making this choice in the household income example:

1. If symmetry of the main body of the data is desired but skewness in the tails is relatively unimportant, we would probably prefer the log transformation. Economists might well favor it for its ease in interpretation.
2. If considerations of symmetry in the tails of the distribution are important, we might prefer the square-root transformation as a fast and easy way to re-express the data.
3. If we want to re-express in a way that tries to balance the degree of skewness of the main part of the distribution against the more modest skewness in the extremes, we might use the fourth root.

We return to this example in Section 4E armed with matched transformations, which facilitate making comparisons. Our purpose has been to introduce the transformation plot for symmetry as an aid in choosing a power transformation. We have also illustrated several transformations, and we have seen one way to compare them through their letter-value displays and midsummaries.

4D. TRANSFORMING FOR OTHER DATA STRUCTURES

Transformation is useful when working with other types of data structures including several batches at different levels, y-versus-x, and two-way tables. We briefly examine each of these in turn.

Transforming for Stable Spread

Data sometimes come to us in several batches at different levels. When the data are amounts or counts, we often find a systematic relationship between spread and level: increasing level usually brings increasing spread. This tendency appears in the largest-cities example of Chapter 3. When the relationship between spread and level is strong, we have several reasons for transforming the data in a way that reduces or eliminates the dependence of spread on level.

1. The transformed data will be better suited for comparison and visual exploration.
2. The transformed data may be better suited for common confirmatory techniques. For example, classical one-way analysis-of-variance models assume constant variance within groups.
3. Other benefits may result from the transformation. The individual batches may become more nearly symmetric and have fewer outliers.

Each of these benefits was realized in Section 3C, at least in part, when the populations of the largest cities were transformed to the log scale.

Chapter 3 presents and illustrates a graphical technique, called the spread-versus-level plot, for finding a power transformation to stabilize spread. Section 8D gives its origins and mathematical properties.

Transforming for Straightness

Data often come to us as measurements on each of two or more variables. We want to find relationships between variables. When these relationships are nearly linear, we often gain advantages.

1. Interpretation is easier.
2. Departures from fit are more easily detected.
3. Interpolation and extrapolation are easy.

Transforming one or both variables sometimes enables us to straighten relationships that were originally curved.

Chapter 5 considers relationships between two variables; Section 8E introduces a plot for choosing a power transformation that can help to straighten the relationship between x and y.

Transforming for Simple Structure

A common reason for transforming a batch of data to a new scale is to simplify the structure of the data. We are doing just this when we straighten a relationship between x and y. The two-way table presents another data structure in which transformation may lead to simplification. Here an additive model, in which a common value, row effects, and column effects add together to account for nearly all of the systematic variation in the table, is easy to understand and explain.

We consider two-way tables and the corresponding diagnostic plot in Chapter 6. We see there that the diagnostic plot helps identify a transformed scale in which an additive model can better summarize the data. This plot is similar to the others we have introduced; Chapter 8 explores these similarities.

4E. MATCHED TRANSFORMATIONS

Linear transformations of re-expressed data present little additional difficulty in interpretation. For instance, if we are willing to work with $T(x)$, then

$$z = a + bT(x) \tag{8}$$

is just as easily interpreted. In this section we show how the proper choice of a and b can aid analysis. We first illustrate this idea with an example.

EXAMPLE: POPULATION OF THE U.S. CITIES

In Section 3A we studied the population of the 15 largest cities in the United States and decided that for large cities it was appropriate to analyze the logarithm of the population. The raw and re-expressed values are shown in Table 4-4.

TABLE 4-4. Population of large U.S. cities in three scales.[a]

	Population	$\log_{10}$(Population)	$200\log_{10}$(Population) $- 300$
New York	778	2.89	278.2
Chicago	355	2.55	210.0
Los Angeles	248	2.39	178.9
Philadelphia	200	2.30	160.2
Detroit	167	2.22	144.5
Baltimore	94	1.97	94.6
Houston	94	1.97	94.6
Cleveland	88	1.94	88.9
Washington	76	1.88	76.2
St. Louis	75	1.88	75.0
Milwaukee	74	1.87	73.8
San Francisco	74	1.87	73.8
Boston	70	1.85	69.0
Dallas	68	1.83	66.5
New Orleans	63	1.80	59.9

[a]Population is expressed in 10,000s and is from the 1960 census.

At first glance, the columns look quite different from each other, but the plot in Figure 4-4 shows that, except for the very largest cities, the logarithm of population is close to a linear transformation of population itself. If the raw population is denoted by x, this means that we can find values a and b such that

$$z = a + b\log_{10}x$$

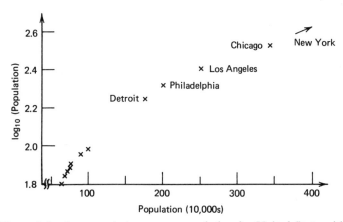

Figure 4-4. Log population versus population for United States cities.

is approximately equal to x, as long as x is not too near one end or the other of the data set. Choosing $a = -300$ and $b = 200$ (by the method given below) yields the value of z shown in the rightmost column of Table 4-4.

Most values are changed very little by the re-expression

$$z = 200 \log_{10} x - 300.$$

That is, z closely resembles x. Only the very large cities—New York, Chicago, Los Angeles, Philadelphia, and Detroit—have been substantially altered, and they have been brought in closer to the rest.

Benefits of Matched Re-Expression

Many of the data-analysis techniques we use are "transparent" to a linear change of scale. This means that if we multiply the data y by a constant b and add the value a, producing $z = a + by$, the results change from r to $a + br$, to br, or are unchanged. For illustration, the median changes to $a + bM$, the fourth-spread is multiplied by b, and a t-test on the data remains unchanged after linear transformation. If data were given in inches, we would not hesitate to do the analysis in feet or centimeters. This is true for location and slope estimates, median polish, and so on. Therefore, if we plan to use one of these techniques on re-expressed data, there is no harm in a further linear transformation.

As the previous example shows, there can be benefits. First of all, we can arrange for the re-expressed data to look, for the most part, like the original data. Only the extreme values change substantially. This will be true as long as (a) the transformation is close to linear near the center of the batch and (b) we match at a point near the center. We usually are already familiar with the magnitudes in the original scale, and we sometimes find matching more convenient than changing our way of thinking. The transformed numbers in the example behave like the log of population, but they look like the raw population figures.

Second, matching emphasizes the changes that are due to transformation. Only the data for Detroit and larger cities were substantially changed—the larger the city, the larger the change.

Third, as we have seen in Figures 4-1 and 4-2, matching helps us to compare the effect of different transformations. In Section 4C, we compared log, square-root, and fourth-root re-expressions of the household income data. The discussion centered on comparison of the midsummaries of the re-expressed data. Since all were expressed in different scales, we compared only patterns and trends.

For comparison purposes, Table 4-5 presents letter-value displays of the original data and matched log, square-root, and fourth-root transformations

TABLE 4-5. **Household income data and three matched re-expressions.**
Letter-value displays with midsummaries.[a]

	Original Data				Log		
M		3480				3480	
F	2412	3678	4944		2204	3453	4702
E	1788	4115.5	6443		1163	3393	5624
D	1517	4400.5	7284		591	3321	6050
C	1248	4799	8350		−89	3219	6526
B	963.5	4978.75	8994		−989	2898	6784
A	727.5	5241	9754.5		−1967	2550	7067
Z	579	5394.5	10210		−2761	2232	7226
Y	345	5510.25	10675.5		−4563	1409	7381
1	114	5494	10874		−8417	−486	7445

	Square Root				Fourth Root		
M		3480				3480	
F	2314	3565	4816		2261	3509	4757
E	1509	3750	5990		1345	3571	5797
D	1115	3852	6589		871	3587	6303
C	688	3995	7301		332	3608	6885
B	182	3946	7709		−343	3433	7210
A	−298	3937	8173		−1028	3272	7571
Z	−641	3900	8442		−1550	3114	7778
Y	−1289	3711	8710		−2629	2677	7982
1	−2220	3301	8823		−4518	1775	8067

[a]Equations of transformations are in the text.

of the data. The matched transformations are:

$$\text{``log''} = 3480 + \frac{\log_{10}(\text{income}) - 3.542}{.0001248},$$

$$\text{``square root''} = 3480 + \frac{\text{income} - 58.99}{.008476},$$

$$\text{``fourth root''} = 3480 + \frac{\text{income} - 7.681}{.0005518},$$

where income is in dollars per year. These each have the property that the median of the re-expressed batch is 3480, and that points close to the median are changed as little as possible. Formulas for the coefficients are given below.

Note that because the transformed values are linear transformations of logs and roots, they can and do take on negative values. Although leaving values near the median relatively unchanged, each of the transformations pulls the upper values in toward the center and pushes the lower values out, sometimes beyond what might be a natural lower bound of zero.

To compare directly the symmetry of the transformed batches, we subtract the median, 3480, from each of the midsummaries in Table 4-5, arriving at the differences in Table 4-6. Figure 4-5 plots midsummary-minus-median in the log, fourth-root, and square-root scales against the values of

$$\frac{(x_U - M)^2 + (M - x_L)^2}{4M}$$

in the original scale from Table 4-2. First, we see that the fourth-root midsummaries are almost exactly midway between the other two sets. This corresponds to the relative position of log, fourth root, and square root on the ladder of powers. Second, we see that the fourth-root transformation is probably close to the optimal power transformation. A power transformation with p between .25 and .5, perhaps .4 as was estimated earlier, would decrease the negative residuals for the extreme letter values at the expense of increasing the positive residuals for the more central letter values. The common scale provided by matched transformations facilitates both of these observations.

TABLE 4-6. **Midsummary-minus-median for the household income data and three matched re-expressions.**

	Original Data	Log	Fourth Root	Square Root
F	198	−27	29	85
E	636	−87	91	270
D	920	−159	107	372
C	1319	−261	128	515
B	1499	−582	−47	466
A	1761	−930	−208	457
Z	1914	−1248	−366	420
Y	2030	−2071	−803	231

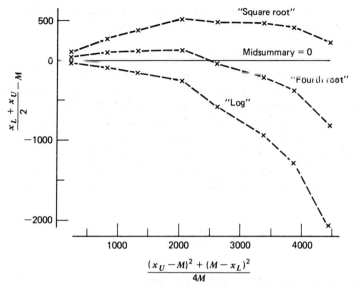

Figure 4-5. Matched midsummaries for the household income data.

Procedure for Obtaining Matched Transformations

Let us denote the original data by x and the nonlinear transformation by $y = T(x)$. We introduce two parameters, a and b, and a further transformation

$$z = a + by = a + bT(x),$$

where we would like to "match" to x.

There are a number of ways to determine a and b. For instance, we can choose two points x_1 and x_2 in the original scale and find a and b so that

$$z_1 = a + bT(x_1) = x_1$$

and

$$z_2 = a + bT(x_2) = x_2.$$

Alternatively, we can choose a point x_0 and require that

$$z_0 = a + bT(x_0) = x_0, \tag{9}$$

and, furthermore, that the derivative of z with respect to x, evaluated at x_0, be 1. That is,

$$\frac{dz}{dx}\bigg|_{x_0} = \frac{d[a + bT(x)]}{dx}\bigg|_{x_0}$$

$$= b\frac{dT(x)}{dx}\bigg|_{x_0} = 1,$$

or

$$b = \frac{1}{T'(x_0)}. \tag{10}$$

This second method relies on the linearity of the transformation near the center of the range of x. We choose x_0 as some central value and force $z_0 = x_0$. Then, because the derivative of z with respect to x is 1, z is close to x to the extent that $T(x)$ is nearly linear in x.

In most problems it will be easiest to identify one point at which we want to match the re-expression and then to let the slope determine the second parameter. Equation (9) indicates that

$$a = x_0 - bT(x_0) = x_0 - \frac{T(x_0)}{T'(x_0)}.$$

Thus

$$z = x_0 + \frac{T(x) - T(x_0)}{T'(x_0)}.$$

For power transformations $T(x) = x^p$, the formula for a matched transformation is particularly simple. Because

$$T'(x_0) = px_0^{p-1},$$

we have

$$z = x_0 + \frac{x^p - x_0^p}{px_0^{p-1}}.$$

The special case of $p = 0$, the logarithmic transformation, is illustrated in the following example, based on the population data.

EXAMPLE (CONTINUED):   POPULATION OF U.S. CITIES

First, we choose x_0 to be 88, the median value of the batch. Other measures
of central location could also be used. The original transformation is:

$$T(x) = \log_{10}(x) = \log_{10}e \cdot \log_e x,$$

so

$$T'(x) = (\log_{10}e)\frac{1}{x} = \frac{.4343}{x}.$$

We calculate

$$b = \frac{1}{T'(x_0)} = \frac{x_0}{.4343} = \frac{88}{.4343} = 202.6.$$

Then

$$a = x_0 - bT(x_0)$$

$$= 88 - 202.6\log_{10}88 = -306.0,$$

so

$$z = 202.6\log_{10}x - 306.0.$$

The simpler transformation

$$z = 200\log_{10}x - 300$$

performs almost as well and was used in the previous example (p. 113).

Matching Expected Values

Although the benefits listed above form the main reason for matching, there
is another consideration. Although the median of a batch of transformed
numbers equals the transform of the median of the raw numbers (as long as
the transformation is monotonic and the sample size is odd), a similar
relation does not hold for the sample mean. That is, $\Sigma T(x_i)/n \neq T(\bar{x})$,
where the x_i are the n values in the batch. Nor is it true that for the random
variable X, $E[T(X)] = T(EX)$, where E denotes expected value. Although
matching cannot eliminate these differences, it can reduce them.

As before, let

$$Z = a + bT(X)$$

with

$$a = x_0 - \frac{T(x_0)}{T'(x_0)}$$

$$b = \frac{1}{T'(x_0)}.$$

Then if we approximate $T(X)$ by the terms up to the quadratic in its Taylor-series expansion around x_0, we have

$$E(Z) = a + bET(X)$$

$$\approx x_0 - \frac{T(x_0)}{T'(x_0)} + \frac{1}{T'(x_0)} \left[T(x_0) + E(X - x_0)T'(x_0) \right.$$

$$\left. + \frac{1}{2} E(X - x_0)^2 T''(x_0) \right].$$

Straightforward algebra yields

$$E(Z) \approx E(X) + \frac{1}{2} E(X - x_0)^2 \frac{T''(x_0)}{T'(x_0)}.$$

We can write $E(X - x_0)^2 = E(X - EX)^2 + (EX - x_0)^2 = \text{Var } X + (EX - x_0)^2$. If we choose x_0 to be near the mean of X, $(EX - x_0)^2$ will be small, so that

$$E(Z) \approx E(X) + \frac{1}{2} \text{Var}(X) \frac{T''(x_0)}{T'(x_0)}. \tag{11}$$

Approximation (11) can be interpreted as follows: If Z is a transformation of X matched at a point near the mean of X, the mean of Z is approximately equal to the mean of X plus a term that depends on the spread of X and the curvature of the transformation. The correction term behaves just as we would expect; matching, which is based on the near linearity of $T(X)$ over small ranges, fails when X is widely spread out or when the curvature of the transformation is very large. Of course, formulas like approximation (11) can be developed for unmatched re-expressions as well. But matched transformations enhance the interpretation.

For the power transformations, including the logarithm,

$$\frac{T_p''(x_0)}{T_p'(x_0)} = \frac{p-1}{x_0}.$$

This identity further simplifies expression (11).

We have seen that matched transformations help us to interpret the effect of transformation in three ways. First, the re-expressed data involve numbers of familiar size, some of which are barely changed. Second, we can clearly see the effect of the transformation on each of the data points. Third, when comparing alternative expressions for a single batch of data, matching facilitates the comparison by putting the results in a common scale. In addition, we have seen how the shift in expected value of a transformed random variable relates to its spread and to the amount of curvature of the re-expression.

4F. SERENDIPITOUS EFFECTS OF TRANSFORMATION

Not all of the five objectives of transformation apply in any one example; they relate to different structures of data—for example, a single batch, a two-way table, or y-versus-x. Still, more than one objective can be relevant for a given data set. For example, when we consider straightness for y-versus-x, we would be pleased if a transformation for (say) y to achieve straightness also gave a more symmetric batch. A similar remark holds when the primary objective of transformation is to stabilize spread for several batches. Experience shows that re-expression to improve a batch in one respect is likely to improve the batch in other respects.

REVIEW OF LARGEST-CITIES EXAMPLE:

Let us return to the example of the 10 largest cities in each of 16 countries. In Chapter 3, the spread-versus-level plot pointed to the log transformation to stabilize the spreads of the 16 batches. A comparison of the boxplots for the raw data (Figure 3-2) with the boxplots for the log data (Figure 3-7) shows the benefits of this transformation:

1. The re-expressed batches show more uniformity of spread. Differences among the spreads no longer show a strong dependence on differences in level.
2. The right-skewness of most of the raw batches is reduced when the batches are re-expressed in the log scale. In particular, eight of the 19 outliers in the raw batches are no longer outliers in the log scale.
3. The details of Figure 3-7 are more visible than those of Figure 3-2. The log scale makes comparisons easier and provides more uniformity of detail in the boxplots.

In sum, transformation has brought improvement in several ways.

Why Does Re-Expression "Work"?

We can easily see that no mathematical principle guarantees that transformation for one purpose will improve batches in other respects. Imagine a data set that consists of several batches at different levels; suppose that each batch is symmetrically distributed about its mean, but that the spreads increase substantially with level. We could construct this hypothetical example in such a way that the log transformation would stabilize spread. But clearly, the resulting batches would be skewed to the left because the original batches were symmetric. A transformation that stabilizes spread would induce asymmetry in the individual batches. Similar mathematical examples can be constructed for other structures of data.

The hypothetical example illustrates the possibility that a transformation that improves a data set in one respect worsens it in another respect. In practice, we have seldom seen this happen. Instead, we often benefit from serendipitous effects such as those illustrated in our previous examples. To appreciate the reasons for our usual "good fortune," we need to consider the question: Why does transformation tend to work well in more than one way?

Let us focus most of our attention on two issues, stable spread and symmetry. Researchers and data analysts have observed that two characteristics,

> Increasing spread with increasing level, and
>
> Right-skewness

often occur together when data are measured or counted. Such data are bounded below by zero; this lower bound provides one explanation for the variation of spread with level. We might expect that batches located far to the right of the origin will be less constrained by the origin and thus will often exhibit greater spread than do those batches closer to the origin. Conversely, batches relatively close to the origin often have less spread, perhaps because of the bounding effect of the origin. This heuristic explanation is consistent with what is usually observed in batches of data at different levels. *Most batches of amounts or counts, when at different levels, exhibit increasing spread with increasing level.*

What about symmetry? The bounding effect of the origin also provides a heuristic explanation of the usual right-skewness present in batches of amounts or counts. Although smaller data values are bounded by the origin and are thus prevented from ranging too far to the left, nothing analogous prevents large values from ranging far to the right. It is then not surprising that batches of amounts or counts have right tails that are longer than their left tails. Finally, we mention a Poisson variable, with mean and variance equal to λ, as an illustration of a random variable having the type of behavior we describe here.

A power transformation, such as log or square root, "works" by changing the scale of data in a nonlinear fashion. For p less than one, these transformations compress the scale for larger data values more than they do for smaller data values. The sketch in Figure 4-6 helps to clarify this idea.

The dashed lines with arrows in Figure 4-6 indicate the mapping by the log transformation of the positive real line to the log scale. The log transformation compresses the scale from 100 to 1000 much more than it compresses the scale from 1 to 10; indeed, line segments of length 900 and 9, respectively, are transformed to segments that each have length 1.

To see why the log transformation can stabilize spread, consider two hypothetical batches:

One with median 10 and fourths at 1 and 100, and
Another with median 100 and fourths at 10 and 1000.

Because the log transformation preserves letter values, we see from Figure 4-6 that the transformed batches, respectively, have

Median equal to 1 and fourths at 0 and 2, and
Median equal to 2 and fourths at 1 and 3.

Thus the transformed batches have equal fourth-spreads.

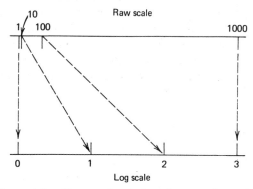

Figure 4-6. Change of scale by a log transformation.

The hypothetical example also demonstrates an improvement in symmetry after transformation. The "original batches" have substantial right-skewness; this is indicated by the positions of the median much nearer the lower fourth than the upper fourth. The "re-expressed batches" have their median exactly midway between the fourths, a feature that is consistent with symmetry. Thus transformation has substantially reduced the asymmetry.

We summarize our discussion of the serendipitous effects of re-expression.

1. Data that are amounts or counts usually display both increasing spread with increasing level and right-skewness.
2. Transformation for stable spread will necessarily compress the scale more for larger values than for smaller values.
3. Transformation for symmetry will also compress the scale more for larger values than for smaller values.
4. Thus transformation for either spread or symmetry will usually help us toward both objectives simultaneously.

4G.   WHEN IS TRANSFORMATION WORTHWHILE?

Transformation, even when it is done correctly and when it brings benefits, can also bring a disadvantage. With a few exceptions, such as those illustrated in Section 4B, re-expression moves us into a scale that is often less familiar. As a result, when we move to a new scale, we may lose some of our intuitive understanding and our ability to make interpretations. Such disadvantages may be of greater concern to the consumers of the analysis than to the analysts. But for most of us it is easier to think in terms of dollars, inches, or years, than it is to consider log-dollars, square-root inches, or one-over-years. Thus when we consider transformation, we should ask whether the benefits justify the effort needed and the inconvenience that may result for our consumer. We need to use judgment in answering this question, and, in a particular context, different people may make different judgments. Still, we can give a few guidelines for deciding when transformation is worthwhile.

Range Considerations

When working with a batch of amounts or counts, transformation is likely to be helpful only when the range of the batch is relatively large. If, for example, the gas mileage for a group of 4-cylinder subcompact cars ranges between 30 and 41 miles per gallon, transformation is unlikely to be worthwhile. With large enough groups, however, it might pay.

A convenient rule of thumb for deciding whether re-expression will change the shape of a batch very much is based upon the ratio

$$\frac{\text{largest data value}}{\text{smallest data value}}.$$

Transformation should be helpful when this ratio is large, say greater than 20, but not helpful when it is small, say less than 2. For the household income data (Table 4-1), the ratio is about 95 and transformation proved helpful. For the largest-cities example (Table 3-3), the ratios range from around 4 to nearly 27 for the 16 batches, and the ratio for the entire data set is approximately 160. Again, transformation provided considerable benefits in this example.

Observe that the rule of thumb is not meaningful for measured data that do not arise as amounts or counts. For example, a batch of temperatures (assumed above freezing) will yield a different ratio, depending on whether the scale is Fahrenheit, Celsius, or Kelvin. When there is no unique "natural zero" that is also a natural boundary, the rule of thumb does not apply, unless we change it to

$$\frac{\text{largest data value} - \text{natural boundary}}{\text{smallest data value} - \text{natural boundary}}.$$

Looking at Residuals

In more complex data structures such as y-versus-x and two-way tables, the residuals from fitting a model tell us much about whether subsequent transformation is likely to be worthwhile. When residuals from a fit are large and show certain kinds of systematic patterns, transformation is likely to be effective in explaining the structure that underlies the data set.

Using Transformation Plots

We have introduced two transformation plots for finding appropriate power transformations: the spread-versus-level plot and the transformation plot for symmetry. We discuss an additional transformation plot in Chapter 6. We explore four transformation plots in greater detail in Chapter 8. In each case, if the slope of a given plot is near 0, transformation is not indicated. If the plot is nearly linear and the slope is quite different from 0, transformation may prove helpful. In this instance, the slope of the plot provides the approximate exponent for a suggested power transformation. When the plot is significantly nonlinear, a power transformation will not do as thorough a job, though if the typical slope is substantial, it is likely to help. Perhaps a more complex transformation is called for, but its very complexity may make it difficult to justify.

Trial and Error

Finally, we note the merit in simply guessing at a transformation and trying it out. Plots and summaries of the re-expressed data set may be compared to those of the original batch. Such comparison helps us to decide whether adopting the new scale is justified.

Other Considerations

We have discussed three general ways for assessing whether transformation is worthwhile; this list of considerations is not all-inclusive. Sometimes custom within a field makes a particular transformation nearly mandatory. Judgments about interpretability of the new scale are very important, as we saw in Section 4B. We may also want to evaluate the sophistication of the users of our analysis in terms of their ability to adjust to a new scale. The size of a data set and the availability of high-speed computing may alter our judgment. Thus a large number of factors will influence our decision about whether to transform, and two analysts will not always interpret these factors in the same way and reach the same conclusion.

4H. SUMMARY

Transformations enable us to express data in a new scale that offers improvements over the original scale of data collection. These improvements may enhance interpretation of the data, or they may promote desirable characteristics of the data in each of our data structures—a single batch, several batches at different levels, y-versus-x, and two-way tables.

Aside from linear transformations, the power transformations (including log) are most useful for data sets consisting of amounts or counted data. Four transformation plots, for the four common data structures, are available to help guide us to a transformation that promotes

Symmetry,
Stable spread,
Straightness, or
Additive structure.

When transformation is likely to be helpful in achieving one of these four goals, it often improves the behavior of the data in other ways too. Although these serendipitous results are not guaranteed by mathematical rules, they are frequently evident in naturally occurring data.

A technique called matching transforms a re-expressed data set linearly in order to make the re-expressed data more nearly comparable to the raw data. Matching enables easier comparison of different transformations; it also enables us to see how a chosen power transformation has substantially altered the original batch.

The remaining chapters of this book use transformations whenever they aid the analysis. Chapter 8 continues our study of transformations at a more mathematical level.

EXERCISES

1. A hypothetical batch of numbers consists of 1, 10, 100, 1,000, 10,000, 100,000, and 1,000,000.

   (a) What power transformation would you recommend?

   (b) What goals of transformation are achieved by this choice?

2. Give a batch of numbers having $n = 6$ values that illustrates that the median is nearly (but not exactly) preserved by a square-root re-expression.

3. Show that the power transformations specified in Definition (1) all preserve the order of any batch of numbers. Give an example to illustrate that order would not be preserved if c were negative when $p = 0$.

*4. Show, using some elementary calculus, that the curves given in Figure 4-1 pass through the point $(1, 0)$ and have common slope 1 at that point. How would the corresponding transformations compare if all values in a batch were between 1.9 and 2.1?

5. What aspects of Table 4-1 suggest that $p = \frac{1}{4}$ might be more effective in some way than the log re-expression? Do the results shown in Table 4-3 tend to support your response?

6. For the household income data, consider a cube-root transformation and then match the results to the raw data.

   (a) What values of a and b in the expression

$$ax^{1/3} + b$$

   lead to a matched transformation?

(b) Give the letter-value displays for the matched cube-root transformation.

(c) How do the trends in the new midsummaries compare with those shown in Table 4-5?

7. Two reasonably accurate thermometers on two bankfronts report temperatures in Fahrenheit degrees and Celsius degrees, respectively. The daily noontime temperatures are recorded for the 30 days of September. Explain why you can easily match the temperatures in the first batch to those in the second batch (Celsius degrees) *without* using the matching process described in this chapter.

8. For the transformed batches described in Tables 4-1 and 4-3, matching is not used.

(a) Find linear transformations that will match each of the three re-expressions to the raw data.

(b) Explain why the comparison of the four matched batches is facilitated.

(c) What conclusions would you now reach about an effective choice of transformation?

CHAPTER 5

Resistant Lines
for y versus x

John D. Emerson
Middlebury College

David C. Hoaglin
Harvard University and Abt Associates Inc.

Various methods have been developed for fitting a straight line of the form

$$y = a + bx$$

to the data (x_i, y_i), $i = 1,\ldots,n$. The best-known and most widely used method is least-squares regression, which involves algebraically simple calculations, fits neatly into the framework of inference built on the Gaussian distribution, and requires only a straightforward mathematical derivation. Unfortunately, the least-squares regression line offers no resistance. A wild data point can easily seize control of the fitted line and cause it to give a totally misleading summary of the relationship between y and x. The three-group resistant line avoids this difficulty. Thus it is much more useful in exploring y-versus-x data.

After a brief description of the way that the resistant-line method divides the data into three groups and achieves resistance by using medians within the groups, this chapter examines the iterative steps that are usually necessary to adjust or polish the line. In their simplest form, these steps of polishing can sometimes produce anomalous results, but a minor modification in the calculations bypasses these problems and leads to further insight into how the method works.

129

A discussion of the background for the three-group resistant line traces the idea of splitting the data into groups. The final sections briefly discuss the ideas of influence and leverage and relate this exploratory method to other alternatives to least-squares regression.

5A. A RESISTANT LINE FROM THREE GROUPS

Customarily, an exploration of how y may vary with x or depend on x begins by plotting y against x. If the pattern of points in this scatter plot is fairly straight, a fitted straight line will summarize how y changes with changes in x. When the plot reveals clear curvature, re-expression of y or of x or of both y and x may be an appropriate way of straightening it. Section 4D discusses the use of transformations for this and other purposes, and Section 8E describes a technique for judging what transformations would be effective. In the present chapter, we assume that the data are, or have been made, satisfactorily straight.

Because the data may include a few points that depart substantially from the primary pattern, the exploratory approach aims at providing resistance so that such data points can generally have only a small effect on the fitted line. Instead, the unusual points will yield large residuals, and these in turn will lead to further probing for explanations of the unusual behavior. After spotting an unusual point in a scatter plot, it is easy to set it aside and fit a line to the rest of the data. In practice, this graphical approach can become laborious. Also, a computerized analysis may simply fit a line, bypassing any visual checks. Thus to increase consistency and ensure protection, the line-fitting procedure should have the resistance built in.

For the simpler task of summarizing the center of a batch of data, the simplest resistant procedure is the sample median, as discussed in Chapters 2 and 10. The exploratory technique for fitting a line (Tukey, 1970) derives its resistance from the median. The basic ideas are to divide the n data points $(x_1, y_1),\ldots,(x_n, y_n)$ into three groups, use medians to form a summary point within each group, and base the line on these three summary points.

Forming Three Groups

We begin by sorting the values of x so that $x_1 \leqslant x_2 \leqslant \cdots \leqslant x_n$. Then, on the basis of these sorted values, we divide the n data points, (x_i, y_i), into three groups—a left group, a middle group, and a right group—as nearly equal in size as possible. When there are no ties among the x_i, the number of points in the three groups depends on the remainder from dividing n by

3. Specifically, we allocate the data to the groups as follows:

Group	$n = 3k$	$n = 3k + 1$	$n = 3k + 2$
Left	k	k	$k + 1$
Middle	k	$k + 1$	k
Right	k	k	$k + 1$

Ties among the x_i may prevent us from achieving precisely this allocation, because we do not break ties. All data points with the same x-value go into the same group. A detailed examination of the treatment of ties would involve a number of special cases, and in some extreme situations it might be impossible to form more than two groups. Instead of enumerating the possibilities, we merely mention that the computer algorithm of Velleman and Hoaglin (1981) provides solutions whenever proceeding to fit a line is justified.

Summary Points

Within each of the "thirds" thus formed, we determine the two coordinates of a summary point by finding first the median x-value and then, separately, the median y-value. We label the coordinates of the three summary points L for left, M for middle, and R for right:

$$(x_L, y_L), (x_M, y_M), (x_R, y_R).$$

Figure 5-1 shows the data points and the summary points in a hypothetical 9-point example. As this figure emphasizes, none of the summary points need be a data point, because the x-medians and y-medians are determined separately. However, all three could be data points, as often happens, for example, when the x_i and the y_i follow the same ordering within each group.

This method of determining the summary points gives the line its resistance. As long as the number of data points in each group is not too small, the median provides resistance to wild values of x, y, or both.

Slope and Intercept

What now remains is to use the summary points in calculating the slope b and the intercept a in the line

$$\hat{y} = a + bx.$$

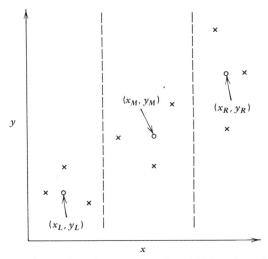

Figure 5-1. Data points ($\times$) and summary points ($\bigcirc$) in a hypothetical example.

(As is customary, the "hat" on y reminds us that the line is the source of fitted or estimated values, both at x-values in the data and at other appropriate values of x.) In this setting, b tells how many units $\hat{y}$ changes in response to a 1-unit change in x, and we obtain this information from the data by working with the left and right summary points:

$$b_0 = \frac{y_R - y_L}{x_R - x_L}. \tag{1}$$

By using (x_L, y_L) and (x_R, y_R) in this way, we strike a balance between (a) the advantage of measuring the change in y over a wide interval of x and (b) the need to have enough data points in the left and right groups for adequate resistance.

When we use the fitted slope b_0 to adjust the y-value of each summary point, the remainder is the intercept value for a line of slope b_0 that passes exactly through that point. The fitted intercept is the average of these three values:

$$a_0 = \tfrac{1}{3}\big[(y_L - b_0 x_L) + (y_M - b_0 x_M) + (y_R - b_0 x_R)\big]. \tag{2}$$

Again, because the summary points are based on medians, a_0 is resistant.

For comparison, the slope and intercept of the line fitted by least squares are

$$b_{LS} = \frac{\Sigma(x_i - \bar{x})(y_i - \bar{y})}{\Sigma(x_i - \bar{x})^2},$$

and

$$a_{LS} = \bar{y} - b_{LS}\bar{x},$$

where $\bar{x} = \Sigma x_i/n$ and $\bar{y} = \Sigma y_i/n$. The impossibility of any resistance is evident in the rigid way that all the x-values and all the y-values enter into b_{LS} and a_{LS}.

Slope and Central Value

The fitting of straight lines in terms of slope and intercept is conventional, but usually artificial. The intercept, which gives a value of y where $x = 0$, may be imprecisely determined; it has little relevance when all the x-values lie far from zero. (In the example to follow, how many private school children do we have at age 0?) Fitting in terms of slope and central value, where the central value is at $x = \bar{x}$ or at $x = \text{median}\{x_i\}$ or at $x = x_M$—or at some convenient nearby round value—is usually more useful. If we work at $x = x_M$ for convenience, then the initial line is

$$\hat{y} = a_0^* + b_0(x - x_M),$$

where b_0 is as before and the central value (also called *level*) is

$$a_0^* = \tfrac{1}{3}\{[y_L - b_0(x_L - x_M)] + y_M + [y_R - b_0(x_R - x_M)]\}.$$

We emphasize this form of the line in the following discussion.

Residuals

Once we have obtained the slope and level for the fitted line, the immediate next step is to calculate the residual for each data point:

$$r_i = y_i - [a^* + b(x_i - x_M)]. \tag{3}$$

As Chapter 7 discusses, plots of the residuals can then reveal a variety of important features and patterns. For the present, we need only emphasize a general property of a set of residuals, both in y versus x and in many more complex situations:

Substituting the residuals for the original y-values (i.e., using (x_i, r_i) instead of (x_i, y_i), $i = 1, \ldots, n$) and then repeating the fitting process lead to a zero fit.

For the straight line, this means that using (x_i, r_i), $i = 1,\ldots,n$, as the data yields zero slope and zero level. In other words, the residuals contain no further straightline behavior to summarize.

An important characteristic of resistant procedures is that they often require iteration. In this book, the 3-group resistant line is the first example. Thus the residuals from the line with slope b_0 and level a_0^* may not have zero slope and level when we attempt to fit a line to them, although the new slope and level would ordinarily be substantially smaller (in magnitude) than b_0 and a_0^*. For this reason, we think of b_0 as an initial value of the slope and a_0^* as an initial value of the level (hence the subscript 0).

Iteration

Ordinarily, we expect that b_0 and a_0^* will need some adjustment. Fitting a further line to the residuals from the initial line gives the adjustments δ_1 and γ_1 to slope and level, respectively. Specifically, we use the initial residuals,

$$r_i^{(0)} = y_i - \left[a_0^* + b_0(x_i - x_M)\right], \qquad i = 1,\ldots,n,$$

in place of the y_i and repeat most of the earlier steps in the fitting process. (The x_i have not changed, so the three groups and the x-medians in the summary points remain the same throughout.)

The adjusted slope and level are then $b_0 + \delta_1$ and $a_0^* + \gamma_1$, and the new residuals are

$$r_i^{(1)} = r_i^{(0)} - \left[\gamma_1 + \delta_1(x_i - x_M)\right], \qquad i = 1,\ldots,n.$$

Now we could try another iteration. In general, we do not know whether we have a suitable set of residuals until we check that they yield a zero fit. In practice, we continue the iterations until the adjustment to the slope is sufficiently small in magnitude—conservatively, at most 1% (or, preferably, .01%) of the size of b_0. Each iteration adds its slope adjustment and level adjustment to the previously adjusted values:

$$b_1 = b_0 + \delta_1, \qquad b_2 = b_1 + \delta_2,\ldots$$

and

$$a_1^* = a_0^* + \gamma_1, \qquad a_2^* = a_1^* + \gamma_2,\ldots.$$

The iterations are usually few enough that the calculations are not very time-consuming; in any event, the resistance justifies the effort.

For some sets of data, this iterative procedure, which has long been used in fitting the three-group resistant line, encounters difficulties. The slope adjustments may decrease slowly or, after a few steps, they may cease to decrease and instead oscillate between two values with the same magnitude and opposite signs. Fortunately, a straightforward modification completely eliminates such problems and allows the number of iterations to be drastically limited. We discuss this in Section 5B, after an example of the resistant line.

EXAMPLE:

As an example for a 1953 discussion, Greenberg gave the ages and heights of two samples of children—one sample from an urban private school and

TABLE 5-1.   Age and height of children in
a private school.

Child	Age (months)	Height (cm)
1	109	137.6
2	113	147.8
3	115	136.8
4	116	140.7
5	119	132.7
6	120	145.4
7	121	135.0
8	124	133.0
9	126	148.5
10	129	148.3
11	130	147.5
12	133	148.8
13	134	133.2
14	135	148.7
15	137	152.0
16	139	150.6
17	141	165.3
18	142	149.9

Source: Bernard G. Greenberg (1953). "The use of analysis of covariance and balancing in analytical studies," American Journal of Public Health, 43, 692–699 (data from Table 1, p. 694).

the other from a rural public school. We reproduce the data from the 18 private school children in Table 5-1 and plot height against age in Figure 5-2. Although the data points do not follow a clear straight line, their pattern is not noticeably curved, and so a fitted line should serve to summarize how height (y) increases with age (x) in this group of children. Relative to nearby points, those for child 13 and child 17 seem to stand out; we will watch these as the process unfolds.

Because 18 is evenly divisible by 3 and the x-values involve no ties, each of the three groups contains six data points. The three summary points are

$$(x_L, y_L) = (115.5, 139.15)$$

$$(x_M, y_M) = (127.5, 147.9)$$

$$(x_R, y_R) = (138, 150.25).$$

Thus the initial slope is

$$b_0 = \frac{y_R - y_L}{x_R - x_L} = \frac{150.25 - 139.15}{138 - 115.5} = 0.4933,$$

and the initial level is

$$a_0^* = \tfrac{1}{3}(145.07 + 147.90 + 145.07) = 146.01.$$

Table 5-2 shows the separation of the data points into the three groups, along with the residuals from this initial line.

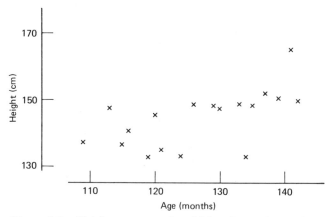

Figure 5-2. Height versus age for children in a private school.

TABLE 5-2. Children's age and height—the three groups and the initial residuals.

Child	Age (x)	Height (y)	$y - [146.01 + 0.4933(x - 127.5)]$
1	109	137.6	0.72
2	113	147.8	8.95
3	115	136.8	−3.04
4	116	140.7	0.37
5	119	132.7	−9.11
6	120	145.4	3.09
7	121	135.0	−7.80
8	124	133.0	−11.28
9	126	148.5	3.23
10	129	148.3	1.55
11	130	147.5	0.26
12	133	148.8	0.08
13	134	133.2	−16.01
14	135	148.7	−1.01
15	137	152.0	1.31
16	139	150.6	−1.08
17	141	165.3	12.63
18	142	149.9	−3.26

To see how the iterations go, we calculate the first adjustments to the slope and level:

$$\delta_1 = \frac{-1.045 - 0.545}{138 - 115.5} = -0.0707$$

$$\gamma_1 = \tfrac{1}{3}(-0.30 + 0.17 - 0.30) = -0.14.$$

We notice that δ_1 is substantially smaller in magnitude than b_0 but still not negligible. Two more iterations bring us to a situation where the process can stop: the most recent adjustment causes less than a 1% change in the slope. The resulting fitted line is

$$\hat{y} = 145.86 + 0.4285(x - 127.5).$$

Figure 5-3 plots the residuals from this line against age (x). On the whole, they are quite satisfactorily flat. The two points toward the right that depart more noticeably—one high, one low—come from child 13 and child

17. (These two residuals are extreme enough to be "outside" according to the rule discussed in Section 2C.) Also, three more residuals for children with ages around 120 months seem a little low. If we had more information, we would try to learn why these children are short or tall for their ages as judged relative to their fellow students. For example, distinguishing the girls from the boys might help.

In this example we can see that the two somewhat unusual data points had little, if any, effect on this line summarizing the bulk of the data. A least-squares line faces more risk of distortion from such points. For these data, the least-squares regression line is

$$\hat{y} = 79.7 + 0.511x$$

or

$$\hat{y} = 144.9 + 0.511(x - 127.5),$$

suggesting that the y-values for children 5, 7, 8, and 17 have helped to twist the line. Indeed, if the y-value for child 13 were not noticeably low, the least-squares line would be even steeper. Because we should usually judge a fit by its residuals, Figure 5-4 plots the least-squares residuals against age. Although rather similar to Figure 5-3, this picture gives some suggestion of a slight downward trend. That is, the least-squares residuals would look more nearly horizontal if we removed from them a line with a slight negative slope.

In this example, the variability of the residuals commands more attention than the difference between the slope of the least-squares line and that of

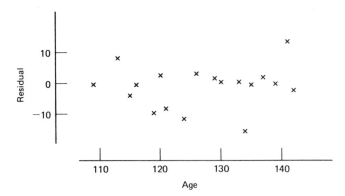

Figure 5-3. Residuals (of height) versus age, after fitting resistant line.

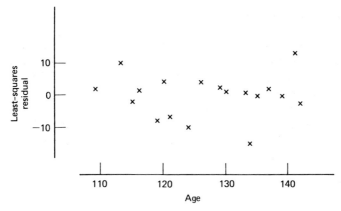

Figure 5-4. Least-squares residuals versus age.

the resistant line. For example, the standard deviation of the least-squares residuals is 7.03, and the standard error of the slope is 0.167, about twice the difference between the two slopes. We can see, qualitatively, how some data points affect the least-squares line more than the resistant line, and it would be straightforward to construct an example that emphasizes the nonresistance of least squares. It is helpful to see, however, that when the data are reasonably well-behaved, the two lines are similar.

5B. IMPROVING THE ITERATIVE ADJUSTMENTS

Section 5A mentioned that some sets of y-versus-x data cause the iterative adjustments of the slope to go astray. We now demonstrate one form of the difficulty with a contrived but dramatic example. Then we discuss how a different way of regarding the iterations avoids the problem altogether.

Oscillation

Consider the following hypothetical 9-point set of (x, y) data, devised by Andrew F. Siegel: $(-4, 0)$, $(-3, 0)$, $(-2, 0)$, $(-1, 0)$, $(0, 0)$, $(1, 0)$, $(2, -5)$, $(3, 5)$, $(12, 1)$. The first six points lie on the x-axis, the last three do not, and a considerable gap separates the last point from the rest. A plot of the data, Figure 5-5, readily suggests that a line with slope 0 would be a reasonable summary.

When we fit a 3-group resistant line to these data according to the procedure described in Section 5A, however, the result is nothing like a

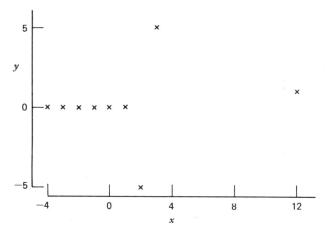

Figure 5-5. Andrew Siegel's pathological 9-point data set.

horizontal line! Table 5-3 shows the adjusted line as it stands at the end of each of the first 10 iterations, and Figure 5-6 plots these slopes against the iteration number. The slope starts at 0.167, decreases to -0.083, and then takes on larger values (both positive and negative) until it alternates between 0.833 and -0.694, neither of which represents a plausible summary of the relationship between y and x.

The residuals from the lines with slopes 0.833 and -0.694, shown in Table 5-4, help us to see the reason for the oscillatory behavior of the slope. In the summary point for the right group, $x_R = 3$, and the median residual comes from either the point with $x = 2$ or the point with $x = 3$, depending

**TABLE 5-3. The resistant line for the hypothetical
           data, as adjusted in each of
           the first 10 iterations.**

Iteration	Adjusted Line
1	$\hat{y} = 0.167x + 0.333$
2	$\hat{y} = -0.083x - 0.167$
3	$\hat{y} = 0.292x + 0.583$
4	$\hat{y} = -0.271x - 0.542$
5	$\hat{y} = 0.573x + 1.15$
6	$\hat{y} = -0.693x - 1.39$
7	$\hat{y} = 0.833x + 1.67$
8	$\hat{y} = -0.694x - 1.39$
9	$\hat{y} = 0.833x + 1.67$
10	$\hat{y} = -0.694x - 1.39$

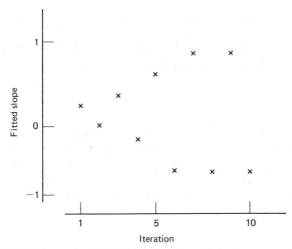

Figure 5-6. Fitted slopes for the first 10 iterations of the 3-group resistant line on the 9-point example.

on the iteration. From Figure 5-5, we would prefer that the median residual come from the point with $x = 12$, but this x-value is sufficiently extreme that the adjustments will not produce this result. Instead, the point with $x = 12$ has either the most negative or the most positive residual in the group.

Of course, this example is quite pathological—one x-value in the right group could be regarded as an outlier and so could two y-values—but similar problems occasionally arise in real data (for example, Exercise 1).

TABLE 5-4. Residuals from the two extreme fitted lines in the hypothetical example.

x	y	$y - (0.833x + 1.67)$	$y - (-0.694x - 1.39)$
-4	0	1.667	-1.389
-3	0	0.833	-0.694
-2	0	0.000	-0.000
-1	0	-0.833	0.694
0	0	-1.667	1.389
1	0	-2.500	2.083
2	-5	-8.333	-2.222
3	5	0.833	8.472
12	1	-10.667	10.722

The difficulty is that the x-median in either or both of the end groups is not a good summary for the point or points that should be involved in the y-median or the median residual. Consequently, the iterative process may converge slowly or it may oscillate, as in the example.

A Safe Iteration

Fortunately, closer study has provided a straightforward way to avoid these convergence problems. The basic idea is simple: If the x-values are the source of the trouble, why not suppress them? It turns out that, after the first step or two, this is easy to do (Johnstone and Velleman, 1982).

Rewriting equation (1) to apply to the adjustment at iteration $j + 1$, we have

$$\delta_{j+1} = \frac{r_R^{(j)} - r_L^{(j)}}{x_R - x_L}, \tag{4}$$

and this will be 0 precisely when $r_R^{(j)} - r_L^{(j)} = 0$. That is, all we must do is find the value of b that yields the same median residual in the right group as in the left group. More formally, we define

$$\Delta r(b) = r_R(b) - r_L(b), \tag{5}$$

showing the functional dependence on b and ignoring the iteration number. We seek the value of b such that $\Delta r(b) = 0$.

Computationally, we begin as before, finding b_0 according to equation (1). We calculate $\Delta r(b_0)$ and then δ_1 according to equation (4). Next we calculate $\Delta r(b_0 + \delta_1)$, the value of Δr for the adjusted slope. (Note that any fitted intercept a would drop out of Δr in equation (5), so we can focus the iterative process on b and leave a until later.) Ordinarily, $\Delta r(b_0)$ and $\Delta r(b_0 + \delta_1)$ will have opposite signs, indicating that the desired value b lies between b_0 and $b_1 = b_0 + \delta_1$. Otherwise, when $\Delta r(b_0)$ and $\Delta r(b_0 + \delta_1)$ have the same sign, we continue to step away from b_0 on the same side as $b_0 + \delta_1$ until we find a b_1 such that $\Delta r(b_1)$ has the opposite sign from $\Delta r(b_0)$.

At this stage, we have b_0 with $\Delta r(b_0)$ and b_1 with $\Delta r(b_1)$, and we know that Δr must be 0 for some b between b_0 and b_1. (The present discussion has not paused to prove this last statement, but we shall see before long that it is correct.) Thus we can readily continue by linear interpolation, obtaining

$$b_2 = b_1 - \Delta r(b_1) \frac{b_1 - b_0}{\Delta r(b_1) - \Delta r(b_0)}. \tag{6}$$

When $\Delta r(b_2)$ is not exactly 0 (or is not yet close enough to 0 for our purposes), we can repeat the interpolation step. To do this, we narrow the interval that we know contains b by using b_2 in place of whichever of b_1 or b_0 had Δr of the same sign as the new $\Delta r(b_2)$.

This new iteration scheme works directly with b to determine where $\Delta r(b) = 0$; after the initial steps involving b_0 and b_1, it makes no use of the x-medians, x_L and x_R.

Uniqueness

So far we have not tried to establish that finding b_0 and b_1 such that $\Delta r(b_0)$ and $\Delta r(b_1)$ have opposite signs brackets the solution to $\Delta r(b) = 0$ between b_0 and b_1, nor have we shown that this solution is unique. We now approach these questions by examining the functional form of $\Delta r(b)$. These techniques and results come primarily from Johnstone and Velleman (1982).

To depict the situation, we use a plot that is closely related to the customary plot of residuals (such as in Figure 5-3). Because the intercept drops out of $\Delta r(b)$, we work only with $y_i - b(x_i - x_M)$; and we plot this quantity against b, instead of against x_i, for all n data points. In effect, we are tracing the value of each residual as a function of b, and we have one line for each data point. Figure 5-7 does this for Siegel's hypothetical example.

Naturally, for even moderate n such a plot will become quite cluttered, so we concentrate on the behavior of $r_L(b)$ and $r_R(b)$, the components of $\Delta r(b)$

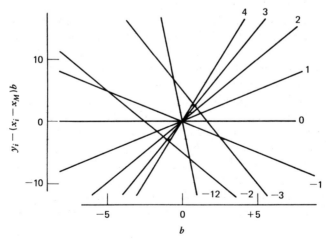

Figure 5-7. Residuals as a function of slope $[r_i = y_i - (x_i - x_M)b]$ in the 9-point example [one line for each data point; label is $-(x_i - x_M)$].

in equation (5). Geometrically, we can easily identify the lines that correspond to data points in the left group. Because the line for (x_i, y_i) in this
plot is $y_i - b(x_i - x_M)$ and thus has slope $-(x_i - x_M)$ and because the left
group contains the smallest x-values, these lines are the most steeply
upward-sloping ones. In the example, the three lines for the left group have
slopes 4, 3, and 2 (because $x_M = 0$ and their x-values are -4, -3, and -2).
Similarly, the lines that correspond to data points in the right group are the
most steeply *downward-sloping*. Often we would simplify the plot by omitting the middle group.

Now, for any given value of b, we can locate $r_L(b)$ and $r_R(b)$ in the plot.
We recall that, by definition (again ignoring the intercept),

$$r_L(b) = \underset{L}{\mathrm{med}} \{ y_i - (x_i - x_M)b \}$$

and

$$r_R(b) = \underset{R}{\mathrm{med}} \{ y_i - (x_i - x_M)b \}.$$

Therefore, within each group, the median value lies on the line that is in the
middle at b (when the group contains an even number of data points, it lies
halfway between the middle two lines).

By varying the value of b, we trace out r_L and r_R as functions of b. We
now see that these two functions are piecewise linear and that a change from
one segment to another coincides with the intersection of two lines within
the group. This property makes it straightforward to plot $r_L(b)$ and $r_R(b)$, as
we do for the 9-point example in Figure 5-8. Because the slopes of all lines
for the left group are positive and the slopes of all lines for the right group
are negative, the piecewise slopes of $r_L(b)$ are positive and those of $r_R(b)$ are
negative, so that $r_L(b)$ and $r_R(b)$ must intersect in a single point. This value
of b is the slope of the three-group resistant line.

Because the iteration scheme in equation (6) uses Δr, we need to restate
these results in terms of Δr. We recall that $\Delta r(b) = 0$ precisely when
$r_L(b) = r_R(b)$. Figure 5-9 plots $\Delta r(b)$ against b and includes the horizontal
line at $\Delta r = 0$ to emphasize the solution. From the equation for the short
line segment, $\Delta r(b) = 1 - 15b$ for $-\frac{4}{9} \leqslant b \leqslant \frac{3}{5}$, we find the solution to
$\Delta r(b) = 0$, namely $\frac{1}{15}$. In effect, the slope is being determined by the data
points $(-3, 0)$ and $(12, 1)$.

Figure 5-9 enables us to see how the original slope-adjustment iterations
failed on this example. The slope values between which they oscillated were
-0.694 and 0.833, and these lie beyond the two ends of the short line

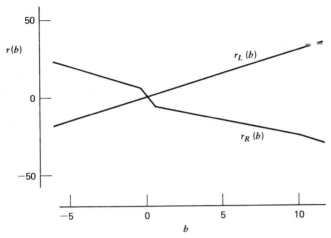

Figure 5-8. Median residual within the left and right groups as a function of slope in the 9-point example.

segment, so that the line segment where $\Delta r(b)$ actually crosses zero remained undiscovered. Also, having $x_R = 3$ did not help either. By using linear interpolation in an interval known to contain the solution [because $\Delta r(b_0)$ and $\Delta r(b_1)$ have opposite signs], the new algorithm ensures convergence—usually at a rapid rate. Thus we have two reasonable choices: (a) agree to make, say, three iterations or (b) agree to iterate to convergence.

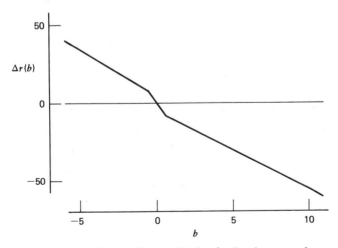

Figure 5-9. $\Delta r(b) = r_R(b) - r_L(b)$ for the 9-point example.

Intercept

This discussion has focused almost entirely on calculating a suitable slope for the 3-group resistant line (as it should, because the slope is usually more important). Still, we need the intercept to complete the fitting process. After determining the slope, we could find the intercept from the summary points as in equation (2). However, just as the iterative algorithm based on linear interpolation deemphasizes the x-values of the summary points, we now prefer an alternative approach. We just use the median of the partial residuals:

$$a^* = \text{med}\{y_i - b(x_i - x_M)\} \tag{7}$$

EXAMPLE:

We conclude this discussion of algorithms for the resistant line by showing how the linear interpolation procedure behaves on the 9-point example. From the data and Table 5-3, we see that

$$b_0 = \tfrac{1}{6} \quad \text{and} \quad b_1 = -\tfrac{1}{12}.$$

Also, we readily obtain

$$\Delta r(b_0) = \text{med}\{-5\tfrac{1}{3}, 4\tfrac{1}{2}, -1\} - \text{med}\{\tfrac{2}{3}, \tfrac{1}{2}, \tfrac{1}{3}\}$$

$$= -1 - \tfrac{1}{2} = -\tfrac{3}{2},$$

and

$$\Delta r(b_1) = \text{med}\{-4\tfrac{5}{6}, 5\tfrac{1}{4}, 2\} - \text{med}\{-\tfrac{1}{3}, -\tfrac{1}{4}, -\tfrac{1}{6}\}$$

$$= 2 - \left(-\tfrac{1}{4}\right) = \tfrac{9}{4}.$$

Now the first interpolation step proceeds according to equation (6):

$$b_2 = b_1 - \Delta r(b_1)\frac{b_1 - b_0}{\Delta r(b_1) - \Delta r(b_0)}$$

$$= -\frac{1}{12} - \frac{9}{4} \cdot \frac{(-1/12) - (1/6)}{(9/4) - (-3/2)}$$

$$= \frac{1}{15}.$$

Calculating $r_L(b_2)$ and $r_R(b_2)$, we find that $\Delta r(b_2) = 0$, and so one iteration of this form has brought us exactly to the solution, whose value we already knew from Figure 5-9.

This result is gratifying after the oscillation of the slope-adjustment procedure (Table 5-3 and Figure 5-6), but we should not always expect such extremely rapid convergence. In this example, b_0 and b_1 both lie in the interval $-\frac{4}{9} \leqslant b \leqslant \frac{3}{5}$ corresponding to the segment of $\Delta r(b)$ that crosses the axis, and so linear interpolation must yield the solution in only one step. Ordinarily, b_0 and b_1 will lie on other segments of Δr, and another iteration or two will be required.

Finally, we calculate the level according to equation (7):

$$a^* = \tfrac{1}{15} \text{med}\{4, 3, 2, 1, 0, -1, -77, 72, 3\}$$

$$= \tfrac{2}{15}.$$

Thus with the proper algorithm for finding the slope, the resistant line gives a very satisfactory fit for this pathological 9-point data set.

5C. BACKGROUND: GROUPS AND SUMMARIES

Although the three-group resistant line represents a separate development, several earlier techniques involved dividing the (x, y) data into groups on the basis of x and then summarizing x and y within the groups. The present section reviews these approaches, paying attention both to historical chronology and to the number of groups and the summary measure.

Some of these methods did not arise as alternatives to least squares in fitting a regression line. Instead, they were developed for fitting a straight line "when both variables are subject to error." Before we examine the methods, we briefly review some statistical models that can underlie a relationship between two variables, x and y.

As a very simple example of a situation in which both x and y are subject to error, we can consider two thermometers, one Celsius and the other Fahrenheit, both measuring the same temperature. The true temperatures follow an exact linear relationship, but imperfections in the thermometers and imprecision in the readings taken from them can introduce errors into both data values. We might suggest this explanation when a temperature display (located, say, on a banking institution) shows Fahrenheit and Celsius temperatures that do not match.

In this discussion we use X and Y to denote the true values of the two variables. We think of these as linearly related:

$$Y = \alpha + \beta X, \tag{8}$$

where α and β are fixed but unknown constants. In simple regression one elaborates the model by allowing for measurement error only in Y; that is, the observation is treated as

$$y = Y + v, \tag{9}$$

where v is symmetric with mean 0 and variance σ_v^2. Then the method of least squares yields estimators of α and β as in Section 5A. If our model is complete, these estimators, a_{LS} and b_{LS}, have the usual optimality properties, including unbiasedness and minimum variance among estimators that are linear functions of the available y_i.

So far, we have not considered the sampling mechanism by which we obtain X. As long as we observe X without error, that mechanism does not matter: either we could fix X and observe y, or we could collect random pairs (X, y).

In general, the notion of "regression" encompasses a broader range of models than the one given by equations (8) and (9). The *regression of y on X* is a function of X, defined by the expected (average) value of y for each fixed value of X. We write this

$$E(y \mid X),$$

adopting the usual conditional-expectation notation, as would be appropriate when X is a random variable. For example, some tables list average weights for men and women as a function of height. As a function of X, $E(y \mid X)$ is not required to be linear, but that is the most commonly used form. Thus the model of equations (8) and (9) is an instance of linear regression:

$$E(y \mid X) = \alpha + \beta X.$$

Another model, more appropriate in many situations, allows v to include other errors in y in addition to measurement error. Still, making this allowance for individual variation has not changed the assumption that we observe X without error.

When we further elaborate the model to accommodate measurement error on X, we have, in addition to equations (8) and (9),

$$x = X + u, \tag{10}$$

where u has mean 0 and variance σ_u^2. Also, we conventionally assume that u and v are not correlated. When X is prechosen (as might happen in a variety of laboratory settings), the relationship $Y = \alpha + \beta X$ is known as a *functional relationship*. When, however, X is random, the term for $Y = \alpha + \beta X$ is *structural relationship*, and one conventionally further assumes that u is uncorrelated with X and v is uncorrelated with Y. From the substantial literature on these "errors-in-variables" models, we need only appreciate that, when X is subject to error, the usual least-squares regression calculations are generally not appropriate. [An exception is the situation described by Berkson (1950).] The difficulty is that α and β, the coefficients we seek to estimate, relate Y to X, whereas we observe not X but $x = X + u$, along with $y = Y + v$.

An extension of the least-squares approach incorporates measurement error in x and seeks to minimize (with respect to a and b) a form such as

$$\sum_{i=1}^{n} \left\{ \frac{[y_i - (a + b\hat{x}_i)]^2}{\sigma_v^2} + \frac{(x_i - \hat{x}_i)^2}{\sigma_u^2} \right\}, \tag{11}$$

instead of

$$\sum_{i=1}^{n} [y_i - (a + bx_i)]^2.$$

Actually working with expression (11), however, requires knowledge of σ_u^2 or σ_v^2 or σ_v^2/σ_u^2; often such information will not be available.

An alternate approach brings in another form of additional information by dividing the observations (x_i, y_i) into groups. If the grouping rule satisfies certain conditions—such as being independent of the errors in the data—then the resulting estimator of β will avoid the shortcomings of the least-squares estimator. Madansky (1959) discusses these and other aspects of the problem of fitting a straight line when both variables are subject to error. Our emphasis in this chapter is simply on the motivation that this statistical problem provides for procedures that use groups. In the remainder of this section we briefly survey the grouping methods of Wald, Nair and Shrivastava, Bartlett, and Brown and Mood.

Wald's Method

For the problem of fitting a straight line when both x and y are subject to error, Wald (1940) first proposed dividing the data (x_i, y_i) into groups. His method uses two groups of equal size. Ideally, all the true values, X_i, in the first group should be smaller than the true values in the second group. In practice, because the X_i are unknown, the grouping would be based on the observed x_i.

Specifically, suppose (for convenience) that n is even and let $m = n/2$. Then, assuming that the x-values are in order, $x_1 \leqslant x_2 \leqslant \cdots \leqslant x_n$, the two groups are

$$(x_1, y_1), \ldots, (x_m, y_m)$$

and

$$(x_{m+1}, y_{m+1}), \ldots, (x_n, y_n).$$

(If it happens that $x_{m+1} = x_m$, the method discards tied x-values at the center before forming the groups. A similar approach is adequate to deal with odd n.) The summary point for each group consists of the averages of the x-values and the y-values in that group. For the slope of the fitted line, Wald uses

$$b_W = \frac{(y_{m+1} + \cdots + y_n) - (y_1 + \cdots + y_m)}{(x_{m+1} + \cdots + x_n) - (x_1 + \cdots + x_m)},$$

the slope of the line through the two summary points. The intercept then comes from the formula

$$a_W = \bar{y} - b_W \bar{x},$$

where $\bar{y} = (y_1 + \cdots + y_n)/n$ and $\bar{x} = (x_1 + \cdots + x_n)/n$, in the same manner as for the least-squares line.

The Method of Nair and Shrivastava

As a computationally attractive alternative to the method of least squares (assuming no error in x), Nair and Shrivastava (1942) introduced the "method of group averages." Ordering the points according to their x-values, $x_1 \leqslant x_2 \leqslant \cdots \leqslant x_n$, they place the first n_L points (x_i, y_i) in the first group and the last n_U points in the second group, and they omit the remaining $n - n_L - n_U$ points. The summary values are the group averages,

$$\bar{x}_L = \frac{x_1 + \cdots + x_{n_L}}{n_L}, \qquad \bar{y}_L = \frac{y_1 + \cdots + y_{n_L}}{n_L}$$

$$\bar{x}_U = \frac{x_{n-n_U+1} + \cdots + x_n}{n_U}, \qquad \bar{y}_U = \frac{y_{n-n_U+1} + \cdots + y_n}{n_U},$$

and both slope and intercept come from the line joining (x_L, y_L) and (x_U, y_U):

$$b_{NS} = \frac{\bar{y}_U - \bar{y}_L}{\bar{x}_U - \bar{x}_L}$$

$$a_{NS} = \bar{y}_L - b_{NS}\bar{x}_L = \bar{y}_U - b_{NS}\bar{x}_U.$$

Nair and Shrivastava also extend this method to polynomials in x. For a polynomial of degree p, which has $p + 1$ coefficients to estimate, they divide the data into $2p + 1$ equal groups, omit the data points in the p even-numbered groups, and use the remaining $p + 1$ group averages to solve for the coefficients.

For fitting a straight line, we must still express n_L and n_U in terms of n. Nair and Shrivastava assume that the $x_1, \ldots, x_n$ are equally spaced, and they then calculate n_L and n_U such that b_{NS} would have the highest possible efficiency relative to b_{LS}. The result is to take $n_L = n_U$ equal to the integer closest to $n/3$. With equally spaced x-values and this choice of n_L and n_U, the relative efficiency of b_{NS} is at least $\frac{8}{9}$.

When both x and y are subject to error, Nair and Banerjee (1943) present simulation results indicating that, in one simple situation, the method of group averages gives more accurate estimates of slope and intercept than does Wald's method.

Bartlett's Method

To achieve greater accuracy than Wald's method, Bartlett (1949) independently introduced the modification of using three equal groups. Like Nair and Shrivastava, he uses the slope of the line joining $(\bar{x}_L, \bar{y}_L)$ and $(\bar{x}_U, \bar{y}_U)$:

$$b_B = \frac{\bar{y}_U - \bar{y}_L}{\bar{x}_U - \bar{x}_L}.$$

For the intercept, however, he requires that the line pass through the point $(\bar{x}, \bar{y})$:

$$a_B = \bar{y} - b_B\bar{x}.$$

Thus Bartlett's method differs from Wald's method only by using three groups instead of two.

For an indication of the accuracy of his method, Bartlett examines the same special case as Nair and Shrivastava: equally spaced values of x, not subject to error. In this situation, where least-squares fitting is appropriate, the efficiency of b_B, relative to b_{LS}, is at least $\frac{8}{9}$, whereas the efficiency of b_W may be as low as $\frac{3}{4}$.

As part of a tutorial exposition of Bartlett's method, Gibson and Jowett (1957) investigate how the allocation of data points to the three groups affects the efficiency of b_B when the x-values come from each of six distributions. Instead of working with samples of x-values, they use the whole true distribution and obtain the optimum allocations shown as proportions in Table 5-5.

The result for the uniform is just what we would expect from the calculations of Nair and Shrivastava and of Bartlett for equally spaced x-values. The rule for the Gaussian, allocating 27% of the data to each of the end groups, coincides with the optimum allocation when one wishes to estimate the correlation coefficient in a bivariate Gaussian distribution from the counts of positive and negative y-values in the lower and upper groups (Mosteller, 1946). This "27% rule" also arises in other contexts. On the whole, the optimum allocations for the four symmetric distributions in Table 5-5 do not differ greatly. Gibson and Jowett focus on the Gaussian distribution and round the allocations, recommending a $1:2:1$ ratio.

TABLE 5-5.   Optimum allocation in the three-group method under six distributions of x.

Distribution	$\dfrac{n_L}{n}$	$\dfrac{n_M}{n}$	$\dfrac{n_U}{n}$	Frequency Function	
Gaussian	.27	.46	.27	$\left(\dfrac{1}{\sqrt{2\pi}}\right)e^{-x^2/2},$	$-\infty < x < +\infty$
Rectangular	.33	.33	.33	$\frac{1}{2},$	$-1 < x < +1$
Bell-shaped	.31	.38	.31	$\frac{3}{4}(1 - x^2),$	$-1 < x < +1$
U-shaped	.39	.22	.39	$\frac{9}{10}\left(\frac{1}{4} + x^4\right),$	$-1 < x < +1$
J-shaped	.45	.40	.15	$e^{-(x+2)},$	$-2 < x < +\infty$
Skew	.36	.45	.19	$\dfrac{x^3}{3!2^4}e^{-x/2},$	$0 < x < +\infty$

Source: Wendy M. Gibson and Geoffrey H. Jowett (1957). " 'Three-group' regression analysis, part I. Simple regression analysis," *Applied Statistics*, **6**, 114–122 (Table IV, p. 121. Reprinted by permission).

The Brown–Mood Line

As an outgrowth of some procedures for setting up tests of hypotheses (primarily in analysis-of-variance situations) in terms of medians, Brown and Mood (1951) describe a method of estimating the coefficients in linear regression models. They consider the general problem involving several explanatory variables, and Mood (1950, pp. 406–408) gives the method for simple linear regression in more detail. More recently, Hogg (1975) has generalized the idea to obtain "percentile lines."

For the simple regression model, Brown and Mood first divide the data into two groups by using the x-median, M_x. The fitted slope b_{BM} and intercept a_{BM} are then chosen to yield zero median residual in each of the two groups:

$$\operatorname*{med}_{x_i \leqslant M_x} \{y_i - a_{BM} - b_{BM}x_i\} = 0$$

$$\operatorname*{med}_{x_i > M_x} \{y_i - a_{BM} - b_{BM}x_i\} = 0.$$

(Including M_x in the left group is arbitrary; the presence of two or more x-values equal to M_x might indeed mean that placing M_x in the right group would make the two groups more nearly equal in size.) To calculate b_{BM}, Mood suggests an iterative procedure similar to the one in Section 5A. He obtains the intercept as $a_{BM} = \operatorname{med}\{y_i - b_{BM}x_i\}$. We would now prefer a modification of the algorithm in Section 5B.

Discussion

To summarize the background material in this section, Table 5-6 lists the methods according to the number of groups (two or three) and the summary

TABLE 5-6. Groups and summaries—the three-group resistant line and related methods.

Number of Groups	Summary Measure	
	Mean	Median
Two	Wald	Brown and Mood
Three	Nair and Shrivastava, Bartlett	Three-group resistant line

measure (mean or median) used within the groups. Viewed in this way, the resistant line fits in nicely with existing methods of long standing.

To compare (a) the $1:2:1$ ratio that Gibson and Jowett recommend for means in three groups and (b) the $1:1:1$ allocation underlying the three-group resistant line, we note a tradeoff. On one hand, keeping the lower group and the upper group well separated tends to produce a more stable estimate of the slope by using a greater range of x-values. On the other hand, putting more data in the two groups increases the stability of the y-summary values. Because the median, although resistant, is ordinarily more variable than the mean for the same number of observations, the equal-groups allocation helps to increase its stability. As a practical matter, however, the allocation for the 3-group resistant line comes more from simplicity and convenience and is not the result of any specific attempt at optimization.

5D. INFLUENCE AND LEVERAGE

The beginning of this chapter mentioned an important drawback of fitting a line by least squares: a single unusual observation can distort the results. For practical data analysis, as well as for comparing least squares to other methods, it helps to judge such effects quantitatively. Two convenient types of measures are the effect of each individual y-value, y_i, on the fitted slope and intercept and the effect of y_i on the corresponding predicted value, $\hat{y}_i$. Both of these involve only straightforward calculations when we fit by least squares. For many other methods, no simple formulas suffice, and research is continuing to pursue quantitative descriptions of this aspect of their behavior. The present section concentrates on least squares.

We recall that the slope and intercept of the least-squares line can be written

$$b_{LS} = \frac{\Sigma(x_i - \bar{x})y_i}{\Sigma(x_k - \bar{x})^2} \tag{12}$$

and

$$a_{LS} = \bar{y} - b_{LS}\bar{x}, \tag{13}$$

respectively. Thus the effect of y_i on b_{LS} depends on how far x_i lies from $\bar{x}$. Specifically, if we change y_i by an amount Δy_i and leave all other y-values unchanged, the result is a new value of the slope, $b_{LS} + \Delta b$:

$$b_{LS} + \Delta b = \frac{\Sigma_{j \neq i}(x_j - \bar{x})y_j + (x_i - \bar{x})(y_i + \Delta y_i)}{\Sigma(x_k - \bar{x})^2}. \qquad (14)$$

Usually we concentrate on the relationship between the changes. Because y_i enters into equation (12) in a simple linear way, we may subtract equation (12) from equation (14) to get

$$\Delta b = \frac{(x_i - \bar{x})}{\Sigma(x_k - \bar{x})^2}\Delta y_i.$$

Thus Δb depends directly on $(x_i - \bar{x})$ and inversely on $\Sigma(x_k - \bar{x})^2$; that is, the further x_i is from $\bar{x}$, the greater the impact on the slope of a given change in y_i. Also, the smaller the variation among the x_k, the greater this impact.

Similarly, we can examine the change in a_{LS} in the face of Δy_i. The result is

$$\Delta a = \frac{1}{n}\Delta y_i - \bar{x}\,\Delta b$$

$$= \left[\frac{1}{n} - \frac{(x_i - \bar{x})\bar{x}}{\Sigma(x_k - \bar{x})^2}\right]\Delta y_i.$$

We note that, although $x_i = \bar{x}$ yields $\Delta b = 0$, the change in y_i still comes through in Δa.

We may rewrite equation (12) as

$$b_{LS} = \Sigma c_i y_i, \qquad (15)$$

where $c_i = (x_i - \bar{x})/\Sigma(x_k - \bar{x})^2$ depends only on the x-values. This version may seem to be merely a cosmetic change, but it meshes with a useful way of approaching more general regression models. Equation (15) holds separately for each regression coefficient in a multiple regression model in the form:

$$b_j = \sum_i c_{ji} y_i.$$

By calculating the c_{ji}, we can learn which observations influence which coefficients in such a model. See Mosteller and Tukey (1977, Section 14C) and Velleman and Welsch (1981).

Our other measure of the effect of y_i is based on $\hat{y}_i$, which we can readily obtain from

$$\hat{y}_i = a_{LS} + b_{LS} x_i$$

$$= \bar{y} + b_{LS}(x_i - \bar{x})$$

$$= \bar{y} + (x_i - \bar{x}) \frac{\sum_{j=1}^{n}(x_j - \bar{x})y_j}{\sum_{k=1}^{n}(x_k - \bar{x})^2}.$$

In this last expression for $\hat{y}_i$, we begin by isolating the contribution of y_j, which appears both in $\bar{y}$ and in the second term:

$$\hat{y}_i = \sum_{j=1}^{n} \left(\frac{1}{n}\right) y_j + \sum_{j=1}^{n} \frac{(x_i - \bar{x})(x_j - \bar{x})}{\sum_{k=1}^{n}(x_k - \bar{x})^2} y_j.$$

In a convenient notation, we have

$$\hat{y}_i = \sum_{j=1}^{n} h_{ij} y_j, \tag{16}$$

with

$$h_{ij} = \frac{1}{n} + \frac{(x_i - \bar{x})(x_j - \bar{x})}{\sum_{k=1}^{n}(x_k - \bar{x})^2}. \tag{17}$$

Among the h_{ij}, the easiest to interpret is h_{ii}, which tells how y_i contributes to $\hat{y}_i$. That is, if we change y_i by Δy_i, then $h_{ii} \Delta y_i$ is the resulting change in $\hat{y}_i$. Thus we see that, under least squares, any change in y_i, even a gross error, has a proportional impact on $\hat{y}_i$. This is the epitome of nonresistance; the impact could easily be arbitrarily great.

We regard h_{ii} as measuring the *leverage* of the data point (x_i, y_i). Equation (17), with $j = i$, shows that the critical issue is how far x_i differs from $\bar{x}$. Thus leverage is entirely a function of the x-values. Further, some limits are built in: $1/n \leqslant h_{ii} \leqslant 1$ (the lower bound is 0 when the model includes no constant term). Helpfully, with an appropriate formula for h_{ij}, equation (16) applies to multiple regression models as well. The h_{ij} constitute what is known as the "hat matrix" (Hoaglin and Welsch, 1978), a valuable diagnostic tool in regression to identify points of high leverage, where an unusual y-value will most disturb the fit.

EXAMPLE:

To illustrate the two measures of the behavior of least-squares fitting, we return to the data of Table 5-1, where $x =$ child's age (in months). Table 5-7 gives c_i and h_{ii} for each of the 18 data points. Because the x-values are already in ascending order, it is easy to recognize the anticipated patterns. The c_i increase from substantial negative to substantial positive values. In trying to judge their size, we must face the absence of any built-in scale or standard. Still, they reveal how much more impact the extreme data points have on b_{LS} than do those closer to the center. And, of course, they remind us that every y-value, no matter how discrepant, has some effect on the value of b_{LS} (with the single exception of one whose x-value exactly equals $\bar{x}$, a situation that does not arise in this example).

For the h_{ii}, we begin to have more of a standard because they must lie between $1/n$ and 1 and because their sum must equal 2 (more generally, the

TABLE 5-7.   Influence and leverage in data
of Table 5-1.[a]

i	c_i	h_{ii}
1	−.0101	.235
2	−.0078	.164
3	−.0067	.135
4	−.0061	.122
5	−.0044	.090
6	−.0039	.082
7	−.0033	.075
8	−.0016	.060
9	−.0005	.056
10	.0012	.058
11	.0018	.061
12	.0035	.077
13	.0040	.085
14	.0046	.093
15	.0057	.114
16	.0069	.139
17	.0080	.169
18	.0086	.185

[a]$\bar{x} = 126.833$, $\Sigma(x_k - \bar{x})^2 = 1770.500$

```
              (unit = .01)
     2          f  5 5
     6          s  6 6 7 7
    (4)        0 · 8 8 9 9
     8         1 * 1
     7          t  2 3 3
                f
     4          s  6 6
     2         1 · 8
               2 *
     1          t  3
```

Figure 5-10. Stem-and-leaf display of the h_{ii} in Table 5-7.

sum equals the number of coefficients in the model). Thus an average h_{ii} in this example is $2/n = 0.111$. In Table 5-7 we see that $h_{1,1}$ is more than twice this size, and in a stem-and-leaf display (Figure 5-10), $h_{18,18}$, $h_{17,17}$, and $h_{2,2}$ stand out a little from the rest. The message is that discrepant y-values in these four data points will cause more difficulty than elsewhere. A look back at Figure 5-2 reveals that the height of child 17 does not fit in well, so that a careful application of least-squares fitting would need to make some allowance for this observation, perhaps by setting it aside.

The book by Belsley, Kuh, and Welsch (1980) discusses leverage and other important aspects of diagnosis in a variety of regression models.

5E.   OTHER ALTERNATIVE METHODS

Section 5C demonstrates that the 3-group resistant line is by no means the first alternative to least squares. One of the earlier lines mentioned there, the Brown–Mood line, offers resistance also. The present section rounds out the picture by briefly describing three further techniques, each of which is intended to provide at least some resistance. These are the least-absolute-residuals line, a nonparametric method, and the repeated-median line. First, we mention a measure of resistance.

Breakdown Bound

An attractive feature of the 3-group resistant line is its ability to tolerate wild points—points that are unusual in their x-value or their y-value or both. To measure this form of resistance, we apply the general notion of breakdown introduced by Hampel (1971).

DEFINITION: The *breakdown bound* of a procedure for fitting a line to n pairs of y-versus-x data is k/n, where k is the greatest number of data points that can be replaced by arbitrary values while always leaving the slope and intercept bounded.

Operationally, we can think of dispatching data points "to infinity" in haphazard or even troublesome directions until the calculated slope and intercept can tolerate it no longer and break down by going off to infinity as well. We ask how large a fraction of the data—no matter how they are chosen—can be so drastically changed without greatly changing the fitted line.

As we have seen in various ways, the least-squares line has breakdown bound equal to 0.

Because the 3-group resistant line uses the median within each group, we find its breakdown bound by taking $\frac{1}{3}$ times the breakdown bound of the median of an ordinary sample. Because the median is the middle value, its breakdown bound is $\frac{1}{2}$, and thus the breakdown bound for the resistant line is $\frac{1}{6}$. (In arriving at this indicative value, we have deliberately glossed over such details as how data points are assigned to groups when x-values are tied or when n is not divisible by 3, as well as whether the sample whose median we seek contains an even number or an odd number of observations. These details are unimportant when n is substantial.) Thus at most $\frac{1}{6}$ of the (x, y) points can become arbitrarily wild without destructively affecting the resistant line. If we spread the wild points over the three groups, we can escape harm, but $\frac{1}{6}$ is the best we can guarantee under the most unfavorable circumstances.

Least Absolute Residuals (LAR)

Minimizing the sum of absolute residuals has a history almost as long as the principle of least squares (Gentle, 1977). To fit a line, we seek b_{LAR} and a_{LAR} such that they minimize

$$\sum_{i=1}^{n} |y_i - a - bx_i| . \tag{18}$$

Unlike least squares, this approach does not yield formulas for b_{LAR} and a_{LAR}. In fact, the slope and intercept are not necessarily unique.

Because the median is the measure that minimizes

$$\sum_{i=1}^{n} |y_i - t| ,$$

we might expect a least-absolute-residuals line to have high breakdown bound. Unfortunately, its breakdown bound is 0. Because equation (18) involves all the x-values and y-values, we could alter x_i to give the data point high leverage and then move y_i far enough to seize control of the fitted line. Even so, such a line will often be less sensitive to moderate disturbances than the least-squares line.

Median of Pairwise Slopes

Another way of applying the median in fitting a line begins with all possible pairs of data points, calculates the slope determined by each pair, and then finds the median of these slopes. More carefully, we assume that the x_i are all distinct, and we define

$$b_{ij} = \frac{y_j - y_i}{x_j - x_i}, \qquad 1 \leq i < j \leq n, \tag{19}$$

yielding $n(n - 1)/2$ slope values. The fitted slope is

$$b_T = \text{med}\{b_{ij}\}.$$

Theil (1950) suggested this method, and Sen (1968) modified it to handle ties among the x_i.

To derive its breakdown bound, we assume that exactly k of the n points are wild. Then the number of wild slopes is

$$\frac{k(k - 1)}{2} + k(n - k).$$

If this number is large enough, b_T will be wild. For large n, we may multiply $n(n - 1)/2$ by $\frac{1}{2}$, the breakdown bound for the median, set the result equal to the above expression, and solve for k to get, approximately, $k/n = .29$. That is, the breakdown bound for b_T is .29.

Repeated-Median Line

To achieve a high breakdown bound, Siegel (1982) devised the repeated-median method. Beginning with the pairwise slopes as in equation (19), this approach takes medians in two stages, first at each point and then across points:

$$b_{RM} = \text{med}_i \left\{ \text{med}_{j \neq i} \{b_{ij}\} \right\}.$$

That is, the first stage takes the median of the slopes of the $n - 1$ lines passing through a given point (x_i, y_i), $i = 1, \ldots, n$; and the second stage takes the median of these n slopes. We note that this method drops the restriction "$i < j$" in equation (19), and so it uses each b_{ij} twice.

For the fitted intercept, the repeated-median method calculates $u_i = y_i - b_{RM} x_i$ and then sets

$$a_{RM} = \operatorname*{med}_i \{a_i\}.$$

Siegel shows that the repeated-median line has breakdown bound essentially equal to $\frac{1}{2}$. We give a heuristic derivation for the case $n = 2k$ and leave the case of n odd as an exercise (Exercise 10). When $n = 2k$, the exact value of the breakdown bound is $(k - 1)/n$. To see this, suppose that $k - 1$ data points are wild and the remaining $k + 1$ are not wild. We define $b_i = \operatorname{med}_{j \neq i} \{b_{ij}\}$. Now, when i_0 denotes one of the nonwild points, b_{i_0} is determined by the remaining k nonwild points and not by the $k - 1$ wild points. On the other hand, the b_i for any wild point must be wild. Thus exactly $k + 1$ of the b_i are nonwild, and these slope estimates determine b_{RM}. Any larger number of wild points would cause b_{RM} to break down. Because the intercept involves only a simple median, we already know that it can tolerate $k - 1$ wild values out of $2k$. Both slope and intercept have the same breakdown bound.

Discussion

Now that we have available methods with a variety of breakdown bounds, how might we choose among them? One consideration is the degree of resistance that a particular application demands. Another issue is the relative precision of the slope estimates, particularly in small to moderate samples, but the quantitative information available at present is not quite adequate for such comparisons. Further empirical studies should enable us to make choices based on both resistance and efficiency.

Finally, assuming that we need at least some resistance and that a breakdown bound of $\frac{1}{6}$ is high enough, we find that increased resistance generally requires more computational effort, especially in the form of sorting. Without plunging into a thicket of technical detail (some of it depending on the particular computer used), we remark that the repeated-median line involves somewhat more effort than the method based on the median of the pairwise slopes. Even though both of these methods are not iterative, the 3-group resistant line, by sorting only the y-values within the left and right groups at each iteration, will require little enough work per iteration and few enough iterations that its overall computational effort is usually substantially lower. For small samples, the differences in computational effort among these three resistant techniques are generally slight.

We have by no means exhausted the methods that have been devised for fitting straight lines. Even so, we have no shortage of techniques that are more resistant than the least-squares line.

REFERENCES

Bartlett, M. S. (1949). "Fitting a straight line when both variables are subject to error," *Biometrics*, **5**, 207–212.

Belsley, D. A., Kuh, E., and Welsch, R. E. (1980). *Regression Diagnostics*. New York: Wiley.

Berkson, J. (1950). "Are there two regressions?" *Journal of the American Statistical Association*, **45**, 164–180.

Brown, G. W. and Mood, A. M. (1951). "On median tests for linear hypotheses." In J. Neyman (Ed.), *Proceedings of the Second Berkeley Symposium on Mathematical Statistics and Probability*. Berkeley and Los Angeles: University of California Press.

Gentle, J. E. (1977). "Least absolute values estimation: an introduction," *Communications in Statistics*, **B6**, 313–328.

Gibson, W. M. and Jowett, G. H. (1957). "'Three-group' regression analysis, part I. Simple regression analysis," *Applied Statistics*, **6**, 114–122.

Greenberg, B. G. (1953). "The use of analysis of covariance and balancing in analytical surveys," *American Journal of Public Health*, **43**, 692–699.

Hampel, F. R. (1971). "A general qualitative definition of robustness," *Annals of Mathematical Statistics*, **42**, 1887–1896.

Hoaglin, D. C. and Welsch, R. E. (1978). "The hat matrix in regression and ANOVA," *The American Statistician*, **32**, 17–22.

Hogg, R. V. (1975). "Estimates of percentile regression lines using salary data," *Journal of the American Statistical Association*, **70**, 56–59.

Johnstone, I. and Velleman, P. F. (1982). "Tukey's resistant line and related methods: asymptotics and algorithms," *1981 Proceedings of the Statistical Computing Section*. Washington, DC: American Statistical Association, pp. 218–223.

Madansky, A. (1959). "The fitting of straight lines when both variables are subject to error," *Journal of the American Statistical Association*, **54**, 173–205.

Mood, A. M. (1950). *Introduction to the Theory of Statistics*. New York: McGraw-Hill.

Mosteller, F. (1946). "On some useful 'inefficient' statistics," *Annals of Mathematical Statistics*, **17**, 377–408.

Mosteller, F. and Tukey, J. W. (1977). *Data Analysis and Regression*. Reading, MA: Addison–Wesley.

Nair, K. R. and Banerjee, K. S. (1943). "A note on fitting of straight lines if both variables are subject to error," *Sankhyā*, **6**, 331.

Nair, K. R. and Shrivastava, M. P. (1942). "On a simple method of curve fitting," *Sankhyā*, **6**, 121–132.

Sen, P. K. (1968). "Estimates of the regression coefficient based on Kendall's tau," *Journal of the American Statistical Association*, **63**, 1379–1389.

Siegel, A. F. (1982). "Robust regression using repeated medians," *Biometrika*, **69**, 242–244.

Theil, H. (1950). "A rank-invariant method of linear and polynomial regression analysis," I, II, and III. *Proceedings of Koninklijke Nederlandse Akademie van Wetenschappen*, **53**, 386–392, 521–525, 1397–1412.

Tukey, J. W. (1970). *Exploratory Data Analysis*, Limited Preliminary Edition, Reading, MA: Addison–Wesley.

Velleman, P. F. and Hoaglin, D. C. (1981). *Applications, Basics, and Computing of Exploratory Data Analysis*. Boston: Duxbury Press.

Velleman, P. F. and Welsch, R. E. (1981). "Efficient computing of regression diagnostics," *The American Statistician*, **35**, 234–242.

Wald, A. (1940). "The fitting of straight lines if both variables are subject to error," *Annals of Mathematical Statistics*, **11**, 284–300.

Additional Literature

Arthanari, T. S. and Dodge, Y. (1981). *Mathematical Programming in Statistics*. New York: Wiley.

Bassett, G., Jr. and Koenker, R. (1978). "Asymptotic theory of least absolute error regression," *Journal of the American Statistical Association*, **73**, 618–622.

Box, G. E. P. and Draper, N. R. (1975). "Robust designs," *Biometrika*, **62**, 347–352.

Brown, B. M. (1980). "Median estimates in simple linear regression," *Australian Journal of Statistics*, **22**, 154–165.

Devlin, S. J., Gnanadesikan, R., and Kettenring, J. R. (1975). "Robust estimation and outlier detection with correlation coefficients," *Biometrika*, **62**, 531–545.

Griffiths, D. and Willcox, M. (1978). "Percentile regression: A parametric approach," *Journal of the American Statistical Association*, **73**, 496–498.

Kildea, D. G. (1981). "Brown–Mood type median estimators for simple regression models," *Annals of Statistics*, **9**, 438–442.

Maritz, J. S. (1979). "On Theil's method in distribution-free regression," *Australian Journal of Statistics*, **21**, 30–35.

McCabe, G. P. (1980). "Use of the 27% rule in experimental design," *Communications in Statistics*, **A9**, 765–776.

Rosenberg, B. and Carlson, D. (1977). "A simple approximation of the sampling distribution of least absolute residuals regression estimates," *Communications in Statistics*, **B6**, 421–437.

Sposito, V. A., Hand, M. L., and McCormick, G. F. (1977). "Using an approximate L_1 estimator," *Communications in Statistics*, **B6**, 263–268.

Sposito, V. A., Kennedy, W. J., and Gentle, J. E. (1980). "Useful generalized properties of L_1 estimators," *Communications in Statistics*, **A9**, 1309–1315.

EXERCISES

1. For each of 12 government agencies, a newspaper report gave the agency's estimate of the number of full-time employees working in public affairs areas in 1976, as well as the costs of these activities (including salaries and operating expenses); Table 5-8 shows these

TABLE 5-8.   Data for public affairs employees in 12 government agencies in 1976.

Agency	Number of Employees	Cost (in Dollars) of Salaries and Operating Expenses
Defense	1486	24,508,000
HEW	388	21,000,000
Agriculture	650	11,467,300
Treasury	202	5,798,235
Congress	446	5,663,174
Commerce	164	5,683,609
Energy Research and Dev. Admin.	128	5,236,800
NASA	208	4,500,000
Transportation	117	2,913,509
HUD	69	2,455,000
White House	85	2,300,000
Veterans Admin.	47	1,313,300
Totals	3990	92,838,927

Source: The Boston Evening Globe, April 27, 1976, p. 2.

data. Fit a three-group resistant line by the original method of Section 5A and the modified method of Section 5B. Discuss the results.

2. For the data in Table 5-1, use the methods of (a) Wald, (b) Bartlett, and (c) Brown and Mood. Compare these to the three-group resistant line.

3. Suppose that, in Table 5-1, the data point $(109, 137.6)$ had been erroneously entered as $(109, 13.76)$. What now would be the equations for the following lines?

(a)   Three-group resistant line.
(b)   Least-squares regression line.
(c)   Bartlett's line.

4. Derive the relationship between Δy_i and Δa given in Section 5D.

5. In vector and matrix notation, the multiple regression model involving n observations on y and the explanatory variables $X_1, \ldots, X_p$ is often written $y = X\beta + \varepsilon$. Use matrix notation to obtain (in terms of X) the constants c_{ij} in the generalization of equation (15). [These are related to the "catchers" discussed by Mosteller and Tukey (1977, Section 14C).]

6. Pursuing equations (15), (16), and (17), calculate the c_i and the h_{ii} for the 9-point hypothetical data given early in Section 5B.

7. Find the repeated-median line for
   (a) the data of Table 5-1,
   (b) the 9-point example in Section 5B.

8. Consider the points $(-2, -2)$, $(-1, -1)$, $(1, 1)$, $(2, 2)$, and (u, v), and sketch an appropriate graph.
   (a) Specify all positions of the point (u, v) for which the 3-group resistant line is $y = x$.
   (b) Specify all positions of the point (u, v) for which the least-squares regression line is $y = x$.
   (c) Is there any point (u, v) for which the least-squares regression line is $y = -x$?
   (d) Specify all positions of the point (u, v) for which the least-absolute-residuals line is $y = x$.

9. Is the repeated-median line for x on y the same as the line for y on x? Explain and illustrate.

10. Suppose that the number of data points is $n = 2k + 1$.
    (a) Give the breakdown bound for the repeated-median line in terms of k.
    (b) Show that, as k becomes large, this breakdown bound has a limiting value of $\frac{1}{2}$.

11. Assuming that the time required to sort n observations is proportional to $n \log n$ (as discussed in Section 1D), develop timing formulas for the three-group resistant line, the median-of-pairwise-slopes method, and the repeated-median line. Include arithmetic operations as multiples of the time required for one addition (or subtraction), one multiplication, and one division.

12. Apply the principle of Bartlett's method to the continuous distribution of x-values F with density f. Assuming that $\text{var}(Y \mid x) = \sigma^2$ for all x, that $x \leq c$ is the left portion, and that $x > d$ is the right portion, show that

$$\text{var}(b_B) = \frac{\sigma^2 \{1/F(c) + 1/[1 - F(d)]\}}{[E(X \mid X > d) - E(X \mid X \leq c)]^2} .$$

Derive the optimum allocations in Table 5-5 [i.e., those based on values of c and d that minimize $\text{var}(b_B)$] for the rectangular and Gaussian distributions.

CHAPTER 6

Analysis of Two-Way Tables by Medians

John D. Emerson
Middlebury College

David C. Hoaglin
Harvard University and Abt Associates Inc.

The data structure known as a *two-way table* or *two-way layout* arises in many settings. In this data structure, each of two *factors* varies regularly and separately from the other, and a value of the *response* variable is observed for each combination of the versions, or levels, of the factors. For example, an individual's life expectancy may vary with sex and with race. We can use the median in a simple iterative fashion to provide a resistant analysis for data of this form. This exploratory technique enables us to summarize any additive dependence of the response variable on the two factors. Also, because the residuals are more directly informative when they come from a resistant analysis, this approach helps us to spot departures from additive structure. The present chapter discusses the use of medians and other methods in analyzing two-way tables.

Section 6A reviews the structural nature of a two-way table, and Section 6B explains and illustrates the basic iterative procedure called median polish. Section 6C compares it to the widely used analysis based on means (as in the classical analysis of variance). Section 6D pursues the connection between fitting by medians and fitting by minimizing the sum of absolute deviations, and Section 6E considers a more general class of iterative procedures. Section 6F examines the degree of resistance of median polish by considering its worst-case breakdown bound, the proportion of bad values that the procedure can tolerate. Sometimes the structural balance of a two-way table is disturbed by the absence of a few data values; Section 6G examines the consequences of such deficiencies in the data. Finally, Section 6H examines the use of a transformation if an additive model is inadequate to summarize the structure of a two-way table.

6A. TWO-WAY TABLES

A *two-way table* is a set of data in which the observations are written

$$y_{ij} \qquad i = 1,\ldots,I; \quad j = 1,\ldots,J; \qquad (1)$$

and displayed in a rectangular array, as shown in Table 6-1. This data structure involves three variables: the *row factor*, which has I versions or levels; the *column factor*, which has J versions or levels; and the *response y*, on which we have $I \times J$ observations, one for each combination of a row and a column. (This intersection of a row and a column is often called a *cell*, so that we are describing a two-way table with one observation per cell. The data may have started that way, or someone may have summarized all the entries in each cell with a single number. We may want to analyze a data set that has more than one observation per cell, but that possibility is not our concern in this chapter.) Only y need be numerical, although either the versions of the row factor or the versions of the column factor (or both) sometimes have numerical labels.

EXAMPLE: INFANT MORTALITY RATES

The National Center for Health Statistics publishes data on infant mortality in the United States and its relation to various socioeconomic factors. Table 6-2 breaks down the infant mortality rate (number of deaths per 1000 live births) according to geographical region of the United States and education

TABLE 6-1.   Format and notation for a two-way table.

i	j		
	1	$\cdots$	J
1	y_{11}	$\cdots$	y_{1J}
$\vdots$	$\vdots$	$\ddots$	$\vdots$
I	y_{I1}	$\cdots$	y_{IJ}

TABLE 6-2.   Infant mortality rates in the United States, all races, 1964–1966, by region and father's education. (Entries are numbers of deaths per 1000 live births.)

| Region | Education of Father (in years) | | | | |
	$\leqslant 8$	9–11	12	13–15	$\geqslant 16$
Northeast	25.3	25.3	18.2	18.3	16.3
North Central	32.1	29.0	18.8	24.3	19.0
South	38.8	31.0	19.3	15.7	16.8
West	25.4	21.1	20.3	24.0	17.5

Source: U.S. Dept. of Health, Education and Welfare, National Center for Health Statistics, *Infant Mortality Rates: Socioeconomic Factors, United States*, Vital and Health Statistics, Series 22, Number 14, Rockville, MD, 1972. DHEW publication number (HSM)72-1045 (data from Table 8, p. 21).

of the father (in years). Here "region" is the row factor, with four levels ($I = 4$), and "education of father" is the column factor, with five levels ($J = 5$). "Infant mortality rate" is the response variable, y_{ij}. "Region" is clearly nonnumerical. "Education of father" may have started out as a simple numerical variable, but in this table it has been grouped into intervals. These correspond to the commonly used classifications "no high school," "some high school," "completed high school," "some college," and "completed college," so that the ordering of the five levels is likely to be the focus in interpreting the contribution of the father's education.

In a two-way table such as Table 6-2, we can attempt to understand the contributions of the two factors by studying the array of numbers, but it is often more effective to break the data down into pieces that have different structures. In applications, these pieces appear in tabular displays, but, in discussing their behavior in general, we represent them by formulas. The formula underlying this chapter is the additive decomposition for a two-way table.

Additive Models

The relationship between the response variable y and the two factors is easiest to summarize and interpret if their joint contribution to the response is the sum of a separate contribution from each factor. When the data depart systematically from this structure, re-expression by a transformation may be desirable. For the present, we assume that any necessary transformation has already been applied. Section 6H illustrates the role of re-expression in removing nonadditivity, and Section 8G gives the mathematical basis for finding transformations that are most likely to be helpful.

The simple *additive model* may be written formally as

$$y_{ij} = \mu + \alpha_i + \beta_j + \varepsilon_{ij}. \tag{2}$$

In this model, μ is an overall typical value for the whole table; we sometimes call it the "common value." The incremental contribution of level i of the row factor, relative to the overall value, is α_i—a "row effect." Similarly, the incremental contribution of level j of the column factor is β_j—a "column effect." Finally, ε_{ij} represents departure of y_{ij} from the purely additive model of $\mu + \alpha_i + \beta_j$; we often try to treat ε_{ij} as a random fluctuation.

Any analysis of a two-way table corresponding to equation (2) decomposes the observed data array y_{ij} into four additive terms: one that is constant over the whole table, one that is constant by rows, one that is constant by columns, and one containing whatever is left over (the *residuals*). We can specify many different ways of doing this; each gives a technique of analysis. In this chapter we are primarily concerned with techniques of analysis that are *resistant*, so that isolated violent disturbances in a small number of cells will not much affect the common value, row effects, or column effects, and, as a consequence, will be reflected in the residuals. The next section describes the simplest such analysis.

6B. MEDIAN POLISH

To obtain an additive fit in the form of equation (2), we can operate iteratively on the data table, finding and subtracting row medians and column medians. For example, we could begin with the rows, calculating the median of each row and then subtracting this value from every observation in the row. Then we would continue with the columns of the resulting table, finding the median of each and subtracting it from the entries in its column. Of course, if a row or column has its median equal to zero, then we will make no change in that row or column. In principle, we could continue repeating the process of subtracting medians—called "median polish"—until all rows and columns have zero median. This means that, having started with the rows and then polished the columns, we would need to check the rows and polish again any that now have nonzero median. And so the iterative process would continue. In fact, it is seldom necessary to use more than a few steps, and (as we discuss in Section 6D) there are technical reasons for not using unrestrained iteration.

A Formal Presentation

To be more formal about this process, we now introduce some notation. The reader who is not interested in an algebraic discussion of the method should skip ahead to the example.

The additive fit analogous to equation (2) is

$$y_{ij} = m + a_i + b_j + e_{ij}. \tag{3}$$

We denote the fit and residuals at the end of n iterations by

$$y_{ij} = m^{(n)} + a_i^{(n)} + b_j^{(n)} + e_{ij}^{(n)}. \tag{4}$$

Because we are describing an iterative process, we state the initial conditions, before the first iteration:

$$m^{(0)} = 0$$

$$a_i^{(0)} = 0 \qquad i = 1, \ldots, I, \tag{5}$$

$$b_j^{(0)} = 0 \qquad j = 1, \ldots, J.$$

As we described the process earlier in this section, we begin with rows and then work with columns. (Whether we do this or reverse the order and start with columns is often a matter of arbitrary choice. As we describe in Section 6D, the two solutions that result need not be the same, but the difference between them is generally unimportant for many purposes, especially those of exploration.) We give below a list of nine equations, numbered (6a) through (6i); they may seem complicated because of the mathematical notation, but their structure is not difficult to comprehend. Equations (6a), (6b), and (6c) describe median polish of the rows, including the row of column effects and the updating of residuals. Equations (6d), (6e), and (6f) give the corresponding polish for columns. Equations (6g), (6h), and (6i) provide for updating the common value, row effects, and column effects. Together, all nine prepare the table for the next iteration, if any. [We could postpone (6b), (6e), and (6g) to (6i) until the last iteration if we chose.]

Throughout, we use the symbol Δ to represent a change, and we assume that n is 0, initially. The steps in an iteration go as follows:

Rows:

$$\Delta a_i^{(n)} = \text{med}\{e_{ij}^{(n-1)} \mid j = 1,\ldots,J\}; \qquad i = 1,\ldots,I; \qquad \text{(6a)}$$

$$\Delta m_b^{(n)} = \text{med}\{b_j^{(n-1)} \mid j = 1,\ldots,J\}; \qquad\qquad \text{(6b)}$$

$$d_{ij}^{(n)} = e_{ij}^{(n-1)} - \Delta a_i^{(n)}; \qquad j = 1,\ldots,J; \quad i = 1,\ldots,I; \qquad \text{(6c)}$$

Columns:

$$\Delta b_j^{(n)} = \text{med}\{d_{ij}^{(n)} \mid i = 1,\ldots,I\}; \qquad j = 1,\ldots,J; \qquad \text{(6d)}$$

$$\Delta m_a^{(n)} = \text{med}\{a_i^{(n-1)} + \Delta a_i^{(n)} \mid i = 1,\ldots,I\}; \qquad\qquad \text{(6e)}$$

$$e_{ij}^{(n)} = d_{ij}^{(n)} - \Delta b_j^{(n)}; \qquad i = 1,\ldots,I; \quad j = 1,\ldots,J; \qquad \text{(6f)}$$

Common Value and Effects:

$$m^{(n)} = m^{(n-1)} + \Delta m_a^{(n)} + \Delta m_b^{(n)}; \qquad\qquad \text{(6g)}$$

$$a_i^{(n)} = a_i^{(n-1)} + \Delta a_i^{(n)} - \Delta m_a^{(n)}; \qquad i = 1,\ldots,I; \qquad \text{(6h)}$$

$$b_j^{(n)} = b_j^{(n-1)} - \Delta m_b^{(n)} + \Delta b_j^{(n)}; \qquad j = 1,\ldots,J. \qquad \text{(6i)}$$

In practice, the steps given by the nine equations above are quite straightforward to carry out by hand on small two-way tables if we make careful use of the space bordering the $e_{ij}^{(n-1)}$, the table of residuals. For ease of bookkeeping, we write down one working table for the steps concerned with rows and a second for those concerned with columns. Tables 6-3 and 6-4 sketch schematic versions of these working tables.

In studying these two tables, we see that Table 6-3 shows the results of equations (6a) and (6b), whereas Table 6-4 shows the results of equations (6c), (6d), and (6e). Continuing with equations (6f) through (6i) takes us back to Table 6-3 with $n + 1$ replacing n and the column headed "med" omitted.

The column and row labeled "prev" in Tables 6-3 and 6-4 are the parts of the previous fit that, in the final step, become the row effects and column effects. As equations (6b) and (6e) indicate, we operate on these in parallel

TABLE 6-3.   Row median polish at iteration n.

i	j 1	$\cdots$	J	new med	prev
1	$e_{11}^{(n-1)}$	$\cdots$	$e_{1J}^{(n-1)}$	$[\Delta a_1^{(n)}]$	$a_1^{(n-1)}$
$\vdots$	$\vdots$	$\ddots$	$\vdots$	$\vdots$	$\vdots$
I	$e_{I1}^{(n-1)}$	$\cdots$	$e_{IJ}^{(n-1)}$	$[\Delta a_I^{(n)}]$	$a_I^{(n-1)}$
prev	$b_1^{(n-1)}$	$\cdots$	$b_J^{(n-1)}$	$[\Delta m_b^{(n)}]$	$m^{(n-1)}$

with the rows and columns of (partial) residuals. When we work by hand, this is convenient, but it may leave one step of adjustment for the $b_j^{(n)}$ to be done after we have otherwise concluded the iterations. Equations (6e) and (6h) center the $a_i^{(n)}$ so that their median is 0.

Sometimes the meaning of one set of versions (or even both) is such that "median = 0" is not the natural centering. When some other centering is more natural, we use it.

The order of execution of equations (6b) and (6i) means that a version of equation (6b) may have to follow equation (6i) and also change $b_j^{(n)}$ and $m^{(n)}$ if we are to have the same centering for the $b_j^{(n)}$. We could equivalently save all centering calculations for the last stage, and the computer algorithm presented by Velleman and Hoaglin (1981) follows this approach.

Using n in the iteration formulas may make it appear that we contemplate a large number of iterations, but this is not so. The "standard" version

TABLE 6-4.   Column median polish at iteration n. (Results immediately follow calculations shown in Table 6-3, and provide for updating.)

i	j 1	$\cdots$	J	prev
1	$d_{11}^{(n)}$	$\cdots$	$d_{1J}^{(n)}$	$a_1^{(n-1)} + \Delta a_1^{(n)}$
$\vdots$	$\vdots$	$\ddots$	$\vdots$	$\vdots$
I	$d_{I1}^{(n)}$	$\cdots$	$d_{IJ}^{(n)}$	$a_I^{(n-1)} + \Delta a_I^{(n)}$
new med	$[\Delta b_1^{(n)}]$	$\cdots$	$[\Delta b_J^{(n)}]$	$[\Delta m_a^{(n)}]$
prev	$b_1^{(n-1)} - \Delta m_b^{(n)}$	$\cdots$	$b_J^{(n-1)} - \Delta m_b^{(n)}$	$m^{(n-1)} + \Delta m_b^{(n)}$

TABLE 6-5. Infant-mortality-rate data and its row medians, ready to begin the first iteration of median polish [new row medians, $\Delta a_i^{(1)}$ and $\Delta m_b^{(1)}$, in parentheses].

Region	\multicolumn{5}{c}{Education of Father}	new med	prev
	$\leqslant 8$	9–11	12	13–15	$\geqslant 16$		
Northeast	25.3	25.3	18.2	18.3	16.3	(18.3)	0.0
North Central	32.1	29.0	18.8	24.3	19.0	(24.3)	0.0
South	38.8	31.0	19.3	15.7	16.8	(19.3)	0.0
West	25.4	21.1	20.3	24.0	17.5	(21.1)	0.0
prev	0.0	0.0	0.0	0.0	0.0	(0.0)	0.0

of this median-based analysis uses $n = 2$. In some instances, the row and column medians of the $e_{ij}^{(2)}$ may not all be 0, but this is not likely to be of great importance. It often happens that the row medians and column medians of the residuals become zero before this stage.

To see how median polish works on a set of data, we return to the infant mortality rates given in Table 6-2.

EXAMPLE: INFANT MORTALITY RATES (CONTINUED)

The initial values for what will become the common value, row effects, and column effects appear in Table 6-5, along with the infant mortality rates. This exhibit corresponds to Table 6-3 for $n = 0$. Table 6-6 displays tem-

TABLE 6-6. Infant-mortality-rate data after first polish of rows [new column medians, $\Delta b_j^{(1)}$ and $\Delta m_a^{(1)}$, in parentheses].

Region	\multicolumn{5}{c}{Education of Father}	prev
	$\leqslant 8$	9–11	12	13–15	$\geqslant 16$	
Northeast	7.0	7.0	−0.1	0.0	−2.0	18.3
North Central	7.8	4.7	−5.5	0.0	−5.3	24.3
South	19.5	11.7	0.0	−3.6	−2.5	19.3
West	4.3	0.0	−0.8	2.9	−3.6	21.1
new med	(7.4)	(5.8)	(−0.4)	(0.0)	(−3.0)	(20.2)
prev	0.0	0.0	0.0	0.0	0.0	0.0

TABLE 6-7.   **Infant-mortality-rate data after first polish of columns**
            [new row medians, $\Delta a_i^{(2)}$ and $\Delta m_b^{(2)}$, in parentheses].

Region	$\leqslant 8$	9–11	12	13–15	$\geqslant 16$	new med	prev
			Education of Father				
Northeast	−0.4	1.2	0.3	0.0	1.0	(0.3)	−1.9
North Central	0.4	−1.1	−5.1	0.0	−2.3	(−1.1)	4.1
South	12.1	5.9	0.4	−3.6	0.5	(0.5)	−0.9
West	−3.1	−5.8	−0.4	2.9	−0.6	(−0.6)	0.9
prev	7.4	5.8	−0.4	0.0	−3.0	(0.0)	20.2

porary residuals $d_{ij}^{(0)}$ and the resulting column medians; this table corresponds to Table 6-4 with $n = 0$. Here and throughout this chapter, we use the convention of rounding $\frac{1}{2}$ to even numbers.

Table 6-7 corresponds to Table 6-3 for $n = 1$; the column medians are subtracted, leaving residuals $e_{ij}^{(1)}$. The exhibit shows the first estimate of the common value and the medians of the current rows. Table 6-8, which corresponds to Table 6-4 with $n = 1$, provides the results of the second polish by rows and gives the medians of the current column entries.

Table 6-9 shows the results of two complete iterations of median polish. Because the columns were most recently polished, their medians are 0; one of the rows does not have a 0 median. This step of polish has produced column effects whose median is 0.4; Table 6-9 also shows the result of an additional step that centers the estimated column effects at 0 and adjusts the

TABLE 6-8.   **Infant-mortality-rate data after second polish of rows**
            [new column medians, $\Delta b_j^{(2)}$ and $\Delta m_a^{(2)}$, in parentheses].

Region	$\leqslant 8$	9–11	12	13–15	$\geqslant 16$	prev
			Education of Father			
Northeast	−0.7	0.9	0.0	−0.3	0.7	−1.6
North Central	1.5	0.0	−4.0	1.1	−1.2	3.0
South	11.6	5.4	−0.1	−4.1	0.0	−0.4
West	−2.5	−5.2	0.2	3.5	0.0	0.3
new med	(0.4)	(0.4)	(0.0)	(0.4)	(0.0)	(0.0)
prev	7.4	5.8	−0.4	0.0	−3.0	20.2

TABLE 6-9. Infant-mortality-rate data after second polish of columns
[new row medians, $\Delta a_i^{(3)}$ and $\Delta m_b^{(3)}$, in parentheses].

Region	Education of Father					new med	prev = old eff
	$\leqslant 8$	9–11	12	13–15	$\geqslant 16$		
Northeast	−1.1	0.5	0.0	−0.7	0.7	(0.0)	−1.6
North Central	1.1	−0.4	−4.0	0.7	−1.2	(−0.4)	3.0
South	11.2	5.0	−0.1	−4.5	0.0	(0.0)	−0.4
West	−2.9	−5.6	0.2	3.1	0.0	(0.0)	0.3
prev	7.8	6.2	−0.4	0.4	−3.0	(0.4)	20.2
new eff	7.4	5.8	−0.8	0.0	−3.4		20.6

estimate of the common value accordingly. The row and column whose
labels contain "eff" provide the estimates of the common value, row effects,
and column effects. The residuals from the fit are available for further
inspection; we would surely want to look at a stem-and-leaf display of the
residuals and perhaps also a boxplot. Figure 6-1 shows the stem-and-leaf
display.

By examining Table 6-9 and Figure 6-1, we can summarize the results of
this analysis of the data on infant mortality rates.

Infant mortality rates are generally highest in the North Central region
and lowest in the Northeast. The education of the father is a stronger factor

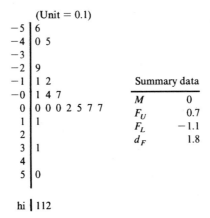

Figure 6-1. Stem-and-leaf display of residuals from median-polish fit to infant-
mortality-rate data.

in distinguishing among these rates than is geography. With a single exception—the high school graduates and the college dropouts—the mortality rates increase as the father's education decreases. In particular, completion of high school appears to exert the greatest single influence in reducing the mortality rates among infant offspring.

An additive model generally summarizes these data rather well. This means that the effects of father's education on infant mortality are roughly the same for different geographical regions. The lack of a pronounced pattern in the residuals suggests that a nonadditive model, if fitted to the table, would probably not enhance our understanding of the infant mortality rates. Still, the large residuals in the South and West hold some interest for us and may call for closer scrutiny. The residual of 11.2 for the least educated fathers in the South is particularly noteworthy.

Background Comments

Using median polish for fitting additive models to two-way tables is a relatively recent development. Along with many other exploratory techniques that resist the influence of a few "bad" data values, median polish was described by Tukey (1970). Mood (1950, p. 405) and Brown and Mood (1951, p. 163) use the same process as part of a procedure for testing interaction in a two-way layout with several observations per cell. More recent textbooks, including those of Tukey (1977), Mosteller and Tukey (1977), and Velleman and Hoaglin (1981), describe median polish and illustrate its use with real data sets.

6C. MEDIANS VERSUS MEANS

Median polish for analyzing a two-way table is a resistant technique: isolated large disturbances in a small number of cells will not affect the common value, row effects, or column effects, and instead will be reflected in the residuals. For this reason, we adopt median polish for exploratory analysis instead of the corresponding analysis using means that forms the basis for classical analysis of variance. This section provides a brief review of the analysis based on means and compares that analysis with median polish.

Analysis Based on Means

To obtain an additive fit in the form of equation (2), we can operate on the data of Table 6-1 by finding and subtracting row and column *means* (or averages). We could begin with the mean of all values in the table and then subtract it from all observations. Next we would calculate the mean of each row and subtract this value from every observation in the row. Analogously, we would find the mean of each column and subtract it from the entries in its column. This method for fitting an additive model to a two-way table does not need iteration (except where roundoff effects matter); it obtains its fitted common value, row effects, column effects, and residuals in one pass.

To be more explicit about this process, we introduce some notation. The additive fit analogous to equation (2) is

$$y_{ij} = \hat{\mu} + \hat{\alpha}_i + \hat{\beta}_j + \hat{\varepsilon}_{ij}, \tag{7}$$

where

$$\hat{\mu} = \frac{1}{IJ} \sum_i \sum_j y_{ij}, \tag{8a}$$

$$\hat{\alpha}_i = \frac{1}{J} \sum_j (y_{ij} - \hat{\mu}) \qquad i = 1,\dots,I, \tag{8b}$$

$$\hat{\beta}_j = \frac{1}{I} \sum_i (y_{ij} - \hat{\mu}) \qquad j = 1,\dots,J. \tag{8c}$$

Analysis based on means is widely used because of its simplicity and intuitive appeal, and because under certain ideal conditions it has some theoretical advantages. The advantages relate to the classical way of assessing the quality of fit of the model to the observed data—the sum of squares of the residuals. In the notation established in equation (7), these residuals are

$$r_{ij} = y_{ij} - \hat{\mu} - \hat{\alpha}_i - \hat{\beta}_j \qquad i = 1,\dots,I; \quad j = 1,\dots,J, \tag{9}$$

and the sum of the squares of the residuals is

$$\text{SSR} = \sum_i \sum_j r_{ij}^2. \tag{10}$$

The additive model that fits the data "best" is often defined as one that chooses estimates for μ, α_i, and β_j so that SSR is smallest. It is well-known that the fit specified in equations (7), (8a), (8b), and (8c) minimizes SSR. This fit is often called the "least-squares fit" to the two-way table of data. It is the basis for two-way analysis of variance (see, e.g., Scheffé, 1959).

Least-squares estimation is optimal in various theoretical senses when the data have an underlying distribution and this distribution has certain special properties; for example, it is often assumed that *the ε_{ij} are fluctuations or errors independently distributed with a Gaussian distribution having mean zero and common variance.* When these or similar distributional assumptions on the errors are not satisfied, other criteria than least squares may be more appropriate. We examine one such criterion in the next section.

The data in a two-way table often do not satisfy the assumptions that give analysis based on means its optimality. For example, one or two gross errors may have crept into the table—that is, values largely inconsistent with the structure underlying the other values in the table. These values can exert a large effect on the fitted values specified in equations (8a), (8b), and (8c): a single value that is markedly different from other values will have considerable impact on any average. We can summarize by saying that analysis by means has very poor resistance to outliers.

To compare the additive fits provided by analysis by means and by median polish, we introduce a simple hypothetical table of data.

EXAMPLE: A HYPOTHETICAL TABLE

We consider a table whose row and column effects are perfectly additive except for a single cell. Table 6-10 depicts a two-way table having a single nonzero entry. This entry has a clear impact on the analysis by means; the results of this analysis appear in Table 6-11. When median polish is used to fit an additive model to the table, the nonzero entry has no impact on the fit and appears as a large residual, as shown in Table 6-12.

The hypothetical example illustrates the primary advantage of median polish over analysis by means: resistance to outliers. To gain this advantage, we are willing to give up certain advantages offered by analysis with means:

Needs no iteration,

Always produces a unique fit,

Always minimizes the sum of squared residuals,

Appears easy to understand and explain, and

Is easy to program on a computer or calculator.

We do not take seriously the advantage flowing from the exact normality of residuals because we do not expect this in practical data.

The difference between the least-squares residuals in Table 6-11 and the median-polish residuals in Table 6-12 exemplifies the general problem of leakage in least-squares fitting. That is, what goes on in one cell—any one

TABLE 6-10.   Hypothetical data for a two-way table.

i	j 1	2	3
1	9	0	0
2	0	0	0
3	0	0	0

TABLE 6-11.   Hypothetical data of Table 6-10 after analysis by means.

i	j 1	2	3	$\hat{\alpha}_i$
1	4	-2	-2	2
2	-2	1	1	-1
3	-2	1	1	-1
$\hat{\beta}_j$	2	-1	-1	$1 (= \hat{\mu})$

cell—affects the fitted values and residuals in all the other cells. When the data value in that one cell is good, leakage does no harm because that cell contributes its fair share to the common value, row effects, and column effects. The difficulty arises when the one cell contains an outlier, because the least-squares method then has no way to prevent the leakage and the entire analysis suffers. The pattern of residuals in Table 6-11 points rather definitely to cell $(1, 1)$ as the location of the unusual value, but the presence

TABLE 6-12.   Hypothetical data of Table 6-10 after analysis by median polish.

i	j 1	2	3	a_i
1	9	0	0	0
2	0	0	0	0
3	0	0	0	0
b_j	0	0	0	$0 (= m)$

of further outliers (in a larger table), some of them in the same row or column as the first outlier, tends to cloud the indication a great deal. Daniel (1978) demonstrates that even highly experienced analysts must resort to rather delicate diagnostic procedures when using the least-squares residuals to locate the outliers in a moderate-sized two-way table. Median polish, on the other hand, generally produces substantial residuals for all the likely outliers.

Analysis by Means for Infant Mortality Rates

We conclude this section by showing the results of analysis by means for the infant-mortality-rate data. Table 6-13, which also repeats the median-polish analysis, shows the tendency of analysis based on means to reduce the magnitudes of the largest residuals by spreading the impact of an unusual data value throughout its row and column. In particular, median polish produces a residual of 11.2 for the lowest-educated fathers in the South; the corresponding residual is only 6.9 in the analysis by means. Because analysis by means must leave residuals that sum to 0 within each row and within each column, quite a number of smaller residuals had to be inflated to compensate for squeezing down the 11.2 and a few other residuals.

TABLE 6-13.   Infant-mortality-rate data after analysis by means.

Region	Education of Father					eff
	$\leqslant 8$	9–11	12	13–15	$\geqslant 16$	
Northeast	−3.0	0.8	1.2	−0.1	1.0	−2.14
North Central	−0.1	0.6	−2.2	1.9	−0.2	1.82
South	6.9	2.9	−1.3	−6.4	−2.1	1.50
West	−3.8	−4.3	2.3	4.6	1.3	−1.16
eff	7.58	3.78	−3.68	−2.25	−5.42	22.82
Analysis by Median Polish (from Table 6-9)						
Northeast	−1.1	0.5	0.0	−0.7	0.7	−1.6
North Central	1.1	−0.4	−4.0	0.7	−1.2	3.0
South	11.2	5.0	−0.1	−4.5	0.0	−0.4
West	−2.9	−5.6	0.2	3.1	0.0	0.3
eff	7.4	5.8	−0.8	0.0	−3.4	20.6

Residuals from
Analysis
by Means   Residuals from
                                                Median Polish

(Unit = 0.1)

Analysis by Means	stem	Median Polish
4	−6	
	−5	6
3	−4	0 5
8 0	−3	
2 1	−2	9
3	−1	1 2
2 1 1	−0	1 4 7
8 6	0	0 0 0 2 5 7 7
9 3 2 0	1	1
9 3	2	
	3	1
6	4	
	5	0
9	6	
	hi	112

Figure 6-2. Comparison of residuals from fit for analysis by means and median polish, infant-mortality-rate data. (The entry after "hi", 112, is a single large residual, 11.2, that is too large to be shown conveniently on a regular stem.)

The overall differences between the two batches of residuals are evident in Figure 6-2. For a more detailed examination of the change in each residual between the median-polish analysis and the analysis by means, we turn to Figure 6-3, where we have plotted vertically the two sets of residuals and linked the two points corresponding to the same cell. A solid line indicates that the residual from the analysis by means is larger (in magnitude) by more than one unit, whereas a broken line indicates that the residual from median polish is larger by more than one unit. Dotted lines mark residuals that have changed by less than one unit.

In general, analysis by means tends to produce fewer residuals with large magnitudes, fewer residuals whose magnitudes are very close to zero, and more residuals of moderate size than does median polish. These properties generally characterize residuals produced by a more classical analysis when they are compared to those produced by a more resistant analysis. We usually prefer the more resistant analysis because it isolates and draws attention to residuals that come from wild values or outliers. Since even carefully collected data sets often contain about 5% wild observations, resistant techniques provide an attractive alternative to more traditional approaches.

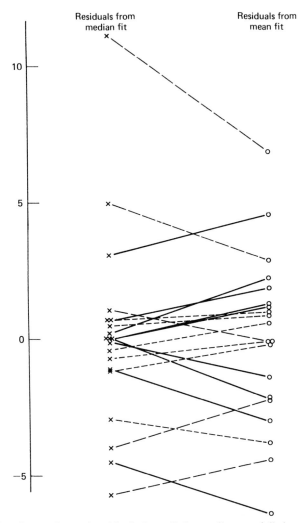

Figure 6-3. Comparison of residuals from fit by medians and fit by means.

6D. LEAST-ABSOLUTE-RESIDUALS FITTING

The previous section emphasizes the dissimilarities between median polish and analysis by means. Median polish is more closely allied with the general method of fitting by minimizing the sum of absolute values of the residuals. Here we explore this method of fitting an additive model to a two-way table and examine its relationship to median polish.

A Location Problem

To clarify the link between medians and least-absolute-residuals fitting, we first briefly consider the problem of estimating the location of a single batch of numbers. Given the data $x_1, \ldots, x_n$, one can choose a single constant c to summarize the x_i according to a variety of criteria. Often the criterion involves minimizing (with respect to c) some function of c and the x_i. A precise formulation specifies a location model for the data:

$$x_i = c + r_i. \tag{11}$$

The task is to select a value of c that optimizes the criterion function.

The least-squares criterion leads to the following minimization problem. Use the data values to choose c so that

$$\mathrm{SSR}(c) = \sum_i (x_i - c)^2 = \sum_i r_i^2 \tag{12}$$

is as small as possible. The familiar solution to this problem is

$$c = \bar{x} = \frac{1}{n} \sum_i x_i. \tag{13}$$

As in the preceding section, the least-squares criterion leads us to analysis by means.

Analogously, the least-absolute-residuals criterion asks us to choose a value for c so that the expression

$$\mathrm{SAR}(c) = \sum_i |x_i - c|, \tag{14}$$

is as small as possible. (Just as SSR is shorthand for "sum of squared residuals," so SAR stands for "sum of absolute residuals.") The solution to this minimization problem is

$$c = \mathrm{med}\{x_1, \ldots, x_n\}. \tag{15}$$

When n is even (say, $n = 2k$), $\mathrm{SAR}(c)$ may take its minimum value at all values of c in an interval. In terms of the order statistics, $x_{(1)} \leqslant \cdots \leqslant x_{(n)}$, these values of c are $x_{(k)} \leqslant c \leqslant x_{(k+1)}$. The conventional definition of "the" median as $\frac{1}{2}[x_{(k)} + x_{(k+1)}]$ in this situation conceals the nonuniqueness, which amounts in practice to a minor inconvenience, not a major difficulty. Roughly speaking, just as the least-squares criterion leads to analysis by means, so the least-absolute-residuals criterion leads to analysis by medians. We have seen this in a very special case, but it tends to hold much more generally.

Median Polish and Least Absolute Residuals

We return to the problem of seeking an additive fit in the form of equation (3) to the data in Table 6-1. A least-absolute-residuals fit to the data chooses values for m, for a_i ($i = 1, \ldots, I$), and for b_j ($j = 1, \ldots, J$) so that

$$\text{SAR} = \sum_i \sum_j |y_{ij} - m - a_i - b_j|, \qquad (16)$$

is as small as possible. Using the analogy with the location model, we might well guess that analysis by medians—that is, median polish—solves the minimization problem in equation (16). This guess is not entirely correct. However, experience with data suggests that median polish often produces a fit to the additive model that is close to being optimal in the least-absolute-residuals sense.

We now examine the connection between median polish and least-absolute-residuals fitting in more detail. Least-absolute-residuals fitting chooses m, the a_i, and the b_j so as to minimize SAR in equation (16). If we begin with rows, median polish first minimizes the inner sum over j with respect to each a_i while holding the b_j fixed:

$$\min_{a_i} \sum_j |y_{ij} - a_i - b_j|, \qquad i = 1, \ldots, I. \qquad (17)$$

(For simplicity here we suppress m and allow it to be absorbed in the a_i.) Median polish then changes the order of summation and minimizes the sum over i with respect to each b_j while holding the a_i fixed:

$$\min_{b_j} \sum_i |y_{ij} - a_i - b_j|, \qquad j = 1, \ldots, J, \qquad (18)$$

where each a_i is the result of the previous step [see equations (6a)–(6i) for explicit detail]. If the order of summation in expression (16) is changed again, the new sums over j may no longer be minimum, and the minimization process—a step of median polish—would repeat. In this way, median polish alternately minimizes by adjustment of row effects (for fixed column effects) and by adjustment of column effects (for fixed row effects). If allowed to iterate, median polish would stop when all rows and all columns of the table have medians equal to zero.

No step of median polish can increase the sum of absolute residuals. Thus the sequence of sums of absolute residuals produced by the iterations of median polish is a nonincreasing sequence of nonnegative numbers. Therefore, this sequence must always converge because it is bounded below by zero: $\lim_{n \to \infty} SAR^{(n)}$ exists.

We can say more if we restrict our attention to two-way tables with an odd number of rows and an odd number of columns. For these, if any row or column of residuals has nonzero median, then median polish actually decreases the sum of absolute residuals. It follows that having all row medians and all column medians of the table of residuals equal to zero is a *necessary* condition for the minimization of SAR. Unfortunately, even for odd-by-odd tables, it is not always sufficient, as the next example, devised by Andrew Siegel, illustrates.

EXAMPLE: MEDIAN POLISH NEED NOT MINIMIZE SAR

Consider the table

$$
\begin{array}{ccc}
1 & 6 & 3 \\
5 & 9 & 2 \;. \\
6 & 4 & 7
\end{array}
$$

If we begin with columns, one full step of median polish gives:

$$
\begin{array}{ccc|c}
-4 & 0 & 0 & 0 \\
0 & 3 & -1 & 0 \\
0 & -3 & 3 & 1 \\
\hline
0 & 1 & -2 & 5
\end{array}
$$

whose sum of absolute residuals is 14.

A full step of median polish starting with rows gives:

$$
\begin{array}{ccc|c}
-2 & 0 & 0 & -2 \\
0 & 1 & -3 & 0 \\
0 & -5 & 1 & 1 \\
\hline
0 & 3 & 0 & 5
\end{array}
$$

and the sum of absolute residuals is 12.

Neither of the fits just calculated using median polish is a least-absolute-residuals solution. If we simultaneously remove, from the last table above, row effects of $(-\frac{1}{2}, \frac{1}{2}, \frac{1}{2})$ and column effects of $(-\frac{1}{2}, \frac{1}{2}, \frac{1}{2})$, we obtain

$$
\begin{array}{ccc|c}
-1 & 0 & 0 & -3 \\
0 & 0 & -4 & 0 \\
0 & -6 & 0 & 1 \\
\hline
-1 & 3 & 0 & 6
\end{array}
$$

after recentering the row effects and column effects. The sum of absolute residuals is now only 11. Independent checking confirms this to be a least-absolute-residuals solution.

This example illustrates the following facts:

1. Median polish starting with rows and median polish starting with columns can lead to different values of the row effects and the column effects.
2. Median polish starting with columns can lead to different values for SAR than median polish starting with rows.
3. Sometimes neither order of steps in median polish leads to a minimum sum of absolute residuals.

Thus this example shows how things can go wrong.

The form of adjustment that we made in the example to go from the rows-first median-polish fit to the least-absolute-residuals fit illustrates why the median-polish process does not always yield the least-absolute-residuals solution. When we restrict ourselves to working only with adjustments in row effects and adjustments in column effects, in alternating steps, some tables of data lead to a position from which no further improvement is possible. Least-absolute-residuals fitting imposes no such restrictions and allows adjustment of any combination of effects at any step.

It would have been a tidy result if median polish always produced a least-absolute-residuals fit, especially if that criterion were widely regarded as the most desirable one for resistant data analysis. It does not occupy that position, however, and we see little reason for concern over the possible difference between a median-polish fit and a least-absolute-residuals one. Experience so far indicates that these differences are not large enough to be important for practical data analysis. From a more mathematical point of view, several questions remain. These include how much, at worst, the median-polish fit can differ from a least-absolute-residuals fit and how one can recognize tables for which the two fits differ.

Some Mathematical Results for 3 × 3 Tables

Although the example above, showing that median polish does not necessarily minimize SAR, involves a 3 × 3 table, the general behavior of median polish is easiest to investigate in such small tables. Four theorems indicate when median polish yields the least-absolute-residuals fit for a 3 × 3 table and what preliminary operations one can perform on the data to ensure this result.

THEOREM 1 (ANDREW SIEGEL): In a 3 × 3 table, whenever the first half-iteration produces an entire row or column of zeros, median polish will converge in at most one and one-half steps to a least-absolute-residuals solution.

We omit the proof because it is a straightforward but tedious analysis of cases.

Thus median polish produces a least-absolute-residuals fit for a 3 × 3 table whenever the data exhibit large row effects or large column effects. Theorem 1 cannot be generalized to 3 × 4 tables, but the case of 3 × 5 tables is open.

THEOREM 2: Suppose that

$$Y_{ij} = A_i + B_j$$

is exactly a sum of row effects and column effects. Then the possible tables of residuals, $\{e_{ij}^*\}$, for

$$y_{ij} + Y_{ij} = m^* + a_i^* + b_j^* + e_{ij}^*$$

are exactly the same as the possible tables of residuals for

$$y_{ij} = m + a_i + b_j + e_{ij}.$$

Accordingly, $\{y_{ij}\}$ and $\{y_{ij} + Y_{ij}\}$ have the same minimum SAR.

PROOF: We put $m^* = m$, $a_i^* = a_i + A_i$, and $b_j^* = b_j + B_j$, leaving $e_{ij}^* = e_{ij}$.

THEOREM 3: Given a 3×3 table y_{ij}, if we take

$$\{ Y_{ij} \} = \begin{array}{ccc} X & X & X \\ 0 & 0 & 0 \\ -X & -X & -X \end{array}$$

for sufficiently large X, then the first step of median polish by columns, applied to $\{ y_{ij} + Y_{ij} \}$, satisfies the hypothesis of Theorem 1. Thus the resulting $\{ e_{ij}^* \}$ achieve the minimum SAR, and the corresponding analysis for $\{ y_{ij} \}$ will have minimum SAR.

PROOF: Apply Theorem 1 and then Theorem 2.

THEOREM 4: In a 3×3 table, if we make all the entries in any chosen row zero by taking out the corresponding parts before applying median polish, then the resulting table of median-polish residuals will have minimum SAR.

PROOF: The first half-iteration of median polish by rows necessarily preserves the row of zeros, so that Theorem 1 applies.

We summarize these four theorems by mentioning that, according to Theorem 1, a row or column of zeros at the end of the first half-step ensures that median polish will yield a least-absolute-residuals fit. Theorems 3 and 4 provide ways of operating on the table so that median polish will perform as Theorem 1 requires. And Theorem 2 shows how the fits that result after applying Theorem 3 or Theorem 4 are related to the original data.

Number of Iterations

In some examples, median polish needs many iterations to produce rows and columns whose medians are near zero. Still, experience reassuringly indicates that a few iterations nearly always give a good fit, in the sense that residuals are small and effects change little with further iterations. (One exception occurs when a table has quite a few missing values; we discuss this in Section 6G.) In practice, two complete iterations are often enough— giving row medians and column medians with small magnitudes compared to the residuals; four iterations will almost always be adequate, surely for exploratory purposes.

Some Background on Least-Absolute-Residuals Fitting

We have indicated that median polish need not strictly solve the least-abso- lute-residuals problem of minimizing equation (16); we have not described a method that will solve this problem. For the interested reader, however, we give a brief outline of the evolution of this criterion, and we provide key references, including some for useful computer routines. Minimizing the sum of absolute residuals was apparently first proposed in 1757 by Boscovich (Eisenhart, 1977), as a method of fitting a straight line to points in a plane. Since then, least-absolute-residuals regression has received the attention of many authors, although this approach to fitting a two-way table has provoked less attention.

Early proposers of methods for minimizing a sum of absolute deviations include Rhodes (1930) and Singleton (1940). However, the development of linear programming and its computer implementation made least-absolute-residuals fitting practical (see Gentle, 1977). Charnes, Cooper, and Ferguson (1955) first indicated that linear programming methods could be used to solve a closely related problem of fitting lines to data of the form y versus x. Wagner (1959) was among the first to provide a linear programming solution to the problem of minimizing a sum of absolute residuals. Since then, many investigators (e.g., Barrodale and Roberts, 1973; Kennedy, Gentle, and Sposito, 1977; and Bloomfield and Steiger, 1980) have provided faster, more efficient, and more general algorithms for solving this problem. We refer to Gentle (1977) for a brief discussion of these developments and for an extensive bibliography on least-absolute-residuals estimation.

Like median polish, least-absolute-residuals estimation need not yield a unique solution; two different sets of estimates for the effects may give the same minimum value for SAR. Harter (1977) gives a history and discussion of the nonuniqueness of least-absolute-residuals estimates; he also discusses constraints that will make the solution unique in a regression setting.

Because exploration needs a clean separation of a two-way table into fit and residuals and because the residuals are often at least as important as the fit or the effects, we are glad to have a resistant method that can isolate large disturbances. At this diagnostic stage, we may have little concern for the potential nonuniqueness of the fit.

Most of the authors mentioned focus on a regression context, although Armstrong and Frome (1976, 1979) direct their attention to least-absolute-residuals fitting of a two-way table. They discuss specialized linear programming algorithms that are highly efficient for two-way tables. They also discuss the nonuniqueness of the least-absolute-residuals fit; they show that alternative solutions frequently exist, except under some stringent conditions, and they discuss some examples of various fits for small two-way tables.

Although an advantage of least-absolute-residuals fitting is that, like median polish, it does not count large residuals heavily, it does suffer from a related disadvantage. Unlike analysis by means, least-absolute-residuals fitting weights very small residuals too heavily. We refer to Mosteller and Tukey (1977, pp. 365–368) for a discussion and illustration of this phenomenon. These authors also propose a modification of the weighting scheme that aims at alleviating this difficulty.

6E. OTHER METHODS OF POLISHING

Other measures of location of the center of a batch exist besides the mean and median. Several of these are described and examined in detail in Chapters 10 and 11. They are more complicated than either the mean or median, but they may combine some advantages of each. In particular, they are all more resistant to outliers than is the mean.

Criteria for Fitting

Section 6C discusses analysis by means, which minimizes the sum of squared residuals [see equation (10)]. Section 6D suggests that we might choose instead to minimize the sum of absolute residuals [see equation (16)]. The expressions SSR and SAR in equations (10) and (16) are examples of criteria for fitting. They provide two ways of summarizing the differences between observed values in a table and fitted values. We are led to different ways of obtaining the fitted values, depending on which criterion we consider.

Other criteria will lead to other procedures for producing fits to the table. Most of these can be defined in terms of a nonnegative function ρ, applied to the residuals from the simple additive model of equation (3). We choose values of m, the a_i, and the b_j to minimize the expression

$$\sum_i \sum_j \rho(y_{ij} - m - a_i - b_j). \tag{19}$$

When

$$\rho(y_{ij} - m - a_i - b_j) = (y_{ij} - m - a_i - b_j)^2,$$

minimization of (19) leads to analysis by means. When

$$\rho(y_{ij} - m - a_i - b_j) = |y_{ij} - m - a_i - b_j|,$$

we get least-absolute-residuals estimation. We give other criteria and their estimation procedures in Chapter 11.

Polishing with Trimmed Means

Any measure of location can be used, in a manner analogous to median polish, to fit the additive model of equation (3) for a two way table. For example, we could use a trimmed mean to estimate row effects and column effects. Aside from this change, the fitting procedure parallels median polish; in particular, it is also iterative.

When using trimmed means as estimates of location, one typically specifies a trimming percentage (say, 25%) and then discards that percent of both the smallest and the largest observations of the ordered batch. For example, a 25%-trimmed mean for a batch of 100 data values is the mean that is computed after the 25 smallest observations and the 25 largest observations are removed from the batch. Chapter 10 discusses trimmed means in detail.

An analysis based on a trimmed mean provides a fit and a table of residuals which are, roughly speaking, between those provided by median polish and those obtained from analysis by means.

Other Fitting Procedures

With the aid of computers, the biweight estimates of location (described in detail in Chapter 11) can also serve as the basis for an iterative fitting procedure for a two-way table. Properly tuned, this method exhibits excellent robustness properties; we postpone further discussion of robust estimation until Chapter 11.

Still another fitting procedure uses the midextreme or midrange. If $x_1, \ldots, x_n$ is a batch of numbers, this location estimate is

$$\tfrac{1}{2}\left(\max_i x_i + \min_i x_i \right). \tag{20}$$

It is of some mathematical interest because it provides the solution to the problem of minimizing the absolute value of the largest residual. In the notation for a two-way table,

$$\text{MAR} = \max_{i,j}\left\{ | y_{ij} - m - a_i - b_j | \right\} \tag{21}$$

is to be made minimum by appropriate choices of m, a_i, and b_j. The criterion in equation (21) is often referred to as the "sup norm"; it is also known as the Chebyshev criterion of best fit. The solution to the analogous location problem for a single batch is the midrange, expression (20); this suggests using the midrange iteratively to approximately minimize MAR, just as median polish approximately minimizes SAR. However, because the resistance of this method is easily seen to be extremely poor, we mention it only to illustrate the breadth of the polishing approach.

Tukey (1972) includes a general discussion of polishing; he considers computational as well as mathematical aspects of polishing techniques for fitting models.

6F.   BREAKDOWN BOUNDS FOR MEDIAN POLISH

The chief advantage of median polish, aside from its suitability for hand calculation, is its resistance to the effects of outlying data values. One way to quantify this resistance uses the notion of *worst-case breakdown bound*.

Hampel (1968, 1971) formalized the notion of the breakdown bound of an estimator in order to characterize the sensitivity of location estimators to gross errors in data. Essentially, one asks what fraction of the observations in a sample can be turned into bad values (for example, by substituting a very large, essentially infinite, number for them) without causing the value of the estimate to change greatly. How large a fraction or how many values —*no matter how unluckily chosen*—can be drastically altered without greatly changing the value of the location estimate?

We could be lucky, as well as unlucky, so there is some merit in the concept of a *well-placed breakdown bound*. Now one asks what fraction of observations in a sample can be bad, *if shrewdly placed so they don't do damage*, without greatly changing the estimate. We can't trust the well-placed breakdown bound as we can the worst-case breakdown bound, but it offers us a limit to optimism to balance the worst-case breakdown bound's pessimism.

In this section we apply the concepts of worst-case breakdown bound and well-placed breakdown bound to median-polish analyses of two-way tables in a straightforward way. In anticipation of this discussion, we first examine a related concept in estimating location from a single batch of data. (Because one value in a sample is like another, we have no real choice as to where a highly exotic value is placed, and the two kinds of breakdown bounds are the same.)

Breakdown Bounds for the Mean and Median

The sample mean offers no protection against gross errors. Because it gives weight $1/n$ to each of the observations in the sample, the presence of only one bad value seriously distorts the value of the mean. The two breakdown bounds for the mean are both 0.

The median, on the other hand, is much more tolerant of gross errors; almost half of a batch of numbers can be bad values without altering the value of the median very much. Specifically, if the sample size, n, is odd, we may write it as $n - 2k + 1$, and the worst-case breakdown bound for the median is $k/(2k + 1)$. And if n is even, say $n = 2k + 2$, then the worst-case breakdown bound is $k/(2k + 2)$. If we introduce the oddness function

$$d(n) = \begin{cases} 0 & \text{for } n \text{ even} \\ 1 & \text{for } n \text{ odd,} \end{cases}$$

then we can unify the two cases by writing the worst-case breakdown bound for the median as

$$\frac{\frac{1}{2}n + \frac{1}{2}d(n) - 1}{n} = \frac{1}{2} - \frac{2 - d(n)}{2n}, \tag{22}$$

so that the relation to $\frac{1}{2}$ is clear.

And what is the well-placed breakdown bound of the median? It is possible to have a larger number of bad values and to balance the signs of the bad values in such a way that the median is not affected. Indeed, it is enough to leave one value alone for samples of odd size or two for samples of even size. The well-placed breakdown bound of the median is $2k/n$ or, in each case,

$$1 - \frac{2 - d(n)}{n}, \tag{23}$$

just twice as large as the worst-case breakdown bound.

Resistance of Median Polish

Because the worst-case breakdown bound for a location estimator is defined by requiring that the value of the estimate not be greatly affected, the most direct way of extending the notion to two-way tables and median polish is to require that no term in the fit be greatly affected. That is, we write the two-way table of data as

$$y_{ij} = m + a_i + b_j + e_{ij},$$

and allow bad observations to affect only e_{ij} and not any of m, the a_i, or the b_j. It is, of course, possible to relax this definition (for instance, to be more concerned about the a_i than the b_j), but doing so is, at least in part, a matter of judgment.

To derive the worst-case breakdown bound for median polish, we examine the configurations in which the bad observations may be placed in the I rows and J columns of the two-way table. In the least favorable configuration, all the bad observations are in the same row or in the same column. This will be most damaging for the smaller dimension of the table. That is, if I is greater than J, then the least favorable configuration places all bad observations among the J observations of a single row. Enough bad observations will affect the median of that row, thereby altering the corresponding row effect. Thus the worst-case breakdown bound, WCBB, for median polish is proportional to the breakdown bound for the median of a sample whose size is the smaller of I and J. By again using the oddness function, we may reduce this to a formula:

$$\text{WCBB} = \frac{\frac{1}{2}\min(I, J) - 1 + \frac{1}{2}d[\min(I, J)]}{IJ}$$

$$= \frac{1}{2\max(I, J)} - \frac{2 - d[\min(I, J)]}{2IJ} \tag{24}$$

The well-placed breakdown bound for median polish of a two-way table is somewhat more difficult to discover. To do so, we let B represent the number of bad observations that median polish can tolerate if the bad observations are arranged in a most favorable configuration. Then B/IJ is the well-placed breakdown bound, so that the following theorem gives the details.

THEOREM 5: The well-placed breakdown bound for median polish of an $I \times J$ two-way table is B/IJ, where

$$B = \frac{1}{2}\min\{J[I - 2 + d(I)], I[J - 2 + d(J)]\}$$

$$= \frac{1}{2}IJ - \frac{1}{2}\max\{2J - Jd(I), 2I - Id(J)\}. \tag{25}$$

More specifically, if $I \leqslant J$, then B is given by the formulas:

1. For I odd, J even, and $J < 2I$,

$$B = \frac{1}{2}IJ - I. \tag{26}$$

2. In all other cases,

$$B = \frac{1}{2}IJ - \frac{1}{2}J[2 - d(I)] = \begin{cases} \frac{1}{2}IJ - J & \text{for } I \text{ even} \\ \frac{1}{2}(IJ - J) & \text{for } I \text{ odd.} \end{cases} \tag{27}$$

Thus $B \geqslant \frac{1}{2}IJ - \max(I, J)$ and $B/IJ \geqslant \frac{1}{2} - 1/\min(I, J)$.

PROOF: We break the proof into three lemmas.

LEMMA 1: $B \leqslant \frac{1}{2}\min\{J[I - 2 + d(I)], I[J - 2 + d(J)]\}$.

PROOF OF LEMMA 1: From expression (22), $\frac{1}{2}I - 1 + \frac{1}{2}d(I)$ is the highest number of bad observations that can be tolerated in any column. Similarly, $\frac{1}{2}J - 1 + \frac{1}{2}d(J)$ is the largest number of bad observations that can be tolerated in any row. The allowable number of bad observations in the table cannot exceed $J[\frac{1}{2}I - 1 + \frac{1}{2}d(I)]$, the result of saturating each column separately, or $I[\frac{1}{2}J - 1 + \frac{1}{2}d(J)]$, the result of saturating each row separately, so that B cannot exceed the smaller of these expressions.

LEMMA 2: For $I < J$,

$$\max\{J[2 - d(I)], I[2 - d(J)]\} = J[2 - d(I)]$$

unless I is odd, J is even, and $J < 2I$, when it equals $2I$.

PROOF OF LEMMA 2: First, if $d(I) = d(J) = 0$, then $\max(2J, 2I) = 2J = [2 - d(I)]J$. Second, if $d(I) = 0$ and $d(J) = 1$, then $\max(2J, I) = 2J = [2 - d(I)]J$. Third, if $d(I) = d(J) = 1$, then $\max(J, I) = J$. Finally, if $d(I) = 1$ and $d(J) = 0$, we must deal with $\max(J, 2I)$, which is $J = [2 - d(I)]J$ unless $2I > J$, when it is $2I$.

LEMMA 3: A configuration of bad values exists for which B achieves the upper bound:

$$B = \min\{J[\frac{1}{2}I - 1 + \frac{1}{2}d(I)], I[\frac{1}{2}J - 1 + \frac{1}{2}d(J)]\}.$$

PROOF OF LEMMA 3: Suppose that B is the maximum number of bad values but that strict inequality holds in Lemma 1. Without loss of generality we assume that the first row has fewer than $\frac{1}{2}J - 1 + \frac{1}{2}d(J)$ bad values, and that the first column has fewer than $\frac{1}{2}I - 1 + \frac{1}{2}d(I)$ bad values.

If cell $(1, 1)$ does not already contain a bad value, we can make the value in that position bad, contradicting the maximality of B.

Suppose that cell $(1, 1)$ contains a bad value. By permuting rows 2 through I (respectively, columns 2 through J), we can assume that the remaining bad elements in column 1 (respectively, row 1) are in the first positions. Thus we suppose there are R bad elements in the first R positions of column 1 and C bad elements in the first C positions of row 1. Of course, our assumptions mean that $R < \frac{1}{2}I - 1 + \frac{1}{2}d(I)$ and $C < \frac{1}{2}J - 1 + \frac{1}{2}d(J)$.

TABLE 6-14. Schematic diagram for a configuration of bad values in a two-way table (the symbol b denotes "bad," and g denotes "good").

i	$j{:}1$	2	$\cdots$	C	$C+1$	$\cdots$	J
1	b	b	$\cdots$	b	g	$\cdots$	g
2	b						$\cdot$
$\vdots$	$\vdots$		$\ddots$	$\vdots$		$\ddots$	$\vdots$
R	b		$\cdots$			$\cdots$	
$R+1$	g		$\cdots$				
$\vdots$	$\vdots$		$\ddots$	$\vdots$	Q		
I	g		$\cdots$				

We illustrate the situation schematically in Table 6-14, using "b" to denote a bad value and "g" to denote a good one.

Let Q be the $I - R$ by $J - C$ subtable in the lower right corner of the two-way table. If Q contains only gs, then the maximality of B is violated, because any element of Q could be changed from g to b while not exceeding $\frac{1}{2}J - 1 + \frac{1}{2}d(J)$ bad elements in its row and $\frac{1}{2}I - 1 + \frac{1}{2}d(I)$ bad elements in its column. Therefore Q contains at least one bad element, which we assume is in position (i, j).

Next we exchange the bad value in position (i, j) with the good value in position $(i, 1)$, and we exchange the bad value in position $(1, 1)$ with the good value in position $(1, j)$. Finally, we replace the good element now in cell $(1, 1)$ with a bad value, thus increasing the total number of bad values to $B + 1$ without leading to breakdown. This contradiction to the maximality of B establishes Lemma 3 and, finally, the theorem.

Now that we have obtained both the worst-case breakdown bound and the well-placed breakdown bound for an arbitrary two-way table, we illustrate these results with an example.

EXAMPLE: BREAKDOWN BOUNDS

Consider a 5×8 table, which has 40 data values. The breakdown bound is

$$\frac{1}{2\max(5,8)} - \frac{2 - d[\min(5,8)]}{80} = \frac{1}{16} - \frac{1}{80} = \frac{2}{40}.$$

TABLE 6-15. Values of B, namely IJ times the well-placed breakdown bound, for $I \times J$ tables.

						J					
I	2	3	4	5	6	7	8	9	10	11	12
2	0	0	0	0	0	0	0	0	0	0	0
3		3	3	5	6	7	8	9	10	11	12
4			4	5	6	7	8	9	10	11	12
5				10	**10**	14	**15**	18	20	22	24
6					12	14	16	18	20	22	24
7						21	**21**	27	**28**	33	**35**
8							24	27	30	33	36
9								36	**36**	44	**45**
10									40	44	48
11										55	**55**
12											60

If 3 of the 40 values were bad (with the same direction) and were all placed in column j, they would greatly affect the estimated effect for column j.

The theorem just proved states that, in a most favorable configuration, the well-placed breakdown bound is determined by [case 1 of Theorem 5]:

$$B = \tfrac{1}{2}(5)(8) - 5 = 15.$$

Table 6-15 gives the well-placed breakdown bounds (times IJ) for two-way tables with $I \leqslant J \leqslant 12$. The boldface values rely on case 1 of Theorem 5, expression (26).

Worst-Case Bound or Well-Placed Bound?

In assessing resistance of median polish to bad cells, should we use the worst-case or the well-placed breakdown bound? Strictly speaking, we should probably be pessimistic and use the worst-case bound in order to be consistent with the notion of a breakdown bound in simpler settings. It is probably quite realistic to guess that bad observations may often go together in the same row or column. If we think that all cells are equally likely to contain bad values and that values are as likely to be bad in one direction as another, the well-placed breakdown bound can help to temper our pessimism. Taken together, the two breakdown bounds define an interval for the fraction of bad values that median polish can tolerate.

6G. TREATMENT OF HOLES

In the preceding sections we have assumed that a two-way table has exactly one value of the response for each combination of levels of the two factors that identify rows and columns. Sometimes a data set is not so simple, in that some values for the response are missing. This can happen for a number of reasons: the data may not have been gathered, a record may be lost, an experimental unit may be damaged, or we may have decided to set aside a data value that we judged to be erroneous. (The last of these situations would occur if we were uncertain about whether a bad value is much larger or much smaller than it should have been. Of course, if we knew that it came from, say, a relatively large value, we would know how to treat it when finding medians. Thus values missing from a table are not always treated in the same way as bad values.) Missing data values produce holes in the table. In general, holes distort a median-polish analysis less than bad values do.

Impact on Analysis by Means

Section 6C explained that analysis by means provides the solution to a least-squares problem that seeks to minimize the sum of squared residuals given in equation (10). When data values are missing, equation (10) will have missing r_{ij}, and its least-squares solution will no longer be as simple as analysis by means. A general least-squares computer program can handle these situations. Alternatively, if no such program is available, special formulas allow us to work around the missing values—essentially by filling them in without increasing the sum of squared residuals. [See, for example, Anderson (1946) and Healy and Westmacott (1956).]

What effects do missing data values have on the analysis by means? They may substantially change the estimated effects. Consider again the hypothetical example in Section 6C. If cell (1, 1) were empty, the original fit shown in Table 6-11 would be substantially altered, and the new analysis by means

TABLE 6-16.   Hypothetical data with missing value,
after analysis by means.

i	j			$a_{i\cdot}$
	1	2	3	
1	abs	0	0	0
2	0	0	0	0
3	0	0	0	0
b_j	0	0	0	$0 (= m)$

would produce a fit whose effects are all 0. The residuals and fitted effects appear in Table 6-16.

The important message, illustrated in the hypothetical example, is that *a missing value or hole in a two-way table can substantially change the additive fit for the entire table that analysis by means provides.* This fact is closely related to the lack of resistance of this fitting method, as discussed in Section 6F.

Impact on Median Polish

An analysis by medians readily proceeds whenever all rows and columns contain at least some data; when we perform median polish, we simply find the median of the available values in a row or column.

How do missing values alter a median polish and the additive fit it gives for a two-way table? If cell $(1, 1)$ were empty in Table 6-10, the fit produced by median polish would coincide with that given originally and shown in Table 6-12. The missing observation does not change the fit.

Suppose, however, that the value in cell $(1, 2)$ were missing instead. If we perform median polish on the table

$$
\begin{array}{ccc}
9 & \text{abs} & 0 \\
0 & 0 & 0 \\
0 & 0 & 0
\end{array}
$$

we obtain the fit shown in Table 6-17. The value for the row-1 effect, a_1, now changes from 0 to 4.5.

In view of the resistance of median polish, it may appear surprising that a single missing observation could cause such a substantial change in the effects produced by the analysis by medians. (One explanation is "local shift sensitivity," discussed in Chapter 11.) This example illustrates that *missing values can substantially alter the breakdown bound for median polish.*

TABLE 6-17. Hypothetical data with missing value, after median polish.

i	j 1	2	3	a_i
1	4.5	abs	-4.5	4.5
2	0	0	0	0
3	0	0	0	0
b_j	0	0	0	$0 (= m)$

Experience with median polish has shown that missing values can sometimes greatly increase the number of iterations needed to remove most of the row and column effects. Tukey (1970, Section 19F) discusses an 8×8 example for which 24 of the cells (37.5%) in a systematic pattern are holes and over 100 iterations are needed to produce a fit whose residuals have row and column medians close to 0. Other resistant fitting methods, especially the square combining table (Tukey, 1970) discussed in a companion volume, can circumvent this difficulty when many data values are missing.

Background Comments

The problem of holes in two-way tables has received attention from many authors since the introduction of analysis of variance. Among those who consider problems associated with missing values are Allan and Wishart (1930), Yates (1933), Bartlett (1937), and Anderson (1946). Healy and Westmacott (1956) discuss computational aspects of the missing data problem. Tukey (1970, Chapter 19), Mosteller and Tukey (1977, Chapter 9), and Velleman and Hoaglin (1981, Chapter 8) discuss the impact of holes on median polish.

6H. NONADDITIVITY AND THE DIAGNOSTIC PLOT

In Section 6A and the development that followed, we assumed that the data follow the additive model of equation (2), except for a few unusual observations. If the data depart substantially and systematically from this structure, a transformation may enable us to proceed in the analysis.

A straightforward plot allows us to judge the extent of any systematic nonadditivity and, at the same time, to see which power transformations (if any) would help to remove it. We begin by calculating an additive fit to the data, so that we have m, the a_i, the b_j, and the e_{ij} as in equation (3). (Of course, we strongly prefer a resistant fit, but the technique is quite general.) For each cell in the table we then define the *comparison value*, cv_{ij}, according to

$$cv_{ij} = \frac{a_i \times b_j}{m}. \tag{28}$$

The *diagnostic plot* consists of the points (cv_{ij}, e_{ij}), one point for each cell; that is, we plot the residual against the comparison value.

If the diagnostic plot reveals no consistent trend or pattern, we can conclude that the data do not depart systematically in such a simple way from an additive model. In practice, an examination of the residuals, e_{ij}, will already have suggested a problem with nonadditivity and will have led us to make the plot, so that we already expect a trend and are concerned primarily about its strength. (A major reason for preferring to base the diagnostic plot on the results of a resistant fit is that one isolated wild value —if wild enough—can cause the plot to suggest systematic nonadditivity if we analyze by means.)

The slope of the diagnostic plot guides us to a transformation or transformations that should help to remove the nonadditivity. If that slope is k, simple powers near $1 - k$ provide useful transformations. Section 8G fills in the underlying mathematical details.

When the plot has a clear slope but powers near $1 - k$ are neither plausible nor appropriate in the context of the data, we can consider adding $k \times cv_{ij}$ to the fit as a further term. We discuss this and more general models in a companion volume.

EXAMPLE: PERCENTAGE POINTS OF CHI-SQUARED

We have all used a table of percentage points for various chi-squared distributions. We know that, as tail probabilities decrease, the corresponding percentage points increase. For a fixed tail probability, an increase in the number of degrees of freedom produces a corresponding increase in the percentage points. Let us see whether the contributions from increasing degrees of freedom and decreasing tail probabilities are nearly additive. If not, then how can the effects of changing tail probabilities and changing degrees of freedom on the percentage points be modeled? In short, we seek insight into the structure that underlies a two-way table of chi-squared upper-tail percentage points.

From a table of chi-squared percentage points, we select the percentage points for five upper-tail percents and six different values of the degrees-of-freedom parameter. The data make up Table 6-18, and Table 6-19 shows the result of a median-polish analysis. The residuals have a "twisted" pattern: those in the upper left and lower right corners are all positive, whereas those in the other corners are negative.

TABLE 6-18.   Upper-tail percentage points for chi-squared distributions.

Degrees of Freedom	p				
	.9	.95	.975	.99	.995
3	6.25	7.81	9.35	11.34	12.84
6	10.64	12.59	14.45	16.81	18.55
9	14.68	16.92	19.02	21.67	23.59
12	18.55	21.03	23.34	26.22	28.30
15	22.31	25.00	27.49	30.58	32.80
18	25.99	28.87	31.53	34.81	37.16

TABLE 6-19. Median-polish analysis of the chi-squared data.

Degrees of Freedom	p					effect
	.9	.95	.975	.99	.995	
3	1.47	0.67	0	−0.78	−1.28	−11.83
6	0.76	0.35	0	−0.41	−0.67	−6.73
9	0.23	0.11	0	−0.12	−0.20	−2.16
12	−0.22	−0.10	0	0.11	0.19	2.16
15	−0.61	−0.28	0	0.32	0.54	6.31
18	−0.97	−0.45	0	0.51	0.86	10.35
effect	−4.57	−2.21	0	2.77	4.77	21.18

The additive fit for the chi-squared table does not capture all the structure in the data; this means that

$$\hat{y}_{ij} = m + a_i + b_j$$

does not summarize the data as well as we would like. To pursue the systematic behavior of the residuals, Table 6-20 gives the comparison values, calculated from the effects and common values in Table 6-19. The diagnostic plot in Figure 6-4 confirms that the nonadditivity is strongly systematic. The linear trend is so strong that we can readily decide on $k = \frac{1}{2}$ as the slope, suggesting a square-root transformation.

TABLE 6-20. Comparison values for chi-squared data: comparison value equals $a_i b_j / m$.

Degrees of Freedom	p					effect
	.9	.95	.975	.99	.995	
3	2.55	1.23	0	−1.55	−2.66	−11.83
6	1.45	0.70	0	−0.88	−1.52	−6.73
9	0.47	0.23	0	−0.28	−0.49	−2.16
12	−0.47	−0.23	0	0.28	0.49	2.16
15	−1.36	−0.66	0	0.83	1.42	6.31
18	−2.33	−1.09	0	1.35	2.33	10.35
effect	−4.57	−2.21	0	2.77	4.77	21.18

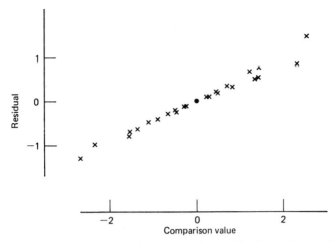

Figure 6-4. Diagnostic plot for chi-squared data. (The large dot at the origin represents six points.)

Pursuing this suggestion, we provide the transformed data in Table 6-21 and the median-polish analysis in Table 6-22. The residuals in Table 6-22 are small, but they do show a slight "twisting" pattern. It is not quite straightforward to compare these residuals with those of Table 6-19 because the values given are in two different scales. (A method of matching discussed in Chapter 4 enables us to partially remedy this difficulty.) The magnitudes of the residuals from the square-root fit form a much smaller fraction of the magnitudes of the various effects than do those from the simple additive model in the original scale. We judge, then, that additivity is a much better description of the data in the square-root scale than in the raw scale.

One reason for exploring the chi-squared table is to discover how the percentage points depend on the probability p and on the degrees-of-freedom parameter ν. We have seen that, to a reasonable approximation,

$$(y_{ij})^{1/2} = m + a_i + b_j + e_{ij} \qquad (29)$$

provides an adequate fit. In different, but still obvious, notation, we could represent this fit as

$$\left(\chi^2_{\nu, p}\right)^{1/2} = m + \alpha(\nu) + \beta(p) + \varepsilon(\nu, p). \qquad (30)$$

This expression of the model raises the possibility that the chi-squared

TABLE 6-21.   Square-root of upper-tail percentage points for chi-squared distributions.

Degrees of Freedom	p				
	.9	.95	.975	.99	.995
3	2.50	2.79	3.06	3.37	3.58
6	3.26	3.55	3.80	4.10	4.31
9	3.83	4.11	4.36	4.66	4.86
12	4.31	4.59	4.83	5.12	5.32
15	4.72	5.00	5.24	5.53	5.73
18	5.10	5.37	5.62	5.90	6.10

percentage points could be approximately specified as a function of v and p. Such a function could then be used (say, in a programmable calculator) to provide values of chi-squared percentage points for various values of v and p. We do not pursue this possibility here, but we refer to the paper by Hoaglin (1977), in which approximate formulas for chi-squared percentage points are derived and evaluated.

6I. SUMMARY

Median polish is an iterative method for fitting an additive model to a two-way table of data. For tables of modest size, it can be performed by hand. A principal advantage of median polish is its resistance to the effects of wild data values and its related ability to deal easily with a few missing data values.

TABLE 6-22.   Median-polish analysis of square-root of chi-squared data.

Degrees of Freedom	p					effect
	.9	.95	.975	.99	.995	
3	−0.03	−0.02	0	0.01	0.02	−1.53
6	−0.01	0.00	0	0.00	0.01	−0.79
9	0.00	0.00	0	0.00	0.00	−0.23
12	0.01	−0.01	0	−0.01	−0.01	0.24
15	0.01	−0.01	0	−0.01	−0.01	0.65
18	0.01	0.00	0	−0.02	−0.02	1.03
effect	−0.53	−0.25	0	0.30	0.50	4.59

Median polish also serves as a quick and approximate method for least-absolute-residuals fitting. Although it does not strictly lead to a solution that minimizes the sum of absolute residuals, in practice it usually gives a solution that is at least qualitatively similar to the SAR solution. Although median polish does not share some of the mathematically appealing properties of analysis by means, it does well enough in practice to be useful in exploring the possibly additive structure of a two-way table. Its breakdown value is relatively high, so that median polish is especially well suited for a preliminary exploratory analysis.

Although we do not discuss them in this chapter, exploratory graphical displays for two-way tables can combine with median polish to reveal more vividly the structure of the fit and the extent of departures from additivity.

When an additive model cannot summarize the structure of a table, another display—the diagnostic plot—enables us to find transformations that can re-express the data in a scale in which the structure of the table is more nearly additive.

REFERENCES

Allan, R. E. and Wishart, J. (1930). "A method of estimating the yield of a missing plot in field experimental work," *Journal of Agricultural Science*, **20**, 399–456.

Anderson, R. L. (1946). "Missing-plot techniques," *Biometrics*, **2**, 41–47.

Andrews, D. F., Bickel, P. J., Hampel, F. R., Huber, P. J., Rogers, W. H., and Tukey, J. W. (1972). *Robust Estimates of Location: Survey and Advances*. Princeton, NJ: Princeton University Press.

Armstrong, R. D. and Frome, E. L. (1976). "The calculation of least absolute value estimators for two-way tables." *1976 Proceedings of the Statistical Computing Section*. Washington, DC: American Statistical Association, pp. 101–106.

Armstrong, R. D. and Frome, E. L. (1979). "Least-absolute-values estimators for one-way and two-way tables," *Naval Research Logistics Quarterly*, **26**, 79–96.

Barrodale, I. and Roberts, F. D. K. (1973). "An improved algorithm for discrete L_1 linear approximation," *SIAM Journal on Numerical Analysis*, **10**, 839–848.

Bartlett, M. S. (1937). "Some examples of statistical methods of research in agriculture and applied biology," *Journal of the Royal Statistical Society*, Series B, **4**, 137–170.

Bloomfield, P. and Steiger, W. (1980). "Least absolute deviations curve-fitting," *SIAM Journal on Scientific and Statistical Computing*, **1**, 290–301.

Brown, G. W. and Mood, A. M. (1951). "On median tests for linear hypotheses." In J. Neyman (Ed.), *Proceedings of the Second Berkeley Symposium on Mathematical Statistics and Probability*. Berkeley and Los Angeles: University of California Press, pp. 159–166.

Charnes, A., Cooper, W. W., and Ferguson, R. O. (1955). "Optimal estimation of executive compensation by linear programming," *Management Science*, **1**, 138–150.

Daniel, C. (1978). "Patterns in residuals in the two-way layout," *Technometrics*, **20**, 385–395.

Eisenhart, C. (1977). "Boscovich and the combination of observations." In M. G. Kendall and

R. L. Plackett (Eds.), *Studies in the History of Statistics and Probability*, Vol. 2. New York: Macmillan.

Gentle, J. E. (1977). "Least absolute values estimation: an introduction," *Communications in Statistics*, **B6**, 313–328.

Hampel, F. R. (1968). *Contributions to the Theory of Robust Estimation*. Ph.D. Thesis, Department of Statistics, University of California, Berkeley.

Hampel, F. R. (1971). "A general qualitative definition of robustness," *Annals of Mathematical Statistics*, **42**, 1887–1896.

Harter, H. L. (1977). "Nonuniqueness of least absolute values regression," *Communications in Statistics*, **A6**, 829–838.

Healy, M. and Westmacott, M. (1956). "Missing values in experiments analysed on automatic computers," *Applied Statistics*, **5**, 203–206.

Hoaglin, D. C. (1977). "Direct approximations for chi-squared percentage points," *Journal of the American Statistical Association*, **72**, 508–515.

Hodges, J. L. (1967). "Efficiency in normal samples and tolerance of extreme values for some estimates of location." In L. M. LeCam and J. Neyman (Eds.), *Proceedings of the Fifth Berkeley Symposium on Mathematical Statistics and Probability*, Vol. 1, 163–186. Berkeley: University of California Press.

Kennedy, W. J., Gentle, J. E., and Sposito, V. A. (1977). "Comparison of algorithms for L_1 estimation in the linear model." Technical Report, Statistical Laboratory, Iowa State University, Ames, IA.

Mood, A. M. (1950). *Introduction to the Theory of Statistics*. New York: McGraw–Hill.

Mosteller, F. and Tukey, J. W. (1977). *Data Analysis and Regression*. Reading, MA: Addison–Wesley.

Rhodes, E. C. (1930). "Reducing observations by the method of minimum deviations," *Philosophical Magazine*, *7th Series*, **9**, 974–992.

Scheffé, H. (1959). *The Analysis of Variance*. New York: Wiley.

Siegel, A. F. (1979). Personal communication.

Singleton, R. R. (1940). "A method of minimizing the sum of absolute values of deviations," *Annals of Mathematical Statistics*, **11**, 301–310.

Tukey, J. W. (1949). "One degree of freedom for non-additivity," *Biometrics*, **5**, 232–242.

Tukey, J. W. (1970). *Exploratory Data Analysis* (Limited Preliminary Edition), Volume II. Reading, MA: Addison–Wesley.

Tukey, J. W. (1972). "Data analysis, computation, and mathematics," *Quarterly of Applied Mathematics*, **30**, 51–65.

Tukey, J. W. (1977). *Exploratory Data Analysis*. Reading, MA: Addison–Wesley.

Velleman, P. F. and Hoaglin, D. C. (1981). *Applications, Basics, and Computing of Exploratory Data Analysis*. Boston, MA: Duxbury Press.

Wagner, H. M. (1959). "Linear programming techniques for regression analysis," *Journal of the American Statistical Association*, **54**, 206–212.

Yates, F. (1933). "The analysis of replicated experiments when the field results are incomplete," *Empire Journal of Experimental Agriculture*, **1**, 129–142.

Additional Literature

Besag, J. (1981). "On resistant techniques and statistical analysis," *Biometrika*, **68**, 463–469.

Brown, B. M. (1975). "A short-cut test for outliers using residuals," *Biometrika*, **62**, 623–629.

Brown, M. B. (1975). "Exploring interaction effects in the ANOVA," *Applied Statistics*, **24**, 288–298.

Draper, N. R. and John, J. A. (1980). "Testing for three or fewer outliers in two-way tables," *Technometrics*, **22**, 9–15

Freund, R. J. (1980). "The case of the missing cell," *The American Statistician*, **34**, 94–98.

Gentleman, J. F. (1980). "Finding the *K* most likely outliers in two-way tables," *Technometrics*, **22**, 591–600.

Gentleman, J. F. and Wilk, M. B. (1975a). "Detecting outliers in a two-way table: I. Statistical behavior of residuals," *Technometrics*, **17**, 1–14.

—— (1975b). "Detecting outliers. II. Supplementing the direct analysis of residuals," *Biometrics*, **31**, 387–410.

John, J. A. (1978). "Outliers in factorial experiments," *Applied Statistics*, **27**, 111–119.

John, J. A. and Prescott, P. (1975). "Critical values of a test to detect outliers in factorial experiments," *Applied Statistics*, **24**, 56–59.

Sposito, V. A., Hand, M. L., and McCormick, G. F. (1977). "Using an approximate L_1 estimator," *Communications in Statistics*, **B6**, 263–268.

Sposito, V. A., Kennedy, W. J., and Gentle, J. E. (1980). "Useful generalized properties of L_1 estimators," *Communications in Statistics*, **A9**, 1309–1315.

EXERCISES

1. A two-way table presents average American life expectancy as the response to two factors: sex (male or female) and ethnicity (Caucasian, Negro, Oriental, or Spanish). Assume that an additive model adequately summarizes the data. Formulate such a model explicitly and interpret the parameters μ, α_i, β_j, and ε_{ij}.

2. Use two complete steps of median polish, starting with columns, to fit an additive model to the infant-mortality-rate data of Table 6-2. Compare your results to those shown in Table 6-9 for median polish starting with rows.

3. Suppose that a two-way table has perfect additive structure, that is, $y_{ij} = \mu + \alpha_i + \beta_j$ for all i and j, where the α_i have mean 0 and the β_j have mean 0.

   (a) Show that analysis by means will provide estimates $m = \mu$, $a_i = \alpha_i$, $b_j = \beta_j$, and $e_{ij} = 0$, thus providing a perfect fit to the table.

   (b) Show that median polish will also provide estimates for which all residuals are 0, but that m, a_i, and b_j need not coincide with μ, α_i, and β_j.

4. Consider the hypothetical two-way table of data

$$
\begin{array}{ccccc}
10 & 2 & 3 & 4 & 5 \\
2 & 3 & 4 & 5 & 6 \\
3 & 4 & 5 & 6 & 7 \\
4 & 5 & 6 & 7 & 8 \\
5 & 6 & 7 & 8 & 20
\end{array}
$$

Give the following fits for this table:
(a) Median polish, starting with rows;
(b) Median polish, starting with columns;
(c) Analysis by means;
(d) Analysis by 20%-trimmed means.

5. Explain how, by changing only two entries in the table of Exercise 4, one could give a table for which all the fits found in Exercise 4 coincide.

6. What can be said about the relationship between median polish and analysis by means for an arbitrary 2×2 table?

7. For an $I \times J$ table, with a nonzero value D in only one cell, derive the common value, row effects, column effects, and residuals that result from least-squares fitting.

8. If $\{y_{ij}\}$ and $\{Y_{ij}\}$ are two-way tables with I rows and J columns and one observation per cell, show that the fit by means for their sum, $\{y_{ij} + Y_{ij}\}$, is the sum of the fits by means for the two tables separately.

9. In the 5×8 table below, the 14 bad observations (denoted by b) occupy positions such that inserting any further bad values without moving one of the existing ones will cause breakdown. However, Theorem 5 guarantees that an arrangement containing 15 bad values is possible. Use the procedure in Lemma 3 to rearrange the 14 bad values and insert an additional one.

	1	2	3	4	5	6	7	8
1	b	b	b					
2	b	b	b					
3				b	b	b		
4				b	b	b		
5							b	b

10. Consider the fit to the chi-squared data in the square-root scale, as shown in Table 6-22.
    (a) Calculate the full table of fitted values for the square-root scale.
    (b) Transform by the square to put each of these fitted values in the raw scale and construct the resulting 6 × 5 table.
    (c) Finally, construct a table of residuals from this fit in the raw scale. Compare this table of residuals to the one shown in Table 6-19.

11. Let $\{x_1, \ldots, x_n\}$ be a batch of n numbers and let the sum of squared residuals be defined by

$$\text{SSR}(c) = \sum_{i=1}^{n} (x_i - c)^2.$$

Show that $c = \sum_{i=1}^{n} x_i / n$ is the value of c that minimizes $\text{SSR}(c)$.

12. Let $x_1, \ldots, x_n$ be a batch of n numbers. Show that the expression

$$\text{SAR}(c) = \sum_{i=1}^{n} |x_i - c|$$

is minimized by taking c equal to the median of the x_i. [Consider first the case with n odd ($n = 2k + 1$). Then, for n even ($n = 2k$), show that any c between $x_{(k)}$ and $x_{(k+1)}$ minimizes $\text{SAR}(c)$.]

13. Let $\{x_1, x_2, \ldots, x_n\}$ be a batch of n numbers, and let $\text{MAR}(c)$ (for maximum absolute residual) be defined by

$$\text{MAR}(c) = \max_i \{|x_i - c|\}.$$

Show that $c = \frac{1}{2}(\min_i x_i + \max_i x_i)$ is the choice of the location estimator c that minimizes $\text{MAR}(c)$.

*14. Give a proof of Theorem 1.

*15. Apply median polish to the following table devised by Andrew F. Siegel:

1	0	$-n$	n	n
0	1	$-n$	n	n
n	n	0	$-n$	$-n$
$-n$	$-n$	n	-1	0
$-n$	$-n$	n	0	-1

Begin by adjusting the rows and then adjust the columns. After one step for rows and one step for columns, compare the table of residuals and the original "data." If n is even (say, $n = 2k$), how many steps will be required to produce a table of residuals all of whose row medians and column medians are zero? Give that table of residuals, along with its row effects, column effects, and common value. Describe the sequence of values of SAR [defined in equation (16)] from original "data" to final fit.

CHAPTER 7

Examining
Residuals

Colin Goodall
Harvard University

Exploratory data analysis places a heavy emphasis on looking at and analyzing residuals. Residuals have, for a long time, helped statisticians with approaches to understanding data and to models. Many residual-using techniques are effective for exploration (Tukey, 1977). This chapter differs from others because many of the methods originated before exploratory data analysis and thus do not stem from either the exploratory or the robustness literature.

Furthermore, this chapter is different because we attempt less mathematical justification. Many of the techniques are already well accepted; other techniques do not require specific justification. For example, graphs convey more, through patterns as well as relative sizes of numbers, than numerical summary statistics. The extra effort involved in looking at residuals is amply repaid by the rich discoveries about the data. The effort required is considerably reduced by the availability of computer packages.

Techniques for batches of residuals take into account the fitting procedure. The batch may be scaled, displayed, and plotted to help identify patterns in the residuals, unusual points, and the residual distribution. We comment on the relationship between error and residual distributions. We plot regression residuals against carriers or the fit to suggest transformations or new carriers to improve the fit. A wedge-shaped residual plot often indicates nonconstant variability. More sophisticated techniques include plots of transformed residuals. The elaboration of a model by fitting to residuals can be a convenient alternative to completely recalculating the fit.

211

7A. RESIDUALS AND THE FIT

Elsewhere in this book we describe ways to summarize a batch of data or a relationship or structure. Frequently, the summary assigns a fitted value to each data value. We then have

$$\text{data} = \text{fit} + \text{residual}.$$

The fit embodies some of the major patterns in the data. To look at the data in more detail, we examine the residuals. Our hope is that the fitting technique puts into the fit as much of the pattern in the data as possible. Therefore, when the same fitting technique is applied to the residuals, the fit is practically zero. We present a range of techniques, some familiar from classical statistics and others suggested by exploratory data analysis, which help to uncover patterns in the residuals.

The fit describes the data—of course, incompletely. We use the residuals to improve the fit in stages, by re-expression (e.g., transforming the data), by the inclusion of additional variables, and by special handling of unusual data points. An informal criterion for goodness of fit is the absence of patterns in the residuals. Other criteria, less effective but more familiar, include a small residual sum of squares and, sometimes closely related, a large value of the multiple correlation coefficient (often stated as a value of R^2). We improve the fit either until it is sufficiently close to the data for our purposes or until we have fitted so many parameters that further patterns in the residuals are likely to be artifacts. Fisher (1973), Mosteller and Tukey (1977), and Box (1980) provide a variety of overviews to contrast with and supplement the preceding simple outline of data description.

Naive Classical and Exploratory Approaches

How we treat residuals reflects our approach to data analysis. For illustration, we contrast the naive classical approach and the exploratory approach. The naive classical approach assumes a model underlying the data, thus

$$\text{data} = \text{model} + \text{error}.$$

The assumption of a particular error distribution helps us fit the specified model in an optimum manner. For example, we arrive easily at estimators of the parameters of the general linear model that will be minimum-variance-unbiased whenever, if ever, the errors are independent, identically distributed, and Gaussian. Confirmatory methods, including significance and hypothesis tests, help check the model and the assumptions. We plot and summarize the residuals to check the error assumptions, and we also use residuals to estimate the error variance on which we base tests.

Exploratory data analysis uses robust and resistant techniques with weaker demands on errors. The techniques are chosen to uncover and magnify patterns in the data. The patterns are not anticipated in a choice of model. Instead, the patterns partly depend on the analytical technique used. Our understanding of the technique tells us in what way the fit is a summary of the data.

In exploratory data analysis we use residual plots extensively to suggest improvements to the fit, to see how the technique acts on the data to give the fit, and to portray the adequacy of the fit. The examination of residual plots is much aided by computer software. In particular, Velleman and Hoaglin (1981) assemble a collection of techniques that appear in an updated version of Minitab (Ryan et al., 1981).

Exceptional Data Points

Exceptional data points have a disproportionate influence on classical fits. They include outliers and, particularly in regression, points with high leverage. Residuals are useful for the detection of outliers because in forming the residual we remove the summary value from the data value. "Outliers" and "outliers in the residuals" are almost synonymous. When we use resistant methods, the fit is almost unaffected by the presence of outliers. Thus the fit better summarizes the majority of the data points, and the residuals more readily betray the presence of outliers.

Methods for fitting to data differ in the way outliers are treated. For example, robust application of classical normal theory may proceed by stages. We first calculate a least-squares fit to the data. We identify aberrant data points, which may include outliers in the residuals, and recalculate the fit without these points. We then compare the two fits. We judge how much of the patterns we have found belongs to the body of the data and how much to just a few points.

On the other hand, one stage usually suffices for a resistant method of analysis. Often, however, resistant estimation itself involves (internal) iteration, as in the resistant line and median polish. A resistant method contains an implicit rule for outliers, which may be to exclude their actual value, although not the fact of their presence, from the fit. For example, outliers help determine the value of the median only through their presence and direction, not through their value.

EXAMPLES:

The examples in this chapter use the data set that McDonald and Schwing (1973) assembled to study the effects of air pollution on mortality while including socioeconomic and climatological variables. Each of the 16 variables was measured at each of 60 Standard Metropolitan Statistical Areas (SMSAs) in the United States. Table 7-1 lists the variables and their acronyms. Table 7-2 lists the SMSAs and the data.

McDonald and Schwing fit all 15 variables to mortality using the ordinary and ridge regression models. They then select subsets of variables by the C_p statistic, recommended by Daniel and Wood (1980), and also by elimination using the ridge trace (Hoerl and Kennard, 1970). Hocking (1976) uses the data set in his review article on variable selection in linear regression. McDonald and Ayers (1978) use Chernoff faces (Chernoff, 1973) to provide initial clusters of variables for a clustering procedure (Beale, 1969) and to identify outstanding data points. They use Chernoff faces

TABLE 7-1.   Socioeconomic, climatological, and air pollution variables in
             60 United States SMSAs.

Variable Acronym	Description (McDonald and Schwing, 1973)
RAIN	Mean annual precipitation in inches
JAN	Mean January temperature in degrees Fahrenheit
JULY	Mean July temperature in degrees Fahrenheit
OLD	Percent of 1960 SMSA population 65 years of age or over
POP/HSE	Population per household, 1960 SMSA
EDUC	Median school years completed for those over 25 in 1960 SMSA
GDHSE	Percent of housing units that are sound with all facilities
POPDEN	Population per square mile in urbanized area in 1960
NONW	Percent of 1960 urbanized area population that is nonwhite
WCOL	Percent employment in white-collar occupations in 1960 urbanized area
POOR	Percent of families with income under $3000 in 1960 urbanized area
HC	Relative pollution potential[a] of hydrocarbons, HC
NOX	Relative pollution potential[a] of oxides of nitrogen, NO_x
SO2	Relative pollution potential[a] of sulfur dioxide, SO_2
HUMID	Percent relative humidity, annual average at 1 P.M.
MORT	Total age-adjusted mortality rate, all causes, expressed as deaths per 100,000 population

[a] "The pollution potential is determined as the product of the tons emitted per day per square kilometer of each pollutant and a dispersion factor which accounts for mixing height, wind speed, number of episode days and dimension of each SMSA." Each SMSA has the same dispersion factor for all pollutants. The pollution potentials are for the year 1963.

further to relate the residuals (from the full least-squares regression) to the variables. Henderson and Velleman (1981) summarize preceding work on the data set and discuss an interactive approach to building a multiple regression model. Whereas McDonald and Schwing argue that outstanding data points provide information on an unusual combination of factors and should not be rejected, Henderson and Velleman insert dummy variables for York, Lancaster, Miami, and New Orleans and reach a better fit.

Unlike the previous analyses, we do not attempt to describe the effects of air pollution on mortality. Our outlook is exploratory. Selecting two or three variables at a time, we illustrate residual techniques and gain some insight into the data set. The pollution potentials benefit from logarithmic transformation, but elsewhere in the data, as we shall see, many patterns in various plots are blurred by a large amount of scatter.

7B. RESIDUALS AS BATCHES

Comparability

First we attend to patterns in the residuals that are consequences of the assumptions and fitting procedure. Simple linear regression provides an illustration.

The model is

$$y_i = \alpha + \beta(x_i - \bar{x}) + \varepsilon_i,$$

where the ε_i are uncorrelated random variables with equal variance σ^2. The least-squares fit is

$$\hat{y}_i = a + b(x_i - \bar{x});$$

Section 5A gives the formula for b. In this form of the model, $a = \bar{y}$. We denote the residuals by r_i:

$$r_i = y_i - \hat{y}_i.$$

In Section 5D we introduce the concept of leverage. For simple linear regression the leverage of the ith point is

$$h_i = \frac{1}{n} + \frac{(x_i - \bar{x})^2}{\sum_{j=1}^{n}(x_j - \bar{x})^2},$$

TABLE 7-2. The air pollution data.

SMSA	RAIN	JAN	JULY	OLD	POP/HSE	EDUC	GDHSE	POPDEN	NONW	W-COL	POOR	HC	NOX	SO2	HUMID	MORT
Akron, OH	36	27	71	8.1	3.34	11.4	81.5	3243	8.8	42.6	11.7	21	15	59	59	921.9
Albany, NY	35	23	72	11.1	3.14	11.0	78.8	4281	3.5	50.7	14.4	8	10	39	57	997.9
Allentown, PA	44	29	74	10.4	3.21	9.8	81.6	4260	0.8	39.4	12.4	6	6	33	54	962.4
Atlanta, GA	47	45	79	6.5	3.41	11.1	77.5	3125	27.1	50.2	20.6	18	8	24	56	982.3
Baltimore, MD	43	35	77	7.6	3.44	9.6	84.6	6441	24.4	43.7	14.3	43	38	206	55	1071.0
Birmingham, AL	53	45	80	7.7	3.45	10.2	66.8	3325	38.5	43.1	25.5	30	32	72	54	1030.0
Boston, MA	43	30	74	10.9	3.23	12.1	83.9	4679	3.5	49.2	11.3	21	32	62	56	934.7
Bridgeport, CT	45	30	73	9.3	3.29	10.6	86.0	2140	5.3	40.4	10.5	6	4	4	56	899.5
Buffalo, NY	36	24	70	9.0	3.31	10.5	83.2	6582	8.1	42.5	12.6	18	12	37	61	1002.0
Canton, OH	36	27	72	9.5	3.36	10.7	79.3	4213	6.7	41.0	13.2	12	7	20	59	912.3
Chattanooga, TN	52	42	79	7.7	3.39	9.6	69.2	2302	22.2	41.3	24.2	18	8	27	56	1018.0
Chicago, IL	33	26	76	8.6	3.20	10.9	83.4	6122	16.3	44.9	10.7	88	63	278	58	1025.0
Cincinnati, OH	40	34	77	9.2	3.21	10.2	77.0	4101	13.0	45.7	15.1	26	26	146	57	970.5
Cleveland, OH	35	28	71	8.8	3.29	11.1	86.8	3042	14.7	44.6	11.4	31	21	64	60	986.0
Columbus, OH	37	31	75	8.0	3.26	11.9	78.4	4259	13.1	49.6	13.9	23	9	15	58	958.8
Dallas, TX	35	46	85	7.1	3.22	11.8	79.9	1441	14.8	51.2	16.1	1	1	1	54	860.1
Dayton, OH	36	30	75	7.5	3.35	11.4	81.9	4029	12.4	44.0	12.0	6	4	16	58	936.2
Denver, CO	15	30	73	8.2	3.15	12.2	84.2	4824	4.7	53.1	12.7	17	8	28	38	871.8
Detroit, MI	31	27	74	7.2	3.44	10.8	87.0	4834	15.8	43.5	13.6	52	35	124	59	959.2
Flint, MI	30	24	72	6.5	3.53	10.8	79.5	3694	13.1	33.8	12.4	11	4	11	61	941.2
Fort Worth, TX	31	45	85	7.3	3.22	11.4	80.7	1844	11.5	48.1	18.5	1	1	1	53	891.7
Grand Rapids, MI	31	24	72	9.0	3.37	10.9	82.8	3226	5.1	45.2	12.3	5	3	10	61	871.3
Greensboro, NC	42	40	77	6.1	3.45	10.4	71.8	2269	22.7	41.4	19.5	8	3	5	53	971.1
Hartford, CT	43	27	72	9.0	3.25	11.5	87.1	2909	7.2	51.6	9.5	7	3	10	56	887.5
Houston, TX	46	55	84	5.6	3.35	11.4	79.7	2647	21.0	46.9	17.9	6	5	1	59	952.5
Indianapolis, IN	39	29	75	8.7	3.23	11.4	78.6	4412	15.6	46.6	13.2	13	7	33	60	968.7
Kansas City, MO	35	31	81	9.2	3.10	12.0	78.3	3262	12.6	48.6	13.9	7	4	4	55	919.7
Lancaster, PA	43	32	74	10.1	3.38	9.5	79.2	3214	2.9	43.7	12.0	11	7	32	54	844.1
Los Angeles, CA	11	53	68	9.2	2.99	12.1	90.6	4700	7.8	48.9	12.3	648	319	130	47	861.8

City															
Louisville, KY	30	35	71	8.3	3.37	9.9	77.4	4474	13.1	42.6	17.7	38	37	193	57	989.3
Memphis, TN	50	42	82	7.3	3.49	10.4	72.5	3497	36.7	43.3	26.4	15	18	34	59	1006.0
Miami, FL	60	67	82	10.0	2.98	11.5	88.6	4657	13.5	47.3	22.4	3	1	1	60	861.4
Milwaukee, WI	30	20	69	8.8	3.26	11.1	85.4	2934	5.8	44.0	9.4	33	23	125	64	929.2
Minneapolis, MN	25	12	73	9.2	3.28	12.1	83.1	2095	2.0	51.9	9.8	20	11	26	58	857.6
Nashville, TN	45	40	80	8.3	3.32	10.1	70.3	2682	21.0	46.1	24.1	17	14	78	56	961.0
New Haven, CT	46	30	72	10.2	3.16	11.3	83.2	3327	8.8	45.3	12.2	4	3	8	58	923.2
New Orleans, LA	54	54	81	7.4	3.36	9.7	72.8	3172	31.4	45.5	24.2	20	17	1	62	1113.0
New York, NY	42	33	77	9.7	3.03	10.7	83.5	7462	11.3	48.7	12.4	41	26	108	58	994.6
Philadelphia, PA	42	32	76	9.1	3.32	10.5	87.5	6092	17.5	45.3	13.2	29	32	161	54	1015.0
Pittsburgh, PA	36	29	72	9.5	3.32	10.6	77.6	3437	8.1	45.5	13.8	45	59	263	56	991.3
Portland, OR	37	38	67	11.3	2.99	12.0	81.5	3508	3.6	50.3	13.5	56	21	44	72	894.0
Providence, RI	42	29	72	10.7	3.19	10.1	79.5	3387	2.2	38.8	15.7	6	4	18	55	938.5
Reading, PA	41	33	77	11.2	3.08	9.6	79.9	4843	2.7	38.6	14.1	11	11	89	54	946.2
Richmond, VA	44	39	78	8.2	3.32	11.0	79.9	3768	28.6	49.5	17.5	12	9	48	53	1026.0
Rochester, NY	32	25	72	10.9	3.21	11.1	82.5	4355	5.0	46.4	10.8	7	4	18	60	874.3
St. Louis, MO	34	32	79	9.3	3.23	9.7	76.8	5160	17.2	45.1	15.3	31	15	68	57	953.6
San Diego, CA	10	55	70	7.3	3.11	12.1	88.9	3033	5.9	51.0	14.0	144	66	20	51	839.7
San Francisco, CA	18	48	63	9.2	2.92	12.2	87.7	4253	13.7	51.2	12.0	311	171	86	71	911.7
San Jose, CA	13	49	68	7.0	3.36	12.2	90.7	2702	3.0	51.9	9.7	105	32	3	71	790.7
Seattle, WA	35	40	64	9.6	3.02	12.2	82.5	3626	5.7	54.3	10.1	20	7	20	72	899.3
Springfield, MA	45	28	74	10.6	3.21	11.1	82.6	1883	3.4	41.9	12.3	5	4	20	56	904.2
Syracuse, NY	38	24	72	9.8	3.34	11.4	78.0	4923	3.8	50.5	11.1	8	5	25	61	950.7
Toledo, OH	31	26	73	9.3	3.22	10.7	81.3	3249	9.5	43.9	13.6	11	2	25	59	972.5
Utica, NY	40	23	71	11.3	3.28	10.3	73.8	1671	2.5	47.4	13.5	5	2	11	60	912.2
Washington, DC	41	37	78	6.2	3.25	12.3	89.5	5308	25.9	59.7	10.3	65	28	102	52	967.8
Wichita, KS	28	32	81	7.0	3.27	12.1	81.0	3665	7.5	51.6	13.2	4	2	1	54	823.8
Wilmington, DE	45	33	76	7.7	3.39	11.3	82.2	3152	12.1	47.3	10.9	14	11	42	56	1004.0
Worcester, MA	45	24	70	11.8	3.25	11.1	79.8	3678	1.0	44.8	14.0	7	3	8	56	895.7
York, PA	42	33	76	9.7	3.22	9.0	76.2	9699	4.8	42.2	14.5	8	8	49	54	911.8
Youngstown, OH	38	28	72	8.9	3.48	10.7	79.8	3451	11.7	37.5	13.0	14	13	39	58	954.4

Source: G. C. McDonald and J. A. Ayers (1978). "Some applications of the 'Chernoff faces': a technique for graphically representing multivariate data." In P. C. C. Wang (Ed.), *Graphical Representation of Multivariate Data.* New York: Academic, pp. 183–19⁻, Table 1, pp. 186–187. Reprinted with permission.

where we abbreviate h_{ii} as h_i. The variance of the ith residual is

$$\text{var}(r_i) = \sigma^2(1 - h_i).$$

Therefore, a pattern is introduced into the residuals through the differing leverages of the data points. The variance of the residual r_i is small when the leverage is large, as occurs when x_i is far from $\bar{x}$. Thus when the model holds, points further from $\bar{x}$ tend to be fitted closer by the regression line.

The pattern in the residuals introduced by the x-values of the data points is removed when we divide each residual by a multiple of its standard deviation. Then the residuals have zero mean and equal variance. The ith *adjusted residual* is

$$r_{ai} = \frac{r_i}{\sqrt{1 - h_i}}.$$

We can estimate σ^2 by the appropriate multiple of the residual sum of squares,

$$s^2 = \frac{1}{n - 2} \sum_{i=1}^{n} r_i^2.$$

The ith *standardized residual* is

$$r_{si} = \frac{r_i}{s\sqrt{1 - h_i}}.$$

Alternatively, we can estimate the variance of the ith residual by using the residual variance from the regression line fitted to all but the ith point. We denote this residual variance by $s^2(i)$ (read "s-squared not-i"). The ith *studentized residual* is

$$r_i^* = \frac{r_i}{s(i)\sqrt{1 - h_i}}$$

and avoids difficulties when σ^2 at point i differs greatly from σ^2 at other points. Behnken and Draper (1972) discuss standardized residuals, and Belsley, Kuh, and Welsch (1980) discuss studentized residuals, both in the context of the multiple regression model.

In Figure 7-1 we display the plot of the residuals from the regression of mortality (y) on population density (x). Figure 7-2, the plot of leverage, shows why the variance of points to the right of the plot is smaller than the rest (the batch of x-values is skewed to the right). In Figure 7-3, the plot of standardized residuals, we adjust the residuals according to the leverages of the points. A careful comparison of Figure 7-1 and Figure 7-3 shows that the rightmost standardized residuals are relatively larger than the ordinary residuals.

Scaled Residuals

Often, when we have a batch of numbers, we first look at the *location* and, secondly, at the *scale* of the batch. To look at the *shape* of the batch, we subtract a summary statistic for location and divide by a summary statistic for scale. A batch of residuals is usually centered at zero. The resulting residuals, with location 0 and scale 1, are the *scaled residuals*. Choices for the scale of the batch of residuals include the standard deviation of the

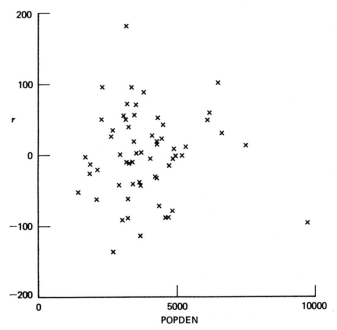

Figure 7-1. Linear regression residuals versus population density.

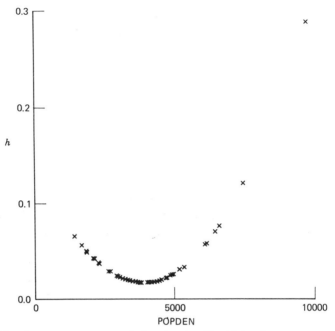

Figure 7-2. Leverage versus population density. The higher leverages correspond to population densities furthest from the mean.

batch, the median absolute deviation

$$\text{MAD} = \underset{i}{\text{med}} \left| r_i - \underset{j}{\text{med}} \, r_j \right|,$$

and the fourth-spread (or the interquartile range).

When we calculate standardized (or studentized) residuals, we divide by $s\sqrt{1 - h_i}$ (or $s(i)\sqrt{1 - h_i}$). The resulting batches have variance approximately 1; thus standardized and studentized residuals are similar to, but more thoroughly adjusted than, scaled residuals. Table 7-3 compares ordinary, scaled, standardized, and studentized residuals from the least-squares regression of mortality on population density. We scale the ordinary residuals using both the fourth-spread (after subtraction of the median) and the sample standard deviation of the batch.

We will seldom want to work with so many types of residuals simultaneously. One or two should suffice, and our choice will depend on how we expect to use them. Unmodified (or raw) residuals, r_i, are in the same scale as the original y-values, so that we can apply subject matter judgment to

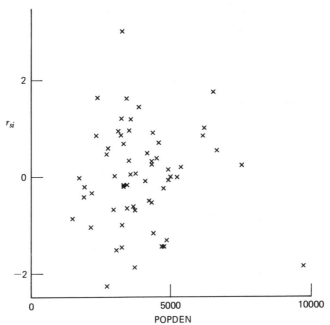

Figure 7-3. Standardized residuals versus population density.

them or simply calculate the $\hat{y}_i$. For most other purposes, we are likely to prefer the studentized residuals, r_i^*, because they are in another familiar scale, they can often be referred (individually) to a t-distribution, and they allow an unusual y-value at a high-leverage point to produce a large residual and thus call attention to itself. This last property of the r_i^* is very valuable because even the standardized residuals offer more limited protection against this shortcoming of least-squares regression, particularly in more complicated models. In the present example, as Table 7-3 shows, the various types of residuals all go along together. In absolute terms, no data point has extremely high leverage, and the ones with the highest leverage do not have unusual y-values.

The scale of the scaled residuals is unity when we use the same scale statistic to scale the original residuals and to measure the scale of the scaled residuals. This leads to simple numerical rules when we consider the shape of a batch of scaled residuals. We discuss a rule for outlier identification. A further example is to compare selected quantiles of the scaled residuals to corresponding quantiles for the Gaussian distribution, as mentioned in Section 2C.

TABLE 7-3. Residuals in five scales: unmodified, scaled by d_F, scaled by standard deviation, standardized, and studentized. SMSAs are ordered according to increasing r_i^*.

SMSA	r_i	Scaled d_F	Scaled s.d.	r_{si}	r_i^*
San Jose, CA	−136.30	−1.688	−2.253	−2.285	−2.375
Wichita, KS	−114.20	−1.412	−1.888	−1.904	−1.950
York, PA	−94.68	−1.169	−1.565	−1.856	−1.897
San Diego, CA	−91.07	−1.124	−1.506	−1.523	−1.541
Lancaster, PA	−88.79	−1.095	−1.468	−1.483	−1.499
Los Angeles, CA	−87.88	−1.084	−1.453	−1.469	−1.484
Miami, FL	−87.79	−1.083	−1.451	−1.467	−1.482
Denver, CO	−79.36	−0.978	−1.312	−1.328	−1.337
Rochester, NY	−71.52	−0.880	−1.182	−1.193	−1.198
Minneapolis, MN	−62.51	−0.768	−1.033	−1.056	−1.057
Grand Rapids, MI	−61.64	−0.757	−1.019	−1.029	−1.030
Dallas, TX	−52.60	−0.644	−0.870	−0.899	−0.897
Worcester, MA	−42.41	−0.517	−0.701	−0.707	−0.704
Hartford, CT	−41.91	−0.511	−0.693	−0.701	−0.698
Portland, OR	−40.81	−0.497	−0.675	−0.681	−0.678
Seattle, WA	−38.25	−0.465	−0.632	−0.638	−0.635
San Francisco, CA	−32.94	−0.399	−0.544	−0.549	−0.546
Canton, OH	−31.84	−0.385	−0.526	−0.531	−0.528
Fort Worth, TX	−25.57	−0.307	−0.423	−0.434	−0.431
Bridgeport, CT	−21.11	−0.252	−0.349	−0.356	−0.354
Boston, MA	−14.78	−0.172	−0.244	−0.247	−0.245
Springfield, MA	−13.57	−0.157	−0.224	−0.230	−0.228
Kansas City, MO	−13.66	−0.159	−0.226	−0.228	−0.226
Akron, OH	−11.30	−0.129	−0.187	−0.189	−0.187
New Haven, CT	−10.89	−0.124	−0.180	−0.182	−0.180
Dayton, OH	−5.86	−0.061	−0.097	−0.098	−0.097
Reading, PA	−5.16	−0.053	−0.085	−0.086	−0.086
Utica, NY	−3.11	−0.027	−0.051	−0.053	−0.052
Syracuse, NY	−1.58	−0.008	−0.026	−0.026	−0.026
St. Louis, MO	−1.38	−0.005	−0.023	−0.023	−0.023
Milwaukee, WI	−0.51	0.005	−0.008	−0.009	−0.008
Providence, RI	2.32	0.041	0.038	0.039	0.038
Flint, MI	2.89	0.048	0.048	0.048	0.048
Detroit, MI	7.98	0.111	0.132	0.134	0.132
Washington, DC	11.18	0.151	0.185	0.188	0.186
Columbus, OH	14.13	0.188	0.234	0.236	0.234
New York, NY	13.56	0.181	0.224	0.239	0.237
Allentown, PA	17.63	0.232	0.292	0.294	0.292
Youngstown, OH	18.91	0.248	0.313	0.315	0.313

the business functions? And again, the answer is a great deal. In addition to the contributions provided by the extension paradigm, this paradigm draws attention to, and provides opportunities for the study of and teaching about the strategic, operational, and administrative consequences of managing national and regional differences. Indeed, how international firms handle the issues of environmental diversity are harbingers of what will probably occur in most "open" societies as a consequence of increased economic, political, and social interdependence.

An "Emerging Interaction" Paradigm of International Business

As we and others have explicitly and implicitly noted (for example, Granovetter 1985; Toyne and Nigh 1997, ch. 1), the social sciences and the business functions are in a period of introspection and are seeking to reformulate the central questions guiding their respective inquiries. Because of this introspection, new paradigms, or new ways of conceptualizing the world, are being sought. In the IB literature, for example, this is most evident in the calls for multidisciplinary research (Dunning 1989), interdisciplinary research (Toyne 1989), the development of more comprehensive theories that require the "coming together" of two or more branches of the social sciences (Casson 1997), the development of a more comprehensive, multilevel, and perhaps holistic, view of IB's domain (Toyne 1997; Boddewyn 1997), and the suggestion that culture will assume a "center stage" role in the economic explanation of firm behavior (Dunning 1997).

In seeking a more comprehensive and holistic definition of IB, we proposed elsewhere a paradigm that views IB as a multilevel, hierarchical system of interactions between two or more socially-embedded business processes and their outcomes (for example, organizations) (Toyne and Nigh 1997, ch. 1). Additionally, these business processes are emergent. That is, they evolve intermittently over time because of the con-

sequences of informed hunch, intuition, and vision (learning coupled with inductive thinking) that can originate at *any* level in the hierarchical system (for example supragovernmental, governmental, industry, firm, or individual). Thus, the "trigger" for all business-related decisions and practices is a combination of three culturally influenced elements or factors: (1) self-interest; (2) the available pool of information, knowledge and experience; and (3) learning (which includes awareness, discernment, interpretation and judgment). In other words, a business process and the various phenomena that are the output of this process change because of the self-interested, bounded learning[13] that occurs as a consequence of interaction with other business processes and their outputs.

So, again, what can a scholarship of IB that is based on an emerging interaction paradigm add to what is already being done by the U.S.-bound business functions? The answer, of course, is a great deal. Unlike the extension and crossborder management paradigms, which are now seen as subparts of a more comprehensive conceptualization, the questions raised and examined as a consequence of this worldview add uniquely to the body of business knowledge. Importantly, it assumes that business is a process that varies from one social setting to another because of the cumulative effects of culture-bound learning. When two or more business processes interact, each will be affected, sometimes radically, sometimes subtlety, but always differently as a consequence of their culture-laden learning processes.

Whether business practice is converging or diverging is of interest only to the extent that such trends may shed light on the learning processes of various "cultures" and the consequences of their interaction with one another. Also, whether the U.S. leads the world in terms of business styles and techniques is only of interest to the extent that the national business leader at any particular point in time may have a role to play in helping other societies and their organizations identify possibly useful business practices (see, for example, Ozawa 1997). However, because of their culture-bound learning processes, emulation will never be exact.

The reason for briefly examining these three paradigms was to set forth in somewhat explicit terms the educational implications associated with the adoption of a particular paradigm. For example, by assuming that IB is merely an extension of domestic business the task of internationalizing the business program can be accomplished by adding an IB survey course and infusing existing courses with judiciously selected IB material. In contrast, when international business is viewed as both the extension of U.S. practice and the management of crossborder peculiarities, the internationalization task requires adding new material and new tools and skill requirements, probably in the form of new courses. Finally, when international business is viewed as involving the interaction of two or more socially embedded business processes, the internationalization effort requires a fundamental change in the entire business educational process that starts with the freshman year and continues through postgraduate work. A few examples of the educational implications of these three paradigms are presented in table 1.1.

WHAT IS AN ADEQUATE LEVEL OF
INTERNATIONAL BUSINESS LITERACY?

The final issue to be addressed here and explored more fully in the following papers and concluding chapter is: What is an adequate level of international business literacy? As IB scholars we would, of course, like to assert that in today's world, an erudite, intellectually learned individual requires more than a nodding acquaintance with the IB literature and its collective and pervasive influence on what a person does as business scholar or practitioner. But such an expectation is frankly utopian. It may even be politically unwise at this time to demand a comprehensive and complete commitment to programmatic reform, since

such a demand would probably result in immediate rejection by most faculties and administrators of business programs. Either the need for internationalization is not yet sufficiently "obvious" to most members of the business academy for a total commitment to be made, or internationalized business programs, particularly at the undergraduate level, have not yet appeared that "captures" the academy's imagination.[14]

A more realistic expectation, given current circumstances, is that most schools should probably seek some "satisficing" level of IB literacy (Simon 1957). Thus, a more pragmatic approach, perhaps, is to ask what minimal body of knowledge, experience, skills and tools can be considered essential for a satisficing level of IB literacy.[15] However, when developing this minimal requirement, the significant educational differences between practitioner programs (for example, B.B.A. and M.B.A.) and scholar programs (for example, Ph.D. programs) need to be explicitly recognized (Toyne 1995).

As shown in figure 1.2, the internationalization of a business school's curriculum can be considered in terms of four levels (or building blocks) of breadth, depth, and focus: global awareness; business contextual awareness; function-specific knowledge; and IB entry level competency.[16] These four educational blocks are briefly explored next.

Global Awareness

As already stressed, what we learn depends on what we have learned, why we have learned it, and how we have learned it. That is, learning is influenced by cultural, historical, situational, and philosophical factors that vary from one locale to another, and from one "culture" to another. Thus, international education in its most general and fundamental sense is both liberal and liberating. Exposure to international phenomena and crosscultural issues opens the minds of students to an entirely different world. Furthermore, by its very nature, international education is inherently interdisciplinary, thus providing at least one reason

Figure 1.2 Building Blocks of an Internationalized Business Education

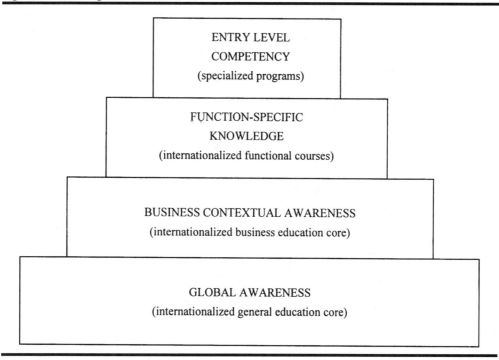

ENTRY LEVEL
COMPETENCY
(specialized programs)

FUNCTION-SPECIFIC
KNOWLEDGE
(internationalized functional courses)

BUSINESS CONTEXTUAL AWARENESS
(internationalized business education core)

GLOBAL AWARENESS
(internationalized general education core)

and one means for overcoming the fragmentation (departmentalization) of students' learning habits.

Thus, the core objectives of any anticipated programmatic reform, such as the internationalization of the business program, should include the goal of providing all business students with a liberal education foundation that is internationally liberating (for example, comparative political systems, comparative legal systems, cultural anthropology).

Business students need a broad perspective on the history and cultures of the several diverse regions of the world as a basis for better understanding the economic, political, social, and technological forces fashioning our interdependent, yet still divided world. Thus, the business school's ability to achieve its internationalization objectives and goals depends, in part, on the successful internationalization of the university core curriculum, generally taken during a student's first two years of college.[17] Moreover, this liberalized and liberating educational also needs to be made a part of more advanced business degrees.

Also, the ability to lead successful and fruitful lives in the internationalized business world of today and tomorrow requires an ability to think well. But as Michael Oakeshott succinctly noted (Fuller 1989, 60–66):

> learning to think is not merely learning how to judge, to interpret and to use information, it is learning to recognize and enjoy the intellectual virtues. How does a pupil learn disinterested curiosity, patience, intellectual honesty, exactness, industry, concentration and doubt? How does he acquire a sensibility to small differences and the ability to recognize intellectual elegance? How does he come to inherit the disposition to submit to refutation? How does he not merely learn the love of truth and justice, but learn it in such a way as to escape the reproach of fanaticism?

This liberal and liberating way of thinking, the body of internationalized knowledge needed, and the academic standards to be sought must be cooperatively established and agreed upon by the various disciplines contributing to a student's education.

Business Contextual Awareness

The implications of the regionalization, even globalization, of business go well beyond the economic and the political; they include the cultural and the social. For the first time in its history, partly because of the North American Free Trade Agreement (NAFTA), the U.S. business education community is faced with the need to develop future managers and business people who have an in-depth appreciation of the international environment of business upon graduation. A parochial business education that seeks merely to provide students with international awareness (such as, how the national economy and its people are connected to the rest of the world) is no longer acceptable or justifiable. In a world of increasing regionalization and globalization, business students in Africa, Asia, the Americas, and Europe need (1) a global appreciation of the vital links between their economy and people and the rest of the world, what causes international business, and how it affects domestic business, economic growth and standards of living, and (2) a regional appreciation of how the international dimension can affect their functional specialty (for example, marketing, finance, human resources, or management). This particular level of awareness is the responsibility of the general business core and can be achieved through: (1) the infusion of appropriate IB subject matter in required courses of the pre-business, professional business core, and major subjects (for example, economics, finance, law, marketing, accounting, and informational systems and labor and management practices); (2) the addition of a well designed introductory, survey course on "International Business"; and, possibly, (3) the requirement that all students must have participated in a study-abroad, or an overseas exchange program, at some point during their academic training.

Function-Specific Knowledge

The globalization of business is a reality that must be faced by the business education community. For example, although most students will not become IB practitioners, all non-IB students should have at least a working, comparative knowledge of the business environments and practices of the members of the regional arrangements (for example, EU, NAFTA), and their implications for domestic competition. This level of knowledge can be gained with the infusion of advanced functional courses with appropriate "regional" material.

IB Entry Level Competency

Academic offerings at this level of global business expertise are typically provided through major concentrations at the undergraduate and master's level programs that provide a focus on a particular region and include language competency at an advanced level, a foreign-based internship or study abroad option, and an area studies component as well as function-oriented global business subjects. It is at this level that the fundamentals of international and transnational functional expertise are developed.

The achievement of these levels is dependent on the acquisition of appropriate resources, the development of appropriate faculty competencies, the development of appropriate attitudes within the school, and the development of intra- and interinstitutional linkages and educational support. Thus, most business schools responding to the globalization challenge have found that they must not only establish objectives and measurable goals for their ongoing activities, they must also undertake new activities. Moreover, those schools that are energetic and proactive seek to individualize their responses and differentiate themselves from other schools' responses in ways that are original and unique.

THE STRUCTURE OF THE BOOK

Although involvement in international business is both educationally and experientially liberating, until quite recently, both IB education and research have been peculiarly national in focus and purpose. Moreover, the philosophical underpinnings of IB education, its content, and even its pedagogy continue to be strongly influenced by both institutional and faculty characteristics. Thus, it was decided to ask nineteen major contributors to IB education—administrators and faculty members—to comment on the challenges and obstacles to a liberal and liberating educational experience (such as, IB education in its broadest sense).

The papers presented in chapter 2 provide an overview to the educational efforts that are occurring in various parts of the world, and sets the stage for a discussion of the merits of alternative approaches to IB education that have developed in different countries. In the first topic paper, Robert R. Locke of the University of Hawaii and Reading University, Great Britain, examines the historical foundations of business education in Western Europe, the United States, and Japan. In the second paper, Leonid I. Evenko, dean of the Moscow Graduate School of International Business, explores the business education challenges confronting the Russian Federation and the other independent states of the territory of the former USSR. In the third paper, William R. Folks of the University of South Carolina reviews the international business educational efforts at his school. The three topic papers are discussed by Robert Grosse, director of research at Thunderbird, and Richard Moxon of the University of Washington from somewhat different perspectives.

The three topic papers included in chapter 3 address the issue of whether IB inquiry should be treated as a peripheral activity artificially fragmented along functional lines, or whether it is time to recognize it as a core activity of an emerging interdiscipline. In the lead paper, Robert G. Hawkins, dean of the Georgia Institute of Technology, suggests that there is no ideal organizational structure and points out the strengths and weaknesses of various alternatives. In the second

paper, Jeffrey Arpan of the University of South Carolina argues that a functional structure is dysfunctional for IB research and discusses the benefits of an IB department. Duane Kujawa, University of Miami, argues in the third topic paper that we need to look more to changes in organizational processes and culture, rather than formal structure, for the solutions to the challenges of internationalization. The commentary by John D. Daniels, University of Richmond, analyzes the sources of the gaps between what IB research should be and what it is. In his commentary, Edwin L. Miller of the University of Michigan emphasizes the important role that academic and faculty leadership plays in instilling an institutional mind set that promotes internationalization.

The three topic papers included in chapter 4 are by David H. Blake, dean at University of California–Irvine, William R. Folks, and Lee Radebaugh, director of the School of Accountancy and Information Systems at Brigham Young University. They examine the benefits, costs, and limitations associated with the various institutional structures currently employed and the central issues that need to be addressed when attempting to introduce the IB dimension at all levels of learning (for example, undergraduate, graduate, doctoral). The commentaries are by Yair Aharoni, rector of College of Management in Israel, and by Bernard M. Wolf of York University, Canada, provide a welcome non-U.S. perspective to these issues.

Chapter 5 includes three papers written by H. David Bess, formerly the dean at the University of Hawaii, Edwin L. Miller, and Duane Kujawa for discussion during a panel session chaired by William R. Folks. A fourth member of the panel was Robert G. Hawkins. The papers addressed the question of how should the IB dimension be institutionally housed in order to simultaneously satisfy its theory-building, research, and educational responsibilities. An introduction to the discussion, and a summary of the discussion that took place during the panel session was written by Folks.

Chapter 6 also includes four papers written by J. Frederick Truitt of Willamette University, Franklin R. Root of the University of Pennsylvania, Paul W. Beamish of the University of Western Ontario, Canada, and John D. Daniels for discussion at a panel session chaired by David A. Ricks, now at Thunderbird. The question addressed by the four panelists was the following: What was the most effective and efficient teaching and learning forms for insuring that the IB dimension is adequately and properly covered in everything from formal degree programs to small, in-house executive training programs?

The final chapter, chapter 7, summarizes the educational and research implications of the various papers and commentaries and suggests future educational and research initiatives for this fledgling field.

Except for chapter 5, we provide an introduction to each chapter that gives our views as to why the topic(s) were selected; what we believe each paper's highlights to be; what we believe the discussants have to say; and finally, to provide our comments and opinions about the significant issues developed in each section. It needs to be clearly understood that neither the reader, the paper authors, nor the discussants may hold similar views.

The references for each paper are combined and listed at the end of each section. This was done to reduce unnecessary duplication of individual citations.

NOTES

1. More complete discussions of the roots of IB education are presented in Fayerweather 1986 and Toyne and Nigh 1997, ch. 1.

2. High school education, wherever taught, tends to reinforce, even complete, a parochial, nationalistic socialization process that starts in the cradle. See, for example, Nigh 1997 for the role of education in building nations and national identity.

3. The various social sciences are also undergoing transformation. For example, economics and sociology are both seeking to redefine their domains of inquiry and the questions they

seek to answer (see, for example, Toyne and Nigh 1997, ch. 1). The implications of this spreading quest for a more holistic and inclusive understanding of human endeavor, of course, is that the current structuring of business schools along functional lines probably will give way to new arrangements. Rather than departments of management, marketing, and so on, perhaps we may have to seek more generalizable (universal) departments such as departments of technology, cultural interaction, information acquisition and assimilation, communication, and so on).

4. Perhaps a distinction needs to be made between training and educating. Training is viewed here as preparing someone for a known "present" or anticipated future (that is, emphasis is placed on acquiring skills and the regurgitation of "temporal facts.") On the other hand, educating a person is to prepare them for an unknown, uncertain future by providing them with a fundamental, liberal, broadly-based understanding of human endeavor, the ability to think critically, and the ability to modify their learned behavior when circumstances change.

5. For example, San Diego State University has joined with another U.S. institution and two Mexican institutions to offer a joint undergraduate degree that requires students to spend their first two years in their home country and their last two years in a host country. Moreover, the students are taught in the languages of the home and host countries.

6. This section draws heavily from Toyne 1995.

7. Although the types of research and teaching are identical at most educational institutions, and thus a two-dimensional matrix could be used in many situations, some institutions may use practitioners and visiting professors to augment their tenure-track faculties. Such business schools could emphasis unifunctional research while offering undergraduate and graduate programs based on integration teaching principles.

8. Essentially, the business functions are interdisciplinary fields (for example, marketing is a synthesis of economics, sociology, and psychology). What is being suggested here and hinted at elsewhere (Dunning 1988 and Toyne 1989) is the redefinition of business functions along lines that more closely approximate what is now occurring within business, and what is being called for by those who advocate a more holistic understanding of business. For a fuller discussion of this point see Toyne and Nigh 1997, ch. 11.

9. Partly because of staffing constraints and partly because of faculty biases and training, the research and teaching approaches adopted and emphasized at particular institutions are often identical.

10. The paradigmic roots of IB inquiry are discussed more fully in Toyne and Nigh 1997, ch. 1.

11. This is based on the widely held belief that business is an economically driven activity that can be examined apart from the social and political dimensions of society. That is, the embeddedness of business in society can be ignored (that is, liberal economics).

12. This last assumption is increasingly questioned as Japanese and European scholars develop a more rigorous understanding of their advancing business practices.

13. Learning is always bounded in that a person or organization has limited information, knowledge, and experience to draw on at any particular point in time.

14. The Master of International Business Studies (M.I.B.S.) program at the University of South Carolina is an example of a graduate-level IB educational model that has been unabashedly copied by other institutions. However, its philosophical underpinnings have not had too much impact on U.S.-bound business degrees (for example, B.B.A. and M.B.A.).

15. Such an approach could open Pandora's box. Since most IB scholars are actually unifunctional scholars who just happen to be interested in international topics, it might be difficult, if not impossible, to get general agreement on what constitutes a satisficing level of IB literacy.

16. A three-level IB literacy approach to the internationalization of business programs was first developed and presented by Jeffrey S. Arpan at John Carroll University and is now widely used (see, for example, Arpan 1993 and Toyne 1992). However, the Arpan model does not draw attention to the influence that prior education has on future learning. For example, it is mute on the general educational core of an undergraduate education. The four-level model described here was developed by Dr. Culpepper at the University of Arkansas.

17. The point being made here is that the student's ability to fully understand the implications of an internationalized business curriculum is amplified when the student has received an internationalized basic university education in the humanities and social sciences. Also, see note 2.

REFERENCES

Alutto, Joseph A. 1993. Whither doctoral business education? An exploration of program models. *Selections* Spring: 37–43.

Arpan, Jeffrey S. 1993. Curricular and administrative considerations—the Cheshire cat parable. *Internationalizing Business Education: Meeting the Challenge.* Ed. S. Tamer Cavusgil. East Lansing: Michigan State University Press.

Bartlett, Christopher A., and S. Ghoshal. 1992. *Transnational Management: Text, Cases, and Readings in Cross-Border Management.* Homewood, Ill.: Richard D. Irwin.

Bartlett, Christopher A., Y. Doz, and G. Hedlund. 1990. *Managing the Global Firm.* New York: Routledge.

Boyacigiller, Nakiye, and N. J. Adler. 1991. The parochial dinosaur: Organizational science in a global context. *The Academy of Management Review* 16(2): 262–90.

Boyer, Ernest L. 1990. *Scholarship Reconsidered: Priorities of the Professoriate,* Princeton, N.J.: The Carnegie Foundation for the Advancement of Teaching.

Casson, Mark. 1997. Economic theories of international business: A research agenda. *International Business: An Emerging Vision.* Ed. Brian Toyne and D. Nigh. Columbia: University of South Carolina Press: 181–93.

Dunning, John. 1997. Micro and macro organizational aspects of MNEs and MNE activity. *International Business: An Emerging Vision.* Ed. Brian Toyne and D. Nigh. Columbia: University of South Carolina Press: 194–204.

———. 1989. The study of international business: A plea for a more interdisciplinary, approach. *Journal of International Business Studies* 20(3): 411–36.

Fayerweather, John. 1986. A history of the Academy of international Business from infancy to, maturity: The first twenty-five years. *Essays in International Business.* Columbia: University of South Carolina Press, no. 6 (November): 1–63.

Feldman, Daniel. 1997. When does demonstrating a difference make a difference? An organizational behavior perspective on international management. *International Business: An Emerging Vision.* Ed. Brian

Toyne and D. Nigh. Columbia: University of South Carolina Press: 446–52.

Fuller, Timothy, ed. 1989. *The Voice of Liberal Learning: Michael Oakeshott on Education.* New Haven: Yale University Press.

Graduate Management Admission Council (GMAC). 1990. *Leadership for a Changing World: The Future Role of Graduate Management Education.* Los Angeles.

Hedlund, Gunnar. 1986. The hypermodern MNC—a heterarchy? *Human Resource Management* 25 (Spring), 1.

Hogner, Robert H. 1997. International business studies: Science, society, and reality. *International Business: An Emerging Vision.* Ed. Brian Toyne and D. Nigh. Columbia: University of South Carolina Press: 115–30.

Nehrt, Lee C. 1993. Business School Curriculum and Faculty: Historical Perspectives and Future Imperatives. *Internationalizing Business Education: Meeting the Challenge.* Ed. S. Tamer Cavusgil. East Lansing: Michigan State University Press.

Nigh, Douglas. 1997. Who's on First?: Nation-states, national identity, and multinational corporations. *International Business: An Emerging Vision.* Ed. Brian Toyne and D. Nigh. Columbia: University of South Carolina Press: 257–68.

O'Reilly, Brian. 1994. What's killing the business deans of America? *Fortune,* 8 August, 64–68.

Ozawa, Terutomo. 1997. Economic theories of international business: Images, economies of concatenation, and animal spirits, dependency vs. emulation paradigms. *International Business: An Emerging Vision.* Ed. Brian Toyne and D. Nigh. Columbia: University of South Carolina Press: 204–20.

Porter, Lyman W., and L. E. McKibbon. 1988. *Management Education and Development: Drift or Thrust into the 21st Century?* New York: McGraw-Hill.

Simon, Herbert A. 1957. *Administrative Behavior,* 2d ed. New York: The Free Press.

Toyne, Brian. 1997. International Business Inquiry: Does It Warrant a Separate Domain? *International Business: An Emerging Vision.* Ed. Brian Toyne and D. Nigh. Columbia: University of South Carolina Press: 62–76.

———. 1995. Internationalizing business scholarship: The essentials. *Internationalizing Doctoral Programs.* Ed. S. Tamer Cavusgil. East Lansing, MI: Michigan State University Press.

———. 1993. Internationalizing the business administration faculty is no easy task. *Internationalizing Business Education: Meeting the Challenge.* Ed. S. Tamer Cavusgil. East Lansing: Michigan State University Press.

———. 1992. Internationalizing business education. *Business & Economic Review* 38(2): 23–27.

———. 1989. International exchange: A foundation for theory building in international business. *Journal of International Business Studies* 20(1): 1–17.

———, and D. Nigh, eds. 1997. *International Business: An Emerging Vision.* Columbia: University of South Carolina Press.

———, and Zaida L. Martinez. 1998. Enterprise Behavior within an International Exchange Context. Paper presented at the annual meeting of the Academy of International Business, San Diego, October.

Wartick, Steven L. 1997. From "Business and Society" to "Businesses in Societies": Comments on the papers by Hogner, Freeman, and Wood and Pasquero. *International Business: An Emerging Vision.* Ed. Brian Toyne and D. Nigh. Columbia: University of South Carolina Press: 165–171.

Vernon, Raymond. 1964. Comments. *Education in International Business.* Ed. Stefan H. Robock and L. C. Nehrt. Bloomington: Graduate School of Business, Indiana University Press.

2

Historical Views on International Business Education

EDITORS' COMMENTS

Until recently, international business (IB) education has been peculiarly "national" in philosophy, orientation, and content. As Robert R. Locke highlights in the first topic paper, there is no widely held consensus at the national level regarding the role played by business education, let alone international business education. Moreover, as described in chapter 1, there is no wide agreement at the institutional level regarding the role played by IB education within the presumably more general business education process. The role played by IB education depends on the IB paradigm adopted by a particular business school's faculty and administration to define the relationship between international business and (national) business, and on the research interests of faculty members. At both levels, national and institutional, the idiosyncratic demand for people trained in business and administration that originates from a combination of cultural, historical, political, economic and social developments has played an important, even decisive role, in the evolution of both business and IB education.

In *Principles of Political Economy,* John Stuart Mill (1965 [1848]) advanced the idea that human progress can be helped by entering into discussion with people who are dissimilar from ourselves in both thought and action. We took this to mean

21

that we can often learn by reflecting on what others assume about us, and how they interpret what we do and how we do it. Thus, we thought it appropriate to ask the authors of the first three topic papers presented in this book to specifically address differences found at both the national and institutional levels.

In the first paper, Robert R. Locke, a historian, presents us with a historical and comparative review of the development of IB education in Western Europe, the United States, and Japan. More specifically, he contrasts the differences to be found in IB education in Western Europe (such as, France, Germany, and Britain), the United States, and Japan, and traces these differences to their treatment of business education in general.

Locke suggests that U.S. business education and, as a consequence, IB education are based on an elite meritocracy philosophy that is strongly influenced by the business-military interaction, particularly after World War II. This business-military heritage, he believes, encouraged the application of scientific methods to management. The U.S. culturally-defined inclination to use hierarchical structures with their inherent bias for specialization also led to a desire for highly specialized professions.

While the U.K. copied the U.S. business education model more than the other countries studied, it did so relatively late and with little resolve. According to Locke, the reason for this late development was due in part to the nature of British elitism; an elitism that is based on a belief that success has more to do with character and intelligence than knowledge. As a consequence of this form of elitism, business education has remained narrowly based since the learning of skills ("training") was intended for lower classes and therefore inferior. This tendency has been reinforced by a commerce and manufacturing elite believe that universities are useless. Thus, since IB education is a result of management education, it too was delayed, and has not received strong support.

France, with a tradition of business and technical education, also copied the U.S. model, but remains weak in research. The reason for this research weakness, according to Locke, is that within the major universities most students are taught by "visiting" faculty who are specialists within their areas of expertise. Thus, there is little in-house incentive to contribute to the development of management and international business as scientific disciplines.

In contrast to France and the U.K., Germany and Japan advanced business education is seen more as a company responsibility. Thus, little value is placed on generic skills.

In conclusion, Locke raises some interesting points that need further development and discussion. For example, as he sees it, the Japanese have the best management in the world and yet have the worst system of academic management education, German firms have always ignored academic institutions when training managers and yet have the most cosmopolitan managers in the world, and U.S. business schools became the "best" in the world while U.S. business lost its international reputation. These differences, he concludes, cannot be explained by a framework that is based on what is offered in U.S. business schools.

In the second topic paper, Leonid I. Evenko also takes a historical, national approach by describing and examining the system of management education in Russia since 1917 under a Marxist regime, and its subsequent, yet to be defined path under a free market regime. Thus, this paper does not focus specifically on IB education, but rather on the Russian system of management education. Importantly, Evenko's paper supports the hypothesis that the thrust and content of management education changes as societies enter into new stages of political and economic development. Evenko echoes educators in other former socialist countries (for example, Hungary) by noting that current training does not prepare leaders knowledgeable in international trade mechanics and foreign business practices. He further argues that these deficiencies must be reduced, if not eliminated.

Finally, in the third topic paper, William R. Folks highlights the development of the Masters of International Business Studies (M.I.B.S.) program at the University of South Carolina, and its relationship to other international activities of the university, and sets forth a number of recommendations based on a review of this particular institution's highly successful experience. As such, the paper is an admirable and thorough case study of a highly successful approach to the internationalization mandate, an approach that assumes IB as a separate and unique discipline. Folks also provides sound reasons for including foreign language skills and international experience through exchange programs and internships in IB programs.

At the same time, Folks's discussion fails to recognize that alternative courses of action have been undertaken (for example, the Hawaiian approach pursued by Bess), and fails to examine the reasons why one approach may be more suitable than another. That is, he ignores the diversity of societal needs for business education highlighted in chapter 1, and made more concrete by Locke and Evenko in their articles. Although he takes note of externalities—target employers, geographic focus, and type of program desired—he appears to assume a monolithic, single-purpose educational mandate, and a single institutional setting, and thus fails to examine the relative merits of the various approaches to internationalization that have been undertaken within the last two decades at institutions around the world (see, for example, *The Economist*, 1991).

The commentaries by Robert Grosse and Richard Moxon complement and advance the issues and challenges raised in the topic papers and addressed in subsequent chapters, since they chose to elaborate on the topic authors' discourses using their considerable experience in teaching and administrative capacities at well known U.S. and foreign schools of business with strong IB programs.

Robert Grosse advances the argument that there is a need to provide a historical description of how "international business became a major field of study in U.S. business schools."[1] For example, while the three topic authors address specific and interesting aspects concerned with IB education, they failed to examine the origins of U.S. international business education. Nor do they look into the future. Grosse further observes that while Evenko presents an interesting description of an educational system long closed to external observers, he does not address the pressing question of how Russia can best provide IB education. Finally, in commenting on Folks's paper, Grosse laments the absence of a history of IB education in the United States to balance the views of Europe and Russia in the other papers. At the same time, however, he considered the decision-making model presented and examined by Folks as an important starting point for a discussion of what would be appropriate IB education in the future.

Grosse concludes by making the following three observations. First, today universities are paying much more attention to IB as an area of study, and that there does not appear to a convergence of IB programs around the world. Second, U.S. educators would benefit quite significantly from discussion with their German and Japanese counterparts regarding the management education most likely to serve the needs of future managers. Also, German and Japanese educators would benefit from discussion with U.S. educators regarding the role of research. Finally, the alternative approaches to IB education that have been developed in different countries need to be examined.

In the final commentary, Richard Moxon notes that each paper should be a reminder to the reader that IB education is localized since it is embedded in an institutional and societal context that determine the kinds of IB programs that are both needed and possible. While Moxon notes that Locke's paper does a laudable job of focusing the reader's attention on the issues of educational relevance and effectiveness by highlighting the need for managers to be adaptable and to continue acquiring knowledge, he criticizes Locke's obvious negative bias towards business education and business educators as unhelpful and erroneous. For example, Moxon notes that IB educators did not exclude the liberal arts and humanities from the curriculum as stated by Locke. Indeed, IB educators have typically been defenders of the importance of the liberal arts, particularly languages. This is because they are well aware of the incisive role played by cultural anthropology, history, and similar disciplines. At the same time, Moxon suggests that Locke does an admirable job of focusing our attention on the issue of future educational relevance and effectiveness.

Concerning the Evenko paper, the most interesting point for Moxon was the reminder that within the Russian system there has always been a tension between ideology and pragmatism, and that in the 1920s and 1930s management education was much more pragmatic than in later years. Another point of interest was Evenko's suggestion that the new Russian "international business leaders" are woefully unprepared, especially in the areas of language, international trade mechanics, and foreign business practices.

While Moxon found the Folks paper to be of interest and value, he added value by critically questioning the underpinning assumptions of the model developed by Folks. For example, Moxon suggests that it would be useful to think about the client needs that lie behind the products listed in table 2.9, and to include students as clients in table 2.10. It would also be useful to provide a methodology for choosing among client segments, and for determining how to compete in each segment that is chosen. Notwithstanding these suggestions, Moxon believes the South Carolina model has provided other schools with several benefits. These include proving that a market exists for M.B.A.-trained international specialists, that foreign language skills

are needed, and that an international experience is beneficial.

Moxon concludes by briefly reviewing the experience at the University of Washington. He remarks on the growth in the number of international courses, the institution's commitment to provide all its business students with an understanding of the global environment of business, and a global vision of business strategy. He also highlighted the fact that their faculty are encouraged to extend their research internationally, and become involved in designing new international courses and programs. Finally, unlike the South Carolina approach that included the creation of an International Business Area, the separate IB group at the University of Washington was eliminated and replaced with an internationally-oriented faculty housed in each of the school's departments.

The discussions presented in the three topic papers and two commentaries provide support for the arguments advanced in chapter 1 and the chapters that follow. For example, as a group, the five papers agree that local (national) environment has a pervasive and critical influence on defining business educational needs, and thus in response, business programs. At the same time there is some disagreement concerning the institutional structure to be used for administering IB business programs. There was also some disagreement concerning the societal role of IB education.

The debate over what constitutes a correct institutional response (structure) is picked up in subsequent chapters. We believe that these differing opinions reflects both institutional and personal philosophical differences concerning business education and the role played by IB (see, for example, Hawkins in chapter 3 and Blake in chapter 4). To some extent it also reflects administrative heritage (such as, the sociocultural history of institutions and groups) that have a pervasive influence on the future initiatives of organizations.

The differing perceptions of the authors regarding the role of business education, and thus IB education, is an interesting subject because it brings into sharp focus the issue raised by Folks and challenged by Moxon. Evenko, for example, appears to view business education as a tool of society. This would suggest that business programs are to be designed, implemented, and maintained for the economic benefit of the state; a thought not much different than that advocated by Robert Reich (1991) and Lester Thurow (1992). Folks, as expressed by his decision-making model, does not appear to recognize a societal need, but only a corporate need. This, of course, can be attributed to the primary reason for the M.I.B.S. program at the University of South Carolina. Moxon, on the other hand, suggests that the individual in the form of the student has a stake in the educational process and its var-ious outcomes. Finally, Locke's important point that business education, and thus IB education, are cultural expressions raise interesting questions concerning the viability of designing business programs that are based on the experiences of other countries. Suffice it to say here, the issues of external stakeholders and the ability and desirability to transfer educational experiences across countries will be addressed in the concluding chapter when responding to the challenges confronting IB education.

NOTES

1. The history of IB inquiry and education has been addressed by Fayerweather (1986), Nehrt (1990), and also by Toyne and Nigh (1997, ch. 1).

■ ■ ■

International Management Education in Western Europe, the United States, and Japan: A Historian's View
Robert R. Locke

I have been asked to present "a historical and comparative examination of the institutional accommodations of the international dimension in business administration education and research" in Western Europe, the United States, and Japan. Although, as a historian, I am an outsider, I accepted this invitation because I firmly believe that the insiders, the people now engaged in international business education, having banished the liberal arts and humanities from their thoughts and curricula in the 1960s, are rather ignorant even about the history of their own disciplines and that a knowledge of history could help the insiders gain perspective about the usefulness of what they are doing. I shall carry out this self-appointed mission in three ways: determine the extent to which International Business Education (IBE) is studied in the countries considered, show how the differences in the state of IBE studies in these countries is a result of historical processes, and explain why an IBE yardstick, like the one adopted at this conference, is of little use when trying to measure the comparative effectiveness of IBE.

THE STUDY OF IBE

To present "a historical and comparative examination of the institutional accommodation of the international dimension in business administration education" seems straightforward enough. First, the number of courses and study programs that have been developed in universities and other institutions of higher education could be counted in various countries. This would include looking at the length of time such studies have been done, comparing the levels of study (undergraduate, graduate, graduate research) as well as the institutionalization of research programs, and measuring the degree of intensity of IBE international exchange programs in research and teaching. Second, the content of IBE could be evaluated, comparing in particular crosscultural educational developments (for example, languages and regional studies) to IBE in the more traditional, largely culture-free, crossborder transactional analysis subjects, for example, marketing, finance, and accounting. The countries with the most courses or programs, with the broadest spectrum of offerings at all levels of higher education, and with the most developed research and teaching would be the winners in the IBE contest.

A recent survey—*International Business Education in European Universities in 1990* —follows this approach (Luostarinen and Pulkkinen 1991, C-116). It shows that Europe compares unfavorably with the United States in the number of institutions offering at least one course in International Business. More U.S. institutions had one or more IB courses in 1980 (80% of respondents) than European institutions had even a decade later (73%). And, of course, the U.S. institutions led the Europeans in terms of the longevity of these offerings.

Schools in the United States, moreover, probably lead schools in Europe in the richness of their crosscultural study programs. This would certainly be true in terms of longevity. The American Graduate School of International Management (Thunderbird), which made foreign languages, history, comparative courses in the humanities, foreign internships, and exchange studies integral to its program, started thirty years ago.[1] Other regional study programs have been around for at least a decade. It is true that, if slow starters, very recently Europeans have made great strides in developing intra-European faculty and student exchanges. Indeed, although U.S. schools have expanded European regional study programs, with the faculty and student exchange agreements, Europe probably leads the United States in this regard. But if viewed from a global instead of a North American perspective, the U.S. regional study programs seem much less parochial than the European. Some U.S. programs are regionally focused, for example Hawaii's Pacific Asian Management Institute (PAMI), Texas (Austin)'s preoccupation with Mexico, the joint M.B.A.-Asian studies degrees offered at Cornell, U.C. Berkeley, Michigan, and Wharton. But others assume a global outlook that is less frequent in Europe, such as the Joseph H. Lauder Institute of Management and International Studies, founded at Wharton in 1983, which grants joint M.B.A.-M.A. degrees to a highly linguistically qualified, internationally recruited group of students. The 1988 Trade Act which authorized the U.S. Department of Education to found sixteen Centers for International Business Education and Research (CIBERs) has added government support to this cultural "globalization," for the sixteen centers that have been created in the past few years—all with one exception in business schools—stress foreign languages, regional studies, foreign business internships, and

Table 2.1

	Europe 1990	USA 1980	USA 1986
Number of institutions responding to survey	231	323	384
Number of institutions with at least one IB course	168	262	382

Luostarinen and Pulkkinen, page 116

study exchange programs on a global scale. European Economic Community programs are, if not exclusively so, much more regionally restricted.

Still, this United States–Europe comparison is not particularly instructive because, as Luostarinen and Pulkkinen show, the disparities in IBE among European countries are greater than that between Europe and the United States. Fifty-seven of fifty-eight U.K. and twenty-six of twenty-seven French institutions responding to their survey offered International Business, while twenty-one out of thirty-six in Germany, nine out of ten in Italy, and eleven out of thirteen in Spain did. Moreover, the gap between the United Kingdom and other major countries increases when a distinction is made between IB programs as opposed to courses (see table 2.2). Luostarinen and Pulkkinen report that the same sort of

disparities emerge when levels of IBE are compared in these countries (see table 2.3). The disparities are reflected, too, in the types and numbers of exchange agreements concluded with foreign institutions (table 2.4) and in the organization of IB research in different countries (table 2.5). Germany not only has less IBE but what it has is concentrated at the graduate level. Undergraduate and M.B.A. IBE in Germany is practically nonexistent.

Britain, then, using this data, emerges as the winner. This is true in almost every category of analysis. But we must not think that the United Kingdom is just another United States. Business education in the United Kingdom, even in an area where they borrowed very heavily from the U.S.—M.B.A. programs, is most un-American. This is true in terms of longevity of business education. The first British business schools

Table 2.2 International Business Programs

	Operating	Planned
United Kingdom	25	18
Germany	3	6
France	16	0
Italy	2	0
Spain	4	1

Luostarinen and Pulkkinen, page 54

Table 2.3

Country	Undergraduate	Graduate	M.B.A.	Doctorate	Total
France	1	26	10	1	38
Germany	0	21	3	0	24
U.K.	50	15	19	3	87

Luostarinen and Pulkkinen, page 54

Table 2.4

Country		Faculty	
	Student Exchange	Teaching	Research
U.K.	38	32	14
Germany	8	2	1
France	42	60	37
Italy	2	1	0
Spain	21	11	2

Luostarinen and Pulkkinen, page 88

Table 2.5

Country	IB Research Number of Institutions	Institutionally not not individually organized
U.K.	36	3
Germany	19	1
France	19	4
Italy	7	0
Spain	4	0

Luostarinen and Pulkkinen, page 64

date from the mid-1960s, hence no academic IBE took place before that time. And it is true in terms of size. In 1990 a premium business school like London only enrolled 180 full-time (f-t) and 60 part-time (p-t) M.B.A. students. In 1989 only 117 f-t and 45 p-t M.B.A.'s were granted. (M.B.A. enrollments in the top five U.S. business school in 1990: 2,620 f-t, 1,010 p-t at Northwestern; 1,575 f-t, 0 p-t at Wharton; 1,600 f-t, 0 p-t at Harvard; 2,260 f-t, 1,160 p-t at Chicago, 670 f-t, 0 p-t at Stanford.) Most British business schools, but not London, moreover, have one-year M.B.A. programs. Hence the time devoted to IBE in a M.B.A. program is considerably less than in the two-year U.S. M.B.A. program. Still, British academic IBE today is more like that of the United States than that of any other major country.

France is the close European second. Luostarinen and Pulkkinen report no IBE doctoral programs there (the one at IN-SEAD does not count because this school is outside the system). Research of an institutional sort, in France, moreover, is rather weak. And there is the problem of what to do about statistics on French M.B.A.-level IBE. This problem arises because of difficulties in comparing French business schools, the *Écoles supérieures de commerce*, with U.S. graduate schools of business administration. The French schools are all preexperience, the best U.S. business schools are not. Moreover, the U.S. schools are graduate, the French "undergraduate" schools, except for the fact that the French "undergraduates" are a highly selected

group of academically precocious youths compared to their U.S. counterparts. Whether one decides that France has M.B.A.-level IBE or "undergraduate" level simply depends on whether one wants to classify the French *Écoles supérieures de commerce* as M.B.A. or undergraduate institutions. Despite these problems, France is clearly second only to the United Kingdom in the scheme of classification used by Luostarinen and Pulkkinen.

Of world-significant European economies, the most powerful, with the greatest volume of foreign trade and a consistently favorable trade and foreign exchange balance, Germany, has proportionally the poorest record in the Luostarinen and Pulkkinen survey of academic IBE. Of German institutions responding to the questionnaire, only 58% have at least one course in IB, as opposed to 96% of the respondents in France and 98% in the United Kingdom. There was almost no undergraduate IBE for Germany, although graduate IBE was moderately strong. Here the same problem arises that cropped up for France. The German first degree in business economics (BWL) is, for the most part, a preexperience degree but it takes five to six years to obtain it. Hence the first-degree holder in BWL has a much greater education than the first-degree holder from a U.S. university. That is why, for want of a better solution, the Germans classify their first degree holders with the U.S. M.B.A.'s. The one highlight in the German picture, compared especially with France, would be research. Luostarinen and Pulkkinen write that in Germany "where the amount of the supply of International Business teaching is rather low, the role of International Business research is quite high such as, 19 (63%) of 30 responding institutions stated that they were doing International Business Research" (209). Nonetheless, as important a university as Mannheim, with 6,000 graduate students in business economics (BWL), has yet to hire faculty specializing in IB (they are currently about to hire a IB professor). Mannheim, moreover, only started a regional studies program last year; few students are involved and there are no Asia options, not even one for Japan.

An extension of the analysis to Japan shows that IBE there is also very weak. Professor Takeshi Yuzawa observes that the faculties of management in Japanese universities average "less than one subject on International Management and Multinationals" in their programs (1988, 74). And "[t]he absence of the U.S.-Style M.B.A. professional management school in Japan's universities" means that these meager offerings occur at the undergraduate level (MacMillan 1988, 132).[2] The weakness of IBE in Japan in effect simply reflects the weakness of academic business education in the country. In this respect people in the United States should not be misled by some recent, U.S.-business-school-inspired developments. There is, as Professor Katsuyuki Nagaoka underscored in a recent paper, still "only one small business school (in Japan) and that is not yet well-known" (1994). And at the undergraduate level, since Japanese university students spend the first two years of their studies on general education, only two years can be devoted to the business major. What is expected of management education during this brief period, therefore, is not an education for an expertise but background management education, namely the passing on of general knowledge about business systems, business firms, organizations, management, some analytical methods, and some elementary knowledge on accounting and computer operations. One does not even pretend to educate managers in Japanese academia.

THE HISTORICAL ANTECEDENTS OF DISPARITIES

Here I want, putting the economic issue on hold, simply to explain why these disparities arose educationally. The answer I give is that IBE is a reflection of how business education came to be done in a country and that this business education in turn reflects peculiar educational history. I shall show why the United States is the leader in the comparative charts; why the United Kingdom, although copying the United States the most, had a late start in IBE and still operates academically, compared to the United States, in reduced circumstances; why France, although with a long business education tradition, copied the United States after World War II, except in matters of research; and why Germany and Japan have copied the United States the least.

Why the United States Is the Leader

People in the United States like to think of theirs as a quintessential democratic country but it is quite elitist. The U.S. elite is a meritocracy based on individual talent; it is broad-based, including government, military, business, and industry leaders in its ranks. All, moreover, share common values that are deeply rooted in the nation. Among them is an instrumental view of education, that knowledge can be applied to problem-solving in a variety of situations. This has led to a great appreciation of intelligence and knowledge in the elite and much knowledge crossover between occupations and professions. Nothing illustrates this better than interaction between business and the military. Ivan Prasker observed, for example, that General Westmoreland

> had an advanced management course at the Harvard Business School in 1954, fitting in so seamlessly among the business executives who were his classmates that one of them later said, "He might have been a vice-president of a corporation." Throughout his career, Westmoreland indicated a passion for efficiency and a fondness for percentages. This meant he

and McNamara who'd not only once taught at Harvard but also emphasized charts and statistics when he was president of Ford, had many of the same instincts and spoke the same language. (1988, 12)

Lest one think that this is a recent phenomenon, it is well to remember how far back the West Point–business knowledge crossover extends. Business historians in the United States are familiar with it. Professor Alfred D. Chandler, Jr., notes that "the United States Military Academy provided the best training in civil engineering in this country until the 1860s," which led not surprisingly to the employment of its graduates in the building of the U.S. railroads (1977, 95). The influence was managerial as well as technical. West Pointers, working for industry, used systems learned at the Point to improve workshop discipline and cost accountability. They were directly involved in working out the "power-knowledge configurations" that brought about the reorganization of factory management and the development of corporate managerialism in this century (Hoskins and Macve 1988, 37–73).

It is best not to exaggerate this education-management connection for the nineteenth century; most managers were shop-floor not college educated. Moreover, the state of scientific knowledge was such that it was of little direct use to managers. Nonetheless the nineteenth century precedent bore fruit. It bore fruit in the business community itself which, as the names of so many business schools attest, generously supported business education. By 1939, 10% of all undergraduates at U.S. colleges and universities studied business; by 1949 the figure had climbed to 17%.

But the really decisive breakthrough came with the introduction of the scientific paradigm (for example, linear programming, mathematical model building, statistics, and decision science) into management education. Once again the military led the way, the operations research teams, spawned by World War II, dramatically demonstrating the efficacy of management

science (cf. Locke 1989, ch. 4). The use of science to solve managerial problems continued in the defense establishment during the Cold War, the most impressive example being the employment of linear programming, developed by George Danzig of the Rand Corporation on United States Air Force contract, to optimize aircraft operations during the Berlin airlift and bombing patterns during the Korean War.

Mathematical modeling and statistical analysis spread in the 1950s from military to civilian governmental agencies—for example, the Census Bureau—and then to the large corporations. At first the techniques were used on transportation and productions problems—for example, queue theory; afterwards they were applied to traditional management fields, such as finance and marketing. The impressive results achieved brought about the impregnation of management education with mathematics and science, initially in the departments of industrial administration of engineering schools—Case Western, Carnegie Tech, and MIT, among others—where the required knowledge of mathematics and statistics was readily available, and then, in the late 1950s and 1960s, their use in business schools prospered. Business schools in the United States radically upgraded admission standards in mathematics and introduced mathematical models and statistics into their behavioral science, traditional functional management (such as marketing and finance), general management, and decision science courses.

Thus the point to be made about how historical experience produced current U.S. hegemony in academic business studies is essentially this: Beginning in the late nineteenth century U.S. manufacturing started to produce good quality, if relatively unsophisticated, products at low prices. The secret was mass production in long, uniform runs, for a market that skilled marketing people developed. It led at times to surpluses and layoffs but, in general, the system worked well as the United States developed "the" consumer society. It also brought about U.S. world military predominance. It obviously impressed Winston Churchill, who glowingly wrote of this "American clear-cut, logical, large-scale, mass production style of thought," which he saw at work preparing the Normandy invasion (Churchill 1952, 95). This style of thought brought the triumph of what Philippe d'Iribarne calls the "classical American form of management" (Iribarne 1989). It called for management with a "high degree of formalization, standardization, and centralization, where managers possessed good conflict resolution skills, good top-down decision-making abilities, good problem-solving, analytical skills, and a capacity to devise good externally imposed evaluation systems." Classical U.S. management was exercised in hierarchical-mode (H-mode) organizations, typical in U.S. manufacturing, "where the whole production process is divided into a series of specialized functions, each standardized, and each work unit is . . . required to perform exclusively the specific job dictated by prior plan" (Locke 1996, ch. 2).

The U.S. system of business education clearly complemented management in H-mode organizations, for the linear programming models, cybernetic feedback systems, statistical analysis, and electronic data processing taught in the academic business school proved very useful for a management where a "high degree of formalization, standardization, and centralization" reigned, where "good top-down decision making abilities, good problem solving, analytical skills, and a capacity to devise good externally imposed evaluation systems" prevailed. The professionalization of elitist management, management by properly credentialed, highly paid experts, resulted. An analogy with the medical profession is not out of place. As the medical doctor studies medicine, this new professional manager studies business administration; as the medical doctor opens consultancies to help sick people, this new manager sets up consultancies to help sick firms. As medical schools do research to im-

prove medical science, business schools do research to establish the theoretical basis of the science of business administration or one of its subdisciplines and to develop the toolkit necessary to doctor business firms.

International Business Education in the United States has from the beginning exhibited this conception of management education. Nothing illustrates this better than the objectives set forth in the prospectus issued for this conference. We are called together "[t]o define the domain, phenomena, and relationships that constitute the field of international business inquiry, and to identify contacts that hold promise for integrating the field and for encouraging theory-building and theory testing in the future." We are asked "[t]o identify opportunities that exist for research in the field, to explore the limitations imposed by, and benefits derived from, various research methodologies and their embedded paradigms, and to suggest directions that seem particularly fruitful for future research." These statements are cast in the accepted U.S. social science jargon and fit well the science paradigm that has dominated U.S. business schools for the past generation. Academics are talking to academics about academic matters.

Because IBE was born into a mature business school community, the first in the world, IBE emerged first in the United States. Since it had to conform to the dictates of this community, it developed fully at all levels of business school education (undergraduate, M.B.A., doctoral, in teaching and research areas, in exchangeships) to become academically respectable. This is why the United States is at the top of the IBE chart.

Why the United Kingdom Copied the United States the Most but Did So Relatively Late and at Low Intensity

The unprecedented hegemony of the United States in the broken world of 1945 made classical U.S. management and the education that went with it an attractive recovery model. but it was a model that the British, because of the nature of their elite, found it hard to adopt. Whereas the U.S. elite was broad-based, the English was not; indeed, there were really two elites in the United Kingdom which differed as profoundly from each other as they did from the U.S. in their outlook towards and involvement in higher education.

One was the political and social elite which had grown out of an aristocratic order in which belief in a society composed of estates not individuals prevailed. Each estate had functions in the social order and attributes necessary to fulfill them. The aristocracy's function was to govern and command. And the attributes the aristocracy assigned to people destined to carry out this function had more to do with character than intelligence or knowledge. Courage, loyalty, and honor—these were essential for good governors, good soldiers, and good colonial administrators. The system of education this elite fashioned to serve its interests sought these ends. English public schools and universities, which meant— aside from the University of London founded in 1839—Oxford and Cambridge in the mid-nineteenth century, based their education on the humanities. Even instruction in them, however, amounted to a poor education since knowledge was incidental to educational purpose. School and college tried to develop leadership qualities by bringing teacher and taught together in a fellowship of "leisure and confidence." They sought to impart "effortless grace, casual assurance" and "light touch in command," qualities of a primarily landholding elite's cultured way of life. The goal was the awakening of the "Gentleman-Ideal" (Gellert 1983, 4). This is why so much emphasis was placed on games: Play was to form the character of young gentlemen destined for higher service.

This education produced the corps of loyal, hardy leaders that served the Victorian Imperium well. But it delayed technical and managerial education in Britain. It was not just that the educational establishment of this political and social elite sought the moral and cultural development of the individual. It was that it made this a class not a national goal. Almost by definition this meant that a different education was intended for the work, commercial, and manufacturing classes and that it would be deemed "inferior." The directing class usually made a distinction between its education and that of the others by calling it "training," a pejorative term signifying the unimaginative, insensitive learning of practical skills (for example, accounting and engineering). Such an attitude encouraged public schools and universities to resist practical studies (hence the refusal of Oxford and Cambridge to accept accounting into their curricula and their obstinate rejection of business schools). It also encouraged the governing elite which had public school and Oxbridge educations to deflect reform movements away form their venerated schools and colleges to nonuniversity institutions or newly created, underfunded, and little respected universities that the government elite avoided.

Britain's second elite counted those who had come to the fore in commerce and manufacturing during the nineteenth century. These were the "practical" men. Had they been in the United States, many would have acquired a "useful" degree, but because the higher education the first elite created was "useless," the commercial and manufacturing leadership would have little of it. Indeed, the negative reaction of manufacturers and merchants to a university ethos which obviously so little served their interests probably accounts for their failure specifically to fund schools of commerce or engineering, as the United States did, as well as, generally, for their strong, persistent anti-intellectual, anti-educational biases.

The periodization of British management education development is, therefore, quite different from that of the United States. Whereas business courses and business schools started at the turn of the century in the United States and grew vigorously thereafter, management education in British academia was, because both elites really opposed it, a nonstarter. It is true that the universities of Birmingham and Edinburgh began to award commerce degrees between the wars, but business and industry did not take up their graduates and other universities did not follow suit (Keeble 1984).

This does not mean that there was no interest in management education in Britain. The war and the increasing centralization of capital in multinational enterprises, the need for creative market and organization strategies, and the growing division between corporate ownership and control which emerged in its aftermath, aroused considerable interest in the subject. Between 1947 and 1950 around sixty teams were sent to the United States from Britain, sponsored by the Anglo-American Council on Productivity, to investigate U.S. practice and six of them were commissioned expressly to look into U.S. management education. These efforts led to the creation of the British Institute of Management (1946) and the National Staff College at Henley (1947).

But, because of the profound distrust that English university traditions evoked in business circles, the enlightened businessmen and management consultants, rejected academic management education. At Henley the methods of education were purely professional; representatives from companies, seeking improved management, met regularly in groups, "syndicates," for a systematic exchange of ideas and experiences. Since academics were not involved and the poorly educated British managers knew little about mathematics or statistics and were incapable of understanding much less propagating linear programming, computer simulation marketing, financial planning or other operational research techniques, the postwar management education movement did not have much to do with the science-based management disciplines developing rapidly on the other side of the Atlantic.[3]

Consequently, the "scientific" approach to management education did not develop in Britain until universities, looking to their thriving U.S. cousins for models,

began business schools in the mid-1960s. Moreover, the advocates of professional management training for and by active managers and the doubting Thomases, who believed that management could not be taught in schools, fought a successful rearguard action. Projected business study growth rates were not attained; M.B.A. programs floundered. In the 1970s the business schools altered their strategy to accommodate the "practical" point of view. Short-term executive programs and company training partnership schemes were inaugurated to bring the schools closer to praxis. This meant that "scientific" management education had to be curtailed. As a result students in British business schools do not, compared to U.S. students, spend much time on campus in coursework and good scientific instruction takes time and effort (think of the grueling two-year workloads at the best U.S. business schools). The practical as opposed to the scientific approach to graduate management education was reinforced, moreover, when the Council for National Academic Awards began to grant the M.B.A. (in the 1980s), for the Polytechnics, where study occurred, stressed sandwich courses in a one-year M.B.A.

Undergraduate business programs in British universities also developed quite differently from programs in the United States. Whereas undergraduate business education had been firmly established in U.S. universities before the M.B.A. boom, in Britain, by contrast, it has taken root only after the graduate business school movement. Academic prejudices against business studies die hard. Of the top twenty subjects studied at British universities (number of students enrolled) in 1966, business did not figure; in 1974 it was twentieth on the list; in 1976 fifteenth, and in 1980, seventeenth (by contrast in German universities business economics was sixth in 1968, fifth in 1970, sixth in 1976, and second [behind law] in 1980) (Locke 1989, 195).

Since the general condition of management education determines the state of academic IBE within a country, English education tradition explains why surveys like that taken by Luostarinen and Pulkkinen put U.K. IBE so far ahead of most European countries. In as much as both British elites had ignored academic business education, there were no enrooted peculiarly British systems of business education in the United Kingdom to stop reformers, when they decided to act in the 1960s, from borrowing freely from the United States. And since they did indeed borrow much from the United States—such as business schools, M.B.A.'s, undergraduate majors, and Ph.D.'s—the IBE that developed in the United Kingdom closely resembled U.S. IBE. Hence the British performance on the comparative IBE charts.

On the other hand, the difference between U.K. and U.S. IBE today can also be explained by the lack of academic management education traditions in the United Kingdom. This is true in the most obvious way. IBE is a much more recent phenomenon in British universities because academic management education is a much more recent phenomenon there. Thus, although the Luostarinen and Pulkkinen survey shows that fifty U.K. institutions had undergraduate IB programs in 1990, had it been done in 1980 few such programs would have been registered.

The differences between U.K. and U.S. IBE, moreover, also exists in ways that these comparative surveys do not reveal. Luostarinen and Pulkkinen note that 19 U.K. institutions have IB M.B.A. programs, the highest number by far of any country in Europe (France is second with ten). But are these programs really comparable to those in the United States? Raymond Thomas, on his retirement (1987) as head of the Business School in Bath, noted that British "University (business) departments . . . are cottage industry units, even at the scale of the largest UK schools" (1987). The comparative disparities are great between British and U.S. schools (in enrollments, in size of faculties [for example, Harvard has 200 full-time faculty, Wharton, 174 on tenure-track, Northwestern 25 teaching finance alone] in physical plant and equip-

ment, in operating budgets, in breath and depth of curricula). The largest U.S. schools (Harvard, Wharton, UCLA, Chicago) are business universities compared to the British schools. Staffs are just not available to present the rich variety of International Management courses available in the United States nor are the students, for the M.B.A. student populations in U.K. business schools are tiny compared to those in the United States. Time constraints of the one-year M.B.A., moreover, also restrict scope and intensity of IB studies. The combined M.A.-M.B.A. offered by the Lauder Institute at Wharton lasts thirty months.

Business schools and their graduates have been fully accepted by U.S. business, which explains their phenomenal growth after World War II.[4] Relative slow growth of M.B.A. programs in the United Kingdom, which resulted mostly from the profoundly anti-intellectual, anti-university tradition in British business and industry, which was itself bred by the anti-utilitarian heritage of English public school-university Gentleman educational ideal, not only limits the numbers of IB M.B.A. students in U.K. institutions but blunts their penetration into higher management.

Why France Copied the United States but Has Weak Research

The French political and social elite and the French business and industrial elite is very different from the English. This is true first of all because there are not two distinct elites but one. it is an elite, moreover, that has almost always been educated in business and technical subjects in institutions of higher education. but there is no reason to belabor the point. The French *grandes écoles* are world famous institutions and I am sure those that are management schools are known to people at this conference. What I wish to do is explain both why the French, despite having their own traditions in business education, copied so much from the United States in the 1960s and why with their great traditions their research in IBE has been so weak.

French universities are very different from those in the United Kingdom. They are state institutions, imbued with a bureaucratic rather than a Gentleman educational ideal. This has been true ever since Napoleon Bonaparte reorganized the system so that it could produce the teachers and functionaries needed in his Empire. Still French and English universities have had one thing in common: neither provided a professional education for businessmen and engineers. As in the United Kingdom management did not become a study option at French universities until the late 1960s. But this does not mean that French managers like British have not been highly educated. During the nineteenth century the French developed those *grandes écoles* to tend to the task.

They, not the universities, are the stars of French management education. This prominence does not rest uniquely on the fact that they were the only place until the 1960s that fledgling managers could be professionally educated. They are stars because of their exclusivity. Whereas universities take all students who have acquired a secondary school diploma, a "bac," the *grandes écoles* admit only those who, after receiving a very good "bac," can pass their competitive entry examinations. Students wishing to attend a *grande école*, therefore, usually spend two years in a special school, after the "bac," preparing for the dreaded entry examination (the *concours*). Those in the *grandes écoles* are, therefore, the most academically accomplished and intellectually gifted students in France.

The bifurcation in French higher education between university and *grande école* is extremely important because of the different approach each takes today to management studies. Although the university programs are new, they are the closest to the United States. All universities, after all, are places of scientific study. Each discipline has a body of knowledge, which is developed through scientific research, and taught. If the discipline is an applied science, this knowledge is used to solve practical problems in the "real" world. The U.S. business school operates this way. Its professors and graduate students do research and teach their discipline to undergraduate and M.B.A. students who go out and apply the knowledge acquired on the job. The French universities, although very different from U.S. universities in many respects, share this research-knowledge view of education. Professors of management and their doctoral students research management and present the results in dissertations, books, and articles. The management students learn the body of knowledge thus engendered and use it at work.

But the French *grandes écoles* of management operate on different educational principles. Throughout their long history they never conceived of their educational role as one of developing knowledge because they did not conceive of management as a "scientific" activity. It was the individual possessed of intellectual brilliance, quick wit, will and energy, not "scientific" management knowledge, that made the good manager. That is why so much emphasis was and is placed at *grandes écoles* of commerce on student selection. In the *grandes écoles* the students were taught by "visiting" faculty, *vacataires:* for example, university professors who taught a course on commercial law or economics, professional accountants who taught courses in their specialties, active managers, or bureaucrats. Since "management" was not studied in universities—since, in fact, no such discipline existed—these imported professors could not focus on how their information applied to management problems. Without permanent faculties and graduate students to do research, the *grandes écoles* of commerce could make no significant contribution to the development of management as a scientific discipline. Consequently, these institutions in the 1960s, after decades of existence, were about as innocent of management science as U.K. and French universities. Like them they had to turn to the United States when, under the impact of the U.S. academic management education revolution, they finally set about repairing their deficiencies.

To understand the state of French IBE, therefore, management educational history has to be taken into consideration. Since management studies are a recent phenomenon, so is IB education. Since they have been heavily influenced by the United States, so has French IBE. The fact that in the Luostarinen and Pulkkinen survey France ranks second to the United Kingdom in the number of institutions with IB programs (87 U.K., 38 France), indicates the extent to which the French have been seduced by U.S. ideas about management education. But the weakness of French research in IBE indicates the extent to which French education differs from that in the United States, for the *grandes écoles* of commerce, which educated the managers, were not research institutions, and the universities, which were, did not until recently research management. Indeed, within the universities management research is still neglected in the most scandalous way.

Why Germany and Japan Have Copied U.S. IBE the Least

Management in Germany and Japan traditionally did not, as in England, undervalue higher education. Managers in both countries are among the most highly educated in the world and have been so throughout this century. Indeed managers sitting on boards of direction of large German corporations often have research doctorates in engineering, law, or business economics (*BWL*). But it is best when looking at Germany and Japan not to use U.S. academic classifications—for example, undergraduate, graduate, M.B.A., Ph.D. To understand how academic education in these two countries differs from that in the United States, one needs to distinguish between pre- and post-experience education.

Roughly speaking the following is true: preexperience education takes place in academic institutions, postexperience does not. There are exceptions to this rule. German managers, for example, often acquire their doctorates after beginning managerial careers. But with few exceptions, once preexperience students depart academic institutions, they never return to them for postexperience management education. That is done in the firm or in extramural nonacademic training establishments utilized by the firms. Consequently there are few German or Japanese M.B.A.'s in IBE because there are few M.B.A.'s in management in either country.

The reason for this postexperience academic management education avoidance can be found in singular German and Japanese views about management and management education. Most people even slightly familiar with Japanese firms know about the employees intense identification with and dedication to their companies. Illustrative stories about it are legion not legend. The employee is not an engineer who works for Toyota but a Toyota man with engineering knowledge and skill, the manager is not a professional employed for his skills but a company man who happens to work in some aspect of what we call management. The difference can be ascertained by paying close attention to Ryutaro Komiya and Kozo Yamamoto's description of the role of "economics in Japanese administration." There are no economists in the Japanese Ministry of Finance, the Ministry of International Trade and Industry, or the Bank of Japan," these scholars explain.

> Most of [their officers in charge of economic affairs] are highly intelligent and capable men. . . . In practice they often play the role of an economists. [But, a]ll are basically generalist administrators, diplomats, or central bank officers; none is a professional economist. They are influential in the economic policy making process not because they apply advance knowledge of economics or economic theory to the issues under consideration but because they have wide experience as generalist administrators and can react promptly to new problems and changing circumstance, mobilize the information, knowledge, and capabilities available among their subordinate staff, and build up a consensus among those concerned

through deliberate persuasion and skillful negotiations (Komiya and Yamamoto 1981, 267).

In Japanese firms management is not conceived of in functional but firm-specific terms. Managers, therefore, do not have professional knowledge and kills that are detachable from the firm: they are not accountants, finance experts, and marketing men who happen to work there. Indeed, as the quote indicates, they are there in management positions as much for their relational as their technical talents, that is, to motivate coworkers, promote group cohesion, and facilitate common effort toward collective goals.

German firms have a somewhat similar attitude toward management. First off, they do not have the word "manager" or "management" in the generic sense, in their language. One manages a shoe factory, or better factory X. The idea that somebody can be brought in from the outside who knows relatively nothing about the actual business to be managed sounds ludicrous to them. To be a good manager at the top, one needs to know the business from the ground up. It is a concept of management that stands in marked contrast to that propagated by the U.S. business schools and their M.B.A. programs. As Professor Heinz Thanheiser observed after studying fifteen West German firms:

> The managers at the highest level, even on the board, were extremely skeptical about the idea of professionalism in management. They did not, then, share the confidence that their American colleagues had in the transfer of "management know-how," confidence which gave them the courage to create the "conglomerates." The German leaders view diversification from a different angle: "We have seriously studied the potential of Sector X (close to us from a technological standpoint) into which we could have easily entered. But nobody on the Board of Directors knows the market, the competitors, the clients. . . . consequently we don't touch it" (Thanheiser 1979, 8). Such views differ greatly

from the viewpoint on diversification and decentralization advanced in Anglo-Saxon writings. (1979, 8)

Since they do not have a generic idea of management neither do Japanese and German firms have one of management education. Germans make a distinction between *berufsfaehig* being (capable of doing a job) and *berufsfertig* (ready to do a job) in their education. To be capable of doing a job, German professors feel that people need a thorough *Denkschulung* (schooling of the mind) in a discipline, say business economics. Germans are, in fact, frequently shocked when they discover how theory naive U.S. M.B.A.'s are in business economic disciplines, for they think that such poorly educated business economists do not have the schooling of the mind necessary for a quick and accurate learning of a profession or occupation after leaving academia. On the other hand, they do not, not thinking of management generically, believe that academic institutions can make people ready to do a job. That is why postexperience education is the province of the firm. And the Japanese feel the same way.

It is important to stress, in order to avoid what Charles de Gaulle would call "Anglo-Saxon" habits of thinking, that the Germans and the Japanese see no necessary antagonism between the practical and the theoretical. In the nineteenth century every manager was trained practically on-the-job everywhere. In England these practical men reacted adversely to the idea of university education, thereby fostering the view that antagonism between shop-trained and university educated was natural. German and Japanese managers certainly did not and do not approve of letting inexperienced theoretical university educated people mind the shop but by the end of the century they began to realize that intellectuality and practicality were not necessarily in conflict.

German universities, unlike English, had always been places for scientific research. Indeed it was impossible to aspire to university status, unless a discipline adhered to the research traditions of German higher education. Originally business and engineer subjects were ignored in the universities but in the late nineteenth century things began to change. Engineering and then business became part of university (*Hochschule*) studies and consequently science research traditions began to fuse with practical on-the-job training. a highly qualified management possessed of both scientific knowledge and practical skills emerged. The German engineers call this combination *Technik* which is not applied science—a subordinate of pure science with all the connotations of intellectual inferiority that the idea implies—but the peculiar combination of knowledge (*Wissen*) and skill (*Koennen*) that makes it something not found in the Anglo-American world.[5] As such, *Technik,* which represents a singular blend of German university and industrial training traditions, constitutes the overarching ideology which brings all levels of management and all types of qualifications together; it provides, along with the community (*Gemeinschaft*) traditions, the cohesive motivational force within firms that contributes mightily to their views of management education. As in Japan that education, although posing no conflict between theory and practice, was brought about by a careful division in educational labors—the university and other *Hochschulen* providing a thorough *Denkschulung* for preexperience people, the postexperience education being done under the guidance of the firm. Together they produced the practically oriented but theoretically sophisticated manager. It is, then, the U.S. effort to let academia play a role in the postexperience education of a new U.S.-invented profession, management, that accounts for the deviation of the German and Japanese IBE from the U.S. model.

WHY IS THIS U.S. IBE YARDSTICK OF LITTLE USE WHEN TRYING TO MEASURE THE REAL EFFECTIVENESS OF IBE?

Everywhere I go I hear that international management is not a field, it lacks the unifying theoretical base necessary for a discipline. But what could this mean? It certainly could not mean that core management disciplines (such as finance and marketing), as opposed to IB, have solid "scientific" underpinnings while IBE does not. People who believe this should spend twenty minutes in a room full of philosophers of science; the former not the latter would come out feeling like complete morons. Statements about solid scientific foundations are nonsense statements. Even the purest natural science has no "solid theoretical foundation." And the social sciences, like those that make up management studies, are even worse off. This sad fact no group of academics at this conference can gainsay.[6] United States' IBE is not a "science"; nor can it or any of the so-called management sciences hope to be one. IBE is, like U.S. management education in general, a cultural expression.

But none of this matters very much. Chemists and physicists are not overly worried because of the theoretical shortcomings of their science. If they can split the atom or find a cure for AIDS they are content. And the same idea applies to management education: if it is useful in bringing about good, competitive management, then it has value. But that is indeed the point. I believe that a reevaluation of U.S. business education and with it IBE has been in order precisely on the grounds of practicality. And I want to make two points in this respect.

First, studies done by the European Parliament show that there is more to educating people than preparing them for a specific functional job or occupation. There are, in fact, good reasons to believe that the ability of a society to cope with economic and technical change has more to do with general literacy and numeracy than with training. Professor Hartmut Wächter reports that

> the time it takes for the content of education to become outmoded and the speed with which it takes place can be positively correlated with its concreteness (closeness to praxis) and negatively correlated with its level of abstraction. It follows, there-

fore, that during periods of sudden social change the basic education needed to prepare us for life is worse the more it is related to concrete practical needs. (1980, 34)

This conclusion is interesting because it means that one should not just attempt to teach what goes on in praxis. It is better, when organizing secondary and higher education, to teach what has been called "key qualifications" (*Schlusselqualifikationen*). Among these key qualifications are, Wächter notes, basic and lateral ones. Basic qualifications are those necessary to learn and lateral ones those that permit the individual, who is properly equipped with basic qualifications, to tap the information that is available in society. The table indicates the nature of the two. Key qualifications are not really qualifications in the sense that they qualify a person for a particular job or occupation in a specific field. During periods of change such qualifications rapidly become managerially outmoded. Indeed, the Institute for Job Market and Occupational Research in West Germany concluded that "it is illusory to believe that quick changes can be made in the training structure, the educational curricula, or in continuing education in order to keep them synchronized with the sudden and constantly changing work world." There is no way, especially during periods of great ferment, to make educational institutions serve directly the specific professional and occupational needs of the economy." Wächter states that

such a response is not really necessary: For most jobs the bosses in a business or firm find that they want people with entirely different qualifications than those imagined when the original position was advertised. Also positions offered can be filled by people of quite diverse training. Mertens [whom Wächter is quoting] concludes that in numerous cases a complete elimination of bottlenecks and surpluses on the labor market occurs, even if calculations about future educational and occupational structures are wrong. The elasticity of the system is important.

Since in Germany at least the system seems to be self-regulating and no real bottlenecks appear, we have to ask ourselves

Table 2.6 Key Qualifications

Basic	
As Educational Goals	**Acquired Education By**
logical thinking	formal logic, algebra
analytical procedures	linguistics, analytical geometry
critical thinking	dialectics
structural thinking	priority order of phenomena
management thinking	organizational theory, economics
conceptual thinking	planning techniques

Lateral	
information acquisition capacity	semantics, information science library science, statistics, media science, foreign languages, language structures, speech, knowledge of symbols, graphics, speed reading, specialized vocabularies, elimination of redundancies and repetitions

Wächter, page 34

why the labor market is capable of adjusting. What in the education makes this flexibility possible? The answer is the acquisition of key qualifications. And if we apply this knowledge to IBE, it would appear that the excellence of IB courses at business schools has much less to do with good international business preparation than a system of education that teaches key qualifications to future managers.

Literacy in the Japanese and German educational systems has always meant learning foreign languages. In German secondary school parlance languages and mathematics not science are considered to be primary subjects (*Hauptfaecher*) because they are the tools with which all knowledge is acquired, they are in effect key qualifications. According to the 1980 survey of German university graduates working in business and industry, all could read English, at least passably, and seven out of ten could also read French. These languages were learned primarily before entering the university; indeed, at university, in many fields, including business economics and engineering, a reading ability in foreign languages is assumed.

This initial advantage, moreover, is exploited in post experience management education. This is true on the job where in large Japanese and German companies technical and scientific foreign language periodicals are circulated to management employees (at Siemens, for example, each employee must initial a cover sheet on each issue to show that he or she has seen it). Special language courses are offered, moreover, regularly in continuing management education. The Esslingen Technical Academy, which is engaged in postexperience education throughout much of Germany, for example, is currently giving two-day courses on "Technical English," "Writing and Telephoning in English," and "Effective Management in English." Such courses do not reveal so much a lack of foreign language knowledge among the managers as its presence, for two-day intensive sessions would hardly be feasible if participants had no prior knowledge of a language.

A comparison between U.S. and German efforts to incorporate languages and regional studies into their university level business study programs, therefore, can be quite misleading. An abundance of foreign language and regional study programs at business schools in fact probably reflect a U.S. inferiority in foreign language education.[7] This inferiority is reinforced because of attitudes encountered in U.S. business. How many U.S. corporations make foreign language periodicals readily available to if not mandatory reading for employees? How many, like the Japanese, send their people on M.B.A. courses in foreign business schools? Few, perhaps none. A recent survey of M.B.A.'s with excellent foreign language competence working in U.S. firms shows that, because "of the English-language bound culture of the corporation," "the utilization of [their] foreign languages in the corporate workplace is limited" (Lambert 1990, 59). The study's conclusion, that recent emphasis on foreign languages in U.S. business schools might be wrong, seems itself incorrect. Since German and Japanese firms would certainly appreciate them, it really amounts to a critique of the lamentable lack of international awareness in U.S. management.

From the firm's perspective, therefore, a more revealing comparison of IBE in these countries would be between secondary schools systems and postexperience firm-sponsored management education. Indeed, the linguist-regional business programs recently adopted in U.S. business schools do not really tell us much about the relative state of international awareness in U.S. M.B.A. students. Although it is assumed in Germany and France that students entering business studies have a good reading knowledge of English and perhaps a second foreign language, no foreign language requirement exists for regular M.B.A. students in the United States and the United Kingdom.[8] Most U.S. M.B.A. students, therefore, remain culturally naive (because learning foreign languages involves more than just learning to talk to people), like the well-meaning Yankee who asks foreign visitors whether they have electricity, running water, or automobiles in their country.

The second point has to do with the motivational aspect of management. In the immediate postwar world, U.S. corporations dominated international business. In the past twenty-five years, however, non-U.S. firms, and in manufacturing, particularly Japanese firms, have been in sharp ascendancy. The Japanese achievement especially has had great implications for traditional U.S. ideas about management. In the West, outstanding management has always been associated with creative individualism. The impressive manager is the high-flyer, the innovative, imaginative, hard-driving, risk-taking, competitive leader who is the opposite of the organizational man. The Japanese have a proverb that the nail that sticks up is hammered down, and the philosophy this expresses seems to pervade Japanese organizations. Hiroko Charles, a Japanese businesswoman (married to an Englishman), who has worked extensively with Japanese and British firms, has pointed out this incongruity: One finds many more capable and intelligent individuals in British firms, but Japanese firms are more effective, risk-taking, quick-learning, flexible, highly competitive collectivities and they are composed of unassuming, modest, indeed rather ordinary organizational men. They seem to have developed a new idea of group as opposed to individual management. And in the competition between individual and group management, the synergy of the latter has proven to be quite effective.

Action by collectivities not individuals requires a form of management that is quite different from classical U.S. management. Whereas, Tadao Kagona et al. point out (1981, 136), contrasting U.S. "mechanistic" management with Japanese "organic," in a U.S. factory the managerial talents essential to success are

1. Good conflict solution skills by the boss;
2. Good individual top-down decision-making;
3. Good use of consultants brought in from the outside as trouble-shooters;
4. Good externally-imposed control systems; and
5. Good problem-solving, analytical skills,

In the Japanese factory, which lacks the formalized, standardized, externally imposed modes of work organization extant in the United States, good management requires

1. Group-oriented consensus making;
2. "Control by sharing of values and information";
3. The cultivation of "relational-skills" in managers; and
4. "Broad consultation before acting."

This collectivist form of Japanese management grew up as a sort of counterculture in a postwar world dominated by the U.S. "mechanistic" form of management. Indeed, the Japanese themselves were greatly affected by U.S. methods. As Nancy McNulty notes, Japanese management "took what could be easily transferred—production management techniques, finance management, some aspects of marketing—[from the United States] and put them to good use" (1980, 17). But she goes on to note that "in areas subject to cultural factors, especially those of organization and personnel management, they stuck firmly to their traditional practices and values."

In the 1960s, while U.S. "mechanistic" management still reigned supreme but the Japanese firms had made their presence felt, the Japanese way of doing things was viewed from the West less as a challenge than a peculiarity. But in the ten-year period 1970–80 peculiarity changed to challenge (from a threatening alien culture) and then subsequently into a new approach to management where "superior Japanese management methods" simply became "superior management methods."

Nowhere has this become truer than in international business management. The Ashridge survey of Global Human Resource

Management offers a recent and convincing example of this (1990). Firms, it reports, now talk about their management needs in terms of the whole organizational culture instead of the skill level of individual managers, for the "mind set and outlook of managers is more important than formal structures and reporting lines [the heart of classical U.S. management] in bringing about a flexible, yet cohesive organization." Quoting the management guru Rosabeth Moss Kanter, the report emphasizes the superiority of "process over structure" because it is not how responsibilities are divided but how people pull together that counts. And it calls for the creation of what Christopher A. Bartlett and Sumantra Ghoshal describe as "transitional" organizations, with "integrated networks which share decision making and where components, products, resources, information and people flow freely between the interdependent units, . . . and dynamic integrative ways of taking decisions through using the commitment of every individual employee to the overall corporate agenda" operate. The "highly formalized and institutionalized central mechanisms" of management have, it stresses, to give way to "co-option," to consensus building with each individual understanding, sharing, and 'internalizing,' the company's purpose, values, and key strategies constitution the most effective means of co-ordination." And the survey's authors claim, quoting Gunnar Hedlung and Dag Rolander, that survival today depends on the firm's "basic strategy," "its guiding principles of conduct" being 'widely shared' throughout its organization wherein 'detailed information' about everything is provided for everybody. The whole organization is expected "to think and act directly on thinking." Flexibility but cohesiveness, process over structure, shared decision-making, integrative thinking, consensus building, global responsibilities, collective motivation, the "firm as a brain"—these are the buzzwords used to describe how the international firm operating in fast-moving interdependent markets

must be successfully managed today. This the vocabulary of managers and collectivist Japanese management ideas has replaced classical U.S. as the universal norm in IB; knowledge of a discipline has been supplanted by group motivational skills in management.

Since under classical U.S. management the U.S. business school became the academic mecca of business education, the question is, how does this paradigm shift effect this connection. Or, in other words, does the U.S. system of academic management education—if it is a U.S. and not a universal norm—serve the collectivist Japanese mode of IBE as well as it did the "mechanistic" U.S. mode? I am not talking just about curricula or teaching methods. Many business schools in the United States have changed both, teaching humanities, foreign languages, emphasizing group learning, team building, and consensus forming, in order to adapt themselves to the demands of "collectivist" management. Rather it is how business schools interrelate as institutions with actual management that the new collectivist Japanese management standard calls into question.

"Management," it is said is a lifetime learning process. That is true but why should the business school have anything to do with it? If we think of organizational loyalties and firm specific skills, then, the role of outside "alien" institutions in management education should be radically diminished. Collectivist Japanese management requires inhouse training, especially when group cohesion and motivation is at stake, for it is not sophisticated management techniques but relational skills that count and these organization-specific methods cannot be learned well in business schools. Even when technique is involved, moreover, it cannot be taught outside the firm as effectively as within. Because academia, almost by its very nature, cannot really duplicate the realities of the firm, the firm's assumption of postexperience management education is a much more effective way to promote integrative group learning and the motivational aspects of good management than doing it in business schools. That, at least, has been the attitude of management in German and Japanese firms.

In fact, business school education, from the viewpoint of the new collectivist management, is actually dysfunctional. The U.S. schools are criticized for teaching "pseudo-skills when [students] might be getting on-the-job experience or acquiring genuine skills" (Samuelson, *Newsweek*, May 14, 1990). Their professors are chastised for "building elegant, abstract models," and their graduates for being "critters with lop-sided brains, icy hearts, and shrunken souls" (*The Economist*, March 2, 1991, 4). The problem is the way that the U.S. business schools, in particular, projects a community disruptive, elitist, classical U.S. management mode, not only by what it teaches but by the way its graduates are received, for example, very high salaries, and treated, for example, fast-track promotions, that undermines the collectivist management necessary, we are now told, to the operational efficiency of the international firm.[9]

The fact that Japanese manufacturing firms have the best management in the world and that Japan has the worst system of academic management education, that the German firms have always ignored academic institutions when training their managers, and still have the most cosmopolitan managers in the world, that the U.S. business school became the "best" in the world while U.S. business lost its international reputation . . . none of this can be explained by a yardstick that simply makes good IBE that which deviates the least from that which is offered in U.S. business schools. But this yardstick, unfortunately, is all too frequently used by business school and if unconsciously by U.S. business school academics.

NOTES

1. More than 20,000 Thunderbird graduates populate the ranks of such companies as Citicorp, BankAmerica, General Motors, American Express, IBM, and General Electric. Many of them having foreign postings or U.S. jobs with International responsibilities. In 1990, 279 companies came to campus to recruit students. Citicorp hired 17 Thunderbird grads, while Exxon and Ecosphere International each hired 4 (Byrne 1991, 193).

2. There are a couple of exceptions: A Graduate School of Business Administration was established at Keio University in 1978, with an M.B.A. program, and the International University of Japan, founded in 1982, has an International Management M.B.A. which graduated its first class in 1990.

3. A questionnaire circulated to the 10,000 members of the BIM in 1978 (45% response rate) shows that only 45% had two or more A levels and only 27% had a university degree.

4. Twenty percent of the top three people in the 500 largest American corporations have M.B.A.s from Harvard, which shows the impact that top American business schools have had on corporate boardrooms.

5. See the many works of Arndt Sorge, Ian Glover, Malcolm Warner, and Michael Fores on *Technik*.

6. See Clifford Gertz's many works. Also works by Larry Laudan who defends the scientific tradition but not by claiming it has "solid theoretical foundations."

7. German, it must be remembered, is increasingly important because of the economic penetration of Germany into Eastern Europe— Poles, Czechs, Hungarians, and Russians are once again learning it.

8. I discuss French strengths in their *Écoles supérieures de commerce* and British weaknesses in language instruction in Locke 1991. Candidates for entry into French schools must pass an oral and written examination in English and one other foreign language for entry. Once admitted, they attend lectures on management subjects in foreign languages (mostly English, although not exclusively), they must do an internship in a foreign language firm and write a paper in the language of the firm—all students, not just those in IBE.

9. One publication, which is consulted universally, exudes this egocentric community disruptive spirit: the Business Week's guide to *The Best Business Schools*. The tenor of the book illustrates what is wrong with the business

school from a collectivist Japanese management perspective. But lest one think the Americans are

alone, just consider the graduates of the French *grandes écoles*.

■ ■ ■

Training and Development of Managers, Entrepreneurs, and Businessmen in Russia: History and Current Trends
Leonid I. Evenko

Of utmost importance in the world economic and social development at the end of the twentieth century is the transformation of sociopolitical and economic systems in the USSR. The seventy-years'-long experiment of applying Marxist and socialist ideas to the practice is coming to its end, with a disintegration of one of the superpowers, affecting the ways of world development and overturning the lives of many nations inhabiting one-sixth of the globe. The phenomena of a few recent years as well as those underway at present on the territories of the states of the former USSR are extremely complicated. They are the result of many factors, and analysts and historians are still to assess and to explain them. But that will undoubtedly take time.

At the same time, Russian economy and society go on functioning. Urgent problems are being solved. A certain inertia exists in the development of economic and social processes in CIS. So in order to manage new opportunities and cope with new challenges, in order to implement a transition from the past to the future, one ought to be able to understand and assess what is being radically changed at present and take steps in the right direction.

An interesting aspect of the current transformations which is of theoretical and practical significance is an evolution, restructuring and developing of the system of training of managerial personnel in Russia and other independent states on the territory of the former USSR. This domain connected with the transformation of knowledge, qual-

ification, cultural level of the people is, on the one hand, of immense importance for the transition to democracy and market economy, but, on the other hand, it is highly sensitive to the changes occurring, responding faster and to a fuller extent to these changes than other areas. Consequently, adjustment processes to these changes run faster here, too.

This article makes an attempt to make a retrospective and partially, perhaps, a forecasting analysis of the developments in this sphere in Russia. In doing so, one should not regard problems of educating, training, and development of personnel to be involved in international business proper independently of the evolution of the entire system of training and development of managers in business, industries, and services. So the analysis undertaken is of a comprehensive nature.

EVOLUTION OF MANAGEMENT TRAINING SYSTEM

Immediately after the October Revolution of 1917 and the Civil War which followed it, the Soviet system undertook the development of its economy in conditions of peace and was confronted with a necessity to effect training and development of personnel able to manage its economy[1]. In all fields, Bolsheviks supported the idea of a class-oriented approach towards solving the problem. And consequently, this could not but tell most decisively on the development of body of managers being extremely class-oriented and politicized in its orientation. At the same time, there were new leaders in Russia capable of a sober assessment of the situation and contributing to a pragmatic approach here. They realized that the revolutionary mottoes and struggle with "class enemies" will one day have to be substituted by routine economic management practices. Those practices were to be effected within the framework of a totalitarian, centrally-planned management of both economy and social life of the country as a whole. So the whole period of socialism prevailing in Russia is typically emphasized by dominating ideology of homogeneously unified and centralized processes of personnel training and developing the relevant state system controlled by the Communist party. No decentralization or initiative outside the official system of management personnel training or training specialists in the sphere on international business could have existed or ever be conceived over the period of more than seven decades.

The initial period was characterized by a dual approach towards personnel training. Firstly, the "Proletarian party" inevitably had to attract intellectuals of the former epoch who, using the phrase of V. I. Lenin, "consisted of 99 per cent representatives of the capitalist class" to its support. However, in compliance with the Bolshevik ideology, these specialists could not be trusted completely, and so every enterprise was typically headed by the leader, referred to as "the Red Director, the Commander of production," who was irreproachably class-conscious but very rarely revealed any professional or otherwise relevant competence. It was for this reason that a decree was signed in September 1920 on setting up workers' departments within schools of higher education targeted at developing new socialist intelligentsia (the famous "rabfacs") providing initial training for workers who had no higher education but were recommended for "a fast-track" promotion in industry and other fields of national economy. In 1925 the first Moscow Courses of "Red Directors" were opened

which were designed to improve qualifications of the executives in their jobs and also to provide a reserve pool for future managers. Started as a year-long course, they gradually switched to a three-year course. In fact, it was a training program compatible to Executive M.B.A. program in contemporary Western practices, though it differed from the latter one in its content considerably mostly because it was ideological and thus was anti-market oriented.

The most interesting phenomenon of the 1920s–1950s in the USSR in this area was the establishment of the so-called Industrial Academies. The first one was opened the end of 1927. Actually, they were nurseries for future presidents and vice-presidents of Soviet industrial firms and other larger organizations and institutions. The term of studies was designed for two years, and those enrolled all had already graduated from technological or engineering schools and had worked for at least five years in industry. The first enrollment included 100 people, with 48 directors of enterprises. During the total period of their existence, between 1927 and 1940, over 31,000 industrial managers were trained there. In its form they were close to full-time M.B.A. courses, but differed from them considerably in their contents. The two foci of training at that time included ideological domination in management and technocratism, the latter being not so alien in Western schools in those days, either.

There was a popular concept at that time, partly fed by "taylorism" as an expansion of mechanized and conveyor type assembly lines, that the source of high labor productivity goes down to application of newest technologies and strict and structured organization at the shop-floor level. Due to this, the bulk of the curricula showed a high proportion of technical innovations, while economic issues and economics at large were confined to studying branch of industries and economics or defining reserves for labor productivity increases in organizing production at an enterprise level. This was further extended by a sizable list of syllabi in ideological subjects: Marxism, political economy, works by Lenin and Stalin, the policy of the Communist party. General education in personnel training in Russia in the Soviet period presented a "mix" of pragmatic knowledge and a thorough "brainwash" of those at the head of enterprises in the centrally-planned economy.

In 1932 the country had 23 Academies functioning, including 11 industrial and 12 branch schools. The latter provided specialized training for transport, coal mining, food processing, and other industries, with the total number of trainees exceeding 9,000. About 10 more Academies were set up in various branches of national economy in the postwar period (the 1940s and the 1950s).

From the point of view of today, this practice can be generally assessed as positive, since it provided specially oriented training centers whose only task was to ensure management training being responsive to special features of the prevailing system and the environment where they were to function as managers. After graduating, the trainees were usually promoted to a higher-ranking position. In the 1920s and the 1930s many of them went for a prolonged internship with leading foreign companies, and acquired relevant experience to be applied in their respective branches of industries.

The system however, was gradually destroyed due to several factors. Firstly, due to the mass repressions of the Stalin's period, physically annihilation was not only representative of prerevolutionary intelligentsia, but many newly trained specialists, for example the red directors, especially those who had had training abroad and had a critical approach to some of the domestic practices or who featured an independent thinking. This had a serious negative impact on the economy at large. Secondly, the concept of training managing personnel for industry was undergoing changes. Starting from 1928 higher technological schools (Technical Universities) began to appear in

the country as the principal facility of training young people as engineers. In compliance with the then prevailing practice of branch management in national economy, these universities also followed the branch orientation: for example, there were specialized technical universities in aviation engineering, auto-vehicle industry, instrumental industry, chemical and food processing industries, and agriculture. This led to a narrow specialization of the graduates, with the universities of a conventional type only ensuring training for future researchers and professors, resulting in a predominantly technocratic orientation of management personnel, poorly supported by inadequate education in liberal arts and based on a general low level of culture. There were very few educational institutions in economics (such as Moscow State Economic Institute, Plekhanov's Institute, or the economic departments of Moscow, Leningrad, or Kiev University) aimed at training economists of a broader type. This resulted in the fact that managers in Russia, even now, have a predominantly engineering background being graduates of an educational institution of the above type.

Gradually the term of studies for engineers was extended to 5–5.5 years. In the number of years it fitted closely to the time pattern of training bachelors and masters in Western universities, but in essence, these programs were different as technological training was predominant while the economic sciences and especially behavioral sciences and general civilization were definitely subordinate. The term of training did not foresee a break in studies between the fourth and remaining 1–1.5 years usually provided for field experience and practical work. "Generalists" now were trained not within the formal system of training but rather practice itself. The most negative outcome of these changes revealed themselves in the closing of industrial academies in the 1950s.

Kosygin's reform of 1965 was connected with an introduction of a more "economically-oriented" methods of management of socialist companies in national economy, it enhanced a new stage in development of the system of training and development of management personnel. Special Resolutions and Orders of the Soviet Government were passed in the period between 1965–67 aimed at improvement both the system of training for management and specialists in different fields of national economy. Since that time, new training and development centers evolved targeted at these fields.

Thus, specialized institutes, similar to Management Development Centres, were set up within branch ministries, central government bodies. They provided short-term training programs—between 3 and 6 months, and frequently shorter courses. Within the period of 1969–84 their number grew from 28 to 79, and the number of their affiliated branches increased from 47 to 134. Interbranch and regional departments for personnel training and development provided services ensuring "generalists." These departments existed at various institutes, including graduate economic schools. The total number of centers which trained generalists was about a dozen, while the number of departments incorporated in graduate schools exceeded 65 in 1969 and grew to 150 in 1984. Specially oriented courses targeted similarly but individually catered were functioning at many enterprises and research institutes and other organizations (their total number growing from 355 to over 750 during the period of 15 years).

The development of a new system of management training and development starting from the 1960s was a way to compensate the disadvantages of the then prevailing system of training at technological universities. On the whole, however, it lagged behind the requirements of the time, like the whole of the system of a centrally-planned economy based on socialist principles of its development.

Table 2.7 provides the structure in the field of management training and development in the USSR in 1988, such as at the moment when it started undergoing radical changes under the influence of a transition of the country to the market economy. Of utmost importance in this system is the top level presenting the interbranch institutes and departments of qualification improvement, with the Academy of National Economy occupying the most important place. These centers are similar to business schools offering Executive M.B.A. programs. They offer annual longterm programs to over 2000 people ensuring basic training in economics and management and providing general training. But the estimated number of managers in national economy of the former USSR was about 7 million, with 10–12% annual rotation ratio. There is, consequently, an annual demand of 88,000 trainees form top executives category. If the number is extended due to the middle-level managers, with 100,000 of them needing an annual retraining arrangement, and 136,000 more of the supervisory-level managers, it is evident that the whole system of training and development ought to accommodate 330,000 people annually. The existing facilities, though, could cover the demands only in a reduced scale. It was for this reason that the most recent years witnessed a development of a fast-growing market offering services in management training and development.

With the personnel for international business taken separately, the changes in the above area are even more acute and more dramatic under the influence of objective requirements. It is worth noting that prior to 1987, when an actual liberalization in foreign economic activities started in the USSR, the demand for the personnel in the relevant field was not so high. The reason for this was "State monopoly of foreign trade"—the principle devotedly adhered to for the 70 years of Soviet power. It was revealed in a ban for individual firms on direct contacts with foreign partners, with the total number of organizations acting as subjects of international activities on behalf of the state being about 150. The system of training the personnel for this sphere

Table 2.7 Management Training and Development Institutions (state statistics, 1988)

Institutions and Industries	Total	Training Centers	Affiliates	University	Programs and Seminars
Manufacturing, construction, transportation, communications, trade	854	86	108	160	500
Vocational training	31	5	26	—	—
Professional high schools	100	—	—	100	—
Universities	157	9	—	129	19
Junior colleges	263	195	—	68	—
Health care	74	16	—	58	—
Cultural institutions	132	8	—	—	124
Agriculture	447	21	8	130	288
Interindustrial institutions	20	20	—	—	—
Business schools	2	2	—	—	—
Independent entrepreneurship	200 (est.)	—	—	—	—
	2080	362	142	645	93

worked traditionally. There were several higher educational institutions providing basic educational programs on the bachelors' level for future employees in international business. They included MGIMO (Moscow Institute of International Relations), a department of the Moscow Financial Institute as well as departments in some leading universities, including Moscow, Leningrad, Kiev and some others. The postgraduate training for practical job in this branch of national economy was almost exclusively confined to the All-Union Academy of Foreign Trade (VAVT), founded in 1931. This academy provides a three-year training program, enrolling annually about 300 hundred specialists with some previous experience in branches and departments such as the Ministry for Foreign Trade and the State Committee for Economic Relations with Third World Countries. Two thousand more specialists used to take

short-term training and development courses at the Institute for qualification improvement under the Academy, their duration varying from a few weeks to a few months. So, this area was controlled by government. The area of foreign economic activities, to a greater extent than anywhere else, was limited to highly concentrated categorized and scarce both in number and in program content.

By its very nature, training there was adjusted to the principle of considering an enterprise as a closed system within a hierarchically centralized pyramid of socialist economy. This was primarily reflected in content of training programs and the patterns of training facilities. Table 2.8 provides information concerning 16 institutes for management training and development in various branches with respect to proportion of individual subjects within the curricula of training people for executive positions in industries. The programs show a certain diversity. Some (for example in electrotechnical industry, bakeries, wood-processing, coal mining) feature a higher proportion of subjects technologically oriented, (accounting for 40% of the total time), management training proper making up from 11 to 67.1%. A more usual distribution of time in these courses reserves from one-quarter to one-third of the total number of instruction hours to these subjects. Courses in economics did not include accounting or corporate finance but a large proportion of time was devoted to the issues of branch economics. The curricula did not include marketing at all. Foreign economic activities was included as a subject into 6 out of 21 curricula, accounting for even fewer instruction hours, since so-called "civil defense" was prescribed as a mandatory course from the regulatory ministry. Most management training institutes definitely underestimated the significance of exposure to experience and field studies. Study tours were provided in only 9 out of 21 programs.

Training personnel for international business, like the whole educational system in the USSR, lacked a real pragmatic orientation. A graduate of the All-Union Academy of Foreign Trade usually had no adequate knowledge of marketing, demand and supply for certain goods in foreign markets, had no real skills in conducting trade negotiations, and were inadequately trained in the methods of economic analysis in open economy. Management courses accounted for only 7% of the total instruction load in the curricula. The trainees of the Academy (age group predominantly 30–35 years old) spent most time on mastering a foreign language (up to 60% of the total time volume), though real practice very often proved insufficient or inefficient to maintain the level of the skills once achieved. Bearing in mind that only about one-third of the graduates got jobs in Soviet agencies abroad, the foreign language skills deteriorated rapidly due a lack of everyday practice, which gives enough reason to state that the whole of the system functioned ineffectively and did not comply with the objective requirements in the area.

There were other serious problems in the area of management training in the USSR. They included archaic, obsolete methods of training (mostly through lectures and tutorials, with a very small proportion of case studies, business games, and similar methods) lack of computers and no access to data banks even of the ministry under which the center functioned. Training facilities were also very poor, providing limited floor space (2 sq. m. for each trainee). The faculty could not do any research because of regulatory limitations, and their salaries were strictly limited. So working for the system of management development centers was less prestigious and less advantageous than working for a university. However, disregarding the disadvantages mentioned above, highly qualified specialists were engaged in these institutes totaling 4,500 full-time faculty and 1,300 part-timers and including 3,200 Ph.D.'s and full professors (Dr.Sc.). Over 50,000 people were invited from industries, management bodies to act as teachers giving lectures and tutorials. It was a way to ensure contacts with routine practices in the course of teaching, bringing instruction closer to life.

On the whole, one may say that during the years of Soviet power in the USSR, a developed and distinct system of personnel training has evolved, adjusted to the prevailing conditions of its time.

CHANGING OF ECONOMIC ENVIRONMENT

How can the essence and the nature of the training system for business and managerial existing in the USSR be characterized and assessed? Theoretically, it should be kept in mind, that humankind during its whole history has evolved only three tools of management, such as affecting other people, different in principle from each other. The first of them is a hierarchy, that is, an organization where the main means of af-

Table 2.8 Proportion of Teaching Courses in Executive Development Programs in Different Industrial Branch Training Center (1988)

Name of training center	Social & Political Sciences	Management	Economics	Engineering & Technology	Law	Informatics	International Business	Civil Defense	Others	Total %	% of Classes	% of Internship	Total	References
1. Chemical industry	2.0	15.8	33.4	17.0	7.5	4.6	7.8	1.3	10.6	100	48.5	51.5	100	Corporate DEO
2. Timber industry (I)	4.6	12.9	22.3	33.3	9.6	8.3	1.8	—	7.4	100	75.0	25.0	100	Corporate DEO
Timber industry (II)	3.8	16.6	23.2	24.0	13.8	4.6	—	7.5	6.5	100	75.0	25.0	100	General managers
Timber industry (III)	3.8	11.0	22.1	31.4	7.3	5.4	3.6	5.4	10.0	100	75.0	25.0	100	General managers
3. Ferrous metal industry	4.6	28.8	43.6	—	6.5	13.8	—	2.7	—	100	50.0	50.0	100	Promotion to general managers and chief engineers
4. Oil and gas industry (I)	4.2	34.7	43.0	—	4.8	10.0	—	—	3.3	100	100.0	—	100	
Oil and gas industry (II)	2.8	67.4	10.1	—	2.8	12.6	—	—	4.3	100	59.0	41.0	100	Promotion to Ministry job
5. Coal industry (I)	3.9	24.6	31.3	29.3	3.2	4.8	—	—	2.4	100	87.5	12.5	100	General managers & corporate technical directors
Coal industry (II)	6.3	18.1	20.9	16.7	4.2	6.9	—	6.9	20.0	100	100.0	—	100	Mining directors
6. Moscow Institute of Management	3.9	37.1	38.0	—	3.6	—	—	—	17.4	100	90.2	9.8	100	
7. Moscow Plechanov's Institute	2.1	22.2	38.5	—	—	12.2	8.7	—	16.3	100	80.0	20.0	100	
8. Energy production industry (I)	4.9	25.3	14.9	10.6	3.0	8.9	—	—	32.4	100	100.0	—	100	Construction company heads
Energy production industry (II)	6.9	36.3	23.8	15.5	3.9	5.8	—	2.3	5.5	100	86.6	13.4	100	Power station heads
9. Radio & electronic industry (Leningrad)	1.5	20.9	14.0	5.5	8.3	12.5	4.1	4.1	29.1	100	100.0	—	100	
10. Radio & electronic industry (Moscow)	6.4	26.0	21.8	17.4	4.2	4.2	—	6.5	13.5	100	74.4	25.6	100	
11. Automatic industry	4.8	32.7	16.6	16.0	4.2	14.6	76.0	1.4	2.0	100	100.0	—	100	
12. Electroengineering industry	3.2	35.5	8.9	36.3	3.2	—	—	—	12.9	100	100.0	—	100	
13. Telecommunications industry	—	38.9	13.0	13.0	11.2	13.8	—	4.6	5.5	100	100.0	—	100	
14. Retail trade	5.5	22.9	47.2	5.0	6.9	6.9	3.5	3.5	2.1	100	100.0	—	100	Department store heads
15. Grain and bread products	5.5	25.6	14.5	24.4	8.9	5.5	—	8.9	6.7	100	69.3	30.7	100	
16. Heavy machinery industry	—	49.1	13.9	13.0	8.3	8.3	—	4.7	2.7	100	100.0	—	100	

fecting is a relation to power: subordination of a person below to the domination of the one above by enforcement, control over compensation, and similar means. The second is culture, such as values elaborated and recognized by society at large: organizations, groups, attitudes, patterns of behavior, and rituals that very often predetermine an individual's behavior and actions. The third of these tools is a free market, a network of fully fledged relations of peers, based on their sale-and-purchase of products and services, on property relations, and on a balance of the interests of seller and customer.

Hierarchical organization, culture and the market are highly sophisticated phenomena. They are a part of social formation of any society. Capitalist economy had been long considered a free market economy, while socialist economy is a planned hierarchical economy. In fact, though, hierarchy, as well as culture and market, as management tools and as manifestations of live economic social systems have always existed in any social formation. We can only speak about priorities in a particular social setting and emphasis. These are the factors determining the essence and the manifestation of the economic organization.

The basis of the soviet administrative-command system was a hierarchy of, so to say, universal character. Everything had its line of reporting, a higher authority and the powers of higher executive authority were unlimited in any way, this was, true, for instance, about the USSR Council of Ministers. But parallel with these practices, the Soviet society has been using a very rigid "hard" culture as a tool of affecting its members. Through ideology, membership in the Communist party which was practically unavoidable for a manager determined at his career development, through the influence of mass media and education. Through officially supported customs and traditions, people in the USSR learned various forms of social limitations and advantages, following acceptable standards of social behavior supervised and approved by

partocracy. They either followed the principles or ran into a conflict with the existing system.

The administrative economic system which has gone through a long period of evolution and development in the USSR was well adjusted and coordinated in its major elements. Since the Stalin days, it made "curbing" society and organizations its principal goals, disregarding all expense—goals that were set from above, to suppress any outward manifestations of incipient conflicts. The whole system was aimed at an aggressive satisfaction of lower social demands and needs of people, the need of affiliation, the need to belong to some group. Despite the rigid suppression of the individual's rights and the myth of universal participation in creating the happy future for the country under the wise leadership, the idea of a highly just society was the basis of the world outlook for the majority of members of this society.

Management and personnel development in Soviet organizations were for decades dominated by and adjusted to the demands of a certain environment, this being the administrative-command system.

The adaptation to this system, not only to its legally organizational and economic mechanisms, but to its policy, ideology and system of values, went on very actively and not quite unsuccessfully in its own way. Fulfilling the plan sent from a higher level of authority, very often at whatever cost, instead of satisfying the needs of a customer; an expansion of enterprises, increase of their production volumes irrespective of its quality and resource consumption, stability instead of dynamics; unification instead of diversification; subordination instead of initiative encouragement and freedom; and using the right of foreign economic relations as an exclusive prerogative of the state and a source of potential ideological threat for managers, those as well as other requirements of the economic system gave rise to specific forms of management allowing to adjust it to specific conditions. Using modern classification, the practices prevailing earlier in Soviet industry can be characterized as bureaucratic hierarchical systems of organization. These requirements preconditioned the system of personnel management, personnel training and development. And now the main task is to overcome the existing prejudice of the "managerial civilization" of a bureaucratic orientation which has proved its historic incompetence and needs a radical, revolutionary restructuring for developing productive forces and ensuring individual human rights.

Market, as a tool of managing economy, as a product of a law-dominated society and the economic realities, is in its very essence capable of this adaptability. Hierarchical organization is a rational means of introducing a stability and regulation into economic and other types of activities. It will not die and will not ruin itself. It will just step into some sectors of economy demanding higher degree of control. It will go deeper down—to the level of individual organizations, while organizations will adapt themselves to the new external, and partly internal environment, and in the depth of their structures, the bureaucratic, mechanistic, systems and management systems will gradually give way to organic, flexible nonbureaucratic structures and systems. Besides, hierarchy and market have their own cultures almost completely opposite in their essence.

We are likely to expect what can be characterized as a "tectonic" cultural shift in our economic and managerial thinking and psychology. It is necessary to orient the consciousness of the managers and employees towards customers but not towards their bosses, towards profits-making rather than towards extravagance, towards entrepreneurs rather than towards bureaucrats, towards an innovator, rather than towards a thoughtless executive, towards pluralism rather that towards unification and standards. The whole of the management system ought to be oriented at motivating employees for hard work and industriousness, and turn investors towards effective use of capital, and managers towards profit

making and rational use of resources. State officials ought to be motivated for ethical and responsible behavior.

At what stage of economic transformations is the economic system of Russia at present and what demands does this impose on the personnel training and qualification upgrading?

As we see it, three stages can be distinguished in implementing the restructuring program (the Perestroika) since 1985.

The first stage in 1985–87 marked in fact only general ideological and psychological, and perhaps partly conceptual preparation for the restructuring proper. At that period the former system of the party and state leadership remained unchanged and the basic principles of economic management using the planning and socialist principle were invincible. At this stage the perestroika concept strongly reminded one of the reform of 1965 conceptually but was aggravated by the failure of a new "acceleration" concept. The most important feature of this period was retaining the rigid party and state control over society at large and the economy.

In 1987, however, the second stage began, marking a real restructuring and featuring openness (glasnost) in the ideological area. This caused a turn of Soviet society towards democratization and had an immense impact as a preparatory stage for a more profound changes. This was also the time of adopting the first liberal economic legislation.

Thus, in January 1987 the government act on joint ventures came and, as time showed later, it turned insufficiently radical to allow new really radical changes into foreign economic mechanism. Nevertheless it implied a revision of an unchangeable principle of the socialist economy in the form of "state monopoly on foreign trade." First new joint ventures started to be set up between Soviet and foreign businesses, in Russia their number growing from a few scores to 2,600 over 5 years. In 1988 the Law on Cooperatives was adopted, which actually opened the way to private property on the means of production in the USSR. It was further enhanced by Law on shareholding companies and partnerships Ltd, adopted in summer 1990. The private sector today includes shareholding associations, cooperatives, leased enterprises and individual private producers and in October 1991 accounts for 13.1% of the industrial output of Russia employing 23.7% of people. This testified to definite positive changes and shifts on the way of Russia towards market economy.

However, at the same time, the period of 1987–91 featured an ever more counteractivities of powerful antirestructuring forces in the leadership of the country. Nickolay Ryzhkov, V. Pavlov, and even M. Gorbachev were not ready actually for real radical transition of the country to the free market. It is not by chance that the "500 days" program, (for all its limitations and somewhat utopian character from the point of view of its practicality) rather progressive in its nature, was sabotaged in fact by the economic leadership in 1990. The political struggle between the democratic forces, the conservative center, the rightwing forces and ethnic national conflicts held the leadership of the country headed by M. Gorbachev from resolute and well thought-through transformations of the country's economy.

The new third stage started in summer 1991, when the aborted putsch of the conservatives brought about the actual destruction of the former state system, causing first a crash of the central power agencies and then the Soviet Union as a single state, leading to eliminating the Communist party and the KGB from the political scene, to the change of people in political leadership of the country. For all the contradictory character of this transformation, the extreme difficulties on the way to a civil society, to a genuinely democratic mechanism of power, towards a market economy, it is quite evident that it was a historically progressive overturn. It signified a termination of an experiment over half a century long of an attempt at socialism in one single state, one of the largest states of the world. At present

there is a definite trend of the peoples of Russia turning to capitalist system of economy with all its production relations and social institutions. Despite the fact that those in favor of the return to socialism sometimes still raise their voice, the majority of the population of Russia irrespective of their fact of their support of the democratic forces—the left or the right—act as supporters of a nonsocialist and moreover nonbolshevik way of development. It is of immense importance and at present it requires a new step in implementing economic reform. The program being implemented at this stage includes the following aspects: financial stabilization, privatization, price liberalization, social setting, income regulation, international economic policy, interrepublican relations, and systematic change. At last the radical (might be even too radical) economic reform has started up. It creates a really new economic environment and sets up new requirements to management and creates new roles in organizations.

NEW REQUIREMENTS FOR LEADERSHIP AND NEW MANAGERIAL ROLES

Along what lines are the requirements to personnel and the system of training, retraining and development changed at the time being?

Suffice it to say that these changes are very serious ones and are radical by nature. They are being realized in two ways. On the one hand, the nature and essence of activities of the trades which have existed in a planned economy and will always exist are changing. On the other, new roles are emerging in a free market economy as a result of new social and economic structures which need new skills and competence unknown before. First of all such changes are concerned with three levels of managers: top executives, middle managers and supervisors. Attached to them are specialists of various kinds who are not involved directly in formal management structure but who play a very important part in management process.

Under the former system, the main strategic decisions on enterprise and its environment, concerning investment, product line mix, and distribution of critical resources, had been taken not on the enterprise level but by the Central or Republican governments—by the Council of Ministers and its staff at the industry branches, managed by the ministries, or functional agencies (the State Planning Committee, the State Material and Technical Supply Committee, the Ministry of Finance, and others). The executive and managerial personnel was precisely within the framework of this system. As a rule they had been promoted to the government from industry, usually from the positions of directors of the largest industrial corporations. In some cases they have a career path inside a certain bureaucratic institution: from a specialist to a top executive (but usually no higher than deputy minister). Soviet chief executive officers and directors (presidents and general managers of corporations and its affiliations, according to Western standards) and experts from industry brought the spirit of live industrial practice into the state bodies. But as a rule the choice of people for promotion was made on the basis of the so-called "nomenclature list" which was under control of Central Committee of the Communist party. As far as the top officials were concerned, this list took into account not only competence but also the belonging to one of the interest groups (close "clique"), which was in power at a particular time at the level of Federation and Union republic, a region or a ministry.

The nature of the Soviet power system as hierarchical bureaucracy and oligopoly led to the fact that the principal part in strategic management in the end was played by two types of executives: bureaucrat (very often with technical background) and partocrat. Bureaucrats and technocrats were engaged in the actual economic activities, whereas partocrats representatives dealt with more general questions—supervision of the society. However on the whole top executives level was formed on the basis of

close overlapping of bureaucrats and partocrats.

It is worth mentioning that this tendency was not pronounced so strongly in the 1930s and 1940s when, from the force of circumstances, rather many young talented and enthusiastic people happened to head governmental institutions, replacing old executives leaving the stage or victims of Stalin's reprisals. But in 1970s–1980s, when these generation got old, the system did not provide an adequate and timely replacement and rotation for them and its stagnation and bureaucratization became especially evident.

In 1990 under new conditions of a transition to a market economy—the former roles of top executives proved to be devoid of any sense. Partocrats as a social strata are leaving the stage. They are being replaced by new politicians, acting in the environment of many political parties in Russia, openness ("glasnost") and democratic institutions. The role of bureaucracies after central ministries and agencies have been eliminated. The economic function of government is rapidly declining. It is limited to regulatory functions and it has nothing within to do with the direct management of the economy. Strategic decision-making, defining the essence of economic development, is being transferred to the level of single enterprises, firms and then in turn is changing the structure of ownership. The foundation of it is emphasizing private capital and independence.

The former leaders are being more and more replaced by a new kind of people—businessmen and entrepreneurs. The businessman is a man who makes money, who is at the head of a new type of enterprise—a capitalist one, existing for the sake of making profit. The entrepreneur takes the risk of a new business, the development of new opportunities. According to ideology predominant at present, they should be the very people capable to be the driving power of new economy and new society in Russia. These social strata do not come out of the blue. They are the corollary of transformation of some social groups existing earlier, the result of social selection in the transition to a market economy system. They are the bearers of expertise and culture existing in the past and now being adapted to new conditions. Who are these people and these strata?

It seems to me one can identify five groups which give rise to businessmen and entrepreneurs in the new Russian society.

The first group of entrepreneurs and businessmen in Russia is formed on the basis of laundering the capital of people who had been engaged in shadow business activities before. According to common Western theory, these could not be entrepreneurs in the former Soviet Union because any entrepreneurship activity was considered criminal. In fact this is not true to life. It is necessary to take into account that in the economy of deficit there had always existed objective conditions for shadow economy and for the last 10 years it was especially typical for the Soviet Union. According to some experts the shadow economy related to about 20% of the Gross National Product of the former Soviet Union. Some 25–30 million people were involved in this kind of activity. Two to three hundred thousand were taken to court every year and were condemned for economic crimes. The shadow economy turnover is estimated from 100 to 300 bln. rubles, depending on the technique of counting: the first place among various sides of criminal deficits revenues is 40 bln. rubles. The second place is given to overreporting of enterprise performance: as fiction, it imputes the basis of which from 30 to 35 bln. rubles were illegally appropriated (in the form of illegally paid bonuses, for example) An illegal home alcohol distillery took the third place at 23 bln. rubles; 6 bln. rubles was acquired at the expense of hidden capacities and the activity of clandestine production liners, the "workshoppers"; direct stealing amounted to 3 bln. rubles; briberies 1 bln.; prostitution gave 700 mln. profit; pornography 300 mln. On the basis of accumulation and redistribution of these assets, the initial accumulation of private

capital was being formed and this capital in the hands of certain groups of economic mafia amounts to billions of rubles. At present it is in the legal turnover and increases at a great pace.

People involved in the shadow economy have all the characteristics of entrepreneurs. It is estimated that two-thirds of them have transferred into legal businesses and one-third remain clandestine, due to the illegal character of their business or on the considerations of tax evasion. Under conditions of radical transition to a market economy, price liberalization and avoidance of state control over the economy, many sources of illegal profits (such as artificially created deficits and performance overreporting) are losing their significance. But at the same time some other spheres of shadow economy, such as briberies for granting business or export licenses and getting permits for real estate acquisitions, are rapidly expanding.

The shadow economy entrepreneurs are the bearers of "hard" culture which effects the mentality not only of these people, but Russian entrepreneurs in general. Their influence is of complex nature. On the one hand they represent rather dynamic component of changes in objectively indispensable directions, but on the other hand many aspects of their business culture (disrespect of law, for example, or social irresponsibility) are negative.

The second group of entrepreneurs are talented young people with no previous practical working experience and life as it is under bureaucratic socialism. They are not overburdened with many prejudices of other social strata and their abilities and skills fitted business activities quite well. It is a so-called "tough youngsters," sometimes they are aggressive, at times they pursue their own business activity but very often they attach to shadow economy people or to foreign businessmen in Russia. Being young, thirsty for life and adventures these young businessmen are aimed more at consumption and wastefulness, westernized life style and outward success than at efficiency and accumulation of capital. Nevertheless they are badly in need of skills and competence and provided they mature, they will probably get cured of adolescent diseases and at least part of them might become true leaders of Russian economy. This group is a raw material for new business culture in Russia and is worthy of a very close attention.

The third group of business people in Russia is represented by people with no connections with shadow economy. They come from normal life conditions existing in the past. It is necessary to bear in mind that in July of 1991 on the territory of former USSR exist about 250,000 cooperatives which produce more than 7% of GNP. The survey showed that from 6,500,000 employees of cooperation, 40% constitute people who were doing management jobs and 30% were engineers and technical specialists. Very often they use their technical background to set up production enterprises. They are well prepared to manage with high efficiency under new conditions of a free market economy.

Their approach to business and their moral values in entrepreneurship are mainly positive. They need training and development and brushing up their business and executive skills and knowledge in market economy conditions, especially in the field of international business.

The fourth group of businessmen are people coming from Soviet establishment who used to hold top positions in the industry. These are former bureaucrats and partocrats converted into businessmen as a result of "nomenclature made privatization." This kind of privatization, which is being carried out for the last two or three years, very often done by some tricks as a legislation evasion put big companies in fact into the hands of top administrative personnel.

Former top executives of state enterprises are well briefed to deal with concrete situations. They have long-lasting unofficial contacts with power structures and they are consistent in securing their interests. They are expertly transforming themselves from one elite stratum of the society to the other. It is hard to estimate their impact on the economy and business culture of new Russia. For many of them past traditions and prejudices become factors of hindrance of transition to capitalist economy. But it is necessary to bear in mind that they had already gone through social selection process and it is quite plausible that new economic system would create for some of them favorable conditions for the development of potential as businessmen.

The fifth group consists of foreign businessmen who dared to pursue serious business in Russia. There might be among them small entrepreneurs who do not consider their activity in Russian seriously and pursue only immediate objectives. But some foreigners are people who can and should play an important role in the development of Russian economy, in setting up business practices in transmission of executive expertise and forming a new business culture.

In general it is possible to state that at present within the framework of Russian economy an intensive process of "the third class" formation is going on. These people need new competence which is acquired by experience and by taking up special programs of training, retraining and upgrading.

At the level of middle management and supervisors very serious changes are also taking place. Some specialists who take part in the process of management also in one way or another need similar professional training in the field of management. The main source of changing attitudes towards management lies in the fact that executives and managers have to work in the environment of the open system, under strong pressures of competition and uncertainties. The typical figures of the past, crucial to the success of an enterprise in a centralized planned economy, were engineers, technicians, planners, accountants, suppliers, controllers and administrators in a broad sense of the word. Their activity was directed inward. They managed to provide for the enforcement of regulations coming from above—from ministries, party committees, directors, and others

At present some of them still maintain their functions, mainly those who belong to "technological nucleus" of an enterprise less subjected to environmental changes. However, along with these people there appear in formation new roles, unknown before. In particular the supply problem, which was crucial for efficient performance of an enterprise in the past, when resource and materials had been distributed from the center, is being gradually replaced by the problem of sales. If an enterprise has a contract, it can function, make a profit, and consequently buy necessary resources in the open market.

Therefore, under new conditions the marketing function is acquiring top priority as well as the function of price-setting which was practically nonexistent in Soviet enterprises in the past. The sale of products is also being organized in a new way, including service, providing guarantees of the delivery and so on. The role of financial workers and accountants is also undergoing substantial changes. In the past, enterprises functioning as large corporations in the ministry system did not work out their own normal financial, credit, and investment policies. Thus, the ministries (like headquarters of large corporations) withdrew their profit, exceeding a "normal level of profitability" at their own discretion. So the expense of profitable enterprises there covered the losses of those who failed to reach normal financial performance. At present under conditions of free market economy this past experience is left behind and financial strategy is becoming most important. An enterprise can go bankrupt easily if not successfully. It is becoming absolutely necessary to analyze the structure of expenditures, the anatomy of profit, and the ways to increase cash flow and to find reserves and means of saving resources. The financial activity of an enterprise is closely connected with strategic planning, investment policy decisions, and analyses of portfolio, as well as with marketing and accounting.

All these functions are fairly new for Russian enterprise.

The functions of engineers and technicians are also undergoing substantial changes. Engineers and technicians have to be oriented on innovations, on exploration of new technical decisions and technologies which make a new product competitive, improve its qualities, and reduce its production cost. It is becoming more and more significant in changing the role of those who work in engineering departments.

The functions of lawyers at the enterprises and other bodies are also changing dramatically. There is a great demand for business lawyers in Russia at present. Because of a shortage of well-qualified lawyers these people are extremely well paid now in contrast with the situation in the past.

During transition into a free market economy a new business infrastructure is being formed and it creates new roles totally or mostly unknown in the Russian economy before—for example, brokers, commercial bankers, advertising people, importers-exporters, and business consultants. In the past in the absence of commodity and stock exchanges, a very large, awkward, and inefficient system of state material and technical supplies (*Gossnab*), distributed resources in a planned manner. State banks served only as distributors and controllers; credit and currency markets were nonexistent. The role of consultants, who pass their highly specialized expertise to those who need it on a commercial basis, is increasing substantially. New data banks and information service networks are being created as well as services in fields such as advertising and public relations.

The same changes are taking place in the requirements to personnel in the field of international business. Under the former system, specialists in this field were concentrated almost exclusively in two governmental bodies—the Ministry of Foreign Trade of the USSR and the State Committee for Economic Relations—mostly in their foreign trade divisions (totaling about 60). There were specialists on international business in Vnesheconombank (Bank for Foreign Economic Activity), in the Customs Service, and so on. The level of training and skills of employees in these bodies were traditionally very high. This rather small detachment was being formed gradually and thoroughly. Practical experience played even far more significant a role than formal education. Strict discipline and control over their relations with foreigners (on the part of KGB) made these people as a rule responsible at their duties; moreover, prestige and economic advantages of working abroad or just going on a business trip abroad served as powerful incentives for good performance and proper behavior.

Under new conditions there has taken place a breakthrough of a dam constructed on the way of international business by a "State monopoly over foreign trade." At the beginning of 1992 the total number of institutions registered as participants in foreign economic activity in CIS countries constituted 45,000, in Russia only 25,000. A new presidential decree in Russia of 18 November 1992 on liberalization of foreign economic activity makes it even easier for Russian enterprises to pursue this kind of activity. This right is automatically granted to any legal person, registered as the subject for business activity. This liberalization of foreign economic activity should be evaluated as domestic positive from the point of view of transition to an open market economy, but at the same time is fraught with numerous problems. One can note a sharp decline in qualification skills and culture of average people who are engaged in foreign economic activity. Those who are trying to do business very often lack proper training with foreigners even if they are top executives of companies or by their positions are involved in international business activity regularly. As a rule they do not speak foreign languages and are completely ignorant of the technique of foreign trade operations. They have very little knowledge about world economy and business legislation. They are not qualified to write contracts or carry out negotiations. Some people substi-

tute the interests of their enterprises with their individual interests, using the possibility to get hard currency as legal or illegal payments for travel to go abroad. These cases are rather sensitive. It is impossible to get rid of them in a short period of time. It will be a reality for some time and will disappear with time on the basis of adjustment to new practices in doing international business, in the process of retraining and under the influence of foreigners.

On the basis of this survey one can come to the conclusion that the system of training, retraining, and upgrading of personnel is facing principally new objectives, requiring its cardinal structural changes. I would like to say that this system is very seriously adapting to new conditions now. Its development is carried out, on the one hand, on the basis of existing personnel training institutions which were mentioned above; on the other hand, on the basis of setting up new structures having similar functions.

NEW INSTITUTES FOR MANAGEMENT TRAINING AND DEVELOPMENT

As it seems now it is possible to define three types of organizations trying to carry out the work with personnel in new conditions.

The first type is a state system or a system being transformed from the former state structures: for example, the recently set up center of educational and scientific organizations in the Committee of External Economic Relations of the Russian Federation. It is supposed to coordinate personnel training in the field of international business; eleven schools of business and upgrading institutes specializing in this field report to it. There has been a decision to work out an All-Russian program to train personnel for market economy under the aegis of the Russian Labour Ministry. The National Committee to certify schools of business is being set up, and those ones that will receive such a certificate are going to get taxes exemptions. In a sense such an approach as that which supposes to maintain the active role of the state in this field has a well-known drawback: that is, inertia and conservatism. However, it is connected with restructuring of the established system and it must be inevitably implemented.

The second type of institutes which carry out personnel training for market economy are newly formed small businesses, cooperatives, or consulting firms which carry out such training on a private basis. As a rule, the initiators of such services are groups of experienced specialists who have a high qualification and deep knowledge and those who are ready to sell these services on an independent basis, not to share profits with unnecessary state-governed structures. A lot of cooperatives and consulting firms are set up at higher educational establishments, research institutes and other enterprises, as their independence allows them a greater flexibility, takes them out of salary regulation of the state, gives them more freedom, becomes a source of additional income for professors, whose full-time jobs are at a higher educational establishment. It also concerns scientists and specialists. This sphere is being developed in an irregular uneven way. Together with successfully made up programs giving real help, there are a lot of purely go-between organizations, which do not have their own serious potential, and which seek short-term financial objectives and their education is of low quality. A really serious problem for such organizations is a problem of material and technical resources. They do not have their own premises, computers and so on. It is most probable that in this sphere an intensive selection is going to take place, and a lot of businesses of this type will go bust, but the best might become real leaders of personnel training in a number of fields.

The third type of institutes which carry out such training, they are joint ventures and foreign firms operating in the Russian market of personnel training services. Some of them are formal on the basis of joint programs together with serious Western partners. Russian-American, Russian-Japanese, Russian-English, Russian-Finnish, and other schools of business, training centers, and even universities have been set up and they really perform transfer of technology in the field of personnel training for management and international business taking into account its adaptation to specific domestic conditions of Russia. It should be borne in mind, admitting at the same time, that such organizations, international in their nature, can play a positive role as a vehicle of transformation of the present-day Russian system.

The most important question here is likely to be financing of such new centers in hard currency, as the present-day conditions in Russian do not allow to adequately compensate foreign experts and professors who are invited to teach here. Success usually is the result of those that enjoy serious financial support from foreign funds.

At the same time there is an active market of Russian management and international business personnel training abroad. Experience shows that to send training groups abroad is quite profitable. But it should be emphasized, that interest of two types are combined here. On the one hand, people really want to receive knowledge and with their own eyes to see the same enterprise function in the West, but on the other hand, when the ruble is going down, a trip abroad for those people is business because a possibility of getting a certain sum of foreign currency for traveling expenses is of great value thanks to the low rate of ruble. That is why top executives very often become participants of such programs. They do not consider that sort of training really useful for themselves and they do not need it, but they sometimes pocket several thousand dollars because of a trip. Besides, Western firms, particularly émigré-founded, which were set up by the former Russians, many of them travel agencies under the pretext of managers and businessmen training, offer services with a lot of entertainment and low standard of training that our visitors do not notice. Their main problem is to enroll people who work at the enterprises able to pay for training in hard currency. Because of such competition they use unlawful actions, for example, paying private persons, which is not allowed, hard currency to form such groups, or offering big bribes to those personnel officers who are responsible to send people abroad to study. As a whole, one may state that at the present time, the system of personnel retraining and upgrading skills in management and international business acquires a number of new traits which did not exist only several years ago.

As a separate subbranch of Russian economy this kind of service of personnel training was the first to become market-oriented, or better to say, a mixed sort of economy. It is important to take this fact into consideration. At present centralized government financing of this process is diminishing and commercial services, when the interested party pays for itself, have become widespread for all institutes, and particularly in postgraduate education. The infrastructure of this market is being developed: a great number of advertising leaflets are sent out, and also informational letters and bulletins; there are announcements on television and in the press and so on. A well-known balance between demand and supply is being achieved in this sphere. The most important positive feature of this process is a quick increase of adaptation of all personnel training system to the concrete needs of different groups of trainees.

Suffice it to say, for instance, now that overall privatization of the main capital in Russia is about to start, a lot of institutes conduct special programs, workshops that make clear the process of privatization, pointing out how to do it correctly and profitably. Brokers' training is carried out in a decentralized way. It is designed first of all for young people eager to try this sort of business. Training of accountants, auditors, financiers, marketing and advertising specialists and qualified salesmen training is in great demand in new conditions. Such programs literally appear within several weeks when they are needed.

However, the commercial basis of this network has now not only advantages, but disadvantages as well. The main thing is that not everything that is profitable is really valuable. In a transition period there is a danger that because of deficit financing, some schools and centers will be obliged to cut their operations and might lose their potential which has been created for years. From a lot of state personnel training institutes people go to private business improving their welfare, but are really doing harm to the standard of training in big centers.

If one evaluates the process going on in the sphere of personnel training for international business, a considerable change in the state of market for the last several years should be underlined. In 1988–89 after the first decrees about the liberalization of the foreign trade activity were adopted, there was a great interest in this sphere occurring everywhere. Many top executives, that is, presidents and vice-presidents of Soviet firms, tried to get such knowledge as soon as possible, and they had to go onto the foreign market to set up joint ventures and have direct contacts with Western partners. This interest has diminished lately, mostly thanks to instability of international business in Russia, hard currency and financial difficulties, and frequent changes of laws concerning foreign trade. As a result there appears a great demand for personnel training from time to time as regards certain specialties. For example, bookkeepers and financiers engaged in international operations, hard currency cashiers, secretaries and administrators with a good command of a foreign language are in great demand. Workshops (tutorials) devoted to international law, cultural specialists of different countries and the like are always popular. There is no doubt about it, that the demand for certain programs depends on the reputation of an institute or a school which offer these programs. However, the ability to feel what the market wants in this field is an important factor in competition and success. One may forecast a considerable demand for such services in Russia in the near future, but it is very closely connected with the stabilization of the whole economic situation.

Personnel training for international business on a longterm basis (for one or two years) is of some importance too. The experience shows, that if in other fields, for example in the field of privatization, it is sufficient to conduct a short seminar, refreshing operating officials qualifications, in the field of international business more fundamental knowledge is necessary. Today the main problem is the problem of learning a foreign language: because of the Soviet economy being closed in the past, most officials and practitioners who have the necessary qualifications and experience have never bothered to learn a foreign language. This is particularly important now that trainees either have a working knowledge of a foreign language, but no professional qualifications, or they are real professionals, but know no foreign languages. The other important aspect is that going out onto the international market for many specialists means study of quite new subjects which they have never taken up before, starting with international law and winding up with, for example, making up contracts, conducting talks, or dealing with intellectual copyright problems. When an official academic basis is absent, training in this field in higher educational establishments becomes more difficult. Finally, short-term business trips abroad are of great importance. Actually such programs should be formed on a "sandwich" basis. Both extremes are not effective: training only abroad or only in Russia. The first is good for young people who have a long road ahead, and those who have a certain background and experience in Russian economy. But teaching case studies in western schools of business is designed for different demands, very often far from Russian scene. Having received such a training, a person may find himself unable to tackle practical problems at home and will prefer to work with a foreign firm. And vice versa, if a training or refresher course takes place wholly in Russia without direct participation of foreign teachers and without a trip abroad, it will be somewhat conservative. But the main thing is that it will not be able to acquire an international mentality and a new culture which is so necessary to take part in international business. For young people training, in particular, M.B.A. programs in the field of international business are a longterm investment in human capital. And it should be realized according to the contents and the forms of organization, to the length of studies. It should differ from forming and realizing of short-term programs and workshops for executives and specialists-practical workers.

But in both cases a drastic change of management training and development system for a market economy, and, in a more narrow aspect, for international business, should be realized on the basis of foreign experience and international cooperation. Russia is expecting help in this field. The return will be sound—Russia's step by step integration into the world economy and access to international markets. Both Western and Eastern firms will gain, and particularly those that will be the first to cooperate with Russia.

NOTES

1. Professor Vladimir I. Matirko (Academy of National Economy), a specialist in the history of setting up and development of the personnel training and upgrading courses system, has kindly offered materials of his paper for this article. The author is most grateful to him.

■ ■ ■

Strategic Dimensions of International Business Education at the University of South Carolina
William R. Folks, Jr.

INTRODUCTION

In 1990, 1991 and again in 1992, *U.S. News and World Report* ranked the International Business area at the University of South Carolina as the top IB area in the United States. Academic studies of IB programs and faculties have consistently placed this university at or near the top.[1] While rankings of this sort are fraught with potential inconsistencies, clearly, on the basis of reputation alone, a review of the strategic development of global activities at the University of South Carolina may provide lessons regarding the dimensions of institutional strategy of benefit to institutions of higher education as they seek to respond to the challenge of relating to an environment where business is conducted globally, rather than locally or nationally.

This paper does not purport to be a complete history of the emergence of all aspects of the international activities of the College of Business Administration at the University of South Carolina. Rather, it seeks to highlight certain key strategic decisions and organizational changes which were critical to the emergence of excellence at the University of South Carolina. It further will focus on the lessons learned from the experience of twenty years of involvement in international business education, and will seek to separate the institutionally unique from the generally applicable.

Nor does this paper purport to be a classic exercise in academic research. Rather, in historical terms, it should serve as a primary source document, from one who, in Dean Acheson's terminology, was "present at the creation." In spring 1972 I taught the first business course offered at the University of South Carolina with an international dimension, BADM 561, International Financial Management, a mixed graduate and undergraduate course providing an overview of international corporate finance, with a class of four students. Since that time I have served as a member of the original task force for the design of the Master of International Business Studies (M.I.B.S.) degree program (1972–75), the first department chair in International Business[2] (1976–85), the chair of the College's Globalization Task Force (1985–88), and the director of the Center for International Business Education and Research (1990–present). In addition, I have been charged with the development of the International Business component of each of the college's strategic plans, including the most recent.

THE STRATEGIC PREMISE OF INTERNATIONALIZATION

Over the twenty years since the university's first tentative international activities, area faculty have developed a working strategic model which we seek to apply in structuring our efforts. That model is explained in greater detail in section 3 of the paper. The model, however, is based on a fundamental premise, that the globalization of business activity is such a significant change in the managerial environment that preexistent activities by business schools, whether in the research, teaching or service area, must be altered significantly to remain relevant.

Business education must constantly change in response to changes in its external environment. The development of cheap and powerful personal computers, for example, has changed the managerial environment, and, consequently, the manner in which business educators must prepare their students. Similarly, the rapid integration of the global economic system in the past twenty-five years and the globalization of business and financial activities have created an equally challenging set of environmental changes and opportunities for managers. Those of us at business schools who are called upon to equip future managers to meet these challenges must also change.

Generically, the process by which business schools respond to the globalization of the marketplace has been referred to as "internationalization." Primarily and properly, for most schools, that response has taken place in the area of curriculum design. To meet the general needs of all managers, many business schools have introduced internationally oriented material into core and major area courses. A more limited number of schools have developed minors, and majors or degrees in international business.[3]

Core curriculum internationalization is an important component of any business school's response to the challenge of global economic integration. The graduates produced by international business degree programs also meet the market's demand for skilled specialists. However, these two responses, however well implemented, do not result in an internationalized business school, nor do they complete the menu of potential responses which business schools can or should make to the phenomenon of the globalization of the economic system. A more comprehensive view of internationalization is needed, in keeping with the multiplicity of activities carried out by business schools.

Internationalization of the business school can best be characterized as the development of a global orientation to all aspects of a business school's endeavors, be it in teaching, research, or service, consistent with the school's overall mission. Thus, strategic premise of the internationalization strategy of the South Carolina's College of Business Administration is as follows:

> The changes required by the globalization of business provide a watershed opportunity for the strategic redirection of business schools. Those schools whose strategic response best meets the challenge of globalization will improve their market position and social value dramatically.

In this paper the internationalization process is characterized as the selection by a business school of a strategic response to global economic integration. I seek to present a dimensional map of the set of strategic responses available to schools. Further, I discuss the strategy selected by the University of South Carolina and its College of Business Administration as a case example of how the internationalization process may be viewed as a strategic response.

Most business schools, including the University of South Carolina, have seen their strategic response to global market integration follow a process of responsive evolution to opportunities. What may appear in retrospect to be the implementation of a consciously designed strategic response probably is a combination of strategic insights and serendipity. However, it is the purpose of this conference and of this paper to extract from our experience more fruitful models of strategic planning, and if the more distant past is pressed *ex post* into such models it is most probably because we at South Carolina have become accustomed to thinking strategically about our entire experience.

Most business schools have not sought to respond with a major programmatic effort, but rather have developed more or less internationalized curricula within existing programs. On the other hand, several schools have engaged in major programmatic efforts, which indicates the development of a coherent strategic response to the globalization of business:

1. The foundation in 1947 of the American Graduate School of International Management (AGSIM), formerly Thunderbird, a school entirely devoted to preparation for international business through the Master's in International Management program.
2. The Master's in International Business Studies (M.I.B.S.) degree at the University of South Carolina, the first large scale commitment of

resources to a separate graduate international business degree at a major business school, initiated in 1974.

3. The development of an internationalized faculty and a wide series of international institutional relationships at New York University, under the leadership of John Fayerweather and Robert Hawkins.

4. The major commitment of Wharton to the international programs of the Lauder Institute, beginning in 1984.

More recently, spurred by the increasing demand for a more relevant M.B.A. curriculum and the strong funding incentives of the Department of Education's International Business Centers initiative, business schools with historical strengths in various aspects of international business have developed a much richer variety of programmatic responses.

THE STRATEGIC DIMENSIONS OF INTERNATIONAL ACTIVITIES AT A BUSINESS SCHOOL

The model for strategic planning which I would offer is derived directly from the Client-Arena-Product Model developed by Ingo Walter for the strategic analysis of financial institutions (Walter 1988, 43–70; Smith and Walter 1990, 587–613). Without pushing the analogy to extremes, financial institutions have the same dual structure which drives educational institutions. Just as financial institutions must provide services to and market to both of their sources of funds (the client as investor) and their users of funds (the client as borrower), educational institutions must also provide service to and market to both the individuals receiving the learning experience (the client as student) and the firms which employ them (the client as employer). In business education, the linkage to the latter is much more direct, perhaps, than it might be in more traditional disciplines.

I begin the explanation of this model with the product dimension. Business schools are primarily involved in education, research and service activities. In more detail, we can subdivide the product markets as shown in table 2.9. Within each product market, each business school has an existing set of educational products.

The globalization of business provides two sets of opportunities for strategic change. First, it most clearly requires adaptation of existing products to the new demands for international educational components from both clienteles, students and firms. Further, new products may be designed with international content which are more attractive than existing products to these clienteles.

The second dimension is the client dimension. Table 2.8 provides a list of the output clienteles of a business school. Business firms are subdivided according to their activity horizon (global, national, regional or local). It is far more difficult to classify the input clienteles of a business school. One might utilize Arpan's awareness-understanding-competency classification, with the clienteles being those who wish to develop their international business capacities

Table 2.9 Product Markets

Education Markets
    Undergraduate Business Education
    Graduate Business Education
    Doctoral Business Education
    Post-doctoral Business Education
    Continuing Business Education
    Integrative Programs for Professionals

Research Markets
    Academic Research
    Sponsored Research

Service Markets
    Economic Development
        Investment Activities
        Trade Activities
        Human Capital Development
    Management Consulting

to the various levels classified here (1991, 2–3). In some ways, however, the boundaries of these groups are in no way fixed, as my experience is that awareness on the part of the student initiates a strong drive for at least understanding if not competency. In further sections of this paper, the client dimension is reserved for the output user of the individuals educated, research produced or services rendered by the business school.

Again, globalization of business causes the needs of the clients to shift and opens the possibility that new clients can be serviced.

The third dimension of the model is the arena dimension. By arena is meant the geographical scope of the activity. That scope can be global, nation-by-nation, or subnational in extent. Changes in geographical scope can be horizontal (adding more nations but continuing to service the market on a nation-by-nation basis), or vertical, in the sense that one takes a program that services a limited subnational region and expands it to a national level.

Within the extensive set of product opportunities, client choices and operating arenas, each business school must identify those which best help it meet its existing mission. However, because of the extraordinary impact of the globalization of business, that mission itself may need redefinition. For purposes of this paper, however, we define the components of an internationalization strategy as the decisions as to which client-arena-product (CAP) combinations the business school should enter anew or in which it should remain active. Following Walter's terminology, these combinations are referred to as cells.

The selection of CAP cells for activity requires a substantial assessment process. A schematic for this process forms table 2.11. Such assessment requires consideration of existing and new clients, the external franchise of the business school, and the social value of the consequences of entering particular cells. Further, the internal capabilities of the business school, in the form of existing faculty capabilities, current product offerings, the capacity of the organization to service particular cells, current relationships on campus with other academic units needed for product development activities, and the current international linkages in place to support international activities, must be assessed, as well.

AN APPLICATION OF THE STRATEGIC PLANNING MODEL: THE DEVELOPMENT OF THE M.I.B.S. PROGRAM AT THE UNIVERSITY OF SOUTH CAROLINA

In 1972, the dean of the College of Business Administration at the University of South Carolina, Dr. James F. Kane, formed a five-person committee chaired by Accounting professor Garnett F. Beazley to explore the college's potential involvement in International Business educational activities. After one year of initial assessment, that committee identified the potential for a specialized graduate program in international business to equip future managers for the challenge of global economic integration. The historical franchise of the university was that of a state-supported Southern university, with a relatively new resident M.B.A. program, an innovative Profes-

---

Table 2.10 Dimensions of Internationalization: The Client Dimension

Global Firms
National Firms
Regional Firms
Local Firms
Governments and International Agencies
Academic Institutions
Accrediting Bodies and Academic Associations

Table 2.11 Building a Strategy for Internationalization

Assessing the Externals
    The Historical Franchise of the Business School
        Government Franchise
        Private Franchise
        Religious Franchise
        Cultural Franchise
    Understanding the International Needs of Existing /Clients
        Corporate Clients
        Government Clients
        Academic Clients
    Identifying New Clients
    Assessing the Social Value and Synergies of Client/Arena/Product Combinations

Assessing the Internals
    Product Content [Curriculum +]
    Faculty Capabilities
    Organizational Structure and Capabilities
    Existing Campus Linkages and Skills
    Existing International Linkages

sional M.B.A. program offered statewide via closed-circuit television, a then small doctoral program and a large and growing undergraduate program. Most graduates remained within the region. Very few large firms had headquarters in South Carolina, and those firms were just beginning a somewhat painful introduction to the challenges of global economic integration. It was the intent, however, of the college, to seek to expand its clientele more widely within the region, and in particular to serve the rapidly developing group of foreign direct investors into South Carolina and the region.

Internally, existing curricula had no real international content, and students appeared to have very limited interest in international courses. However, there were a number of faculty, primarily represented on the committee, who had substantial academic training in or personal experience in international business activities. The college had a very weak departmental structure with rotating "Program Directors," an administered tenure and promotion system with limited faculty input, and a strong and powerful administration, a situation which was mirrored in the university's central administration. Cooperation with other areas of the university was limited, but collegial, and the College of Business Administration had no international linkages.

Recognizing, however, that strategic advantage could be gained by responding to the apparent growth in global business activity, the committee recommended and secured faculty and administrative approval for the introduction of new product representing the following client-arena-product choices:

Client: Regional firms becoming active in international trade, and foreign firms investing in the region.
Arena: Executed in Germany and Colombia, but with a global longterm perspective.
Product: A fully internationalized Masters program with language, area studies and internship requirements.

IMPLEMENTING THE INTERNATIONAL PLAN

Once the choice of specific client-arena-products cells has been made, there are at least six planning components which require attention:

1. Product redesign and/or product development.
2. Product marketing to identified clients/arenas.
3. Faculty development.
4. Organizational development.
5. Development of external linkages.
6. Funding support.

Each of these components will be discussed in some detail with reference to the Masters in International Business Studies program at the University of South Carolina.

DEVELOPMENT OF THE MASTERS IN INTERNATIONAL BUSINESS STUDIES CURRICULUM

Many of the components of the Masters in International Business Studies curriculum, such as intensive language study and area studies, conceptually predate the M.I.B.S. program. Others, such as an internationally integrated graduate business curriculum and an overseas internship, had been conceptualized but never fully implemented. The primary contribution of this program has been the technology of combining all of these complementary educational components into a single learning experience.

The M.I.B.S. program, as originally designed in 1973–74, consisted of seven different activities:

1. Intensive Summer Foreign Language Program. In 1974, German and Spanish were offered, with French and Portuguese being added in 1975, according to the original program design. The curricula of these summer courses were developed by the various language departments at the University of South Carolina. The courses required seven hours of class daily, Monday through Friday, for a ten-week period in the beginning summer of the program. Further, an English-language track for foreign nationals was initiated in 1975,

with specialized English language coursework offered by the English Program for Internationals. Within two years, the English language capabilities of foreign national applicants enabled the college to drop the English language coursework.

2. Unified Business Program. The core business material normally covered in the first year of an M.B.A. program and coursework in International Trade Theory, International Monetary Economics, International Finance, International Marketing, International Accounting, and International Management, were integrated into approximately 450 classroom hours of instruction over the fall and spring semesters of the first year of the program. With some subjects, such as marketing and finance, integration of the international dimension was accomplished by actual class-by-class interweaving of international and "domestic" topics. Other international material was integrated by positioning of the specific coursework in the appropriate sequence. Over the first ten years of program operation, two major changes took place. First, to meet accreditation standards, a segment in Management Information Systems was added. Second, because of changes in personnel and the difficulty of providing class-by-class integration, both marketing and finance were split into "domestic" and international segments and integration accomplished by positioning the segments appropriately. An exception to the general breakdown of class-by-class integration was the coalescence of Macroeconomics and International Monetary Economics, utilizing an open-economy macroeconomics approach.

3. Language Maintenance Program. Immediately after the intensive

summer language programs, students began their business program. To maintain language skills, 2.5 hours of language activity were scheduled per week in the fall and spring semesters. Over time, this period began to be used for the development of commercial and business vocabulary in the language of study.

4. Area Studies Program. Each student being prepared for overseas activity in Europe was provided with an area studies course, initially extending over a five week period in June and early July of the program's second year. This course, carrying six credit hours, was staffed by specialists from the Departments of Geography and Government and International Studies. A similar course was offered on Latin America, and, when the foreign national track was initiated in 1975, a course on the United States was added.

5. Overseas Language. Recognizing that the language training offered would not be sufficient to secure the level of competency needed to operate on internship, a specialized M.I.B.S. language course was developed and offered in late July and August of the program's second year. These courses utilized language training facilities, in countries where the language was spoken and where our students would do internships. No USC faculty were used in these courses, which were administered by the College of Business Administration.

6. Overseas Internship. The unique feature of the M.I.B.S. program, and its primary marketing point to prospective students, is the overseas internship. The original concept, which still motivates the existing internship program, was to provide some business experience in an environment in which the student would be required to use the business, language, and acculturation skills developed in the earlier stages of the program. The internship began September 1 and was completed by late February of the program's second year.

7. Summary Coursework and Directed Study. Upon return to the South Carolina campus approximately March 1, students would take a six-credit-hour internationalized policy-strategy course, and a three-credit-hour directed study course, one-on-one with faculty. These courses would be completed by early May, when the students would graduate. Within five years, primarily due to increased enrollments, the directed study course became a topic specific seminar, with instructors meeting with study groups on a regularly scheduled basis. In 1985 the curriculum of the final semester was altered. The policy-strategy course was reduced to three credit hours, and a series of ten international business electives were introduced. Students were required to take two of the ten electives. Most of the electives evolved from the topic specific seminars.

In 1982 three-year tracks in Arabic and Japanese were introduced. The first and third years closely paralleled the existing M.I.B.S. program, while the second year involved study at a university located in the foreign region. Where USC did not have faculty resources in foreign language, existing intensive language programs at other sites were utilized. Since 1985 two-year Italian and Russian and three-year Korean and Chinese tracks have been implemented.

In its curriculum, the M.I.B.S. program subscribes to a number of philosophical perspectives. Its objective is to provide its graduates with a modern graduate business education, with the international dimension of business integrated throughout, either through integrated subject curricula or proper positioning of material. Its graduates should have mastered at least one language other than their own, and have utilized that language in an internship in a business environment where that language is predominant. They should have the experience of living in a foreign culture, and of preparing for living in that culture. The end product should be a well-educated, globally oriented, business professional with an experience equivalent to a first overseas assignment, and with the ability to understand and replicate that experience if called upon to do so. In general, the objective of the curriculum is to provide this capability to those who have had no prior area studies, language, or business education.

The fundamental technology of the M.I.B.S. program has stood the test of eighteen years of changes in the global economy. Much of the current frenetic changes to adapt the traditional M.B.A. curriculum to secure greater global content and a multicultural perspective were anticipated almost a generation earlier by M.I.B.S. The extensive experience of our administrators and faculty with this type of program now represents an internal strength of the College of Business Administration, a strength which is being applied in a number of innovative new ventures.

MARKETING THE M.I.B.S. PROGRAM

As with all academic programs, in M.I.B.S. there have been two dimensions to program marketing. With respect to prospective students, the initial approach was based on the employment environment of the 1970s and the traditional role of the graduate professional school as providing more career specific training after a liberal arts or technical undergraduate degree program. M.I.B.S. was primarily marketed to recent graduates of liberal arts programs who were seeking a professional business

degree which built on area studies or foreign language expertise developed in the undergraduate program. Business experience was valued but not actively sought. The initial promotional strategy of the program involved creation of a college newspaper campaign featuring overseas scenes and the phrase "Picture yourself in International Business." Subsequently, large mailings of brochures and posters to faculty in area studies and foreign language, and to members of the Academy of International Business proved effective, as students would consult closely with faculty in these areas as to graduate study opportunities with an international dimension. By 1985 the program had grown to an average of approximately 140 students entering per annum; average GMAT was 565, with GPR's averaging 3.2 from a reasonable mix of schools. Approximately 40% of the students were from the Southeast, and another 25% were foreign nationals.

The original arena selected for placement of M.I.B.S. graduates was a regional one. Existing university placement facilities were deemed sufficient and no specialized placement capability for the program was developed. The first graduates were primarily placed in the Southeast in firms becoming involved in international activities. However, as the second class included a substantial number of foreign nationals and a geographically diverse group of U.S. nationals, the program began to place on a national basis. U.S. multinationals came to view the M.I.B.S. program as an alternate source to Thunderbird (AGSIM) for foreign nationals and globally oriented U.S. nationals. Further, the internship proved to be a very effective placement tool, with some 10–30% of the students finding employment with their internship firm yearly. Because of this interaction with the internship acquisition process, by 1980 the M.I.B.S. program administration had assumed responsibility for the placement of M.I.B.S. students, running a separate placement operation with separate interview schedules. While the mid-February return of the students from internship hindered somewhat the interviewing activity of the M.I.B.S. students, through 1985 the program typically had greater than 95% placement within six months of graduation, with average salaries higher than regional M.B.A. programs but lower than "top ten" M.B.A. programs.

FACULTY ACQUISITION AND DEVELOPMENT

One of the fundamental needs of an International Business program of the magnitude of M.I.B.S. is a faculty competent in the subject matter being taught. M.I.B.S. is particularly challenging because it requires that competence be drawn from a wide number of sources on campus.

M.I.B.S. foreign language instruction was developed by faculty in the foreign language departments of the university. The summer intensive programs became the summer teaching assignments of a large number of the faculty, supported by numerous instructors and teaching assistants. As the language faculty of the university had a primary interest in literature and a research orientation, certain faculty elected to limit their involvement in this program. Over time, a number of foreign language faculty were hired specifically to develop M.I.B.S. curricula, including some specialists in the business dimensions of a given language. Several of these specialists failed to receive promotion and tenure, given the research direction of their department. The grafting of an intensive summer language program onto a research-based faculty has proven to be an ongoing managerial challenge, but one which has generally been met successfully.

Area studies faculty had a less significant adjustment to make to teaching in M.I.B.S., and the outstanding scholars of the two departments concerned have generally taught repeatedly in the M.I.B.S. program.

Within the business program, existing faculty were identified who had the ability to teach the international components of the program. Where such individuals were not available, the various departments were charged with recruiting international specialists. In this manner, within the first two years, international specialists in marketing and management were added. In 1976 a separate International Business department was created, and all international faculty (with the exceptions of accounting and economics) were placed in this area. The Economics and Accounting departments recruited additional international specialists, and the International Business faculty grew from an initial group of five in 1976 to eleven by 1985. The increase in staff enabled the program to grow to two sections of the Unified Business Program.

Many of the faculty of the Unified Business Program were members of other departments teaching the "domestic" components of their discipline. Through providing support for foreign travel and research for these faculty, the International Business area sought to increase their international expertise and ability to relate to M.I.B.S. students in the classroom. External projects were sought to provide additional exposure; as an example, in 1975 seven of the Finance Department faculty were provided summer compensation to teach two or three weeks in an executive program in Rio de Janeiro. Over time, a body of faculty with international interests were developed to support the international expertise of those in the International Business Department.

In the early years of the program, a substantial number of visiting faculty were used in program activity. Use of visiting faculty diminished as the size of South Carolina's International Business faculty grew.

ORGANIZATIONAL STRUCTURE

As I will argue in another paper developed for this conference, one of the primary obstacles to the development of a successful international strategy is the relatively inflexible interdisciplinary structure of the academic environment in general, and in particular the seeming resistance to programmatic change where there is a strong departmental structure (Folks 1992). The organizational environment which bred the M.I.B.S. program was a remarkably flexible one within the College of Business Administration. In a period of rapid enrollment growth and faculty expansion, positions were set aside to support the initiation and expansion of M.I.B.S. In the 1970s the departmental structure within the college would be classified as relatively weak, with, at least initially, limited faculty input relative to tenure and promotion, centralized planning at the dean's level, highly open channels of communication across departmental lines, and little or no history of a politicized departmental environment. Such a structure favored interdisciplinary efforts and risk-taking, precisely the requirements to launch a venture such as M.I.B.S.

As discussed earlier, the original M.I.B.S. program was drawn together by an interdisciplinary task force. Approximately six months prior to the program's launch in 1974, the dean and the task force retained a consultant, Dr. Allen B. Dickerman, who had operated a program for foreign managers at Syracuse University and who had extensive international contacts. As the program was launched, Dickerman became a faculty member of the college, in the Management area, and was named director of the M.I.B.S. program. In this capacity, Dickerman reported to the director of Graduate Studies, who had responsibility for the administration of all graduate programs in the college. Dickerman's responsibilities included recruiting, admissions, student services, arrangement of overseas language programs, and arrangement of internships. Although, as one might suspect, the success of M.I.B.S. has generated many "fathers," Allen Dickerman's outstanding contributions in the development of all of these difficult activities laid the groundwork for the future success of the program.

During the first year of the program the task force (enlarged to include faculty teaching in the program) served as an executive committee, overseeing the operation of the program and providing policy guidance to the director. Although the task force continued to work effectively in this regard, the inordinate time requirements and the difficulty of having policy decisions made by such a large group convinced the dean that the new product had been successfully launched and that the "new product" committee could be disbanded. Consequently, the program was administered in 1975–76 by Allen Dickerman as director.

In 1976 the program began its third year of operation with a third organizational structure. As indicated in the previous section, an International Business Department was organized, with the author as chair. In addition to providing an organizational base for the international faculty of the college, the area was charged with the design and staffing of the M.I.B.S. academic courses. Dickerman and the newly hired associate director of the M.I.B.S. program, Dr. Robert J. Kuhne, were assigned as faculty members of the new International Business Department. Over the period 1976–78, as the program grew in complexity, supervisory responsibility over program activities was gradually transferred from the Graduate Division to the International Business Department, and budgetary responsibility for M.I.B.S. related activities placed with the department chair. The M.I.B.S. program became a part of the International Business Department administratively, although the ultimate responsibility for admissions and student services remained with the Graduate Division, and numerous other departments continued to contribute faculty to the program. Over time, faculty from the International Business area played an increasing role in the administration of segments of the program, taking over internship administration as Dickerman neared retirement age. By 1983 International Business tenure track faculty members administered all aspects of the M.I.B.S. program, including final placement of M.I.B.S. students in positions.

The period of intensive faculty involvement in the administration of M.I.B.S. lasted for only a few years. Beginning in the fall of 1983, non–tenure track program administrators were employed to administer the internship and placement process, which were gradually handed off from the faculty over a two-year period. Direct faculty control over admissions and retention of students continued for a longer period, although by 1986 all but the most difficult situations were the responsibility of a director of Student Services. However, the International Business faculty retained strong oversight over the program's operation, providing planning input and continued support for solving program problems as they developed. Further, the budget for program activities was included in the International Business Department's overall budget, and the department chair had complete flexibility to allocate resources between normal departmental activities and activities in support of the M.I.B.S. program.

The reduction in faculty involvement in administration was implemented for a number of reasons. First, the intellectual challenge of developing systems to administer the unique admissions, program operation, internship and placement activities of the M.I.B.S. program had been met, and these systems could be given to professionally trained administrators who would be held accountable by those in the department who had developed them. Secondly, the success of M.I.B.S. created numerous additional opportunities which required substantial faculty input if they were to be yield advantage to the college and university. Further, college tenure and promotion decisions had been decentralized to the department level, and the criteria changed to reduce the emphasis on service to the college and provide greater emphasis on scholarship. This change reflected also the strategic decision by the International Business faculty to take advantage of the success of the M.I.B.S. program to build a strong research capability, with a large Ph.D. program and a high standard of research productivity. Such productivity could not be achieved if department faculty spent significant periods of time involved in program administration.

Yet the continued success of the M.I.B.S. program is founded on the decision by Kane in 1976 to create a separate International Business faculty, and the subsequent period of faculty involvement in the direct administration of the M.I.B.S. program. Providing a separate department provided a signal to other departments that the program was held in high regard by the administration of the college and university, and provided the program with equal access to resources and equal influence in the councils where decisions affecting college strategy were taken. The college's two planning exercises over the period, the 1982 and 1990 plans, were done on a department and division basis; the International Business faculty was able to exert strong influence on the M.I.B.S. program's direction and the direction of the college during those periods.

Having a department, and placing the academic program under that department, provided the program with strong champions. Given the relatively closed structures of the modern U.S. university, programmatic initiatives must have an initial champion within the organization simply to receive a hearing. Further, that support must be continuous, and the linking of the success of a group of faculty to the success of a program clearly provides that continuity of support. What is impressive about the historical development of the M.I.B.S. program is the fact that top administrative support for that program has not wavered over its eighteen years of existence.

Equally impressive is the continued development of creative changes to enable the program to adjust to the increased globalization of business. M.I.B.S. has constantly adapted in subtle ways, not just in its curriculum, but in the way in which its students are recruited, placed on internship, and positioned for employment. It is often overlooked that the management of international educational programs is truly a global business activity, and the source of global business expertise on a campus is its internationally trained faculty. Collecting that faculty into a unit, and drawing on that expertise for program management, may not insure success, but very few, if any, business schools have been successful internationally without providing a similar vehicle for applying expertise to solving the problems raised by international activity. For this reason alone, business schools which seek to operate internationally should consider creating an International Business faculty unit. There are, of course, more important reasons for doing so, including, in my view, the fact that International Business is indeed a separate domain for inquiry, but that is an issue which others are addressing in other ways in this conference.

DEVELOPING EXTERNAL LINKAGES

There were two primary external linkages needed for the development of the M.I.B.S. program. First, selected institutions in foreign countries to provide the overseas language training had to be identified. These institutions were also utilized, in some cases, to provide for administration of the local components of the internship. The initial development of the institutional relationships for the program utilized a mixture of academic institutions, for profit training institutions, and eleemosynary public service institutions. In more recent years, as new language tracks have developed, local universities have been selected to provide language training, and the involvement of the local partner in internship administration has been substantially reduced.

From the first, those involved in the M.I.B.S. program have been intent on providing internship experiences of the highest quality. For this reason we have elected to keep a high level of control of the process of identifying internship opportunities, assigning students to internships, and evaluating student performance. From the earliest days of the program we have visited companies who might offer internships to our students, seeking to collaborate in the design of the internship experience, and continually monitoring the performance of students while on internship. Students are interviewed regard-

ing the type of internship they would prefer. After consultation with the students, those whom we believe appropriate for particular internships are proposed to the host company for acceptance. Students usually are actively discouraged from contacting firms regarding internships directly, but are asked to channel contacts through the faculty or professional staff member responsible for the track. Placement of students on internship is an exacting task, requiring continued balancing of the interests of all the students in the program, the needs of all the firms providing internships, and the continuing needs of the program for future internships.

Initial internship contacts were with foreign firms investing in South Carolina and with local country and regional managers of U.S. multinationals operating in the internship country. In the 1970s, with little national reputation, the University of South Carolina found resistance in approaching personnel executives of major multinationals in the United States. Further, we found that local managers had sufficient authority to offer internships without reference to headquarters, and were more cooperative if they did not perceive the intern as being an imposition by headquarters. As we expanded language tracks, and began to work with the same multinational in a number of different countries, U.S. personnel executives, who were being approached by us for foreign national internships in the United States, became far more receptive to taking interns overseas. The excellent track record of our early graduates also enabled us to shift some of the internship acquisition effort to corporate headquarters.

FUNDING INTERNATIONAL ACTIVITIES

As it currently operates, the M.I.B.S. program provides substantial positive incremental cash flows to the University of South Carolina and to the College of Business Administration. After three years of operation, M.I.B.S. had begun to generate a substantial portion of its associated out-of-pocket expenses. Study of its funding offers numerous instructive lessons as to how a successful international program can create resources which allow for further growth and resource generation.

In 1974, the M.I.B.S. program was approved by the South Carolina Commission on Higher Education on the condition that the university would be provided no incremental resources earmarked specifically for the program. Further, foreign travel funding from state sources was limited to $1,000 per trip. In order to launch M.I.B.S. successfully, Kane utilized private sources of funds from the USC Business Partnership Foundation to provide for the funding of foreign travel for the M.I.B.S. program. As the USC Summer School was operated on a pay as you go basis, with summer tuition for the M.I.B.S. foreign language courses being placed at the disposal of the Department of Foreign Languages and Literatures for payment of instructors, a clearly designated revenue stream was available to support language instruction. Business instruction in the first year was provided by reassignment of faculty from other programs and the continuing growth of the college.

The ability to develop internship opportunities and supervise students while on the internship depended on the continued provision of foreign travel money. As part of the internship, firms were asked to provide a stipend to students through the University of South Carolina. Approximately 10% of each stipend was withheld and utilized for travel to the internship country. In addition, once the International Business Department was established in 1976, a modest budget, designed to cover faculty travel and other normal departmental expenses, was provided from university funds.

By 1981, the market position of the M.I.B.S. program was such that a surcharge for the extraordinary expenses associated with the program could be levied. A matriculation fee of $600 per student was levied, which provided a substantial portion of the program operating budget. In 1985, the fee was increased to $1,000; with an incoming class of 140, the program generated

$140,000 in fee income, which, coupled with approximately $75,000 in internship stipend fees, provided the entire operating budget of the program and the International Business Department.

The pronounced market success of M.I.B.S. enabled the university to increase the matriculation fee rapidly in the latter half of the 1980s. Currently, with an entering enrollment of 180 per annum, the matriculation fee is set at $7,500 per student ($3,750 for South Carolina residents). The total level of tuition and fees charged to students is approximately $13,000 for out-of-state students for the two years of the program. Since the program's initiation in 1974, there has been no in-state/out-of-state tuition differential at the graduate level in the University of South Carolina. Beginning in 1992–93, out-of-state tuition will be approximately twice that of in-state tuition, raising the total tuition and fees over a two year period to approximately $21,500. With student generated revenues of approximately $3.5 million (adjusting for in-state students), the program will provide a satisfactory cash flow to the university and college and is on a sound financial footing.

BUILDING ON THE M.I.B.S. PROGRAM: OTHER INTERNATIONAL ACTIVITIES AT THE UNIVERSITY OF SOUTH CAROLINA

Although the M.I.B.S. program provides the primary growth mechanism for international activities in the College of Business Administration, the college has sought in numerous ways to take advantage of the opportunities which the operation of a program such as M.I.B.S. generates. The primary effort has been to develop programs which complement M.I.B.S. without compromising the market identity of the program.

Activities in Education Markets

At the undergraduate level the South Carolina faculty has made a strategic decision not to offer an undergraduate international business major. Our view is that the proper education of future managers of global firms requires the combination of activities included in the M.I.B.S. program, executed when the student is sufficiently seasoned to make a contribution to the internship firm and benefit from the graduate level of instruction in M.I.B.S. USC provides a limited menu of undergraduate International Business electives and summer overseas programs. Scarce International Business faculty resources are deployed primarily at the graduate level.

At the graduate level, in addition to M.I.B.S., the college offers a resident M.B.A. program which services a regional student clientele. While it is appropriate to offer these students an exposure to international business, it is more important that the M.B.A. program remain distinct from M.I.B.S. Consequently, a separate group of four international business electives are offered to M.B.A. students in the Fall semester of their second year; M.B.A. students are required to take at least one international business elective. These electives are prerequisite to the advanced international business electives also open to M.I.B.S. students, and many M.B.A. students take a second international elective.

Initially, the M.B.A. student body had a minimal foreign student enrollment. To remedy this deficiency, the college has negotiated a number of agreements with European business schools allowing students from those schools to spend one semester at the university taking graduate business courses at full fees. These agreements are supplemented by a number of exchange programs which allow up to ten M.B.A. students to study overseas for one semester. Further, special arrangements with five-year business programs at select European business schools have been made to provide for completion of the USC M.B.A. program in a fifteen month period. In 1991–92, more than 50 foreign graduate students participated in various exchange and agreement programs on the Columbia campus.

In 1980 the International Business faculty developed a doctoral concentration with in International Business with an over-seas residency and foreign language requirement for U.S. nationals. Currently the doctoral program has grown to 18 degree candidates; some 52 candidates applied for three entry positions into the program in 1992. A minor is offered to support doctoral candidates in other disciplines. In 1990, in conjunction with the Finance Department, a doctoral major in International Finance was initiated, which combines the doctoral coursework of the Finance Ph.D. concentration with two international finance and one international business theory doctoral seminar. The doctoral concentrations were developed to provide well trained scholars and teachers in the disciplines of international business and international finance, which were and are in demand in the academic market. Strategically, development of a strong doctoral program was a major part of the strategic move in 1983–85 toward greater research productivity of the faculty.

The lack of international content in most doctoral programs coupled with the demand for faculty capable of teaching about a global business environment provided the strategic market opening which led to the Faculty Development in International Business (FDIB) program. FDIB is a two week postdoctoral program offered to business school faculty seeking to develop an in-depth teaching and/or research capability in the international dimension of their functional discipline, and was initially developed and managed by David A. Ricks of the International Business faculty. From a relatively modest beginning, FDIB trained 120 faculty in summer 1991, and is so successful that two sessions are being offered in summer 1992. An Eastern Europe postdoctoral program was offered in 1991 for faculty of U.S. business schools, and will be repeated. In addition, a program designed to help business school deans and other administrators develop an international activities strategy is also offered each year.

From time to time the college has offered managerial education programs in international business, either publicly through the university's continuing education center, or contractually for firms. The small local market in South Carolina and fierce national competition in this market has limited our success.

Activities in Research Markets

The shift in emphasis of faculty activities from program administration to research, which began in the early 1980s, has resulted in an extremely active research faculty. Strategically, the college sought to complement that research emphasis through hiring practices and through concentrating faculty research publication in peer refereed journals. A key aspect of this strategy was the college's successful effort to be given editorial responsibility for the *Journal of International Business Studies*, arguably the outstanding academic journal in the field. The successful editorial work of David A. Ricks, Brian Toyne, Jeffrey S. Arpan and Douglas W. Nigh, contributed positively to the external image of the University of South Carolina.

Two research centers were developed to do sponsored research in International Business in the 1980s. On January 1, 1990, these two centers were consolidated in the Center for International Business Education and Research (CIBER), one of sixteen federally funded centers designed to be national and regional resources. CIBER currently receives $427,000 in federal funds, matched more than 50% by university funding, and is responsible for new educational program initiatives and a number of major research projects. Because existing programs in International Business at South Carolina met many of the statute's requirements, and because of the large number of research active faculty, a significant proportion of CIBER's budget supports those aspects of international business research, such as foreign travel, questionnaire translation, and assemblies of international scholars, as exemplified by this conference.

Activities in Service Markets

Because of the reputation of the university for excellence in curriculum design, based initially on the design of the M.I.B.S. program, faculty in the college have had the opportunity to participate in a wide variety of curriculum design projects at other institutions of higher education. The college was selected by AID to provide technical support to the Universidad Catolica Madre y Maestra in the Dominican Republic in the development of the country's first M.B.A. degree, an $8,000,000 project spread over eight years. The college also was selected in 1990 by the United Nations Development Program to serve as the U.S. business school in a joint British, Canadian, and U.S. effort to develop an international master's degree curriculum for the University of International Business and Economics (UIBE), the educational unit of the Ministry of Foreign Economic Relations and Trade of the People's Republic of China located in Beijing.

The college has declined to participate in a number of similar ventures. Each accepted venture was chosen because it provided significant faculty development opportunities and was congruent with strategic decisions being made at the time. The Dominican Republic venture was initiated as the college began to diversify its M.I.B.S. operations in the Spanish track away from Colombia, in response to deteriorating security for U.S. nationals in that country in the 1980s. The relationship with UIBE was sought as a prelude to the development of a M.I.B.S. Chinese track, which would include an internship within the People's Republic of China.

THE 1990 STRATEGIC SHIFT— REDESIGNING M.I.B.S.

In the latter half of the 1980s the M.I.B.S. program encountered significant new challenges. While the general client-arena-product combination, the graduate business education for future managers of global firms regardless of headquarters location, remained the same, the reduction in managerial workforce experienced in many firms and the desire of personnel executives for individuals with prior work experience and a functional specialization made the traditional approach of providing a generalized business education to liberal arts undergraduates less valued in the marketplace. Over time, under the leadership of Jeffrey S. Arpan, appointed department chair in 1985, admissions patterns were shifted to emphasize prior work experience and students were urged to utilize their elective courses to create two course concentrations in a functional field.

In 1989, an interdisciplinary committee chaired by the author was formed to redesign the business components of the M.I.B.S. program. The report of this committee in 1990 recommended a major product redesign to accommodate the changed environment facing the program. This redesign was approved by the faculty and implemented for the class entering the program in 1991. It included the following steps:

1. The Unified Business Program was redesigned. The core material on business was separated (except for marketing and finance) from the core international material. Undergraduate business majors were exempted from the initial business course and began the program in the fall with the international core. Those exempting the foreign language and nonbusiness major foreign nationals took the initial business course in the summer and began the fall with the international core. Most of those taking the business core in the fall and the international business core in the spring will be students who either have a technical background or who are in one of the less commonly taught languages, and would number no more than one-third of the students.

2. The set of international business electives has been expanded, and the number of noninternational business electives offered also enlarged. Some two-thirds of the stu-

dents would take seven electives, four in the spring semester of the first year of the program and three in the spring of the second year. Other students take at least three electives. Currently, informal area concentrations are being designed to provide guidance to students regarding development of functional expertise.

3. The area studies, overseas language and internship components of the program now begin 30–45 days earlier, to allow the students to return to campus in early January. Early return broadens the availability of electives and puts the student into the U.S. job market significantly earlier.

4. Admissions policies were changed to emphasize prior business experience. For the class entering in 1992, some 85% will have at least two years of relevant business experience.

5. The program would be expanded to 180, with further expansion possible if resources become available. The new program has proven to be a much more attractive one to prospective students. The average GMAT of students admitted to enter in 1992 will be approximately 630, with a higher average GPR from undergraduate institutions of greater selectivity in their admissions.

Coupled with the curriculum changes was a major organization shift, effected in 1990. The management of the M.I.B.S. program was separated from the management of the International Business Department, and a collegewide Faculty Executive Committee was created to provide faculty oversight and policy guidance. Day-to-day administration of the program became the responsibility of a nonacademic director of the M.I.B.S. program. In order to create effective organizational linkages with the admissions and student management functions, M.I.B.S. administrators were assigned to the Graduate Division of the College of Business Administration. Budgetary responsibility for the program now rests with the Graduate Division, and the International Business faculty have a separate operating budget.

While on-campus administration of the program has improved under the new organizational structure, and the increased number of internships needed to operate the program at the increased size have been arranged, other aspects of the reorganization have perhaps been less effective. The college needs to provide an effective placement capability for the M.I.B.S. program; under the reorganized structure the management of the program has limited resources to devote to corporate relations, which organizationally are located elsewhere in the college. Further, financial resources are readily shifted among a number of competing resident graduate programs. In a general environment of reduced state funding for higher education, funds generating programs may be called upon to support those programs which do not generate funds.

One of the reasons for the separation of the International Business faculty from the direct administration of the M.I.B.S. program was to provide broader college ownership of its flagship program. Although some non–International Business faculty have played a prominent role through the program executive committee, they are only now developing the technical expertise of international operations needed to resolve the difficult issues which arise in managing such a complex global program. Distancing the International Business faculty from program management has clearly resulted in diminished championship of the program by anybody.

However, through the continued support of the program by International Business faculty, who are responsible for approximately 40% of the business teaching in it, development of international expertise among a select few faculty in other departments, and greater understanding of program complexities by top administrators, M.I.B.S. has begun to evolve a new set of working relationships which promise greater success in the future. Symbolic of this set of relationships is the creation of a policy formulation group for all international activities of the college, the Globalization Committee, which includes all associate deans, the director of the Graduate Division, the chair of the M.I.B.S. Faculty Executive Committee, and four International Business faculty who have administrative components in their duties.

SUMMARY AND CONCLUSION

Some general principles can be developed from the experience of the University of South Carolina's in its twenty years of international activity.

First, business schools must desist from thinking about internationalization only in curricular terms, but must instead develop a concept of internationalization as a strategic response along all dimensions of activity.

No international activity should be undertaken unless it is congruent with the mission of the given business school, unless the administration of the school is seeking to change the school's mission. Development of the M.I.B.S. program was a part of a conscious decision to change the college's mission.

The impact of the globalization of business cannot be escaped; it requires the redesign of almost all academic products, but particularly the M.B.A. and undergraduate curricula. The fact that such change is required offers the opportunity to break out of restrictions imposed by the past experience and reputation of the given business school.

Internationalization can be contextual; that is, other strategic choices can be the primary impetus in the strategy of the school, and international activities take place in the context of that strategy. The University of

South Carolina has chosen to make internationalization one major impetus in its longterm strategy. Other schools have established significant reputations by taking other major developments in the business environment and applying the same strategic concepts to improve their market position. An example of this approach is the University of Tennessee, where an emphasis on Total Quality Management has led to numerous overseas seminars on the subject by their faculty.

The administrative structure used to manage international activities should complement the strategic choice of the school. However, whatever strategic and administrative choices are made, they must respect the fact that there is a separate domain of study known as international business.

As implementation of international activities requires changed faculty behavior, and as the type of behavioral change needed is not frequently rewarded in the incentive system of most major business schools, the reward structure in the business school must be examined and probably must be altered to induce changed faculty behavior.

There are substantial economies of scale available in international activities, and strategies should be designed to take advantage of them. It is far easier to negotiate and manage two agreements bringing twenty students to campus for one semester than it is to negotiate and manage ten agreements bringing two students each to campus for one semester. It is far easier to add ten internships in a country where one already has twenty than to begin ten internships in a new country.

Linkages between or among schools should never be created unless they serve some strategic purpose.

Finally, substantial market opportunities still exist as the globalization of educational activity begins to parallel the globalization of business activity. International activities can be expensive, and mistakes can be costly. However, as the experience of the University of South Carolina so clearly shows, properly designed activities which meet a true market need may also be rewarding to the business school, the university community as a whole, and to faculty who develop the skills to provide that service.

NOTES

1. For examples of ranking systems see Ball and McCulloch 1984 and Nehrt 1987a. For a detailed discussion of their merits, see Douglas 1989 and Nehrt 1989b.

2. The College of Business Administration at the University of South Carolina has no formal departments or departmental chairs. Faculty are grouped into program areas for tenure and promotion purposes, and each area is headed by a program director appointed by the dean. For purposes of this paper, the phrase "department" refers to the program area, and "department chair" refers to the program director.

3. The most recent survey of U.S. business school efforts at internationalization is the Academy of International Business sponsored survey of Thanopoulos [1986]. A new extensive survey is currently underway, sponsored by the Academy of International Business, the American Assembly of Collegiate Schools of Business, and the Center for International Business Education and Research at the University of South Carolina.

■ ■ ■

HISTORY OF INTERNATIONAL BUSINESS
EDUCATION: THE U.S. VIEW
Robert Grosse

International business education is a phenomenon of the post–World War II period. Before that time the international dimension of business administration was largely examined under the heading of international trade or commerce, focusing on exports and imports rather than overseas subsidiaries and other contractual forms. So the discussion of international business education is limited to a time frame that most of us have experienced first-hand.

There is no doubt that the greatest concentration of firms that established overseas affiliates in the immediate postwar period (or continued with such activities since before the war) were based in the United States. Thus, it is not surprising, as noted by Robert Locke, that U.S. universities were the earliest to place emphasis on international business education. And it is not surprising to see that the global association of IB teachers began in the United States (as the Association for Education in International Business in 1959, and subsequently as the Academy of International Business beginning in 1972).[1]

The history discussed here is presented from an "institutional perspective" in that it is not a history of international business, but rather a history of the study of international business. Also, this history focuses on the study of international business activity that emphasizes firms with foreign affiliates, so it is limited for the most part to the last half-century. Our interest is not to look too far backward to understand the study of international aspects of business in times when trade was the primary form of dealing with overseas clients and suppliers, but rather to concentrate on recent times. In sum, our focus is on the late twentieth century and on the ways in which universities and business schools treat international business studies.

From these terms of reference it is quite interesting to see how the discipline of international business has developed during the past 30 years. The largest number of schools teaching courses that are designated as international business clearly are concentrated in the United States. A common claim in European business schools is that they offer internationalized courses across the range of functional areas, since business in European countries has been largely internationalized for many years, certainly since the opening of the Common Market in 1958. This point really needs to be examined in detail, because similar claims have been used by U.S. business schools to avoid dealing with the need to internationalize.[2] What is clear is that the number of IB courses and programs is much more limited in Europe than in the United States—and that at the same time business is much more international in Europe than in the United States.

THE DEVELOPMENT OF IB EDUCATION IN THE UNITED STATES

At the risk of too greatly simplifying the process through which international business became a major field of study in U.S. business schools, I think that the following points need to be made for the purpose of comparison among countries.

First, international business teaching as it was initially conceived came out of the study of international economics, particularly international trade. When Steve Robock, Lee Nehrt, Ken Simmonds, and others at Indiana University began to define the discipline, the roots were fundamentally based in economics. The professors such as these few, plus Jack Behrman, Jean Boddewyn, John Fayerweather, and a handful of others, were far out on the fringe of business studies. Their emphasis on the study of multinational corporations was given a great boost by the publication of Raymond Vernon's article on the international product cycle in 1966. While their social science base was economics, the new IB analysts were interested in studying all facets of the multinational firm.

Second, the internationalist movement in U.S. business schools grew continuously if not spectacularly during the 1960s. While schools were becoming somewhat receptive to the idea of teaching about international business, there was no model for putting such an idea into practice. Courses were introduced at various schools in the two most popular functional fields: international marketing and international finance. As well, some other courses such as a survey of international business, and management of international operations of multinational corporations, were introduced in some schools. The field grew fitfully both institutionally in U.S. business schools and organizationally in the Association for Education in International Business.

Third, it was not until 1974, when the AACSB's guidelines began to require business schools to include a global perspective in their curricula (Fayerweather 1986, 18), that interest in the field really spread to all parts of the country and to all levels of schools. As shown in table 2.12, constructed from the AIB-sponsored International Business Curriculum Surveys by Terpstra (1970), Daniels and Radebaugh (1974), Grosse and Perritt (1980), and Thanopoulos (1986), the growth rate of teaching in international business was greatest in the late 1970s and early 1980s.

Another boom may be underway today, with the greatly heightened awareness on the part of U.S. business that a global view is needed to compete successfully against foreign as well as domestic rivals. This remains to be documented in the next AIB Curriculum Survey.

The treatment of international business in business school curricula initially developed as a series of applications of functional area studies to crossborder phenomena. That is, an international marketing course was typically added to the offerings of a marketing department, and an international finance course likewise was placed in a school's finance department. No international business departments (except the Thunderbird School in Arizona) existed in the 1960s, though some pockets of specialization (such as at Indiana University) did develop.[3]

It was only in the 1970s that international business became a field sufficiently defined as to produce the debate on how to infuse international content into business school curricula, viz., through internationalizing functional area courses or through institution of international courses in each functional area. This debate has lasted through two decades and shows no sign of lessening today. Several of the papers in this volume treat the issue in great detail. The reality is that by now international business has been accepted adequately as a discipline, such that this kind of debate can take place and be resolved in ways that (hopefully) build international knowledge and thinking into business schools across the country.

The point of this commentary is not to inadequately paraphrase pieces of John Fayerweather's (1986) history of the Academy of International Business, but rather to highlight some of the key institutional turning points that underlie the development of the IB discipline in the United States. There should be no question that the initial launching of IB as a field of study resulted from the efforts of a visionary few academic analysts. Likewise, the acceptance of the field in the United States came from the business school accrediting body's recognition that a very large part of US business was competing in a global context, and that schools needed to prepare students to think about the relevant concerns in that context.

Perhaps in the past the insularity of the United States, with its huge internal market, called for measures that were unnecessary in Europe, with a dozen smaller and more internationally integrated markets. Perhaps the globalization of business in the 1990s

Table 2.12 Percentage of U.S. Business Schools Offering International Business Courses

AIB Curriculum Survey authors	year	percentage of schools offering IB
Terpstra	1970	n/a
Daniels and Radebuagh	1974	64%
Grosse and Perritt	1980	81%
Thanapolous	1986	98%

will reduce the future need for specific focus on international business studies, since they perforce will be incorporated in the functional field courses. Such an eventuality seems quite unlikely, given professors' tendencies to focus narrowly on their own disciplines and only under duress to open their horizons to additional complications such as globalization. This problem surely will remain in Europe as well as in the United States, so the continued development of IB studies appears certain as a counterbalance to the specialization in other areas.

LOCKE'S DISCUSSION OF EUROPEAN IB EDUCATION

Locke's description of IB education in European countries offers quite a useful set of examples to place in contrast with the US experience. His overall perspective places the US achievements in a greatly reduced role, by his argument that US firms are being outperformed by German and Japanese rivals, and hence that the German and Japanese schooling must be more appropriate for current business conditions.

But moving beyond his superficial comparison of systemic inadequacies, Locke does note an instructive point concerning German and Japanese education in international business. By showing that schools in these two countries place much more emphasis on technical education before a student enters the workplace, and on in-company education as a vehicle for advanced study, he has posed a very interesting model for discussion. Do the U.S. and British (as well as other non-German European) schools need to consider an alternative framework for teaching IB, which would emphasize teaching inhouse courses or seminars at companies?[4]

This point is fascinating, since it seems to argue in favor of more undergraduate education in business (and international business in particular), and then graduate training primarily through short courses aimed at specific firms. The way of the M.B.A. is again in question! Perhaps we should not go too far in exploring this issue, since our goal is to examine international business education and not all business education. Nevertheless, the issue is raised, and we should at least consider a response from the IB discipline that supports a direction for business schools in general. Let us leave that response until examining the other perspectives presented by Evenko and Folks.

It seems that a point of reference is required to compare business school/university training in international business at present, in order to draw more concrete conclusions about where systems converge and diverge, and also about where each country's schools may wish to move in the future. Taking the Thanopoulos study of 1986 as such a reference, we see that only 75% of non-U.S. schools offered international business courses in that year, compared with 98% of U.S. schools. The Japanese comparison could not be drawn, because of the very small number of Japanese respondents to the survey.

The U.S. business schools tended to offer IB course predominantly at the undergraduate level, whereas the European schools tended to focus more on the graduate level. Locke points out that these responses may be improperly categorized, since the European schools follow a different system of university-level education that may confuse advanced undergraduate and masters courses. Locke's explanation of the various national systems of business education in the United Kingdom, France, Germany, and Japan, is quite valuable for purposes of advancing our discussion, since I expect that most of us are not familiar with more than one or two systems beyond our home country's one.

Beyond presenting this description of business education in other countries, Locke does not advance the discussion at all. He offers no information about international business teaching other than the Finnish study. Therefore, our analysis must move beyond the paper to make comparisons. In the discussion of general business programs,

Locke does emphasize the bilingual or multilingual nature of business students in Europe. This is a refreshing change from the common situation in the United States, and one that we U.S. professors envy quite seriously. It leads to the conclusion that U.S. schools will have to pay even more attention to "interculturalizing" our students, since the society does not accomplish this for us before the students arrive at our doorstep.

Locke defines his target for IB education as training managers who will succeed in international business competition. This narrow focus must be expanded in our consideration of alternative systems, since the goals of IB education include training of students at all levels of higher education as well as advancing knowledge of how the world of business functions. These multiple goals argue in favor of a portfolio of IB educational offerings, from research centers to executive programs to undergraduate introductory courses. This broader context should be the one in which comparisons and contrasts are made and where our search for improving IB education leads.

EVENKO'S DISCUSSION OF RUSSIAN IB TRAINING

Evenko's treatment of the Russian experience in business education is truly enlightening in its scope of coverage over time and across the sources and types of education that may provide IB content. He follows the path of IB education in the Soviet Union through the decades, until Perestroika of the late 1980s and the dissolution of the country into separate republics. He tends to emphasize overall management education, however with frequent reference to the aspects of that education that relate specifically to international business.

On the subject of IB teaching specifically, Evenko notes that Soviet Industrial Academies from the 1920s through the 1950s trained managers, some of whom would receive assignments to international business functions such as purchasing foreign products or dealing with needed foreign financ-

ing. He states that in the 1920s and 1930s many trainees went for extensive internships with leading foreign companies, acquiring experience relevant to their respective industries. This process was halted with Stalin's repression, during which very little international exposure was available to people who would lead industrial enterprises.

According to Evenko, it was only in the late 1960s that "economically-oriented" management education became accepted in the Soviet Union. From that time until Perestroika, Soviet managers were able to obtain education in management techniques, but no specific focus was placed on international business education. He explains that this was not a significant problem for the country, since demand for people with IB knowledge was limited to the few government divisions responsible for foreign trade.

Apparently, international business/trade education was available to the personnel involved in foreign trade from the All-Union Academy of Foreign Trade, founded in 1931. Undergraduate programs were available for these few people through the Moscow Institute of International Relations. Nevertheless, Evenko faults the programs for failing to provide any training in marketing, trade negotiations, and other key areas.

The move to open-market IB education and management training in general began in 1985 with President Gorbachev's declaration of economic restructuring. It became a reality in 1987 with the adoption of liberal economic legislation (for example, permitting joint ventures). And according to Evenko, the movement was only completed when the Soviet state was destroyed and the independent republics formed. The new educational system is being formed and reformed now, so that full description of its characteristics is not yet possible.

Evenko points out that there are three sources of IB education now operating in Russia: the Committee of External Economic Relations that offers courses in IB; consulting firms that sell training on a private basis; and joint ventures in which foreign partner firms train their Russian counterparts in the context of the joint venture companies. He muses that the fundamental teaching resources are coming from abroad, and that the key bottleneck in the system is financing the purchase of such teaching services, since the Russian economy is so weak and foreign exchange reserves are essentially nil.

This paper provides a fascinating look at a situation of opportunity for IB education in a newly democratic economy—but it leaves the question as to how best to provide that education to others.

THE SOUTH CAROLINA MODEL AS SUGGESTIVE OF A "U.S. SOLUTION"

Folks's paper does not deal in any detail with the history of IB education, except at the University of South Carolina. This history itself is minimized, with the expressed goal of looking to the future to design international business education programs at universities in the United States. While this ultimate direction is quite useful, it would have been helpful to see some comparison with IB education at other U.S. universities.

Folks's suggested direction of IB education is really an institutional one, that is, it seeks to prescribe to individual institutions the appropriate path to "globalize" activities, depending on the institution's target employers, geographic focus, and type of program desired (for example, undergraduate, graduate, M.I.B.S., or other). These decision-making considerations provide an important taking-off point for discussion of appropriate IB education in the future, and should be pursued in that context. They are not discussed further here.

CONCLUSIONS

The main conclusions to be drawn at this stage of our thinking on the history of IB education are that business schools/universities are today paying much more attention to IB as an area of study than a decade ago or earlier, and that there does not appear to be a convergence of IB programs around the world. There may be, from the

analyses of Locke and Folks, even heated debate about the appropriate direction to move from where we are today.

Given the efforts of functional area disciplines such as marketing and finance to internationalize their teaching, at least in nominal terms, international business educators will need to redouble their efforts to push real IB research and teaching. This issue is especially notable at U.S. business schools, but may be a feature of business schools in other countries as well. The continued development of the field really requires new leaders who are willing to work on the fringe of business administration research and teaching to pursue the educational goals that we as a discipline define for ourselves.

Also, it appears from Locke's analysis that the U.S. educators could benefit quite significantly from discussion with their German and Japanese counterparts as to the kinds of management training that are most likely to serve the needs of future business managers. By the same token, it appears that the German and Japanese educators could benefit from discussions with their U.S. counterparts about research that can advance knowledge in the field. The lack of fora in which Japanese, U.S., and German educators meet to discuss these kinds of issues is itself notable. We need to take concrete steps to force this learning process to take place.

It appears that the stage is set to discuss the alternative strategies for IB education that have developed and are developing in different countries. We should not be diverted into too broad a discussion of overall management education, unless that choice is made explicitly. The opportunity is here today to define a clear path for IB education as it contributes to overall management education. All we need to provide is the leadership.

NOTES

1. See Fayerweather 1986 for factual details on this issue and on the development of IB education in the United States in general.

2. That is, U.S. businesses often make the claim that they infuse International content into

functional area courses. This is generally akin to saying that domestic textbooks such as financial management and marketing management are 'internationalized' because they include a final chapter on International aspects of the subject—despite the fact that almost no teachers ever use that chapter in their courses.

3. A few major international business research centers also began at that time. For example, the Harvard Multinational Enterprise Project was begun in 1966 and produced a series of books and articles that continued into the 1980s.

4. This concept hits home especially directly for me, after I have just spent a 10–day period teaching and coordinating an advanced management program for the 60 top executives in the newly-reorganized Argentine national oil company. This program was designed and has been used in previous years for country managers of MNEs operating in Latin America. By offering the program in-house for the management team of this firm, we outside academics were able to bring some new ideas to the table and to serve as a sounding board for the managers. We had to be indoctrinated into the newly-defined mission of the firm and to the goals placed by the CEO and Board of Directors. With those basics in place, we taught applications of International business for the use of this specific firm. It is far from certain that this means of teaching is the best vehicle for matching academic thinking to corporate needs, but it is a method that follows the Japanese and German models to a significant degree.

■ ■ ■

HISTORY OF INTERNATIONAL BUSINESS FROM AN INSTITUTIONAL PERSPECTIVE: COMMENTS ON THE PAPERS BY LOCKE, EVENKO, AND FOLKS

Richard W. Moxon

All three papers remind us of the particular institutional and societal contexts in which our activities are embedded, and which influence what kinds of international business educational programs are needed and possible. The Locke paper is a comparative examination of IB education in the context of the business education traditions of several countries. The Evenko paper looks at the influence of one country's ideology and political and economic context on management education. The Folks paper promises to examine the influence of a particular institution's resources and goals on the development of an international business educational program. Thus, while the three papers are very different, a common thread unites them and may help us to draw some general lessons.

The Locke paper is ambitious, attempting as it does to provide a historical perspective on international business education across five countries. This paper is thought-provoking, and makes some points that should encourage us to reexamine what we are doing. But before commenting on these very positive aspects of the paper, let me offer a few critical comments.

The author, as a historian, takes delight in making fun of (rather scathingly at times) the scholarly pretensions of business educators who, according to him, lack historical perspective and are second-rate intellectuals at best. Most of this is harmless fun, but I found myself resenting some of the caricatures he used. Specifically, international business educators did not banish liberal arts and humanities from their curriculum. These studies have indeed gradually been downgraded in business schools, although they are now making a modest comeback. Professors teaching IB, however, have typically been the defenders of the importance of liberal arts courses in the business curriculum. Our interdisciplinary preferences simply were overwhelmed by the "discipline-based" economists, psychologists, and statisticians.

Locke also makes a point of saying that U.S. IB educators, and this conference in particular, seem to believe that counting IB courses or IB programs is a good way of ranking countries with regard to international education. But this is Locke's straw man. This conference has not adopted any yardstick, and few U.S. IB educators would accept such a measure. Most would agree, for example, that an institution like IN-

SEAD is among the most internationalized in its teaching and research programs, but has few courses specifically focused on international operations or strategy. He particularly criticizes this tendency to equate quality with the number of courses or programs in a recent study by Reijo Luostarinen and Tuija Pulkkinen, *International Business Education in European Universities in 1990*, sponsored by the European International Business Association, and published by the Helsinki School of Economics and Business Administration. But this criticism is misplaced, as the study has not been endorsed by any official body, and can hardly be said to represent a consensus of business school educators.

Locke does a good job in describing the historical development of business education in different countries, but I was disappointed that he did little in describing the history of international education in business schools. He basically says, "since business education developed this way, this meant that international business education was like this." He neglects the historical development of international business itself, and what influence that has had on education. The most obvious example is the growth of the multinational corporation, a predominantly U.S. phenomenon in the postwar world, and having a strong influence on the development of international business education in the United States.

Moving to the more useful aspects of the Locke paper, he does an excellent job of focusing our attention on the issue of educational relevance and effectiveness. He tells us not to worry so much about whether IB is a discipline or not, as he argues that no business subject has a very solid theoretical foundation. Rather, we should worry about relevance; can IB education result in effective and competitive management? Here his doubts have a ring of truth. He argues that in a volatile world, the ability of businesses and managers to adapt is crucial, and that the most relevant education is that which provides a foundation for a person to continue acquiring knowledge. He argues for focusing on key qualifications, and sees especially mathematics and languages as the most important subjects. He implies that U.S. IB education puts too much emphasis on business and not enough on these basic skills. Locke's point is a good one, but I would argue that U.S. business schools are already responding to these criticisms. Foreign language training and other efforts to overcome the charge of "naive American managers" are getting a lot of attention today. In this sense, Locke may be preaching to the converted.

Locke's other provocative point is that maybe universities are not the places to provide the kind of education that managers need. Much of the needed education is industry-specific and organization-specific, with the key to successful management being relational skills and industry-specific knowledge. He argues that business schools tend to breed arrogant generalists, who think they can manage anything anywhere. Again, U.S. business schools are recognizing these challenges and responding to them. There is a strong emphasis on building relational skills, and more attention to developing industry-specific courses and executive programs.

Locke concludes his paper by noting that Germany and Japan have good managers but poor management education systems, while U.S. business education has gained its reputation; U.S. business lost competitiveness. The implication is that the U.S. system has it all wrong. But this may be another caricature, without much evidence to support it. Japanese management is not perfect. Much of the service sector in Japan is hopelessly inefficient and "customer-unfriendly." While Japanese manufacturing firms are world-class competitors, there is a dark side to Japanese management—the demands on individuals, and some aspects of the *keiretsu* system and business-government relations. And Japanese firms have been far from paragons in the management of their foreign operations, often exhibiting insensitivity to foreign cultures. The Japanese, in response to some of these problems, are now beginning to develop new systems of management education.

Things are changing in Germany and the United States as well. German firms may no longer be ignoring academic institutions, as executive education is gaining in popularity in Germany, and some innovative German business institutions have recently been created. And while U.S. industry has lost ground in many sectors, it is simplistic at best to lay a major share of the blame on business schools. Furthermore, many U.S. companies are exceptionally strong world-class competitors, and their management the envy of Europeans and Asians. The Locke paper is purposely provocative, and inspires us to think carefully about the role of IB education. But in exaggerating to make a point, it loses some of the bite that it could have had.

The Evenko paper does not really focus on IB education, but instead on the whole system of management education in Russia. The paper does a good job of describing the system as it has evolved. It is not surprising that the system is characterized as heavily centralized and hierarchical, and that the education emphasizes technical and ideological subjects, with many programs organized along industry lines. Little emphasis has been placed on general economics, and none on marketing and finance, reflecting a system which gave managers responsibility for production volume, with little need to consider consumer needs, the efficient use of resources, or tradeoffs of any kind. Specialist training has been emphasized over generalist education, although more generalist programs were developed in recent years. Most interesting to me was the reminder that within the Russian system there has always been a tension between ideology and pragmatism, and that in the 1920s and 1930s management education was much more pragmatic than in later years, incorporating foreign internships for managers, for example. It is also interesting to be reminded that the Russians have had programs similar in structure to our M.B.A. and Executive M.B.A. programs, albeit with a different curricular emphasis, for many years.

The paper touches only briefly on international business education. It notes that the centralization of foreign trade meant that the need for IB managers was limited. Most of the emphasis in the education of the foreign trade bureaucrats was on foreign languages and the mechanics of foreign trade, not on understanding management or foreign economic systems. Ironically, when Evenko comes to evaluating the current needs of IB education in Russia, he notes that the new Russian "international business leaders" are woefully unprepared, being especially deficient in languages and in knowledge of international trade mechanics and understanding of foreign business practices. The author's conclusions about what is needed for these leaders to function effectively in IB are not surprising. Basically, he is calling for a modern business curriculum plus foreign language and the study of foreign economies.

The paper is interesting to read, as it opens a world that most of us know nothing about, but the history and conclusions regarding management education needs are hardly surprising. The paper reminds us that IB education is part of the larger economic, political and ideological context of society, but beyond that I find little that is surprising in this paper.

The Folks paper attempts a far less ambitious task, that of drawing general lessons regarding the design of IB education from the experience of one institution, the University of South Carolina. Reading the history of the development of the M.I.B.S. program, and of the broad range of international activities into which this effort has developed, reminds us of the admirable record of institution-building that has been accomplished by our hosts at this conference. From a little known business school with no international business courses, and in a section of the country where few international corporations are headquartered, the university's international business programs have emerged as nationally recognized and highly ranked, and are one of the models for what is now being developed at many of the country's leading schools. My subsequent criticisms of this paper, and my reservations regarding "the South Carolina model," should be interpreted in the light of my admiration for what this institution has accomplished.

The Folks paper promises to develop a framework for assessing strategic alternatives in international business education, and then to use the South Carolina example to illustrate the power or usefulness of this framework. What I found in the paper, however, was an underdeveloped framework of no real power or persuasiveness, a long historical description of the development of South Carolina's M.I.B.S. program, and a few general conclusions only loosely connected to, and weakly supported by, that history. Sprinkled into the paper were many useful words of advice, but the reader hardly comes away with strategic guidelines.

First, a few comments on the strategic framework. The client-arena-product (CAP) idea is useful in helping us to be explicit about the "product-market segments" in which an institution may choose to compete. I have only minor quibbles with how this framework is applied in the paper. In table 2.9 we see a list of products defined along conventional lines, but it could also be useful to think about client needs that lie behind these products. This might lead to more creative thinking about what business we are in and what needs we are trying to satisfy. In table 2.10, I am bothered by the exclusion of students from the client dimension. The chart implies that we are only concerned about the employers of our students. Especially given the advertising strategy of M.I.B.S., which is targeted at students with slogans like "Imagine Yourself in International Business," I find it hard to believe that the M.I.B.S. program was really designed in response to the articulated needs of corporate clients. The marketing orientation implies that we should be thinking about both the perceived needs of students and those of business. In this same table, I do not see academic institutions and

accrediting bodies as clients, but rather as institutions which may influence the design of product offerings. Finally, the arena (A) side of the CAP framework is not developed in the paper at all.

Just identifying the relevant product-market segments is a long way from having a strategy framework, however. What is next needed is a methodology for choosing among segments, and for figuring out how to compete in each segment that is chosen. Here the paper does not provide much guidance. Table 2.11 mentions the historical franchise of the school, and calls for "assessing the internals," both of which provide indications of institutional strengths and weaknesses. The table also mentions "understanding the international needs of existing clients" and "identifying new clients." But the paper does not go beyond this obvious "SWOT Analysis," and in fact fails to illustrate how South Carolina performed such an analysis. We do not know what alternatives were considered, which factors weighed most heavily in the decisions and what were the key tradeoffs that were made. For those of us trying to learn something from South Carolina's experience, a clear explanation of these tradeoffs would be much more interesting than an overly broad "strategic framework."

Next, some comments on the "South Carolina model." Most academics teaching in the IB field were probably skeptical of the M.I.B.S. program design when South Carolina introduced the program. While it had some new features, especially the required international internship, it seemed quite similar to the Thunderbird model. It featured a separate degree, emphasized language training for students with little or no language background, and was housed at an institution not known for its international studies programs. Most IB professors were convinced that Thunderbird was not a serious academic program, and South Carolina seemed to be following the same model. Many felt that a separate international degree would lead to a dilution of the M.B.A. core, that accepting students with little language background would mean that many graduates would not be truly proficient in the language by the end of the program, and that South Carolina would have trouble attracting a sufficiently talented group of faculty and students. The result would be a large number of graduates who would have trouble finding challenging international jobs with top corporations.

While skeptics still abound, many have come to recognize the strengths of the South Carolina program. Virtually all of us are putting increased emphasis on foreign language ability and on opportunities for international experiences through exchange programs and internships. At my own institution we have completely overhauled our international specialization within the business school to emphasize language, area studies and foreign experiences instead of encouraging students to take a lot of IB electives. Many of us have realized that however good a job we do in internationalizing our M.B.A. core curriculum, there will also be a need to train international specialists in programs like that of South Carolina. The Lauder program is well known, and several of the federally-funded CIBERs are creating such programs, albeit much smaller than South Carolina's. And we have realized that South Carolina has succeeded in attracting students with solid academic credentials.

But some of our initial doubts remain, and are in fact reinforced by the changes in M.I.B.S. described in the paper. In some ways at least, the M.I.B.S. program is evolving toward the type of program most academics favored from the beginning. The quality of the students is increasing, the level of language ability of incoming students is higher, the curriculum is putting more emphasis on core M.B.A.-type skills and knowledge. The "South Carolina model," by proving that a market exists, has inspired many of us to overcome institutional inertia and design new international programs ourselves. As these programs have developed, and competition increases for South Carolina, the M.I.B.S. program itself has evolved. The original M.I.B.S. in-

spired us to be more innovative, and provided one model for IB education, but M.I.B.S. itself will need to continue to evolve to be a healthy survivor in the future.

Finally, a few comments of my own on the "history of international business education from an institutional perspective." What strikes me about this topic, and about the whole theme of this conference, is that this is just the newest phase of what has been a continuing debate for many years—how the international dimension can best be accommodated into business education. Several conferences on this topic were held in the 1960s, and I suspect many were held before that decade. And despite the fact that "internationalization" or "globalization" is the new mantra of business schools, the international dimension has been incorporated into business education virtually from its beginning in the United States. What has changed most over the years is the topical emphasis and pedagogical philosophy of the international curriculum, not the relative attention paid to the international dimension.

A brief review of the experience of my own institution is revealing. Our School of Business Administration admitted its first class in 1917. One of the six first-year required courses was "Resources of the World," an economic geography course with a global perspective. Foreign trade was one of the ten majors available to students, and as early as 1920 the School had twelve Chinese students in its first exchange program. The curriculum emphasized foreign trade practices and transportation, and this practical approach to trade continued into the 1950s. The international group was naturally linked organizationally to the faculty in marketing and transportation.

By the early 1960s, our program had entered a second phase in which the curriculum focused not so much on trade, but instead on international corporate activity. Foreign area analysis was introduced as part of the curriculum, and gradually evolved into regional courses focusing on the dynamics of the international business environment—European integration, the

East-West interface, and developing countries. This curriculum reform of 1960 laid the basic foundation on which the international curriculum is based today. During the 1970s and 1980s the number of internationally-oriented faculty grew, a major project was launched to encourage faculty research on the Pacific Rim region, and new international executive programs were created. A separate group of "internationalists" was maintained during this time, and was somewhat uncomfortably housed in the marketing department, but international courses were being offered in several other departments of the school.

The latest phase in the development of our international programs began in the late 1980s, and was crystallized by the creation of a Center for International Business Education and Research. CIBER has become the catalyst for an expanded program of international activities—teaching, research and outreach—and more importantly, for the incorporation of the international dimension into the mainstream of the school's programs, culture and organization. The school is committed to educating all students to understand the global environment of business, and to have a global vision of business strategy and leadership. This is reflected in required courses on global markets and competition (note that these are not typical IB survey courses) in all degree programs, the incorporation of international material into several core courses, and a wide range of international electives. At the same time, the school is developing in-depth programs for international business specialists, who are expected to not only understand international business issues, but to demonstrate proficiency in a foreign language and to have meaningful experience in a foreign culture.

Faculty in all fields are being encouraged to extend their research internationally, and many professors are participating in the design of our new international courses and programs for students. So while international programs have always been important at our institution, we are now at the point where the international dimension is being incorporated into a wide range of courses and programs. This is reflected in our new organization, in which the separate international business group is being eliminated, and internationally-oriented faculty are housed in each of the school's departments, with CIBER acting as a catalyst and coordinator for international programs.

My conclusion for IB education in the United States is optimistic based on these papers, this conference and the historical evolution that I have observed. Despite frustration by some of us committed "internationalists," if we consider realistically the societal, business and institutional context in which we have had to operate, the international dimension has not been neglected in the past, and is not being neglected now. Innovations in international business education are being made, and we are responding to the needs of business and students. In fact, in many cases we are leading where business is yet to follow. United States' business schools have problems, as does IB education. But constant reexamination of their mission and programs is a hallmark of these schools, and we are at it once again, as is evident in this conference.

REFERENCES

Arpan, Jeffrey S. 1993. Curricular and administrative considerations—the Cheshire cat parable. *Internationalizing Business Education: Meeting the Challenge.* Ed. S. Tamer Cavusgil. East Lansing: Michigan State University Press.

Ball, Donald A., and W. M. McCulloch, Jr. 1984. International business education programs in American schools: How they are ranked by members of the academy of International business. *Journal of International Business Studies* 1 (Spring/Summer): 175–80.

Byrne, John A., et al. 1991. *Business Week's Guide to the Best Business Schools.* New York: McGraw-Hill.

Chandler, Alfred D. 1977. *The Visible Hand.* Cambridge, Mass.: Harvard University Press.

Churchill, Winston S. 1952. *The Second World War,* v. London: Cassell.

Douglas, Susan P. 1989. The ranking of masters programs in International business—comment. *Journal of International Business Studies* 1(Spring): 157–62.

The Economist Intelligence Unit. 1990. *The Quest for the International Manager, A Survey of Global Human Resources Strategies* (Ashridge Management Research Group).

Fayerweather, John.1986. A history of the Academy of International Business from infancy to maturity: The first 25 years. *Essays in International Business.* Columbia: University of South Carolina Press, no. 6 (November):1–63.

Folks, William R., Jr. 1992. Institutional barriers to internationalization. Perspectives on International Business: Theory, Research and Institutional Arrangements. Center for International Business Education and Research, University of South Carolina, Working Paper.

Gellert, Claudius. 1983 *Vergleich des Studiums an englischen und deutschen Universitäten.* Munich: Bayerisches Staatsinstitut für Hochschulforschung und Hochschulplanung.

Hoskins, Keith W., and R. H. Macve. 1988. The genesis of accountability: The West Point connections. *Accounting, Organizations and Society*13 (1):37–73.

Iribarne, Philippe d'. 1989. *La Logique de l'honneur: Gestion des entreprises et traditions nationales.* Paris: Editions du seuil.

Kagona, Tadoa, et al. 1981. Mechanistic vs. Organic management systems: A comparative study of adaptive patterns of U.S. and Japanese firms. Reprint from Kobe University's *Annals of the School of Business Administration,* no. 25.

Keeble, Shirley, 1984. "University Education and Business Management from the 1890s to the 1950s: A Reluctant Relationship," Ph.D. dissertation, London School of Economics and Political Science.

Komiya, Ryutaro, and K. Yamamoto. 1981. "Japan: The officer in charge of economic affairs," 263–67, 167, in Alfred w. Coats, ed. 1981. *Economist in Government.* Durham: University of North Carolina Press.

Lambert, Richard D. 1990. Foreign Language use among International Business graduates. *Annals of the American Academy of Political and Social Science* (September): 511.

Locke, Robert R., 1989. *Management and Higher Education Since 1940.* New York: CUP.

———. 1991. Mastering the lingo. *The Times Higher Educational Supplement,* 5.4.

———. 1996. *The Collapse of the American Management Mystique.* Oxford: OUP.

Luostarinen, Reijo, and T. Pulkkinen. 1991. *International Business Education in European Universities in 1990.* Helsinki: Helsingin Kauppakorkeakoulun julkaisuja.

McNulty, Nancy G. 1980. *Management Development Programs, The World's Best* (Amsterdam).

Mill, John Stuart. 1965 (1848). *Principles of Political Economy.* Ed. W. J. Ashley. New York: A. M. Kelly.

Nagaoka, Katsuyuki. 1991. Japanese business education. Paper read at the Anders Berch Symposium, 16 September, at Uppsala: Motala Grafiska AB.

———. 1994. "The Japanese System of Academic Management Education." 138–54. in Lars Engwall and E. Gunnarsson, eds. 1994. *Management Studies in an Academic Context.* Uppsala: Motala Grafiska AB.

Nehrt, Lee C. 1987. The ranking of masters programs in International business. *Journal of International Business Studies* 3(Fall): 91–99.

———. 1989. The ranking of masters programs in International business-reply. *Journal of International Business Studies* 1(Spring): 163–68.

———. 1993. Business school curriculum and faculty: Historical perspectives and future imperatives. In *International Business Education: Meeting the Challenge,* ed. S. Tamer Cavusgil. East Lansing: Michigan State University Press.

Prashker, Ivan. 1988. *Duty, Honor, Vietnam, Twelve Men of West Point.* New York: ArborHouse/William Morrow.

Reich, Robert B. 1991. *The Work of Nations: Preparing Ourselves for 21st Century Capitalism.* New York: Alfred A. Knopf.

Samuelson, Robert J. 1990. What good Are B-Schools?, *Newsweek* 14 May,

Smith, Roy C., and I. Walter. 1990. *Global Financial Services*. Grand Rapids: Harper Business.

Thanheiser, Heinz. 1979. Stratégie et planification allemandes. *Revue française de gestion* 21 (May–June): 6–13.

Thanopoulos, John. 1986. *International Business Curricula: A Global Survey.* Academy of International Business.

Thomas, Raymond. 1987. "Management Education and Development: Innovation or Déjà Vu?," June, in manuscript.

Thurow, Lester. 1992. *Head to Head: The Coming Economic Battle Among Japan, Europe and America.* New York: William Morrow.

Toyne, Brian. and D. Nigh. 1997. *International Business: An Emerging Vision.* Columbia: University of South Carolina Press.

Wächter, Hartmut. 1980. *Praxisbezug des wirtschaftswissenschaftlichen Studiums* (September) Expertise für die Studienreformkommision Wirtschaftswissenschaften.

Walter, Ingo. 1988. *Global Competition in Financial Services.* Cambridge, Mass.: Ballinger Publishing Company.

Yuzawa, Takeshi. 1988. Thesis on the history of Japanese Business and Commercial Higher Education, parts of which were given to the author, quote from p. 74.

3

Institutional Structure and International Business Research: Functional or Dysfunctional?

EDITORS' COMMENTS

For some time now it has been recognized that the focus and conduct of scientific inquiry is partly conditioned by the structuring of the scientific process, for example, the physical and organizational location of the activity in question, the activity's level of financial support, the reward/sanction system, and the hiring and retention of researchers. Within academic circles, the structuring of business research has generally been along the lines of the traditional business disciplines, such as accounting, finance, marketing, management. In this chapter we asked our authors to address several questions related to the implications of such a functional approach for the development of international business research.

We first acknowledge the recent trends in international business theory, research, and practice and ask: Given such recent trends, is the functional approach the most effective today? Should international business research be treated as a peripheral activity fragmented along functional lines or should it be recognized as a core activity of an emerging interdiscipline? Do the benefits of the functional approach to business research in general and international business research in particular outweigh the costs? Are the international business research questions and approaches developed with a functional approach adequately meeting the needs of scholars and practitioners in their quest to understand international business phenomena?

The five authors who address these and related questions in this chapter have a wealth of experience as researchers, as teachers, and as academic administrators. They have had to wrestle with the organizational issues related to international business research not only on a personal basis, but also in leadership positions for various institutions. Their contributions draw heavily on their collective decades of experience.

In the lead paper, Robert G. Hawkins first warns that there is no ideal structure at all times for all institutions. He goes on to highlight the importance of the objectives of IB research and links these to various organizational structures. Turning his attention first to the objectives of IB research, Hawkins discusses what IB research is supposed to accomplish along the following categories: description, understanding and explanation, assessment and advocacy, base for curriculum and teaching, and advancement and notoriety. Of particular note is Hawkins's discussion of the often unexamined values that underlie the conclusions of our normative research. Since those conducting IB research come from a variety of disciplinary backgrounds, they do not necessarily share the same sets of standards or objective functions, leading to some stress within IB groups when it comes to advocacy positions.

Hawkins next discusses four alternative organizational structures for international business research: IB lodged within each functional field; IB as a separate department, center, or unit; IB-defined project or program; and matrix, IB plus a functional field. He reviews the advantages and disadvantages of each and provides a major business school as an example of each alternative. Of particular note here is the role that the professor's group identity and allegiance plays. Finally, Hawkins looks at the fit between research objectives and organizational alternatives and concludes that the IB as a separate department seems to facilitate the highest number of IB research objectives, while IB within the functional fields seems to provide the least support for achieving the multiple IB research objectives. Importantly, Hawkins acknowledges that the cost and complexity of each alternative must be taken into account—an analysis that Hawkins does not do in any depth.

In the second paper, Jeffrey S. Arpan concludes that a functional structure is generally dysfunctional for international business research. He first examines the adequacy of IB research and finds the quantity to be far less than the importance of IB phenomena warrants. This situation is particularly true for IB research that does not fit neatly in a traditional business function. While IB research gets passing marks for quality, Arpan has strong reservations about its usefulness to practitioners and policy-makers.

Arpan's explanation for this state of affairs incorporates a number of factors: lack of faculty awareness or interest, lack of faculty training, greater difficulty in publishing IB research in "top-tier" journals, and most importantly of all, the current organizational structure of most U.S. business schools. The picture he paints of "the functional-specialist" departmental structure emphasizes the extent to which it hinders the type of collegial interaction and cross-fertilization necessary for multidisciplinary and international business research." In this structure, the training of new scholars, the publication of research, and the rewards for scholarly activity all tend to be carried out in ways detrimental to the development of international business research. Arpan's remedy involves changes in doctoral programs, faculty development programs, faculty reward systems, and departmental structures. He closes with a evaluation of the benefits of an IB department, particularly those related to the quantity and scope of IB research.

In the third paper, Duane Kujawa argues that the trend towards internationalized management will force all business schools—big or small, old or new—to deal with their own international business "imperatives." He examines the environmental pressures that influence organizational structures. These include demographic and general economic trends, as well as changes in students' career preferences and businesses' education expectations. The upshot of these trends is the likelihood that business schools will not change existing functionally oriented departmental structures and add a new international business department. Rather business schools will deal with the these IB imperatives by creating certain facilitating, integrating positions, for example, associate dean for IB programs, creating certain crossdepartmental task forces and committees, and changing certain processes related to funding activities and reward systems. Kujawa differs from Arpan in that Kujawa is optimistic about the possibility of schools instituting such changes and their implementation having a positive effect on IB research and education. Arpan would acknowledge that such changes are possible, but just not likely, unless an IB department is formed.

In his commentary, John D. Daniels asks if there any identifiable problems concerning the state of academic business research and if so, does structure have a significant impact on them? Using a modified gap analysis and the image of IB research flowing out of a pipeline, Daniels analyzes where the "leakages" occur. His paper is full of thoughtful (some provocative and some practical) recommendations to reduce the research effort gap, the international applicability gap, the dissemination gap, and the target market gap. For Daniels, the key to getting more international business research done lies not in changes in formal organizational structure (IB department versus other alternatives). Rather like Kujawa, Daniels suggests we look more to changes in organizational processes and culture. Nonetheless, one can identify throughout his paper examples of the negative effect of a functional specialist structure on IB research, most notably in the

areas of doctoral training and the dissemination of IB research.

In his commentary, Edwin L. Miller argues that internationalization of faculty and institutions demands aggressive and skilled administrative leadership. Miller sees as absolutely critical the role of the dean (and under some circumstances, the dean's designated representative) in restructuring the school, adjusting reward systems, and instilling an institutional mindset that promotes internationalization of the school and its faculty.

In looking at all five papers as a group, it's clear that all five authors agree that organizational structure (taken in its broadest sense to include organizational structure, processes, and culture) does have a significant impact on the state of IB research today. When it comes to the value of different alternative formal organizational structures and particularly the value of a separate IB department, the five authors disagree to some extent. Kujawa, Daniels, and Miller see the departmental structure as less important and put more emphasis on designing and implementing appropriate reward systems, information systems, crossdepartmental integrating mechanisms, and institutional mindsets. Arpan argues that you probably will not get those desirable organizational processes and culture without an IB department.

We believe that you do not necessarily have to have an IB department in a school in order to promote IB research of the quantity, quality, and scope that we all desire. The key lies instead in what a "department" means within a school and in the nature of the relationships among the departments and their members. In some schools, departments based on functional disciplines are the center of their members' professional universe, their primary identification and target of allegiance, the primary determinant of professional rewards; interpersonal interaction, mutual understanding and respect across departments is low. In such a school, an IB department is needed if IB research (and education) is ever going to amount to anything. On the other hand, in schools where departments are less central, where crossdepartmental interaction, understanding and respect is higher, an IB department is not necessary.

Most instructive in understanding the effect of departments on IB knowledge generation is a 1969 paper by Donald T. Campbell, "Ethnocentrism of Disciplines and the Fish-Scale Model of Omniscience." Campbell identifies "the tribalism or nationalism or ingroup partisanship in the internal and external relations of university departments, national scientific organizations, and academic disciplines" as a main obstacle to the development of a comprehensive, integrated social science (1969, 328). Such ethnocentrism of the disciplines leads to a state of knowledge in which we have a redundant piling up of highly similar specialties within disciplines (and departments) and large interdisciplinary gaps—a far cry from the more uniform distribution of knowledge across all possible areas, which Campbell calls the fish scale model of omniscience. Noting first that present disciplines are arbitrary composites shaped by in large part by historical accident, Campbell carefully describes the how organizing academic specialties into departments leads to certain specific decisions in departmental chair selection, rewards, training, and curriculum with certain outcomes in ingroup/outgroup identification and scholarly communication and competence. The end result is very strong pressures leading to departmental separateness and interdepartmental gaps. "In my professional career at four universities I have repeatedly seen men who were of exceptional creativity and competence, and who were absolutely central as far as social science or behavioral science was concerned ..., be budgetarily neglected and eventually squeezed out because the departmental organization of specialties made them departmentally peripheral" (Campbell 1969, 335).

Since international business as a scholarly field of study is interdisciplinary, we see departmental structure issues as serious ones for the progress of the field, since all too often the IB specialist may be peripheral within the department. Yes, the strong departmental dynamics can be overcome by a school with the concomitant positive effects on its international business research and education performance, but these dynamics need to be recognized as the strong, unrelenting current that they are.

■ ■ ■

Organizational Structures for International Business Research
Robert G. Hawkins

INTRODUCTION

This paper addresses alternative organizational arrangements, largely within universities, for conducting research in international business and its subfields. It does not conclude that one or a subgroup of potential organizational structures are optimal or preferred. There is no ideal organizational structure for all institutions in all times. Rather, the paper attempts to identify the importance of broader institutional characteristics as a factor in the preferred organizational structure for research; it also suggests that the objectives of the research activities should be another important determinant in choosing the organizational structure.

The paper has two major sections. The first deals with the primordial issue of the objective of international business research. What is it supposed to accomplish? Section

III describes some alternative organizational structures for international business, in academic settings, and their characteristics with respect to international business research. Some examples or models of organizational structure related to research are identified, along with some advantages and disadvantages of each.

The final section provides some brief conclusions, linking the objectives of IB research with organizational structures.

OBJECTIVES OF IB RESEARCH

The question of the purpose or objectives of research in international business is not a trivial one. The answers vary among institutions, across disciplines, and across individuals. There are no appropriate or inappropriate objectives or purposes of scholarship relating to international business—its beauty is in the eyes of the beholder. This section attempts to identify some of the more standard objectives or rationales for international business research to provide a backdrop for organizational structures relating to that research.

To test opinion (or discomfort) with international business research, some serious questions should be answered—for the individual and the institution. What is the research attempting to accomplish? This leads to the derivative question of what do we include in "research"? Or is originality, or primary information, a benchmark of "real" research?

Obviously, the appropriate answer to these questions depends upon the institutional objectives, and perhaps the individual's objectives as well, of the scholarship. Answers to the questions of what is IB research supposed to accomplish can be identified with more or less precision.

Description

A major objective of research in any field is to describe, and perhaps analyze in the sense of breaking it into component parts, the subject of the observations. In international business, this has typically in-

volved the individual firm or organization, originally the exporting and importing company, the bank that does foreign exchange and international banking activities, or the individual managers who engage in international activities. By the 1960s, the focus moved exponentially towards multinational, transnational, or global companies. What do they do? What is their profile?

This type of research need not be confined to individual private enterprises. Description and analysis of the activities of public and international organizations, such as the IMF, World Bank, national central banks, or international policy-making bodies are legitimate subjects for this international business research focus. By the late 1960s and 1970s, developing countries, as national units, were being described and profiled; strategies, and international investment policies were placed increasingly under the microscope of the international business researcher.

A slightly different dimension, also within the category of description, involves comparative analysis, involving two or more descriptions, on a focused topic or groups of behaviors of nations or companies. Much of this dealt with personnel and human resource practices among countries, the style of management decision-making, and the role of government in the international business sector. All of this research falls under a generic objective to increase the understanding of the individuals or organizations under study, that is, to better understand and be able to describe meaningfully the world of international business as it exists. Clearly much of the research in this category also begins to address the question of "why and how" does it exist this way, incorporating the methodologies of the various social sciences and historical analysis into these descriptive studies. But this research objective generally seeks to provide a meaningful description of complex phenomenon in understandable terms, rather than being predictive or prescriptive.

Some may question the usefulness of this descriptive research. But in fact, this type of descriptive researches is extremely useful in providing the factual base upon which we build causative, prescriptive, and reductive models and recommendations. This research provides the contextual base upon which much of the curriculum in international business has been developed and designed. This research has served as a major touchstone against which companies compare their own policies and plans, organizational structure and style, or approaches to international markets. Further, the academic community has contributed greatly to an understanding within both business and government of the actual practices of those very same governments, international organizations, and individual companies. This, in turn, establishes the context in which policy managers carry out their own agendas.

Descriptive research need not be mundane or soft. Indeed, the application of relatively sophisticated empirical techniques, including regression and other forms of multivariate analysis, has been applied in a number of studies which are essentially descriptive in nature, attempting to sort out commonalities among characteristics when the characteristics have multiple dimensions. Over the past two decades, our meaningful knowledge of multinational companies, of economic policy practices, and of international financial relationships, have greatly benefited from the application of multivariate empirical techniques.

Descriptive research also frequently involves case studies which provide an additional, and in some sense an analog, dimension of description of international business phenomena in organizations. Case studies are particularly useful in describing the complexity and the context of administering or applying policies in a discontinuous and ambiguous international environment.

Thus, pride can be taken in the descriptive dimensions of international business research over the past few decades. Through it, international business researchers have more accurately and efficiently described and analyzed the actual situation in international business, and in the process eliminated some of the preconceived notions, illusion, and misconceptions of researchers in the traditional functional fields. The facts must be presented before other researchers or decision makers can be confused by them.

Understand and Explain

A second fundamentally important objective of international business research is to establish cause and effect relationships between variables or phenomena, and thereby set up testable hypotheses or ask meaningful questions of why international businesses behave the way they do, why governmental policies have particular impacts, or why nations or other groups behave in particular ways. This objective relies upon hypothesized and observed relationships between and among variables. In the international business arena, this research seeks to help understand the evolution and nature of international business transactions and to explain how and why they change in particular ways.

The literature to explain and understand international business phenomena cuts across a broad range of methodologies. The earliest research in the field evolved from economic analysis and was originally embedded in international economic theory. Much of the international finance and macroeconomics research was based upon economic relationships. The field then moved on to incorporate a broad range of research on organizational structure, management behavior, and governmental behavior, seeking to explain why structures or decisions were as they appeared. By employing a wide variety of social-scientific and behavioral variables, this research links one phenomena (or characteristic) with other variables (characteristics or events).

Research in this area ranges from the limited and simplistic, seeking to establish the effect of one variable on another and explore the simple relationship, to complex simultaneous systems involving a multitude of variables and relationships. For example, some studies examine the simple relationships between types of technologies and firms' international organizational structure; or the impact of exchange rate changes on exports of a particular country. At the other extreme is the search for a general paradigm to explain international business or the behavior of multinational corporations These models involve a multitude, or at least several, variables which interact in a very complex manner. An early initiative of this type was the product life cycle theory of international trade and investment associated with Raymond Vernon and others at Harvard; a prominent recent effort is the "eclectic theory" associated with John Dunning and several of his colleagues. Thus far, the quest for a single paradigm for internationalization and the behavior of multinational companies has not been highly satisfying in producing a robust and tractable model.

The IB research which focuses on "understanding and explaining" does precisely that. In general, it is positive, not normative in nature. But as will be noted below, the end use of much positive research in the international arena is to provide a basis for making policy prescriptions or normative statements.

Assess and Advocate

A variation or extension of the "explain and understand" type of research is that which is normative. This scholarship, which covers a wide variety of studies, establishes hierarchies of "better and worse," with respect to such phenomena as types of behavior, company or government policies, and market or organizational structures. It extends beyond establishing relationships and measuring their coherence or size, which is where positive analysis would stop.

In order to establish a hierarchy of better and worse, a concept of the "standard" or ideal is necessary; that is, an objective function which lays out relationships which yield utility or desirability. Such standards or objective functions come in many varieties, perhaps the most pervasive is the standard of the competitive market. Indeed, the standard of perfect competition is frequently used to establish explicit or implicit models against which to judge relationships about which normative (or advocating) conclusions may be made.

Perhaps the most extreme advocacy model rests on the assumption in much of the finance literature that the maximization of shareholder wealth is the "standard" or ideal against which company policy is judged. It is within this context that much of the international managerial finance literature and research is carried out. Specific company policies or practices can be judged as better or worse from the analysis, based upon whether they improve or reduce shareholder wealth.

Other examples are found in analyses of national development policies or international trade policies and their effects. In most such analyses, especially in the Anglo-Saxon literature, the competitive market becomes the standard against which actual policies are compared. In both instances, the analysis turns to advocacy, in ascribing superiority to policies which move the situation closer to perfect competition.

This type of analysis is not confined to those relating to the competitive economic paradigm. They extend to the organizational design and business strategy literature in the international context as well. For example, many studies accept as an implicit standard or desired goal the improvement of employee productivity/efficiency within enterprises. Thus, incentive systems of international organizational structures which improve employee productivity/efficiency is judged superior to those which reduce it. And in much of the marketing literature, the analyses go beyond description and understanding to accept the assumption that activities which increase sales, reduce the cost of sales, or more fully satisfy customer desires is better rather than worse.

While this may be a cumbersome way of saying that much of our research and its conclusions is value laden, it is relevant for a discussion of appropriate organizational structures for international business research. The several disciplines which make up "international business" do not necessarily share the same sets of standards or objective functions for the normative or advocacy dimensions of their research. This can sometimes make for stressful bedfellows from diverse disciplines who find themselves in the same organizational IB unit.

A Base for Curriculum and Teaching

An overlapping objective of international business research, and an important one in the evolution of the "field", is to provide the intellectual and factual basis upon which courses and teaching materials are developed. As will be noted, this may be an explicit objective of the scholarship of some faculty members or departments.

The materials to support the education process have and will include research that is descriptive; that which establishes linkages among international business variables in a positive, nonassessive vein; and also materials which advocate and are prescriptive in nature. The desired mix of such materials will vary among faculty members, and perhaps from department to department.

The proliferation of journals, texts, and monographs in international business and related fields over the past twenty years is almost beyond belief, in historical perspective. This suggests that this important objective of international business research, of providing an adequate base for wider understanding and education, has been fulfilled—some would say with a vengeance.

Advancement and Notoriety

In international business as in most academic pursuits, one objective of scholarship is to attain the required level of publications and acceptance in the field to qualify for academic promotion and tenure. Viewing this objective of research cynically, it proliferates marginal and sometimes inconsequential writings, journals, and books. It may pull otherwise effective faculty members into frustrating and counterproductive writing and publishing activities.

But another side of this objective is positive. Published and reviewed scholarship is an external and objective indication that the scholar is keeping up and participating in the field, or at least some part of it. Exposing scholarship to the review and opinion of others also gives an external and relatively detached view of the quality and importance of the work. While publishing for its own sake may be questionable, publishing—regardless of the organizational structure for international business—will clearly be one of the elements of legitimacy in the eyes of other organizational units within the school or the university.

ALTERNATIVE ORGANIZATION STRUCTURES

The several organizational and individual objectives in research in international business outlined above will be one influence in setting a school's organizational structure for the international business faculty and its activities. But certainly not the only objective. The history and evolution of international business within the school is likely to be a determining force, as are the personalities and diverse agendas of the individual faculty leaders who were there at the inception of international business programs.

This section examines a few among a wide spectrum of possible organizational structures and how they may relate to the research mission in international business. The wide range of organizations is very difficult to describe. At the one extreme, it may be the small school which has one faculty member interested in international business research topics who patiently conducts his or her individual research program without much attention or support from the institution. Historically, such individuals tended to be in the management (business strategy) department/group, or less often in the marketing group.

The other extreme is a full-fledged international business department, with a departmental chair, a separate identity, and with the primary allegiance of the individual faculty members to international business department. Faculty members may participate in other departments and departmental activities but international business reigns supreme with respect to recruiting, promotion and tenure, and judging and evaluating research productivity and acceptability. The gradations from the full-fledged department are several, including subdepartments within the larger business policy/strategy departments, or occasionally as subgroups within the marketing department.

Table 3.1 identifies four generic types of organizational structure for international business research. These are intended to be illustrative rather than definitive. Associated with each type are admittedly incomplete lists of advantages and disadvantages.

Table 3.1 Alternative Organizational Structures for International Business Research

Type of Organizational Unit	Advantages	Disadvantages	Examples
IB lodged within each functional field	Tight link with function field	Difficulty in cross-disciplinary research and collaboration	Stanford Texas
		Lack of critical mass and cohesion of IB	
International business as separate department; center; unit	Conducive to cross-disciplinary research	Detached from functional field; cross of rigorous peer review	University of South Carolina
	Creates an identity critical mass	Viewed as second-class citizen	
		Another administrative unit in the school	
IB-defined project or program	Assembles faculty from several disciplines	Lack of continuity	Harvard
	Faculty assessments left to functional departments	No true IB identity of administrative influence	
	Attractiveness for external funding		
MATRIX: IB plus functional field	All of the above	Dual allegiances of faculty members	New York University
		Managerial complications	
		Possible double hurdles for advancement	

Finally, an example of a major business school which approximately fits each type of organizational structure described is indicated.

IB Within the Functional Fields

This type of organizational structure subsumes the international dimension under the various functional disciplines within the school. In most instances, if the school is a large one, there will be formal departments in marketing, finance, and similar disciplines. The international business dimension in each of those functional fields will be handled by one or more faculty members conducting research and teaching in it. Many schools have less formal departmental lines, but are likely to have divisions or other faculty units (sections) with administrative responsibilities for curriculum, faculty recruiting, and student advisement within the functional field. Faculty members with international interests are likely to be listed in several or all of those groupings.

Obviously the main advantage from a research perspective is that the international business research, and faculty members conducting it, are very tightly linked to the functional field. This, in the view of many, elevates the rigor of peer review, the purity of the methodology and acceptability of the research. It should assure a strong functional-field methodology base, and improve the acceptability and the credibility of the research in the functional field mainstream.

A subsidiary advantage is that integration of international business within the functional fields avoids the administrative costs and complexities of another department or organizational unit. There is precision in the advancement and quality hurdles which a faculty member must meet. These may involve appraisals from international business faculty members form other departments in the school, which may influence indirectly the assessment.

The disadvantages of having international business research scattered among faculty members in several functional fields are obvious. In such an environment, it is difficult to accomplish research on topics or projects having several disciplinary dimensions. Individual faculty members must bypass the vertical lines of organization to collaborate with researchers in another administrative unit or department of the school. This complicates proposals for research funding, and it complicates administration of research funding once received.

Perhaps more important, the dispersion of the international business group across several functional fields leaves the impression of a minor level of institutional commitment to international business, and the lack of a critical mass to project the research programs and teaching activities to audiences within and outside the university. The difficulty of creating an identity or focus for international business research or programs is magnified by having faculty members dispersed across various departments which represent their primary allegiances.

Casual observation suggests that smaller schools, but not exclusively, tend to spread the international business research faculty across several departments—not on purpose, but as their institutions have evolved. Among these might be included relatively small, but prestigious institutions such as Stanford, Carnegie-Mellon, and Dartmouth. But one might include in this group substantial institutions, some of which have minor variations on the functional field organization, including the University of Nebraska, the University of Texas, and some other institutions of substantial size.

IB as a Separate Department or Unit

The other extreme is a school which establishes a department of international business, and assembles in it faculty members with primary orientation in international business and a secondary orientation in the functional field. In several instances, this involves faculty members whose functional field training spans the spectrum among economics, finance, marketing, accounting, and business policy and strategy. In the extreme cases, the international business department has its own budget, behaves as a traditional academic department including responsibility for some or all of its research funding, and initiates faculty recruitment and promotion/tenure processes.

From the exclusive perspective of international business research, this form of organization has several major advantages over other organizations. Foremost is that this organizational structure provides a convenient assemblage of faculty members trained in various disciplines whose primary allegiance is to international business; thus crossdisciplinary research is facilitated—perhaps even essential. At the same time, as-sembling of at least a few faculty members from various disciplines means that a critical mass of faculty will be involved in helping to define the mission and research objectives of the unit. If this is successfully done, it enhances the prospect of serious external funding for the research of the "IB department." A subsidiary advantage is that the department unit also enhances the international business influence in curricular matters, and may serve as the focal point for the school's initiatives in internationalizing its programs and curriculum.

Project or Program

Another type of organizational structure occurs when faculty members from several functional fields assemble as a program team to carry out a research project or establish an educational program in the international business area. The primary mission of the faculty group is to pursue and complete the research or program while most administrative matters are handled by the traditional, functional units. The project may have a loosely defined administrative role, may be given a separate budget account, and pursue external sources of funding, and even serve as a faculty advocacy group for international business components of the curriculum.

Perhaps the Harvard Business School's Multinational Enterprise Unit of the 1960s–1970s is the most visible example of this flexible type of arrangement. The primary allegiance of the faculty remains with the functional field department. But the unit does define an identity for the international business research group and provides its administrative focus and support. It has the advantage of bringing together faculty members from different functional fields to pursue multidisciplinary or interdisciplinary research. It preserves the integrity of the functional-field administrative organization while providing a "home" for the international business aspects of a subset of the faculty.

The disadvantages are also clear. Such project- or program-based groupings may well disappear when the project is completed. They are unlikely to have serious influence in the longer term organizational structure of the school or on the role of international business in the curriculum or in the research priorities of the school. And the project' influence in such critical faculty matters as promotion and tenure, and recruitment, may be limited, thus constraining the group's success in legitimizing international business within the overall priorities of the school.

Matrix Structure

A fourth generic organizational structure involves a matrix relationship between international business and the traditional functional fields. In this type of structure, international business is defined as a department (area or unit) on a co-equal basis with the functional field departments. It has a chair, a budget in certain circumstances, and a level of coresponsibility for faculty recruitment and promotion and tenure.

The individual faculty member has a dual allegiance. He or she is a member of the international business department and a functional field department at the same time. The faculty member undergoes performance reviews by both groups, in most instances.

This departmental structure clearly has an advantage of assembling faculty members from several disciplines interested in international business, developing cohesion and potential for multi- and interdisciplinary research on international business topics, and developing a sense of identity of the international dimension in the school. It further has an advantage of developing a critical mass which will be helpful in securing external research funding.

But the advantages are balanced against several difficulties. The chair of a "matrixed" IB department faces significant complications in securing internal resources, in coordinating faculty assignments, and in assuring consistent criteria for promotion and tenure. The matrix has the further disadvantage for the individual faculty member of dual allegiances with the functional field and with the IB unit. He or she may be torn and confused over which to place first and how to manage his or her role in the two departments.

New York University is the most systematic example of the matrix form for IB. The success of this type of unit, including at NYU, requires a coherent and accepted mission and philosophy about international business (and IB research) to which all or most of the IB faculty adhere. It also requires a permissive and cooperative attitude by the faculty and chairs of the functional field departments. Without it, the necessary coordination will be incomplete and conflicts elevate. Without these, the matrixed IB unit could be relegated to second-class status in the school. Upon any sign of this, individual faculty members will recognize the functional field as the superior choice in the dual allegiances.

Evolution of Organizational Structures

It should be reiterated that the four generic types of organizational structure described above are not necessarily exhaustive, nor do they describe any specific school organization. They are generic types towards which specific organizational arrangements gravitate for handling international business in the university context. The more interesting issue is why and how particular departments have evolved towards particular types of organizational arrangement.

One may speculate that the size of the school may have an influence. Large schools appear more likely to be able to sustain and accept an international business unit/department which is viewed as equivalent to the units in the functional fields. The same can be said for the matrix organization, which appears to fit into or require a large institution. Smaller schools appear to leave the international business faculty lodged within the functional units, or to have them assemble for particular international business research projects or program design and development. Size also will be correlated with the preexisting organizational structure within the school; large schools will more likely be departmentalized; small schools tend to be loosely organized without formal departments.

A second influence is likely to be the age of the school and the length of time the school has been interested in and committed to international business research and programs. Almost always the school enters the international business arena on a selective and ad hoc basis, leaving the international business faculty members lodged within the traditional functional fields. As interest mounts and size increases, one of the other forms of organizational structure emerges, depending upon various elements in the school's profile. Thus, older schools which are larger, and which have had longstanding significant interest in international business, will be more likely to have a department or matrix of departments involving international business than a recently formed school with a recent commitment to international business.

There may be exceptions, of course. Those exceptions often follow from whether the leadership of the school and its faculty have decided to focus on international business as one of its top priorities. This frequently is reflected in the initial organizational structure of a new school, so that international business is visibly represented in that structure as a reflection of its priority in the school's mission.

The diverse influences on organizational form, and the diverse forms of organization for international business found in U.S. universities, poses a further interesting question. Does the organization structure matter? Causal observation produces examples of great successes in international business research in organizations in each of the generic forms. That success is based upon the quality and commitment of the individual faculty members, which may be irrepressible by even the most hostile organizational structure. Perhaps the most that can be expected from alternative organizational forms is to reduce or limit the organization as a hindrance in realizing the IB research capabilities and objectives of the faculty. But as will be noted below, certain organizational forms may promote or facilitate certain types of success in international business research, depending upon its objectives or focus.

AN OPTIMAL ORGANIZATION STRUCTURE FOR IB RESEARCH?

The foregoing suggests that the organization of the international business group within any university depends upon a multitude of historical and special factors, only some of which may involve the focus of the international business research and programs. It also identified several objectives and types of international business research, and their linkages with the functional fields. This section suggests some tentative conclusions. At a minimum, it supplies some relevant questions to be asked about the school's research mission in setting an appropriate organizational structure for IB research.

A linking of research objectives and the generic organizational forms is shown in table 3.2. The column heads show various research objectives and types. The Xs in the matrix indicate a compatibility, or strong compatibility, between each organizational structure and the particular research characteristics shown in the left column. An absence of Xs indicates a relatively poor fit between the pursuit of that objective and that particular organizational form.

To reiterate a disclaimer, the success of international business research depends largely upon the quality of the faculty members and methodologies involved. Thus, the level of compatibilities shown in the table are to be viewed only as tendencies observed by the author, rather than involving any mechanistic relationship.

All forms of organization can and should support descriptive research about international business. It is likely, however, that the tendency for organizational structures which assemble faculty members from several functional disciplines will be more focused on description and relatively less on causality and prescriptive analysis, than those which retain a tight functional field relationship with faculty members.

Likewise, a research focus which gives high priority to establishing and testing relationships among international business variables; that is, the "understanding and explaining" objective, as well as the descriptive and normative types of research in international business, are likely to have high priority in functional field–based faculty relationships. This is particularly true in the economics and finance components of the IB research, where more attention is given to analytic rigor and empirical validity than, perhaps, to description. Indeed, it seems reasonable to expect that there will be less emphasis on normative and prescriptive types of research in organizational structures which assemble faculty members from several diverse disciplines, among which research methodologies and normative standards may conflict. Contrarily, if a school has a particular methodological or analytical focus, the IB research is likely to fit more closely the testing of the "ideal" assumption.

Casual observation suggests that this relationship may have some validity, as seen in schools which retain very discipline-based organizational groupings with a focus on analytic rigor in its international business research. Carnegie-Mellon University and the University of Chicago are examples.

As to the objective of international business research as a tool to build the pres-

Table 3.2 Research Objectives and Organizational Fit

Objectives of Research	Functional Field Department	IB Department	IB Project	Matrix
Describe	x	xx	xx	x
Explain	xx	xx	xx	xx
Normative, Advocate	xx	x	x	xx
Build image		xx	x	xx
Faculty advancement	xx	xx		x
Encourage cross-disciplinary work		xx	xx	x
Attract research funding		xx	xx	xx

tige and external image of the international business group or the school, it seems that the functional field–based organization structure is not likely to be optimal. This is because the critical mass of faculty may not be achieved, individual faculty members will try to associate their research with the functional field rather than fitting it for the journals and literature in IB, and the "large IB project" with significant external funding is less likely to be forthcoming.

At the other extreme, a cohesive international business department or matrixed IB department are more likely to have the critical faculty mass, be able to attract research funding, and to exercise internal influence in the school's priorities so as to create significant external visibility. The same potential exists, with somewhat less continuity, in the project-oriented organization structure which also assembles faculty members from divergent functional fields. The lower potential for the latter stems from the uncertainty about continuity and its focus on a particular dimension of international business research or programs.

Assessing the capability of the four organizational forms to provide legitimacy, support, and acceptance of the international business research of faculty members is necessarily subjective and uncertain. It seems to follow, however, that a functional field–based departmental structure is more conducive to legitimacy and predictability in defining the research hurdles for promotion and tenure. Thus, a strong relationship is shown in table

3.2 for the organizational structures involving the functional field–based form, despite the fact that this may tilt the researchers efforts towards functional field legitimacy and detract from his or her focus on the international dimension. Likewise, a legitimate international business department should enjoy superior position in defining, implementing, and judging the research hurdles for the individual members. The matrixed IB department may supply the definition of acceptable research for the IB faculty member, but that faculty member also must be acceptable to the functional field groups as well. Our assessment is that the project-based organizational form will be less supportive, given the fact that the faculty member will be mainly judged by a functional field group which may or may not consider the IB project or program of adequate legitimacy and quality according to the criteria of their functional field.

One of the objectives of international business research in general may be to encourage crossdisciplinary or multidisciplinary analyses of international business phenomena. Indeed, the broad nature of the international business field suggests that this is a legitimate priority for many institutions, and including the Academy of International Business. Three organizational forms fit more comfortably with this as an objective than others. Obviously, the lodging of the international business faculty member exclusively in a functional field department creates a barrier which must be overcome to participate in multidisciplinary research projects.

At the other extreme, the international business department and international business project organizational forms, by definition, bring together faculty members from more than one discipline. Likewise, the matrixed department also assembles faculty members from several functional fields, though each retains a strong linkage with his or her functional field as well.

Finally, one of the objectives of international business research (and the visibility associated with it) is to attract external funding for faculty members, departments, and schools. The reverse is also true: significant funding for international business academic activities will clearly enhance the research effort and agenda in international business.

Again, the functional field–based organizational structure may provide the lowest level of attractiveness for externally funded research for the international business dimension of that field. Alternatively, the department-based and project-based organizational forms, with their critical mass and representation of several functional fields, should have more attraction for supporters of research in general, and IB-focused research in particular.

The advantages of a recognizable unit, such as a matrixed IB department, or an international business unit or center, for attracting research funding for IB projects have probably been enhanced in the past decade as various federal and, in several states, state agencies have turned their attention towards globalization of the education process. United States Department of Education–sponsored international business centers represent one manifestation. Funding by state education agencies, such as in New York, for export promotion centers, international business and foreign language projects and programs is another indication. Such governmental initiatives tend to stimulate multidisciplinary projects, increase the desirability of an organizational unit such as a department, and in other ways encourage organizational change towards structures which cut across functional field disciplines.

Looking down the rows in table 3.2, it appears that the organizational structures which would facilitate the highest number of objectives of international business research is the free-standing, international business department, with the matrixed IB department showing a close second. In third place, because of the relative lack of continuity and separate organizational identity, is the international business project. The least conducive or supportive of the potential multiple objectives of IB research is that which leaves the IB faculty member with a primary or sole allegiance in the functional field department.

This hierarchy of organizational structure for supporting IB research objectives ignores the complications and costs which particular forms carry with them. Any school and university must consider these tradeoffs. Paramount among these is the cost and complexity of an additional IB department, or even any department, especially in a small institution. Even for a medium- or large-sized institution, the assembling of an international business group of faculty in a separate department carries with it additional budgetary costs. But it carries a more basic risk. Will a separate IB department achieve true acceptance and legitimacy among the functional field colleagues; will it increase the danger of second-class citizenship and less influence in critical administrative processes—compared with the other organizational forms? The matrixed IB department also places additional burdens and uncertainty on the individual faculty members who must march to the two groups of drummers in the functional field department as well as the international business unit. This paper suggests that no single answer exists for all institutions. Certainly, the quality and volume of IB research is high and accepted. As the disciplines have "internalized" IB dimensions, legitimacy has risen while the need for a separate identity for IB may have been reduced. But for a multitude of schools, the organizational dilemma is an important one.

How any one institution chooses its organizational unit will depend not upon this paper, but upon its history, the perceptions and philosophies of the faculty leaders in international business in juxtaposition with other faculty leaders in the school, and upon the priority which the school's administration places on international business in general, and international business research in particular. Hopefully, this paper provides a typology and hierarchy to permit the schools and their faculties to pose the questions and tradeoffs in a more systematic way.

■ ■ ■

Organizational Structure and International Business Research: Hamlet's Dilemma
Jeffrey S. Arpan

Too many business faculty members today face a Hamlet-type dilemma. To wit: "To internationalize or not to internationalize (one's research), that is the question. Whether 'tis nobler in the mind to suffer the slings and arrows of outrageous colleagues or to take arms against their sea of criticisms and by opposing, end them."

They face this dilemma when they are surrounded by colleagues who undervalue the importance and complexity of international research and who, by nature of the peer review, promotion and tenure process, decide their professional fate. And since this situation predominates U.S. academia, today's Hamlets remain outnumbered by their adversaries.

This opening analogy provides a glimpse of my conclusion: functional structure is generally dysfunctional to international business research. I hasten to add, however, that several nonorganizational factors also impede international business research in academic institutions, and that having an international business department does not guarantee superior international business research in quality or quantity.

Taken collectively, the impediments result in what I consider to be an embarrassing and inexcusable paucity of meaningful (useful) international business research conducted by U.S. business school faculty.

DEFINITIONS AND POINTS OF DEPARTURE

One of the goals of the conference for which this paper was prepared was to define more precisely the very nature of international business and hence international business research. Therefore, it would have been premature, if not presumptuous for me to have done so before the conference and especially unilaterally. However, for this paper it is important for me to specify how I define international business research. Otherwise, I could not coherently argue the related issues, nor should others.

The definition I use is an operational one—operational in the sense of applying to this paper. In this sense, I define international business (IB) research to encompass inquiries about all aspects of business crossing national boundaries, including both firm-level and public policy aspects, and theoretical and applied varieties. However, as will be argued subsequently, the precise definition of IB research is not very important to the central issue of this paper, because so little IB research is done at all, no matter how narrowly or broadly it is defined.

It is also germane to mention at this point that my comments and observations pertain largely to the United States and larger universities. Their applicability to other countries and smaller institutions is best left to subsequent debate. Finally, the specific issue addressed in this paper is part of a larger question concerning the overall adequacy of IB research (for example, quantity, quality, and importance) relative to the importance of IB to business and other major world phenomena and issues. My discussion begins with this larger question.

THE ADEQUACY OF IB RESEARCH

By virtually any measure, international business has become increasingly important to the world economy and to all nations. International trade, investments, and other forms of international capital flows have never been higher or more pervasive than in the current period. On a global basis, international trade increased from $51 billion in 1948 to $328 billion in 1972 to over $2.5 *trillion* in 1990. Total foreign direct investment cumulative outward flows by industrial countries increased from $70.5 billion in the 1960s to $302.3 billion in the 1970s to $561.6 billion in the 1980s. For the United States, foreign trade as a percentage of GNP has more than doubled and foreign direct investments in the USA increased from $3 billion in 1950 to over $400 billion in 1990; US foreign direct investment abroad increased from $46 billion during the 1960s to over $121 billion in the 1980s. The upshot is that virtually no country, industry, or company is insulated today from international competition or otherwise unaffected by developments in the global economy.

It is typically assumed that business faculty conduct research primarily on important, current phenomena. If true, then the significant increases in international business activity and its importance should have resulted in a commensurate increase in the quantity of research concerning international business. It has not. While there has been an increase in research concerning international business, it has not come close to paralleling the increase in international business activity or its importance. Hence either

the faculty assumption is wrong or other factors intervene. I believe it is the latter.

If the quantity of research is inadequate, what about the quality? To be sure, quality in research is an elusive concept to define precisely. One widely accepted yardstick is "defendability," for which the sophistication of the research design and the analytical tools employed are common surrogates. Using this definition of quality, IB research has clearly increased in quality. For the most part, it has moved sequentially from being essentially descriptive to simplistically analytical to rigorously analytical. Tight methodologies and powerful statistical techniques are now employed in most IB research published in "scholarly/academic" journals. Those that do not are seldom published. Another "quality" dimension is the research's contribution to the development or enhancement of theory. Again, I conclude that IB research has increased in this definitional dimension of quality, particularly during the last decade. Hence, the quality of most published research is probably adequate, at least as defined above. Whether the research being published is of any practical interest or use is another question, and will be addressed later in this paper.

What about the adequacy of the scope of IB research? From research primarily focused on exporting and international trade theory, IB research has broadened substantially. It now encompasses most functional/operational aspects of the multinational enterprise (MNE) and their related theoretical aspects as well as topics relating IB to economic development and national sovereignty, North-South and East-West issues, and more recently, international competition and competitiveness. The enormous variety and diversity of topics researched is clearly evident at the annual meetings of professional organizations such as the Academy of International Business, the American Marketing Association, the American Accounting Association, and the Allied Social Sciences Association.

However, IB research by business faculty published in the scholarly academic journals has been of much more narrow focus—dealing largely, if not almost exclusively, with international aspects of single function–based topics. The main reason is that almost all of the "scholarly/academic" journals in business are single function based—for example, *Journal of Marketing Research*, *Journal of Financial Management*, and *Accounting Review*. Hence, interdisciplinary and public policy related research has far fewer publication outlets.

Finally, a few observations are in order about the adequacy of the value of IB research being undertaken. Like quality, value can be highly subjective. Like beauty, it is often "in the eyes of the beholder." In terms of research, value can be defined as the research's contribution to knowledge of the field or discipline (that is, the academic/esoteric value), to practitioner understanding, skills, or efficiency/ effectiveness (the pragmatic/applied value), or to the understanding and competency of public-policymakers and the electorate of the country (the civics/citizenry value). It is my assessment that IB research has moved away from the pragmatic toward the esoteric, and has seldom been directed toward public policy. Hence concerning value: academic/esoteric—probably adequate; practitioner/pragmatic—increasingly inadequate; government/public policy—very inadequate. Overall: inadequate.

Could it be because most social science research, and, by definition, empirical research, lag the phenomena being investigated, that an increase in research about international business would lag the increase in international business per se? If so, could this explain why IB research is currently inadequate, and that it's only a matter of time before it becomes commensurately greater? It's possible, but doubtful. I believe the explanations are found elsewhere.

EXPLANATIONS

There are a number of possible explanations for the inadequacy of international business research undertaken by faculty.

One possibility is that business schools have little interest in, and place minor emphasis on research (compared, for example, to teaching or service). If true, then the insufficiency of IB research could derive from a paucity of research in general. Possible, but doubtful. While business school faculties initially were not paragons of research, this can not be said of the majority of them now.

Most business school faculty currently are evaluated, ranked, promoted and tenured in large part (if not primarily) on the quality and quantity of their research publications. "Publish or perish" is now as applicable to business faculty as it is to their colleagues in humanities and social sciences. Business school legitimacy and rankings are increasingly measured by faculty scholarship. As a result, there has been a proliferation of "research professorships" within business schools and, in general, a steady increase in the emphasis and importance placed on faculty research. Therefore, the explanation for the inadequacy of international research must be found elsewhere.

One possibility is the lack of opportunities for research. Even if the business school environment emphasizes research, it is difficult to do research if teaching and administrative loads are substantial. In addition, if the subjects of research are unwilling or uncooperative, it is difficult to conduct research. Could either condition explain the inadequacy of international business research? Again, it's doubtful. Internally, teaching loads have generally decreased, (at least de facto teaching loads) and funding for research has increased substantially, such as research fellowships and "summer pay" for research. Externally, among the business and government communities the attitude toward research remains conducive and supportive. Probably more than ever before, there is easier access to information of virtually all varieties. Finally, given the number of countries, companies, and issues to be researched on a worldwide basis, it is difficult to support an argument that there are insufficient subjects and issues to research. If not a lack of emphasis or opportunities for research, then what?

One possibility is the lack of publication outlets for research, or at least international research. With respect to the former, doubtful; with respect to the latter, possible. Overall, the proliferation of research has lead to a proliferation of journals, both "scholarly" (academic oriented) and "professional" (practitioner oriented). Hence today there are more publication outlets than previously. However, those which are considered to be top-tier academic journals remain single-discipline based, almost without exception. Furthermore, the number of consensus top-tier academic journals has remained fairly constant—seldom more than two or three per discipline. Finally, the academic journals have moved toward inclusion of only highly quantitative, statistically rigorous research.

Getting one's research published in the top-tier journals today typically requires research that is sufficiently narrowly focused within a single discipline to lend itself to quantitatively powerful research techniques and be somewhat timeless because the publication lags have become elongated. Most international business research does not fit these characteristics. Hence most international business research does not get published in the top-tier journals. For this reason and because most faculty prefer to undertake research that has a high probability of being published in a top-tier journal, many do not undertake international research.

Another possible explanation is a lack of faculty awareness or interest in the international dimensions of their primary discipline, or in international business topics and trends that transcend single disciplines. This is more than a possible explanation; it is a probable one. Albeit increasing, most business school faculty still do not attach much importance to international dimensions or topics. They are not in their immediate frames of reference. They are not areas to which most faculty have been exposed, let alone schooled or trained. They are, at best, peripheral to faculties' vision. Things on the periphery tend to remain on the periphery.

Even more explanatory is the phenomenon that when faculty do become interested in conducting international research, they usually recognize they have insufficient training to do it. As reported in several studies during the last few decades (for example, Nehrt 1987, Kuhne 1990) most business faculty have never taken a single international business course, and most doctoral programs do not require any international business courses be taken. Hence existing and future faculty are "unschooled" in international business.

Further, international business research is arguably more complicated than domestic research. Different languages, concepts, constructs, values, business practices and a host of differing socio-economic-political contexts must be overcome. Collection of primary data is more difficult and expensive; utilization of secondary data, when available, is typically more hazardous. Few business faculty receive or acquire the requisite research training to deal adequately with these complications. Lacking compensatory interest or motivation, faculty facing such inadequacies shy away from doing internationally oriented research, just as they shy away from adding international dimensions to the courses they teach or (teaching their subjects from a global perspective).

An existing inadequacy could be remedied by interacting with those who possess the needed training/expertise, such as one's colleagues. But since most of one's colleagues suffer from the same inadequacy . . .

Finally, there is the possible, nay most probable, explanation for the inadequacy of international business research: the organizational/structural aspects of business schools.

STRUCTURE/STRATEGY

After decades of research, we still cannot say for sure whether strategy determines structure, or structure determines strategy. Maybe it does not really matter that much.

Nonetheless, I believe that structure influences behavior as well as strategy, and primarily explains why so little international business research is conducted.

Virtually all business schools continue to be organized along single discipline, "functional" departments (for example, finance, marketing, and accounting). Certainly not unique to business schools, the departmental structure reflects a historic emphasis on subject mastery and hence specialization within academia. To a large extent, it is mirrored in business. Under the functional-specialist mode of structure and operation, one continually seeks to further define and refine one's area of specialization; to develop ever more expertise on a specific subject, ad infinitum. To do this, one typically minutely examines increasingly more minute aspects of a subject, sometimes ad nauseam.

Far too infrequently is there examination or integration of other fields. Far too infrequently are there professional interactions and collaborations with scholars in other disciplines. As a result, knowledge typically advances at the margin, rather than in quantum leaps. Far too often, the downsides of intermarriage become realities of departmentally based faculties. Far too often functional structure hinders the type of collegial interaction and crossfertilization necessary for multidisciplinary and international business research. The functional-specialist (departmental) structure also physically separates and isolates faculty from those with other specializations.

The functional-specialist structure also effects career preparation. Doctoral candidates preparing for academic careers are trained by already specialized faculty, who in turn extol their protégés to become even more specialized. For example, several years ago the USC business faculty approved a change in the doctoral program which eliminated minors and replaced them with cognates in order to achieve even greater specialization in a student's major area. The same faculty steadfastly resisted a policy change that would have required all Ph.D.

students to take an international course even though the faculty approved an international course requirement for all master's-degree students. Their resistance was largely on the grounds that for doctoral students an international course would not be as useful as an additional course in students' functional major or cognate area.

Like their predecessors, most recently minted Ph.D.'s throughout the United States, lack international exposure and training at a time when the curriculum in which they teach is supposed to be getting more internationalized and the environment in which their students will live and work is becoming more globalized. Without receiving international exposure and training, faculty are less likely to value appropriately its importance. As a result, they are also less likely to internationalize their research and courses, and find it more difficult to do so.

Reinforcing, if not driving the functional specialist structure is the promotion and tenure systems at most universities. Promotion and tenure processes begin at the departmental level, where one's specialized expertise is judged initially by one's similarly specialized colleagues. Especially at the larger and "top-tier" universities, the dominant criteria for promotion and tenure is one's research publication record, and within this criteria, one's publication record in "scholarly" journals.

The determination of which journals are scholarly, not surprisingly, is made by functional specialists who have typically published in, or are reviewers for, such journals. Hence the *Journal of Marketing Research* is considered the best scholarly journal by marketing faculty, and the *Accounting Review* is considered the best scholarly journal by accounting faculty. Since these journals almost exclusively publish articles that are single function–based, faculty who seek publication in such journals conduct research that is similarly single function–based. If all international business topics could be subsumed in functional areas, then the single-function orientation of the top-tier scholarly journals in and of itself might not contribute to the inadequacy of international business research. However, many international business topics cannot be subsumed within a single functional area, and hence lack a top-tier scholarly publication outlet. (The same problem is faced by researchers seeking publication of crossdisciplinary research of any kind, not just international business.) In addition, as discussed below, the top-tier scholarly journals do not publish much international research even when the topics can be subsumed in a single functional area. For example, fewer than 3% of articles published in the *Journal of Marketing Research* during its multidecade history have dealt with international marketing issues or phenomenon.

There are several plausible reasons why the top scholarly journals so seldom publish internationally oriented research. Two alluded to above are that less international research is conducted and that much of it is too broad-based for the highly specialized journals. Another possibility is that much of the international research submitted does not meet the journals' acceptance criteria. More than possible; quite probable. During the past several decades, articles published in the top scholarly journals have become increasingly quantitative in nature, and to a large extent, methodology oriented, if not driven. In some journals, it appears that the research methodology and theoretical implications have become more important than the topic or its relevance to "the real world."

To some faculty, this change represents an evolutionary upgrading of the sophistication of research—a necessary and beneficial increase in the defendability of conclusions. To others it represents a trend toward pseudo intellectual sophistication—a movement away from the real world toward the ivory tower; toward research that is statistically meaningful but meaningless to the real world; to the study of only those things that are easily measurable, rather than of what needs to be studied. Conceivably, both perspectives are true. Yet increased quantification of research typically poses greater

problems for international (versus domestic) research due to the greater number of uncontrollables and number of factors that are difficult, if not impossible to quantify, such cultural attributes or influences. As a result, international research is less likely to be accepted for publication in top scholarly journals.

Finally, promotion and tenure considerations typically focus on the number of scholarly publications and on single authorship. When combined with all that has been said above, this process fosters the conducting of small-scale, single discipline, short-time-frame projects that are methodologically "tight" and which concern highly measurable phenomenon—attributes not characteristic of, or conducive to most international business research.

If promotion and tenure play such important, influential roles in the behavior of faculty, and if the promotion and tenure process hinders international research, then tenured full professors should be conducting more international research. They should, but they seldom do. Most remain internationally naive and untrained. Others become highly involved in administrative and service activities, lessening the time available to do research of any kind.

This said, tenured full professors often do broaden their fields of interest from their originally narrow fields of specialization, such as moving from accounting or finance towards "control" or "strategy/policy." They also often become more involved in consulting or executive education/management development, putting them more in touch with the real world and real world issues and practices. These phenomena open avenues for more international research, particularly if xenophobia and other obstacles can be overcome!

REMEDIES

Despite rumors to the contrary, faculty are not so different from the rest of humanity. They eat, sleep, breath, and respond to stimuli. Most of their behaviors result from conditioning. To change faculty behavior, faculty stimuli or conditioning must be changed. I have argued above that the inadequacy of international business research by faculty stems from inadequate conditioning and stimulation. I argue below that remedying the inadequacy requires changing faculties' conditioning and stimulation. To do this, changes will be necessary in business school structures, programs and processes.

Future Faculty

Nehrt's and others' studies paint a dismaying picture of the international business training being given to recent Ph.D. students in business—or more accurately, the lack of international business training they're receiving. Less than 10% take even a single course in international business; even fewer receive foreign language training; and even fewer participate in a "foreign" experience. And at least one specific study dealing with the integration of international material into business programs showed that there is significantly less in doctoral programs than at masters and undergraduate levels (Arpan et al. 1993). As a result, future business faculty are receiving virtually no international training, and very little international exposure except for interaction with foreign students and some foreign faculty. Thus, it is reasonable to expect that they will not conduct a great deal more international business research than the faculty members they will be joining, unless some changes are made in their doctoral programs.

Several needed changes include requiring all Ph.D. students to take at least one international course in their functional area, one course on international research techniques, and if they have not taken one previously, a survey of international business course. They should also be required to demonstrate minimal proficiency in at least one foreign language before graduation. Other remedies include requiring them to write at least one case dealing with international business or becoming involved in at least one international business research project. They also need to be provided funded opportunities to study, teach, or conduct research abroad during their program. Finally, they should be encouraged to begin developing a regional expertise (for example, Asia or Latin America) during their doctoral studies.

Existing Faculty

As noted earlier in this paper, the vast majority of current business school faculty have little interest or expertise in undertaking international business research. As a result, their institutions and administrators must provide increased encouragement, opportunities and incentives for existing faculty to increase their international awareness, understanding and research expertise. All the remedies listed below are worthy of consideration and implementation:

Faculty should be offered teaching load reductions and additional funds for international travel, data collection, data bases and foreign collaboration.

Administrators should prioritize travel funds for faculty presenting internationally oriented papers or participating in international conferences.

Sabbaticals should be linked to the internationalization of the faculty involved.

More funds should be made available for faculty to attend seminars and symposia on international research and the international dimensions of business.

Arrangements should be enhanced for international faculty exchanges and for other internationalists (academics and practitioners) to interact with faculty.

Criteria for remuneration, promotion and tenure must include international aspects of professional activities such as research.

In assessing research, factors other than the number of single authored articles in scholarly journals must be accorded greater emphasis. Such factors should include the overarching importance of the topic being researched, the complexity of the topic, the number and magnitude of obstacles that had to be overcome (versus assumed away) integration and linkages with other disciplines and schools of thought. If a departmental structure is to be left intact, ways must be developed to foster faculty interaction across departments. An international business center or institute is one way increasingly being utilized throughout the country. A matrix organization where "international" is one of the areas constitutes another alternative.

Clearly another option within the confines of the departmental organization is to establish a separate IB department, with all the rights, privileges and responsibilities thereto appertaining. While few in number, their number is growing. The debate surrounding this structural alternative is sufficiently important to warrant elongated treatment below.

The International Business (IB) Department

For decades, many faculties have debated the desirability of centralizing international business faculty expertise an international business department versus dispersing and integrating it among and within functional departments. The arguments often parallel those for a corporation's organization with respect to its international expertise and activities (as exemplified in Howard Perlmutter's classic article, "The Tortuous Evolution of a Multinational Corporation"). An issue more restricted to academia, however, is whether or not international business is a distinct, separate discipline or a dimension/extension of other disciplines. It is not my purpose here to resolve conclusively this debate, or to rehash comprehensively the arguments on each side. Rather, I offer some observations about the issue as it pertains to IB research and, to a lesser extent, IB education.

In my opinion, most academics conduct research primarily for promotion and tenure reasons. To be sure, some research is done for other reasons: for knowledge's sake, to satisfy one's own intellectual curiosity, to unselfishly contribute to the advancement of the field of knowledge, to

further a social or political cause and to improve one's ability to teach. Some research is also done for fame and fortune, for one's ego and to further one's national and international reputation and recognition. But without research being important for promotion and tenure, I doubt seriously that half as much of it would be done. If so, then for IB research to increase, then IB research per se must be counted (weighted) more substantially in promotion and tenure decisions.

In this light, one factor clearly in favor of an IB department is that IB research is weighted more heavily in promotion and tenure decisions by the faculty of that department. Therefore, with an IB department there is cadre of faculty whose promotion and tenure depends to a significant extent on IB research. This is not likely to be the case in any alternative organizational structure. In fact, the opposite is more likely: faculty in functional departments who do too much IB research often (if not typically) experience promotion and tenure problems within their own departments, even if they are well known as good IB researchers by their international business colleagues at other institutions. The primary reasons are that their functional department peers do not value or understand IB research, and/or do not rank as highly the journals in which the IB research is published. Matrix organizations or dual appointments are not panaceas either. Satisfying promotion and tenure criteria for two separate and typically disparate sets of faculty peers is a Scylla and Charybdis situation few faculty, and even fewer junior faculty, can negotiate successfully.

Another research advantage of an IB department is a greater facility to conceptualize and undertake larger scale, multidisciplinary research projects. Most IB departments consist of faculty with different functional specializations but who share a commitment to international business. Research coordination, professional interaction and developing a spirit of camaraderie are facilitated by being housed together (organizationally and spatially). IB centers and institutes can attempt to replicate these advantages, but without a cadre of continually interacting faculty whose promotion and tenure are determined by international activities, it is more difficult for centers and institutes to be as effective.

A potential research disadvantage of an IB department is that faculty in other departments accede to the IB faculty the responsibility for IB research, and hence not get involved themselves. This can also happen with regard to internationalization of the curriculum. Friction and rivalry may develop between IB and the other departments over international course offerings, research projects, and other international activities.

While these potential disadvantages are not to be taken lightly, in my judgment they are outweighed by the advantages of an IB department. More international research gets done; more broad-based, integrative research is undertaken; the department's existence demonstrates the institution's commitment to IB education and research to faculty, students, business, government, and funding sources; and specialized international business degree programs are more easily established and operated. It would be ideal if all business schools had deans, department chairs, and faculty in all disciplines who were well versed and strongly supportive of international research, and if promotion and tenure decisions were based significantly on the quality and quantity of research undertaken that was internationally relevant. However, I would estimate that this situation applies to less than 1% of business schools in the United States. Based on my experience and observations of business schools, especially in the United States, having an IB department is the most effective way of obtaining maximum international research and curriculum development.

Case in Point: The University of South Carolina (USC)

Until early in the 1970s there was virtually no international business courses, research or faculty at USC's School of Business. Consequently, there were no international business degree programs nor an international business department. Nor were there many foreign students or internationally oriented students studying business at USC. Today there are over twenty regularly offered international business courses at the undergraduate and graduate levels. There is a specialized masters degree program in international business studies (M.I.B.S.) with over 350 students—one of the largest of its kind at an AACSB accredited institution. There is also a joint M.I.B.S./J.D. degree program, and a proposed M.I.B.S./M.S. in Engineering degree program. There are international business majors and minors in the Ph.D. program (one of the largest international Ph.D. programs in the country) and also a specific international business/international finance dual major in the Ph.D. program.

USC's international business department houses a dozen tenure-track faculty, five of whom have been ranked among the 50 most prolific researchers in international business during the 1980s. There are also at least a dozen faculty housed in other departments who have significant international interests and are annually involved in international research and teaching. Each year over 150 foreign students attend degree programs in the business school, the majority under a series of exchange programs arranged by the international business faculty.

There is also a Center for International Business Education and Research administered by the IB faculty (one of the five original national centers designated and funded by the U.S. Department of Education). The grant to establish and operate the center, and over half a dozen other grants awarded to USC by outside agencies to expand and enhance its international activities were conceptualized and written by IB faculty. An international business advisory board provides continuous input concerning international activities. From 1984 to 1992, USC was the copublisher and editorial home of the *Journal of International Busi-*

ness Studies, the official journal of the Academy of International Business and the number-one ranked academic journal in its field. USC's IB faculty also conduct business school "internationalization" seminars on a regularly scheduled basis at USC and frequently elsewhere. Its international business department is ranked #1 in the nation, as is its M.I.B.S. program.

In short, in less than two decades USC has become an acknowledged powerhouse in international business education and research, despite having an unlikely geographic location and a university and business school that were not considered outstanding in virtually any respect. To be sure, there were many individuals involved in making such an achievement possible, and certain synergies facilitated continued development. However, it was arguably the establishment, growth and development of USC's international business department that enabled this transformation to occur more than any other single factor.

The internationalization of USC's business school began very modestly in the early 1970s with the establishment of a faculty committee to investigate how international business might play a role in the school's future growth and development. The investigations and brainstorming of this committee resulted in the establishment of the M.I.B.S. program in 1974. Prior to 1974, there were fewer than half-a-dozen international business courses being taught, and about that many faculty who had even minimal academic training in international business or the international dimension of their functional area. Establishing the M.I.B.S. program necessitated the development of a radically new curriculum, the offering of several new courses in international business, international studies and foreign languages, the arranging of six month overseas internships for all M.I.B.S. students, and the establishment of administrative staff and operandi to make it all work. Some of the faculty for the new IB courses had to be imported from other institutions (that is, visiting professors) as

well as several of the senior program administrators.

It quickly became clear that more IB expertise and a change in organizational structure were going to be necessary. As a result, an international business department at USC was established formally in 1976, comprised of four tenure track faculty and one distinguished lecturer (a former Mobile Oil executive). Of the four faculty, one had been an assistant professor in the management science department at USC for several years (but whose dissertation had been in the area of international finance); one was a full professor hired from Syracuse with considerable international administrative experience (who also formerly had been a consultant to the CBA committee that conceived the M.I.B.S. program); one was an assistant professor with a Ph.D. in international business who had been hired by the marketing department one year earlier; and one was a newly minted assistant professor with a degree in international business who was hired specifically to work in and for the M.I.B.S. program.

During the first few years, M.I.B.S. students were much fewer in number and of considerably lower average quality than those who were to follow during the 1980s and 1990s. This was not a surprising result, given that the program was new, radically different and untested, and the faculty, college of business, university and city were not very well known or highly regarded. However, the then dean of the college, James F. Kane, held fast to what he believed was a viable niche in the academic market, and a needed and valuable product in the business market place: international business expertise. As a result, over the next ten years he hired eight other international business faculty—all of whom had specialized in international business during their academic training. The first several hired during the expansion were already well known in the field and hence added to the legitimacy and reputation of the international program. Their presence also facilitated the hiring of subsequent faculty, all of whom joined as junior faculty.

The increased size, expertise and commitment of USC's IB faculty caused or facilitated many things. One was the continuous expansion and enhancement of the M.I.B.S. program. Another was a dramatic increase in the amount, variety and significance of IB research undertaken. Another was obtaining a steady stream of grants for further internationalization of educational and research programs, and conceiving several new ways of obtaining additional funds. Another was being selected as copublisher and editorial home of *JIBS*, and later, as a U.S. Department of Education national center for international business education and research. Another was the establishment of international business as a major and minor in the Ph.D. program, and the significant expansion of international content and courses throughout the entire curriculum. Still another was the increased internationalization of faculty in other departments via joint research, grants, and other types of interaction. Another was the successful offering of faculty and curriculum internationalization programs for faculty at other institutions.

Could this have happened without the establishment, growth and development of the IB department? Possibly. Would it have happened? Almost certainly not. Without an IB department, it is possible to add international content to formerly domestically focused courses, offer some specialized international courses, have an international business major and/or minor, obtain grants, establish a IB center, even become a federally designated and funded center, and develop a reputation for international business expertise. However, I do not believe these things happen as quickly, thoroughly, comprehensively, synergistically or as enduringly as they do with an IB department.

It is rarely sufficient to have an IB champion or even a group of IB specialists scattered throughout the faculty as change agents. It more typically requires a group who view themselves as internationalists first (and functional specialists second), who are committed above all else to international research and education, and whose promotion, tenure and professional reputation are determined primarily by what they accomplish in these areas. Put more succinctly, it takes an international business department.

■ ■ ■

On the Institutional Environment and International Business Research
Duane Kujawa

It is my intention to examine the issue of "institutional structure and international business research" within the broader context of the "institutional environment," that is, the university and business school settings (especially regarding budgetary, program, and enrollment matters), the expectations of those hiring business school graduates (especially regarding curriculum impact), students' interests, administrative leadership concerns at the business school level, and faculty interests. The goal of this analysis is to identify the direction and potential impact of the changing academic institutional environment on the conduct of international business research.

This analysis rests on the simple assumption that the institutional environment affects budgets, strategies, programs, curricula, and personnel, and, therefore, institutional structure and research. This analysis is also concerned with the question "Where are we headed?" rather than with "Where are we now?" Nonetheless, any prognostic effort such as this must begin by establishing a baseline. The discussion turns briefly then to the question of "Where are we now?"

INSTITUTIONAL STRUCTURE AND INTERNATIONAL BUSINESS RESEARCH: THE PRESENT SITUATION

Arpan's "Organizational Structure and IB Research: Hamlet's Dilemma" (this chapter) presents a view shared by many that today's institutional structures affect IB research in essentially dysfunctional ways. Arpan's observations and reasons, and perhaps a few others, can be summarized as follows.

The quantity of IB research has not grown over recent years even though International trade, direct foreign investment, and other measures of IB activities have recorded significant growth. The academic quality of IB research has improved substantially over recent years. In the topical sense, the scope of IB research has increased considerably over time, but seems to have succumbed to the structural environment in most business schools; that is, the research is single function–based (for example, marketing, finance, or accounting) and appears in narrowly focused journals. The value of IB research is more evident in areas important to academic scholars concerned with a field of knowledge, rather than to managers, public-policymakers, and other practitioners whose needs are more applications oriented. For these latter groups, the value of IB research has, in fact, been inadequate. "Deficiencies" in the quantity, scope, and value of IB research are not the result of deficiencies in the value of research in business schools, in the funding and release time provided for the conduct of research, or in the general availability of publication outlets. Deficiencies in the quantity, scope, and value of IB research likely result (1) from a narrow definition by faculty of the "top-tier" academic journals—almost all of which are single discipline–based; (2) from a lack of faculty awareness or interest in discipline-oriented or interdisciplinary research which is international in nature; or (3) from insufficient training (at the Ph.D. level) to conduct IB research (which is inherently more complicated than domestic research). Because of these deficiencies, faculty are faced with a low probability for success in their IB research, assuming they cross over the threshold and develop an interest in IB research. Structure itself is seen then to be at the core of the problem, that is, the functional, discipline-defined departmentalization of business schools. This structure results in specialization and parochialism. It blocks interdisciplinary research and the development of new and more broadly based paradigms. It results in "inbreeding" and biases the promotion and tenure-awarding process towards success in research that is evidenced by publishing in discipline-oriented journals. Some academics may well disagree with some of Arpan's observations here, while others may well agree with many of them. The latter see business schools as more myopic than forward-thinking in terms of structure and the implications of structure regarding curriculum and faculty competencies, including research. The reasons for this are complex, to say the least. Nonetheless, structures result from pressures and are affected by pressures. Our discussion now turns to some of these pressures, and to their potential structural impact.

ENVIRONMENTAL CHANGES AFFECTING INTERNATIONAL BUSINESS RESEARCH

The environment affecting international Business research is multifaceted. It includes the university setting, the business school setting, financial matters, curriculum matters, the needs of firms recruiting business school graduates, and faculty interests. These different perspectives are complex in themselves, and generally overlapping. To illustrate, financial matters influence institutional structure, which in turn affects programs and faculty interests. Potential employers of graduates also affect programs, as do faculty. Programs impact finances.

Nonetheless, this section will look at several key elements in the environment affecting international business research in fairly discrete fashion. A subsequent section will pull together some of the more important influences and attempt to identify their more aggregated impact.

Demographics and the Economy

Demographic and economic trends affecting the university environment appear unfavorable. On the negative side, college-level enrollments are expected to level off, if not decline, at the undergraduate level while remaining about steady at the graduate level (Green 1991). This has been caused by several demographic changes. One of these, the "baby bust," is highlighted by the fact that the number of high school graduates has been declining since 1989. In fact, the number has declined from about 27 million in 1989 to 24 million in 1992—a drop in excess of 10%. By 1994, it is estimated the number of eighteen-year-olds will have dropped by 25% compared to the 1980 figure. A second factor is that the racial and ethnic mix of college-age students has been changing of late, as the participation of minorities (who are generally less affluent than Anglos) in the eighteen-year-old pool is expected to grow from about 13% to 20% between 1980 and 1995. The general economic recession of the past few years has also had an important effect on the affordability of university-level education.

These recent trends have resulted in tremendous economic pressures on universities. The private universities have experienced these pressures directly, while many public universities have had to respond to declining state government financial support. With the latter, enrollments have been cut, or growth plans stalled. Faculty salaries have declined in real terms in some instances. The financial support for faculty research (especially for summer research support) and for new program development has been cut in many other cases.

Many private universities have responded with cost-cutting measures too and the lowering of admission standards. In addition, some, my own included, have closely examined opportunities for developing new, interdisciplinary academic programs involving faculty and courses from several different schools. But, there have been all too few successes as departmental interests have resulted in formidable barriers to change—many of the same barriers which were discussed by Arpan (1992).

The conclusions here are several. The university environment is a financially difficult environment. To sustain faculty research resources, class sizes, especially in public institutions, have grown in many instances. "Turf battles" have arisen in other cases as specific faculty groups have fought pressures for downsizing, or for "diluting" their department's academic offerings in the guise of building new, interdisciplinary programs. Promotion and tenure decisions have also turned more negative in this environment. New faculty hiring has declined in many instances.

Changing Career Preferences Affecting Business Schools

Adding to the negative pressures resulting from demographic and economic changes, career preferences of those entering college since the latter 1980s are turning against careers in business. The Higher Education Research Institute at UCLA reports, for example, that since 1988 the percentage of college freshmen planning to pursue business careers has declined from 23.6% to about 18%—suggesting there may well be a decline in business schools' enrollments of nearly 25% (Astin 1991). Moreover, preferences for careers in elementary and secondary education have been growing since the early 1980s—especially among females (whose growing preference for business careers was responsible for much of the growth in business school enrollments during the mid-1960s to the mid-1980s). Some contend this does not bode well for management education at either undergraduate or graduate levels.

Similarly, the continuing decline in engineering enrollments since the early 1980s means fewer undergraduate students, who find engineering difficult, are switching to business schools. It also means a traditionally rich source of students for graduate business programs is declining. In contrast, albeit in a minor way, undergraduate liberal arts enrollments are increasing slightly lately. The pool of these students interested in graduate business degrees is also likely growing slightly.

For our purposes, the conclusions here are several. For one, the negative financial effects of the demographic and economic trends discussed earlier are being intensified in business schools by the generally negative changes in the career preferences of potential business students. Another is that all the pressures for looking more carefully at faculty promotion and tenure decisions are being intensified. Pressures to hire new faculty have declined substantially in the current environment. This makes programmatic and structural changes more difficult to accomplish because the composition of the faculty resource base remains fixed—at least in the short-run. But this does not mean pressures for programmatic and structural changes are declining.

Changing Pressures and Changing Directions

Strategic and operating changes occurring in the corporate community have a substantial impact on business schools for several reasons. One is that business schools, as professional schools, are sensitive to the trends affecting the employment prospects of their graduates. Business school faculty concerned about relevance in their research are also following such changes and seeking to understand their rationales and prospects for the future. Also, business school students are concerned about these changes because they want to be prepared in their academic activities to meet the ongoing professional challenges these changes portend. The net result is that, more often than not, and more often "better sooner than later," business schools are responding to pressures to change or modify their academic offerings in the face of "demand-pull" pressures exerted by companies and (likely prospective) students.

Many studies on the pressures for and likely directions of programmatic changes in business schools have been conducted by organizations such as the American Assembly of Collegiate Schools of Business (AACSB), the European Foundation for Management Development (EFMD), and the Graduate Management Admissions Council (GMAC). Their results are perhaps best typified by what I consider to be the most comprehensive of these studies—the AACSB-commissioned study by Porter and McKibbin, *Management Education and Development: Drift or Thrust into the 21st Century?* (1988).

The Porter and McKibbin study, after carefully examining the history of the development of business schools' curricula, *inter alia*, and the trends in faculty research interests and in corporations' interests in business school graduates, found that business schools needed to be increasingly concerned with developing students' interpersonal and communications skills, understanding of ethics, entrepreneurship capabilities, knowledge of external environmental factors, and understanding of the globalization of business activities (Porter and McKibbin 1988). In terms of the "globalization" perspective, their study clearly identified the potential for major academic impact of the powerful trends that were already underway regarding the global expanse of markets (both resource and product), the internationalization of competition, and the facilitating, if not encouraging, supportive activities favoring the development of democracies and free markets by governments worldwide.

One need not look too distant to confirm the global commercialization of what heretofore were often seen as domestic markets. The emergence of the U.S. multinational enterprises in the 1960s, the growth of the European multinationals in the 1970s, and the explosive, trans-Pacific emergence of Japan's *Kaisha* in the 1980s have led to a world economy dominated to a considerable extent by what one observer has christened the "triad powers" (Ohmae 1985). Coming into the 1990s, we have witnessed the demise of the Soviet Union, the emergence of democracies and market-based economies in Eastern Europe, substantial progress in European integration via EC '92, the birth of the European Economic Area (EEA), substantial privatization of industry and services, especially in Latin America, negotiations for a North American Free Trade Area, and so on. Coming into the 1990s, we are also witnessing the maturing of "transnational" enterprises—companies operating internationally with resources and activities globally dispersed, yet specialized, so that both efficiency and flexibility can be achieved within an integrated, interdependent, worldwide network (Bartlett and Ghoshal 1992). We also see explosive growth in international trade, the globalization of international financial markets, new forms of increasingly utilized global strategic alliances, rapid growth in international tourism, and changing roles for governments seeking to enhance the international competitiveness of their firms and of their internal markets.

The significance, pervasiveness, and distinctiveness of all these changes have, in turn, led to profound changes in management practices, the aspirations and needs of those preparing for professional careers in such a changing, globalizing environment, and the educational programs required from academe. We are seeing, for example, newly emerging academic degree programs featuring internationalization of the curriculum, skill development program enhancements to prepare students for dealing crossculturally and in leadership positions, expanded foreign language training, and, especially in Europe, foreign internships where students have the opportunity to deal with real international business issues while presenting themselves to a prospective employer in a hopefully favorable light.

Faculty traditionally bear the responsibility for developing and implementing academic degree programs. Keeping in mind that educational programs must be built upon faculty competencies, and, to be meaningful at the university level, faculty competencies must be grounded in the faculties' research orientation and research productivity, the discussion now returns to the more specific impact of the institutional environment on international business research.

EFFECTS OF THE CHANGING ENVIRONMENT ON RESEARCH

With but one single exception, the various changing environmental factors discussed thus far portend neither good nor bad as far as the institutional environment of international business research is concerned. I say this because I have no a priori preferred model that concludes one institutional arrangement is better than another regarding the ability of any such arrangement to foster international business research. The environment is what it is. And the single exception just noted bodes well for international business research. It is the overarching, not-to-be-denied trend that management today is becoming increasingly internationalized. This trend is being confronted by business schools big and small, old and new, locally and across continents. It is affecting faculty young and old, quantitative-oriented and qualitative-oriented, functionally focused and interdisciplinary. I personally feel this trend is so strong, all the other environmental factors will not negate or offset it; they will instead simply affect the manner in which business schools' implement their international business imperatives. Few, if any, business schools will be able to exempt themselves from this trend. And those that do, I submit, will likely atrophy in the longer run.

The Impact of Demographic and Economic Factors

An earlier discussion concluded the present environment as affected by demographic and economic changes was a financially difficult environment. Research in general, let alone research in international business, generates individual and organizational benefits which are longterm. Research budgets are "investment" budgets. Research is also expensive. In difficult financial times, those responsible for establishing university budgets, especially legislators in publicly funded institutions who are somewhat removed from day-by-day concerns about university operations, may well see research funding as temporarily expendable. Similarly, administrators and trustees at privately funded universities that are tuition-driven face pressures to sustain the teaching mission—thus forcing budget cuts on services and/or research support.

In some cases, financial difficulties have directly resulted in systemwide mandates that require fewer sections of a class being offered, larger class sizes, and less diversity in class offerings and degree programs. In other cases, faculty and school-level administrators may have been able to deflect budget cuts directed at research support by deliberately reducing section, class, and program offerings. As noted earlier, in all cases faculty success in securing tenure and promotions will likely be less frequent, and new faculty recruitment less pressing. "Turf battles" may likely characterize relations among faculty in the different departments.

In these respects, building new interdisciplinary programs in international business—which is the keystone for expanding international business research—may well be opposed by the faculty and difficult to accomplish. This kind of environment also mandates that international business programs, if they are developed, have to be implemented with existing faculty who, given what Arpan and others suggest, are likely not professionally competent in international business. This may well serve to extend the implementation phase for such programs as faculty competencies in international business are being developed via new-founded interests in international business research. It will also likely serve to have faculty research in international business be

more functionally focused than interdisciplinary. Some observers see this as a problem.

The Impact of Changing Career Preferences

At first glance, the fact that students are turning away from being management majors can be seen as magnifying the financial difficulties characterizing business schools' demographic and economic situations. A "first level" of analysis may well lead one to such a conclusion. Fewer students mean less money for operations. This is true in both public and private institutions. But, I think there is more to it than that.

A "second level" of analysis reveals changing career preferences are a structural phenomenon, and the potentially effective responses by business schools should thus require strategic repositioning as well as investment budgets. Moreover, structural change is not a difficult situation for faculty to understand and respond to. If enrollments are declining in a traditional program, such as the "regular" M.B.A. or the undergraduate B.B.A. programs, business school faculty are certainly open to suggestions on how to offset declining enrollments in these programs via the establishment of new programs aimed at different market segments. This was precisely the situation at my own university when the new IB dual degree program (consisting of an M.B.A. degree with a functional major, a Master of Science in International Business degree, and program enhancements including skill development seminars, foreign language training, and an overseas internship) was developed by and approved by the faculty. This was done in the face of (slight, but financially significant) declines in the enrollment in the business school's undergraduate programs. The structural problem of declining enrollments in general actually served to enhance the IB program's attractiveness.

Changing career preferences are also affecting the development of international business programs in another important way. They are encouraging interdisciplinary courses and program enhancements. Structural change means business schools need to be "closer" to their customers. The new international business programs are much more comprehensive than the traditional, functionally oriented programs. In the IB programs, foreign language competency is more than relevant, it is essential. International business means doing business crossculturally. Sociology, anthropology, international relations, political science, and history, for example, become increasingly relevant to such programs—as do crosscultural communications and comparative legal systems. To summarize then regarding both the demographic/economic factors and the changing career preference factors, the demographic/economic factors situation offers solutions which are not consistent with those of the changing career preference factors. The former suggests downsizing and retrenchment; the latter innovation and restructuring. The international business interests—including those concerned with research in international business—are promoted more effectively via the latter alternative.

CONCLUSIONS

All the preceding discussion leads to several conclusions regarding institutional structures and international business research.

Internationalization of academic programs and of faculty is mandated by the professional markets business school serve. University budgetary constraints and this need for program and faculty "restructuring" suggest internationalization will result by working with and adapting existing resources (that is, faculty). Since the restructuring spans across the entire business school, all the faculty and all the departments will likely be affected. This, in turn, suggests institutional structures other than those using an international business department to spearhead the internationalization process will likely be favored. One alternative used by several business schools involved appointing an associate dean for international programs to serve as a catalyst and facilitator (with financial resources) to assist de-

partments and faculty in internationalizing. This person also provided leadership in programmatic developments to strengthen international content. Another alternative is the more extensive use of faculty committees and task forces with specific charters. At my business school, for instance, we have a "faculty development committee" which reviews and funds faculty proposals for international research support, international travel, and foreign language training.

Looking at how all this affects international business research, several observations seem pertinent.

Given the strategic origins of the shift towards internationalization in business schools, faculty interest in international business research will certainly expand. Indeed, some of the "die-hard," anti-internationalist faculty may even see the light. (Enlightened self-interest is a wonderful motivating factor!) The *quantity* of international business research will thus likely increase. As more faculty get into IB research and more intellectual power is applied to the issues and problems addressed in IB research, the *academic quality* of the research should also likely increase.

Expanded interests in the international aspects of heretofore traditionally functionally defined, departmentally based research will likely result in commonalities of research interests across the departmentalized structural environment which characterizes so many business schools. This phenomenon, coupled with "market-based" pressures for interdisciplinary IB programs, may well lead then to substantial expansion in the *scope* of international business research.

Given this movement away from purely functionally based research, one might also expect IB research to be less directed towards academic audiences and more towards the identification and solution of problems confronted by managers and policymakers. In this sense, the *value* of future international business research may well be enhanced. Similarly, faculty may themselves begin moving away from narrow definitions of top-tier academic journals and *include* interdisciplinary journals, such as the Academy's *Journal of International Business Studies*, in their top-tier groupings. Indeed, I have already seen evidence of this in several business schools.

Moreover, given the present pressures for change, faculty training to develop international business research competencies will likely be accelerated. But, this training focuses on existing faculty, not necessarily just Ph.D. students. There is every reason to believe this training will be utilized in a time-efficient and effective manner. Existing faculty are already functionally competent and seasoned regarding research design and methodology. They should be able to expand already existing competencies into the international arena quite easily. In time, departmentalized, internationalized faculty will be demanding a stronger international orientation in their own doctoral students.

In the overall scenario developed and presented in this paper, structure is not the *core of the problem* where departmentalization is seen as sustaining specialization and parochialism, promoting in-breeding, and standing in the way of interdisciplinary programs and research. Rather, I see an evolving structure, one that is changing in response to the strategic needs of the "academic enterprise." These needs embrace internationalism in all its applications and ramifications. Structure is not a barrier, per se, to international business research—as it may well have been in the past. Today I see structure as increasingly encouraging international business research while influencing the content, usefulness, and pervasiveness of this research in important ways.

■ ■ ■

INTERNATIONAL BUSINESS RESEARCH AND STRUCTURE

John D. Daniels

Finding something appropriate to say in a discussion of papers can sometimes be more difficult than the preparation of the papers themselves, particularly if one agrees substantially with what has been presented. However, I have had a different problem. Without receipt of two of the three papers, at first, I thought it would be difficult to have anything to say; however, on second thought, I realized I have known all three panelists for more than twenty years and feel I can be fairly certain of what they would have said in the papers I did not receive. For example, I have been on the same faculty with two of them and almost with the third. Two of us were doctoral students together, and two of us coauthored a paper. I will forego analogies about leopards and old dogs by simply saying that, basically, I think I know their unchanged positions from twenty years ago.

As we examine the relationship between international business research and institutional structure, we must ask two key questions. First, are there any identifiable problems concerning the current state of academic international business research? If not, then there is little point in examining the connection between research and structure. Second, if there are research problems, does structure have a significant impact on them? If not, we should try to identify what the causes are and then address those causes.

In order to answer the first questions, it is important to examine why academic business research should be undertaken at all. There are basically two reasons: to understand better how business is and should be conducted, and to provide theories to explain why. Both of these reasons presuppose that there are and/or should be dynamic environments and business reactions that necessitate an ever-evolving field of knowledge. The research output should also relate to the internal and external constituencies served by the business schools. In other words, the knowledge created through research must be transferred to these constituencies if it is to satisfy the rationale I have listed for the conduct of research. The transfer process may be handled orally or in publications. It may be handled directly by the researcher or indirectly by colleagues who report findings in their own writing and teaching efforts. Through these processes the work should eventually have an impact on the conduct of business, either through knowledge that business organizations or their regulatory bodies gain in publications, training programs, and consultations; or through the hiring of students who have gained knowledge in their coursework.

In applying these concepts of research adequacy, there are several linkage points at which there may be structural impediments. I have shown these as a modified gap analysis in table 3.3. This table and my explanation of it may also be thought of as a pipeline, through which there is an input flow at point 1 that is substantially higher than the tap outflow at point 6. For example, the potential for academic research that has an impact on the international conduct of business is largely constrained by the number of faculty members. Leakage begins at point 1 as some faculty undertake little or no research of any kind and continues at point 2 to the extent that projects have no demonstrable international applicability. Point 3 illustrates that of the international research projects begun, some portion is lost because of not being completed or because of nonacceptance for publications and conference presentations. Of that which is published or presented at meetings (point 4), not all is read or heard by everyone with potential use for the information. However, even within the audience that does read or hear, some will find the re-

search results irrelevant to their own needs and will, for example, not transfer the knowledge into their own teaching, research, or action programs (point 5). Finally, point 6 shows that even among the studies which are acted upon by business organizations, regulatory bodies, and faculty members, some will have negligible or no impacts on the conduct of international business. This raises the questions of whether the output flow from the pipeline is sufficiently high? If the answer is yes, there may still be more efficient means of reaching this end. If the answer is no, there is a need to reduce leakages at appropriate points.

EFFORT GAP

To stress more research in general assumes that more research will be undertaken and that some portion of the increase will reach the end of the pipeline. Emphasis at this point seems consistent with the overall enhanced importance of research within business schools over the last two decades; however, some recent events seem to harbinger a changing trend. These include reports by various commissions that have decried an overemphasis on research, AACSB's placation of a group of potential runaway-schools through recognition of teaching missions, and state budgetary authorities' questioning of how much time professors spend in the classroom.

In spite of some current opinion to downplay research, I feel that more research emphasis in general may yield some positive results for improving the adequacy of inter-

Table 3.3 Modified Gap Analysis to Examine the Adequacy of International Business Research

Gaps	Symptoms
Effort	Faculty doing no or insufficient research
International Application	Research is domestically focused
Dissemination	Work is not published or presented at meetings
Target Market	Articles and speeches miss part of constituency
Application	Knowledge receivers do not act
Impact	Action has no positive outcome

national business research. My comments will relate only to North America inasmuch as I know too little about activities and missions elsewhere. Furthermore, I shall direct most of my comments to business programs, although I realize that there may be substantial opportunities to increase the international business related component of research by faculty in such fields as geography, political science, anthropology, and foreign languages.

There seems to be a prevailing current opinion that almost all business faculty are heavily engaged in research activities. My observation is that this is not the case, but it is difficult to find precise figures. For example, various people at AACSB have told me that there are between 1,200 and 1,300 degree-granting business programs in the United States, of which only about 270 are accredited. Some business research from the nonaccredited programs takes place and makes it through the aforementioned pipeline; however, I estimate that this is close to zero. Among the accredited programs, there are vast differences in research emphasis as well, both in quantity and quality; consequently, very few schools account for a disproportionate share of the studies that are undertaken, get published, have international content, are read, and have some impact. These are the approximately 50 schools that claim to be among the top ten business programs in the country. In summary, only about 20% of the business programs are accredited and only about 20% of these do most of the research; consequently, research is heavily concentrated in about 4% of business degree-granting institutions. From a structural standpoint, these schools differ from the population of business schools as a whole in that they hire a greater portion of their faculty from other programs within this inner circle, tend to have larger doctoral programs, offer more substantial perks such as released time and summer grants for research, and pay demonstrably more attention to top-tiered journals and research reputations when evaluating faculty for promotion and tenure.

However, the deviation in research activity goes beyond institutional differences. Even within the 4% of schools that I have identified, I estimate that not more than half of faculty are currently active researchers. This is because there are some nontenured faculty who never learn how to play the research game, some tenured faculty who have effectively retired from research, and some other faculty who have moved primarily into administrative activities. Conservatively, only 2–5% of all business faculty must account for well over half the research that fits my definition of being adequate. This is a skewness that is at least as disparate as Brazil's income distribution, and it is one which, like income distributions among and within countries, is probably worsening rather than improving.

Before I address what might be done structurally, let me say that the present market situation offers some hope of moving additional business programs upward in quality, which could result in having more than the present 50 schools within the top 10. The current glut of new business Ph.D.'s should result in more placements from top-tiered doctoral research programs into the LDCs (lesser developed colleges) of business programs. The slowing of business school growth should also enable these same upwardly aspiring programs to enforce more stringent research requirements within their retention decisions.

Structurally, I have some suggestions that are not really proposals, rather they are brainstorming thoughts. Some may seem like heresy, but will hopefully generate some meaningful discussion.

Reduce the Number of Academic Administrators

Within the business schools, we teach that organizations should not try to make sales managers out of their best salespersons; however, we too often take the best, or at least competent, academic researchers and make administrators of them. Therein, they forego or reduce their research output. They often are willing to make the moves because there are structural limits to increase their financial compensation from the university any other way. Proponents for using academicians as administrators argue that outsiders do not understand the eccentricities of a university environment (does anyone?) and point to substantial anecdotal evidence of their faux pas. In response, I would argue that (a) one can probably gather at least as many anecdotes about the ineptness of academicians as administrators, (b) there are many examples of nonacademicians who have had high success, (c) professional administrators are usually less expensive, and (d) money saved in using professional administrators could be used for more research incentives. I would also argue that faculty spend too much time in committees because of faculty governance objectives. Not only do the committees take time from research, most of their decisions are neither better nor different than those that would be made by an administrator acting alone.

Move Toward "Free Trade" in the Allocation of Research Grants

Most of us at an international meeting extol free trade; in fact, most of our business school colleagues have similar views. Yet, when we set rules within our own business schools on who can receive research grants, we impose many competitive restrictions, such as on rank and place of employment. Some of these restrictions help to widen the schism between the "haves" and the "have nots," both within and among universities. Might we consider experimentation by opening up some competition from outside research proposals? These proposals could come from other colleges, other universities, or outside the academic environment. Additional competition might force faculty members in "have" schools to upgrade their efforts, especially where they have come to expect grants regardless of effort or outcome. Competition might also result in expanding the pool of people who even try to undertake research projects. Such experimentation does not imply immediate free

trade, rather the replacement of an embargo with a limited quota.

Eliminate Tenure

I have already alluded to some faculties' effective retirement from research after receipt of promotion and tenure. The original rationale for tenure was to provide job security so that senior faculty could pursue longterm research projects without being held accountable for output on a year by year basis. I would argue that the tenure process has not generated the large and longterm research results. Furthermore, the goals could be pursued through the alternative of multiyear contracts. More recently, arguments for tenure have been based on academic freedom. However, in the United States, court rulings on first amendment rights would seem to mitigate a need for tenure based on academic freedom.

Any move away from tenure raises the question of whether qualified faculty could be attracted. The present market situation may make more faculty members willing to accept the multiyear appointments in lieu of tenure track positions. Furthermore, the more prestigious business schools would likely have less of a hiring problem. If they were to take leadership, other schools could follow more easily. A second question concerns equity. In answer, there is much precedence for gradual phaseouts of other fringe benefits. A third question is whether a substitution of multiyear contracts in place of tenure would prolong the research efforts of senior faculty members. The senior faculty would be more accountable; however, their accountability may or may not include an emphasis on research output.

Keep Expensive Human Resources Busy on Scarce and Expensive Activities

My observation through the years is that universities handle budgetary problems largely by cutting back on the less expensive human resources and support services. One reason for this is structural. The tenure system does not easily allow for adjustment of faculty lines; consequently, faculty changes generally lag market conditions considerably. The results are counter-intuitive and counter to what is taught in business schools, that is, to keep expensive resources busy. There is a tendency to end up with too little in support services relative to the number of faculty positions, leading either to idle faculty time and/or the diversion of faculty time to activities that should be handled through support services. These activities include typing, retrieval of library and film materials, running of computer programs, filing, and the handling of routine administrative requests. Not only does the lack of slack in support services result in an inefficient use of expensive faculty resources, it diverts time away from primary research functions.

INTERNATIONAL APPLICABILITY GAP

We are all aware that only a small percentage of the business research undertaken has international application. I hesitate to estimate what portion that is. However, I do want to address the issue of why and why not so little is international and suggest some means for reducing the gap.

International Business Department Structure

Does an international business department make a difference? There seems to be a popular image that newly-minted Ph.D.'s arrive at their first university posts equipped and eager to undertake international research projects. However, these junior professors quickly abandon their international orientations in order to pursue domestically focused studies because they (a) can be completed and published more easily and (b) will be treated more favorably by functional department review committees that prefer people to do similar work to what they, themselves, are doing. Many people who accept this image argue the necessity of having a separate international business department. Although I am the first to support an international business department for many reasons, the research argument is a very tenuous one for making a case.

First, I cannot think of a single anecdote of the so-called turncoat internationalist. Nevertheless, I admittedly can think of people whose international orientation may have been a factor in their not being hired or in their nonretention. However, I can also think of people whose international orientation may have been a factor in their not being hired or in their nonretention. However, I can also think of people whose specific functional, regional, or methodological orientation was a factor in their not being hired or retained within an international business department. Second, we now have a great deal of information on companies' organizational placement of international operations and should realize that there are both success and failure stories associated with any structural type. Likewise, I suspect that if we were to agree upon a large sample of meaningful international business studies, we would find that some originated from people in separate international business departments, some from functional departments, and some from various mixed or matrix structures. I doubt that there would be any significant difference by structural type. Third, company case studies are rife with examples of people who achieve their international objectives by working within a variety of structures, for example, Corning Glass Works International. Therefore, I do not recommend structural changes for departments as a means of addressing problems in international business research. People who are really competent and committed to do international business research will find means to do it.

Organizational Culture

Schools obviously differ not only in research emphases, but also in the degree to which international is encouraged and undertaken. Some differences are due to regulatory impediments, such as rules preventing the use of state funds for foreign travel. However, other differences seem to be due to some yet to be explained organizational culture. In this respect, it would seem that we might draw on the many demographic and behavioral studies that explain international differences among business organizations, such as why some export and others do not. From these studies it should be apparent that one important factor is a positive international experience of a key decision maker. Another is a positive response to a fortuitous foreign inquiry. Still another is a reaction to a changing competitive situation. It seems reasonable to expect that these same behavioral factors would influence actions of business school faculty as well. These factors may help to explain differences among academic institutions and among departments within the same business school. Any attempts to encourage international research must contend with these behavioral factors, not just with rationally oriented arguments.

Research Training

Lee H. Nehrt has pointed out in his research that very few new Ph.D.'s in business have studied the international aspects of their functional specialties. This shortcoming has been addressed primarily from the standpoint of teaching; however, it also affects the research dimension. New Ph.D.'s usually have a need to publish very quickly and rely on content and methodology from their dissertations.

There is simply very little time to retool. The same methodological and content drawbacks occur for research in general, such as people who do questionnaire surveys for their dissertations tend to do questionnaire surveys throughout much of their careers and people who examine consumer satisfaction tend to follow that subject for many years. Therefore, a major bottleneck that prevents junior faculty from pursuing international business research is at the doctoral levels. A greater emphasis on international must be addressed within doctoral programs. Of course, a structural impediment is that most Ph.D. programs in business are handled within highly autonomous functional departments. Unless this autonomy is broken down or unless some international supporters are added to the departments,

they perpetuate their hiring practices and orientations to doctoral programs. Their output becomes the hiring input for other institutions; and the situation is perpetuated further.

Another problem may confront faculty members who wish to make a transition from purely domestic to internationally oriented research questions. They may lack skills in the nuances required for sampling, questionnaire preparation, accessing data bases, and interpreting results in an international context. Or, they may be so overly polycentric that they are afraid to undertake international projects. As international business specialists, we have undoubtedly contributed to polycentrism by promoting mystiques that international research takes longer, is more difficult, and is more expensive. I do not accept these mystiques, although I do accept that international research projects may necessitate different approaches. On this point, I wish to make two recommendations. First, the field needs, as is available for most other areas, a comprehensive book on how to conduct international business research. This should concentrate on both the differences and similarities to domestic research. Second, we should offer seminars for faculty to retool them on how to do international business research in much the way that we now offer programs to help them add international dimensions to their teaching.

DISSEMINATION GAP

Some work, regardless of field, never gets finished or never gets published. I have no idea of whether this is a bigger problem for international business than for other areas. There are certainly as many or more outlets for international business research than for any other type of studies; and the methodological requirements are no more stringent. However, there are two potential problems that may plague publishability in top-tier journals. I alluded to one problem in the previous section. As researchers move from a domestic to an international focus, they may overlook some of the research adjustments necessary for making valid conclusions; and this problem goes unnoticed until too late. A second problem relates to the crossfunctional nature of the field and the crossdisciplinary base for much of the research, the result being more difficulty in accessing all relevant literature. Therefore, some studies add too little to the existing literature to be easily publishable. I discussed this problem as one of "reinventing wheels" in my AIB presidential address at the 1990 Toronto meeting. At that time I suggested state-of-the-art presentations for annual meetings and am pleased that the Fellows have responded positively. For my teaching paper in this conference I have discussed the need for an IB abstract service akin to what *International Executive* used to publish. Such a service should have benefits for making studies more publishable.

TARGET MARKET GAP

Ultimately, if research studies are to be disseminated into the classroom, they must first be read (or assigned) by instructors. However, we have very little information on how many and which journals are perused by faculty for different types of courses. I can only speculate that faculty pay attention to work in a very narrow spectrum of journals and that these journals are the ones considered to be top-tier within their own areas. If my speculation is correct, then much of the international business research is seldom examined by the functional specialists whose courses we would like to see internationalized. In other words, a professor who teaches advertising is likely to read only mainstream marketing and advertising journals and miss the international research on advertising that appears either in crossfunctional journals, such as the *Journal of International Business Studies*, or in internationally specialized marketing journals, such as *International Marketing Review*. This situation places more burden on functional textbooks to convey international

business materials; such as a text on advertising. So far, these texts have generally not integrated international materials very well. Might we consider means of targeting specific courses with a newsletter list of sources of recent international materials that are likely to have been missed by most instructors?

I must speculate merely about a second area as well—that faculty members pay more attention to the so-called gurus in their areas than to other authors, regardless of the quality or newness of what they say. In other words, if an already established leader were then to pursue international research projects, the domestically oriented rank and file would be apt to pay heed. I believe that this phenomenon helps to explain why research and teaching within strategy have become more international than research and teaching in other areas of business. Michael Porter was already a recognized leader in the strategy area when he began addressing international issues; and his reorientation has been followed by others in strategy. If other accepted academic leaders were to become "reborn internationalists," we might also see more infusion of international content into other domestically oriented functional courses and research. We might thus explore ways to use accepted academic leaders as opinion leaders to bring about change. Perhaps we could cosponsor joint association meetings, invite some of these leaders to make plenary presentations at international business conferences, or enter limited joint venture arrangements among associations to publish special journal issues in order to reach a broader clientele and to gain more acceptance of the research in international business.

There is also a potential audience for international business research outside of academia; however, I feel safe in saying that very few people in business or government read any of the articles that appear in the academically-directed journals. Some international business studies appear in academically-housed practitioner journals, which reach more, but still a small part, of the outside constituency. Some of the studies from academic and practitioner journals are eventually covered in newspapers and magazines, but my impression is that the reporting is considerably delayed and covers very few of the academic studies. Afterward I do not know how much is read, but probably not a great deal. Part of the problem is structural inasmuch as the academically-directed journals are often the only ones that are viewed favorably for faculty review. For example, articles in practitioner journals, such as *Business Horizons* and *Harvard Business Review*, are treated primarily as service rather than a research component within promotion and tenure reviews at some institutions. Therefore, there is almost a negative incentive for academicians to communicate their research results beyond the ivied walls of academia. On a positive note, I understand that the University of Colorado–Denver is planning a journal which will rewrite academic international business research for a practitioner clientele.

APPLICATION AND IMPACT GAPS

I do not wish to repeat what I said in my Toronto speech about the lack of relevance of much of the international business research that has been published since one may read those comments in the *Journal of International Business Studies* (Second Quarter, 1991). I do not imply that international researchers should be apologists; we probably do no better or worse than other business researchers. Nevertheless, the questions of relevance is undoubtedly a factor in whether readers act on what they read.

Since Professor Jeffrey S. Arpan began his paper with a paraphrase of a Hamlet dilemma, I shall close with a direct quote from Hamlet that I believe has applicability to the questions of relevance and impact:

Polonius: What do you read, my Lord?
Hamlet: Words, words, words.

■ ■ ■

SOME OF THE DETERMINANTS OF INTELLECTUALLY EXCELLENT INTERNATIONAL BUSINESS RESEARCH; OR ORGANIZATION STRUCTURE IS NOT THE ANSWER

Edwin L. Miller

The key to high quality, relevant international business research is the combination of the scholar engaged in rigorous research, the dean whose academic and administrative leadership is capable of creating conditions supportive of international business research and a faculty that is committed to integration of an international business perspective throughout the academic and scholarship programs. It is the scholar who is well grounded in the theory and body of knowledge in his or her discipline, sensitive to the problems and challenges of the world of practice and capable of designing, conducting and completing rigorous research projects. It is the dean whose vision and leadership skills are applied in such a way as to create an environment in which the scholar and teacher can achieve his or her goals best by directing their efforts toward expanding their professional competence to include an understanding of the international dimensions of the field. It is the mindset of the faculty to commit itself to the imperative of incorporating an international dimension throughout the academic and research programs and the institution's overall orientation and perspective. Organizational structure, however it is defined, is not the predominant variable responsible for driving the quality, quantity or adequacy of international research. Intellectual excellence, faculty development and renewal, academic leadership and its support of a global or international perspective of the institution's teaching and research activities are of extreme importance.

I will comment on Jeffrey Arpan's paper and, perhaps, offer an alternative opinion for extending and enhancing the intellectual substance of international business scholarship. I have some strong opinions regarding the nature of international business research, and I would like to offer some steps that can be taken to stimulate high quality scholarship in the international business area. I realize that some of my points will be at odds with those of Arpan's while others will tend to overlap, and that is as it should be. Rather than point a finger or lament about the reasons for the supposed disinterest in international business scholarship, I would like to stress the fundamental importance of the individual scholar and the role that he or she plays in the transformation of a field of scholarship. I'm reminded of the statement "that if one buys a carton of skim milk then one should expect to find skim milk, nothing better and nothing worse." In the same way, if an international business group is composed of weak or mediocre scholars, one should expect weak or mediocre scholarship to be the hallmark of their work as well as the reverse condition: strong, well-grounded scholars will produce strong, well grounded scholarship.

Let me begin by saying that I am unprepared to comment on the quality, quantity, and adequacy of international business research in fields other than international management and its subfield: international human resource management. As I reflect on the state of scholarship in those areas over the last decade, I must say that I'm encouraged by the rays of light that are beginning to appear on the horizon. International management is a field of research and teaching that is in transition as it moves from anecdote to theory grounded research and from descriptive to empirical analysis. Internationally oriented articles are being published in the leading refereed journals, and they are appearing with increased frequency. The scholarship is excellent in terms of the issues being studied and the application of appropriate research methodologies. A review of the *Academy of Management Journal*, the *Academy of Management Review*, and the *Journal of International Business Studies*, to name a few journals, indicates the field is attracting the attention of increasing numbers of management scholars. Their work is successfully negotiating the ever increasing demands of the referee process of the discipline oriented journals and the application of real world problems.

INTELLECTUAL EXCELLENCE

What is the distinguishing characteristic of the emerging body of international management and subfield of international human resource management literature? For me, it is intellectual excellence on the part of the scholars and the work they are producing. Nothing is more important, and nothing less should be demanded of scholars engaged in any field of legitimate scholarship. The stature of the field will be measured by the dual criteria of the quality of knowledge and ideas produced by scholars and by the impact that these intellectual contributions have on the worlds of academics and practice. Surprisingly, much of the research has not been limited to faculty members grouped in an international business department. I realize the traps that quickly appear as one extrapolates from a limited sample to the whole. However, I will accept that risk in order to sharpen the argument. Without intellectual excellence, the quantity and adequacy of international business research become nonissues, and the scholarship associated with the field will be condemned to the trash heap.

How does one go about stimulating the quality, quantity and adequacy of the flow of research in international business? Is it by organizing an international business department to be staffed by faculty members who have banded together to teach and do research on international business related subjects? Perhaps! Arpan has told us that the establishment of an international business faculty group is the way to go. That is one alternative, and it is an appropriate suggestion—if those faculty members are excellent scholars well grounded in their disciplines, focusing on internationally oriented issues and utilizing appropriate research methodologies. The results should be similar to those reported in Arpan's paper. However, I'm cautious about accepting organization structure as the ideal solution for overcoming the "anemic condition" of international business oriented research.

Let me share some of my concerns about Arpan's recommendation. First, I'm afraid that an organization design that leads to the concentration of faculty who consider themselves to be international specialists in a department of international business is courting academic disaster and intellectual suicide. Such a design can easily lead to encapsulation and isolation of the international faculty. International business is an applied field, and it needs continuous infusions from the basic business disciplines. There is far too much of a smokestack mentality in business schools, and this recommendation will exacerbate the problem. The barriers that separate faculty groups from each other are far too high to scale and too dense to penetrate, and the consequences are discipline isolation, faculty inbreeding and intellectual diminution. It is time to destroy these barriers dividing disciplines and applied fields.

Second, I consider Arpan's suggestion as one that encourages a mindset that differentiates international from domestic perspectives and ultimately influences scholarship and teaching programs. Such a differentiation seems to be the direct opposite of organizational design trends and the global perspective that is emerging among corporations and leading business schools. I thought that we, as business educators, have been challenged to press toward the goal of successfully integrating an international dimension into the academic and research programs resulting in a different way of thinking about the business and management process. No longer will there be a distinction between global and domestic priorities. Unfortunately, strict departmentalization reinforces the mentality that allows disciplines to insulate themselves from each other.

Third, if there is an international business faculty I recommend that it be considered to be a business school resource. That faculty with its expertise should be encouraged to participate in the various discipline oriented research seminars, engage in joint research projects and provide advice on the design of internationally oriented research projects and course development. It is essential that the departmental boundaries be easily penetrated and faculty members, regardless of discipline orientation, be encouraged to consider internationally oriented topics as viable sources of scholarly inquiry. Crossfertilization is a necessary condition for professional growth and development.

Fourth, interaction among faculty researchers and those skilled in the tools needed to conduct rigorous international business research is essential to the establishment of an intellectual climate that is conducive to internationally oriented research. The key to faculty interaction will be the dean's ability to provide the intrinsic and extrinsic rewards to those scholars, regardless of discipline orientation, who engage in international business related research activities and as those who incorporate an international dimension in the classroom.

FACULTY DEVELOPMENT

Faculty development represents the key to intellectual excellence and it is a tool for stimulating high quality, important and relevant international business research. Professional development and renewal are processes that must continue throughout one's career, and they should be a concern of the individual scholar, academic departments and school or college of business administration. The international business scholar must maintain competence in one of the basic business disciplines, apply the body of knowledge to internationally related problems and design research studies that utilize appropriate research methodologies to international oriented projects.

For the scholar, well established in his or her discipline and specialty, the challenge of maintaining one's professional competence becomes a challenge of the first order because (1) there has been a knowledge explosion within the basic disciplines and the practice of management; (2) international and global perspectives have become a challenge for scholars interested in expanding theory or understanding the practice of management; (3) researchers who are successfully publishing their findings of internationally oriented research studies are incorporating complex and rigorous research methodologies and data analysis. I'm afraid an international business faculty that is insulated from the basic disciplines of business administration will be unaware of changes and the consequent result will be atrophy.

At this stage in the development of international business as a field of scholarly inquiry, and as a mindset pervading the intellectual environment of business schools, an international business department probably provides an intellectual anchor and a relatively safe harbor for such scholars. Banding together can help establish an environment that will encourage the faculty members to take risks with respect to research subjects, consider alternative publication outlets for their research findings and encourage interactions with scholars whose educational competencies are different from theirs. I would expect the internationally oriented faculty groups to be among the most entrepreneurial and active of the various faculty groups. Tangentially Arpan's thesis deals with these issues, and based on my experience, such an organizational design can prove to be successful, and I can appreciate his reservations about the fear of ultimate disappearance of internationally based research if the scholar is placed in a basic discipline oriented faculty group.

RELATIONSHIP WITH THE WORLD OF MANAGEMENT PRACTICE

I strongly recommend that business school faculty members, regardless of their disciplines, consider the multinational corporation be considered as an important re-source for enhancing the intellectual excellence and the quality and relevance of scholarship. The multinational corporation can provide an important laboratory for testing theory and understanding current and evolving management practice.

Managements of multinational corporations have been coping with problems of structure and management styles, and their grasp of the problems as well as their resolutions to them are far advanced from those proposed by international business scholars. It is a sad commentary to discover that scholars are frequently offering solutions to yesterday's problems. No wonder management considers some international business scholarship to be irrelevant because it is not timely and applicable to management's needs.

Researchers have much to gain by interacting with executives and managers engaged in international business. The field requires the constant infusion of new ideas and approaches from those firms who are competing with others in the international marketplace. The impact of regional pressures upon corporate strategy development, models of organization structure, communication and control mechanisms, management styles and human resource management practices are among some of the day-to-day issues and problems that multinational corporations have to cope with. These real-life experiences present a rich vein of practical knowledge and insight that must be transferred to internationally oriented management scholars, and ultimately into educational course materials designed for students preparing for careers in business as well as present managers participating in executive and management development programs.

The corporation competing in the global marketplace is the laboratory for testing new models, developing new techniques for successfully coping with the intense competition and becoming sensitive to the management problems. At this point in time, researchers must come to accept the notion that an important stimulus for the advancement of thinking and practice in the international area may flow from the corporate setting rather than academe. In other words, we as academics will be the learners and the collaborators with management rather than their teachers. By means of continual interchange with the multinational corporation, researchers will be working on problem-driven research issues, reflecting on what they had found in their research projects, encouraged to publish their findings in refereed journals and striving to advance theory and construct models.

Intellectual excellence remains the measure of the international business scholar, and that criterion can be applied to research that has a practical orientation as well. What are some of the steps that the academic researcher may exercise in order to have an impact on the quality of research? First, scholars can learn a great deal by becoming familiar with the current management practices of those firms doing international business. What does this entail? It means that the scholar must be sensitive to larger organizational, social, legal and economic issues associated with the management of firms engaged in international business, and more specifically it requires the researcher to become familiar with the organizational problems confronting management. Second, when a scholar is engaged in a company-supported research project, it is important that the researcher come up with suggestions and viable propositions that can be implemented. It maybe difficult to justify an academic researcher who is limited to providing theoretical, and perhaps unrealistic or impractical solutions to real problems. Third, the researcher has an important role as an educator responsible for helping to prepare tomorrow's cadre of managers as well as current managers engaged in the management of the firm involved in international competition. The scholar must be capable of blending theory and practical knowledge with understanding because it is the best of the two worlds, and he or she can contribute significantly to the international business research agenda of this decade.

ADMINISTRATIVE AND ACADEMIC LEADERSHIP

The influence of the dean's leadership cannot be discounted in the process for the establishment of conditions that will encourage the incorporation of an international perspective throughout the institution. The administrative leadership must be instrumental in the establishment of conditions that will promote the success of an institution's internationally oriented research program. I noted above the dean can encourage interaction among faculty researchers to reinforce the importance of international business related research. Such interaction can contribute to the development of a mindset that integrates global and domestic priorities.

The dean needs to address several factors that can contribute to a research program that will enhance the intellectual substance of international business scholarship, and these are: (1) internationally oriented research becomes a necessary consideration for favorable promotion and tenure decisions; (2) availability of financial and nonfinancial resources necessary to support the research process; and (3) the development of an intellectual climate that will be supportive of faculty interested in pursuing international business related research. These variables can combine in such a way as to impact the mindset of the faculty and emphasize the importance of international business in the research and teaching programs.

CONCLUSION

In this paper I have attempted to identify several of the variables that influence the growth of internationally oriented scholarship. Throughout this paper I have stressed the importance of intellectual excellence of the individual faculty and the significance of faculty development and renewal. They are critical variables in the equation, but they are not the only variables. From my perspective, the challenge facing the academic and administrative leadership is the following: the creation of an environment such that the individual faculty member can produce internationally oriented research that has relevance to theory construction and or the practice of management. How the leadership goes about the creation of an environment that is supportive of international business scholar will be reflective of its creativity and commitment to internationalization of the institution's educational and scholarship mission. In conclusion let me restate my earlier position: as scholars, we are in the knowledge business, developing new knowledge, transmitting that knowledge and applying it to real world problems and conditions. Longwinded discussions about the ideal departmental structure pale before the imperative of intellectual excellence, academic leadership and the recognition that the scholar and teacher is in the knowledge business.

REFERENCES

Campbell, Donald T. 1969. Ethnocentrism of disciplines and the fish-scale model of omniscience. In *Interdisciplinary Relationships in the Social-Sciences,* eds. Muzafer Sherif and Carolyn W. Sherif, Chicago: Aldine Publishing Company.

Arpan, Jeffrey, William R. Folks, Jr., and Chuck C. Y. Kwok. 1993. *International Business Education in the 1990s.* St. Louis, Mo.: AACSB.

Astin, Alexander W. 1991. *The American Freshman: Twenty-Five Year Trends.* Los Angeles: Higher Education Research Institute, UCLA. (Comments in the text on career preference trends are based on data and information presented in this article.)

Bartlett, Christopher A., and S. Ghoshal. 1992. *Transnational Management.* Homewood, Ill.: Richard D. Irwin.

Green, Kenneth C. 1991. Stalking the M.B.A. class of 1999. *Selections* 8(2): 29–35.

Ohmae, Kenichi. 1985. *Triad Power: The Coming Shape of Global Competition.* New York: The Free Press.

Porter, Lyman W., and L. E. McKibbin. 1988. *Management Education and Development: Drift or Thrust into the 21st Century.* New York: McGraw-Hill Book Company.

Kuhne, Robert, 1990. Comparative analysis of U.S. doctoral program in International Business. *The Journal of Teaching International Business* 1(3/4): 85–99.

Nehrt, Lee C. 1987. The internationalization of the curriculum. *Journal of International Business Studies* 44(Spring): 83–90.

4

Institutional Structure and International Business Education: Functional or Dysfunctional?

EDITORS' COMMENTS

In this chapter we focus on the transfer of knowledge and explore the implications of alternative institutional arrangements for IB education and teaching. As in the case of IB research, the increased internationalization of the world of business puts pressure on the schools of business to internationalize their educational programs. We recognize that research and education activities mutually influence one another and will have some comments below about such significant interactions.

We asked the authors in this chapter to examine the benefits, costs, and limitations of the various institutional structures currently employed in IB education. We were particularly interested in understanding the central issues that need to be addressed when trying to introduce the international business dimension at all levels of learning (for example, undergraduate, graduate, doctoral).

As in the previous chapter, we are fortunate to have a set of authors with extensive experience in leadership positions in business schools that have been wrestling with the challenges of internationalization. As a group, these authors provide not only the perspective of U.S. institutions, but also those of Canada and Israel.

In the lead paper, David Blake identifies one key factor that exerts a strong influence on the development of an institution's programs and arrangements for international business education—that is, the institution's conceptualization of international business as a field of academic study and business practice. Blake sees international business as having a dual nature. On the one hand, international business is a separate discipline with its own Ph.D. programs, journals, and professional associations; business has its experts in international business, international finance, international marketing, international accounting, and so on. On the other hand, international business is the context that affects every function of business and management and the way they should be taught, learned, and practiced.

Blake discusses three pedagogical views of international business: the multidomestic, the international business specialist, and the global. He makes the important point that in some schools, one view is dominant; in others, a blend of all three exists. Consequently, there is a tremendous diversity across institutions in the conceptualization of just what is international business, with the resulting diversity in institutional objectives for IB education and institutional structures.

Blake reviews the alternative missions adopted by schools for IB education and the strengths and weaknesses of different institutional structures. He sees the diversity of approaches and structures as a positive aspect of the development of IB education, primarily due to the possibility of crossinstitutional learning. Like Ed Miller in chapter 3, Blake contends that success in internationalization of a school depends crucially on administrative and faculty leadership—particularly in developing a coherent vision of international business within the institution and the specific objectives that provide direction for IB education.

In the second paper, Randy Folks assesses the institutional obstacles to the development of international business education initiatives and judges them worse today than twenty years ago. Barriers to internationalization include the training and supply of fac-

ulty, conflict with business school research objectives, organizational structures and incentives in business schools, and funding dilemmas.

Like David Blake, Folks finds the main source for these barriers lies in the existing institutional conceptualizations of international business as a field of study and practice. Again we find two different perspectives: IB as an academic discipline in its own right and international as a context within which business (and the traditional business functions) are carried out. Folks argues that disregarding the existence of a core body of theoretical and empirical material in the separate discipline of international business and adopting only the latter of the two perspectives above is the fundamental conceptual error that generates formidable obstacles to the internationalization of business schools.

Folks shows how this conceptual error leads to seeing internationalization only as a curricular challenge rather than as a needed strategic response to a major environmental change. Further, this misconceptualization leads to institutional arrangements in which faculty with international functional expertise are grouped with faculty in the specific discipline. In the previous chapter on IB research, we also heard much about the pluses and minuses of such functional specialist structures. Here Folks argues that effect of such a structure not only retards IB research, but also IB education and the internationalization of the institution.

In the third paper, Lee Radebaugh addresses many of the same issues raised by David Blake and relates them to the experience of his university in the internationalization of the business school. Radebaugh notes that his business school has not developed a mission statement for international business—as Blake argued was all too often the case today.

Radebaugh suggests that schools evolve through three phases in internationalizing (mostly ignore it, partial ad hoc efforts, and a more comprehensive strategic response). He links his school's internationalization efforts to Blake's multidomestic, international specialist, and global approaches and shows the relevance of all three perspectives for his institution. He goes on to add an interesting discussion on the importance of strategic alliances—both inside and outside the university—for the promotion of internationalization of the business school.

In his commentary, Yair Aharoni agrees with Blake and Folks on the importance of academic beliefs and norms in the internationalization of the business school. He takes the three authors to task for concentrating so heavily on the experience of U.S. business schools and introduces some insights concerning European approaches to international business education and research.

Aharoni finds the record of progress in the United States in international business education and research to be sub par. His explanation for this revolves largely around the view of international business as a multidisciplinary field and the difficulty any multidisciplinary effort in U.S. academia faces in contending with the disciplinary structure of universities. (See editors' comments in chapter 3 concerning the ethnocentrism of the disciplines and its effect on IB.)

What's to be done? Aharoni thinks that the solution of a separate IB department (advocated by Folks) is just not feasible for most schools. Rather only shifts in the professional norms of the traditional functional disciplines can help most schools. This is indeed a daunting challenge for those of us interested in the development of IB research and education. How do we go about convincing the opinion leaders of the various functional disciplines that the international dimension of their discipline is an integral and necessary part of the professions, as Aharoni suggests? An important point here by Aharoni is that much of the challenge of internationalization of the business school lies beyond the scope of a single school or dean and rather must be dealt with at a national level.

Finally, Aharoni insists that there is still a place for international business specialists to carry out the core IB research and teach the core IB courses. Thus Aharoni like Blake and Folks covers both aspects of IB's dual nature: academic discipline in its own right and context within which functional disciplines operate.

In his commentary, Bernard M. Wolfe provides a Canadian perspective as well as some information on internationalization in Europe. Like Radebaugh he illustrates many of the ideas in this chapter with examples from his own institution's internationalization experience. We see the importance of leadership and funding to making progress in internationalization. Particularly noteworthy is his point on the important links between undergraduate and graduate international business education. Increasingly students are using their undergraduate education to prepare better for an internationalized masters-level business education.

The discussion in this chapter on institutional arrangements and IB education blends very well with that of the previous chapter on institutional arrangements and IB research. Not surprisingly we find many of the same effects of a functionalist departmental structure on both IB research and education. For the research-oriented university, research performance dominates other considerations in hiring and rewarding faculty. In such cases, teaching suffers. When the ethnocentrism of the disciplines and departments is added, any multidisciplinary educational effort in general, and international business education in particular, suffers.

We find it interesting that the authors in this chapter find the key to progress in internationalization of educational programs and institutions lies largely in how we conceptualize, or think about, international business as a scholarly activity and business practice. Blake's ideas about IB's dual nature; Folks's emphasis on the misconceptualization of IB as a fundamental barrier; Aharoni's views on the need to change academic beliefs and norms, particularly in how the functional disciplines view the international dimension. We too believe that how we think about IB as an academic field of study is important and direct the reader to the other book to come out of the conference, *International Business: An Emerging Vision*, for a pertinent discussion of this important issue.

In addition, we find quite interesting Aharoni's idea that the nature of the problems in IB education and the likely solutions lie at the national level. Given the interdependencies among schools of business in the development, employment, and evaluation of faculty, a single school does face serious constraints in how it responds to the challenges presented by the increasing internationalization of the world.

Finally, we are also taken by Aharoni's and Wolf's idea that we need to expand the scope of our deliberations of internationalization to include non-U.S. perspectives. Educational institutions, like business organizations, are embedded in their societies and reflect the particular path of that society's development. It is exactly such cross-national interaction that is likely to generate a better understanding of why internationalization is approached as it is in the United States (or some other country) and to provide a wider range of experience to draw on, as educational institutions struggle to respond to the internationalization of the world.

■ ■ ■

Organizing for Education in International Business
David H. Blake

INTRODUCTION

For the most part, schools of business are organized by disciplines even though in some schools formal departments are eschewed for interest groups or some other principal of organization. Regardless of school structure, faculty usually identify primarily with their discipline-based professional field, for example, marketing, finance, or accounting. However, the organizational structure for international business varies widely across schools like other relatively new and somewhat underdeveloped fields such as business and society.

There are several reasons for this variation in organization for education in international business. First, the field is still emerging; yet it must fit into business schools with well-established disciplines which are the primary basis for faculty organization and identification. Second, because of its relatively underdeveloped stage, schools and often faculty within schools have different and changing perspectives about the role of international business in the education of their students. Third, in many schools, the number of faculty knowledgeable in the field is relatively small, resulting in the placement of the international faculty in an area or department largely for convenience.

Another and very important factor is the dual personality of the field. On the one hand, international business is a separate discipline with its own Ph.D. programs, journals, and professional associations. This disciplinary approach somewhat reflects the business world where there are business experts in international business or the international nature of a specific function. For example, there are academic and business experts in international trade, international finance, exporting and importing, licensing, international tax issues, and many others.

However, on the other hand, international or global business is the overarching and all-embracing context that affects all of the functions of business and management and the way they should be learned, taught, and practiced. In this sense, much of business is international or global, and the field is not just one more function among many.

Thus, almost by definition, every department, faculty, and course should teach their theories and concepts in ways that contribute to a greater understanding of the global nature of business and management. This perspective acknowledges that specific international technical expertise is necessary, but for the general business executive such detailed technical knowledge may not be necessary and is definitely not sufficient. The pedagogical implication of this view is that all faculty and courses need to be globalized. Further, this perspective should coexist with but cannot be replaced by the specialized expertise of faculty who view themselves as international business scholars.

These two very different yet fully compatible views contend with each other in the determination of how schools are structured for education in international business. This conflict is not inherent to the two approaches. Instead, it is the result of first a failure to understand the dual nature of international business and second a conceptual misunderstanding that the functional approach to international business is a surrogate for the even more important global context of business and management.

Some programs offer international business as a separate discipline or function and seek to produce graduates who are technical experts in the function of international business. Other schools are less concerned with educating students for careers as international experts, but instead they are trying to ensure that all their students are thoroughly grounded in the fundamentally global nature of business and the management theory and practice that will lead to success in a global business system. Other schools are attempting to achieve both objectives. And there are others still that have no consistent plan or approach.

Given the differences just discussed, there is no "best" way to organize and place international business within management education. In this paper I will not attempt to argue that one way is better than another, nor are my comments the result of an exhaustive survey of current practice in the United States. However, the discussion that follows is based on several personal biases: first, as a teacher and researcher I am committed to a full understanding of the global nature of business and the implications for management, companies, states, and the international system; second, business schools should prepare students to operate in a business system which is increasingly global with the knowledge and understanding appropriate for this global system; not just in the future but now; and third, as a dean for more than eighteen years, I have attempted with varying degrees of success and failure to globalize curriculum programs, and, most difficult of all, faculty.

This paper will first discuss different philosophical approaches to educating management students (especially at the graduate level) for international business. Then, a number of archetypal structures found in business schools will be examined relating them to various perspectives or missions and to the strengths and limitations of each.

EDUCATING STUDENTS FOR INTERNATIONAL BUSINESS: A VARIETY OF MISSIONS

Schools of management and business have adopted a number of different approaches to educate their students for effectiveness in international business. These different approaches depend upon how a school defines (or does not define) the field of international business for pedagogical purposes which in turn is somewhat based upon the understanding of the faculty and administration about the nature and scope of international business. The confusion about the meaning of the field discussed above and its implications for management education lead to intraschool inconsistency, ambivalence, and often ineffectiveness as the lack of clarity of understanding gives rise to a mismatch between academic mission and implementation.

142

Pedagogical Views of International Business

To some, international business means how to engage in business in other or "foreign" countries. This basically geographical orientation is sometimes referred to as the multidomestic approach, that is, business is conducted domestically and differently in a number of different countries. Given this view, students need to understand the complexities of doing business in different states, including such things as market, business and accounting systems; relationships to governmental, legal, political, and social systems; and cultural, religious, ethnic, and language systems.

This view of the field tends to cause faculty and the curriculum to focus on the differences among the individual states being studied with sharp contrasts drawn between business practices and the business climate in the home state where the school or corporation is located and other states. The difficulties of doing business in these countries are stressed, implicitly or explicitly, and apply to the full range of international business activities including exporting and importing, licensing and franchising, as well as direct foreign investment.

A second approach recognizes that there is a need for specific international business functional expertise, enabling managers to become international business specialists. Graduates of this type of program would seek to use their technical knowledge about international business in a specific or several functional areas. Graduates might manage international currency transactions, implement international tax strategies, develop an international transportation system, or chart an export drive.

Here the focus is not on doing business in other countries but rather developing specific technical skills applicable to the conduct of international business. A specific variation would be a focus on exporting. Graduates are hired for their technical knowledge that helps a company in its international plans and transactions. They are placed where this expert knowledge is needed or best fits.

A third and entirely different understanding of international business focuses on the rapidly developing global and regional economic business systems. This view recognizes that there is an international or global business system. (I will use the term "global business system" and "global business" to differentiate this perspective from the first which I will call "multidomestic business" and the second which I will label "international business specialist.") With this approach, students and faculty need to have a thorough understanding of the complexities and opportunities associated with the interconnections existing in the global system. Further, they need to know the theories, concepts, tools, and techniques that will enable them to be effective at the global level.

While the multidomestic approach is a part of the global perspective, for companies must still operate in different states, the global approach is more systemic and encompassing than the more state-oriented view. Global business can be pursued from anywhere in the world while the multidomestic view looks at operating from and within discrete states. Global business treats state boundaries as merely part of the business environment, but the multidomestic approach tends to define business by those boundaries and their many implications.

An analogy based on U.S. experience helps to illustrate the difference between the two approaches. The multidomestic perspective would focus on doing business in the individual states of the United States much the way it was prior to the Civil War whereas the global perspective would concentrate on the United States as a whole. More recently, the change from a Union of Soviet Socialist Republics to multiple independent and increasingly disconnected states makes the point going in the reverse direction.

The global approach is based upon the conviction that a well-educated business person must understand the connections and opportunities that exist worldwide and the consequences for domestic as well as international business. Fundamental comfort with the nature of the global system and the managerial knowledge that enables an executive to operate successfully in this broad venue are key educational objectives.[1]

The international business specialist approach educates for the manager who has or will have specific international responsibilities requiring technical knowledge. A significant portion of this person's career will be in that particular international specialty. The curriculum and program are designed to provide the student with the appropriate international functional expertise. The global approach, though, educates for effective management in a global system. For example, while the intricacies of international tax matters may not be understood in technical detail, the graduate with a global approach will have general and broad functional knowledge appropriate to a company that is at least affected by and most likely involved in business on a global or regional basis. The manager's knowledge base will be global in scope with a similar perspective to match. Business problems are framed and diagnosed in global terms first and foremost and then reduced to the specialist or local level if needed.

In order to highlight the differences between these three approaches, I have probably drawn sharper distinctions than is necessary. Further, some schools blend all three of these perspectives just as effective corporations must. However, observation suggests that most business executives and business faculty tend to equate international business with the multidomestic or international business specialist approaches rather than with the global view. The result is that international business becomes a specialized functional area or has a limited geographical focus potentially competing for resources and course time with other more established business disciplines. Moreover, there is the implication, false I think, that in order for faculty to teach "international" they must become international business scholars and specialists themselves.[2] This misses the key point: all business and management disciplines are largely conducted in and therefore must be taught as part of the global context.

School Educational Objectives for International Business

Some schools have adopted clear statements about how they approach the task of educating students for international business. More often such a statement is lacking, or the school's implicit approach provides little guidance to faculty or students about what can be expected in the educational process. Confusion regarding the school's understanding of international business coupled with the lack of specificity about its pedagogical objectives often lead to a lack of direction for international business education and consequently how faculty should be organized to achieve these goals.

A few schools have determined that their educational programs are designed primarily to prepare the graduate to live and work in a specific country or countries. General business education is supported by international business courses, but students also take language training and country- or region-specific work in the history, politics, and culture of their chosen countries. Internships abroad contribute to the student's country specific knowledge and their hope that a position will be found that will enable the student to work in that or a similar "foreign" country. A few schools are well-known for their focused masters programs in international management or business. Several others offer similar programs but within the context of a more traditional M.B.A. program.

Many more schools offer international business as one of several functional specializations like finance, marketing, or MIS, or they may focus primarily on international business as a functional specialty. These programs are seeking to educate the international business specialist who will use the knowledge gained to develop a career in the international business field specifically. With the exception of core coursework, most of the courses are international functional courses in areas like marketing and accounting with some general international courses.

A small number of other schools are attempting systematically to educate students for managerial effectiveness in the global business environment. Their graduates are not expected to live and work overseas, at least initially, but they are expected to operate comfortably within the global business system wherever they are located be it overseas or in a domestic locale. These students are not expected to know all about a specific host state or states (though there may be state- or region-specific courses) or to be fully adept in all the technicalities expected of an international business specialist. They are expected, though, to have the skills to manage effectively in the context of a global business system. More importantly, these students develop a global mind set that enables them automatically to approach business or managerial problems in a global framework. This international exposure may be achieved through a required international business course or two, a required "constrained" international business elective or electives, or, less frequently, the conscious integration of international material into the core courses.

Most schools have a haphazard blend of each of these objectives, more the result of accident than planning. A typical curriculum would be some traditional discipline-based international business courses such as international finance and possibly a more generic international business policy or environment course. Further, depending upon faculty interests and capability, various international electives may be offered about doing business in specific countries (Brazil) or regions (the Middle East). However, there often is not a clear approach or philosophy that guides what the curriculum should achieve and what students will learn. In addition, few schools have really understood the necessity of developing the global mind set in all students.

Probably the great majority of schools and especially their faculty have largely ignored the issue of what their students should be learning about the international nature of business. For some, the issue is of no real significance and is merely a passing fad. For others, there is so little understanding of the global nature of all business that courses are taught largely in ignorance of the global context. Perhaps a last chapter or last lecture is devoted to the subject, but even then it is considered to be an aberration, not a basic part of the course. For other schools, lip service is paid to education for international business without the substantive follow through.

ARCHETYPAL INSTITUTIONAL STRUCTURES

Given the varying understandings of what is meant by international business and the differences in educational missions for international business education, it is not surprising that there are different institutional structures and combinations that have been developed. Each results in differing sets of opportunities and limitations, though the absence of clear direction may result in far more limitations than opportunities in actual practice.

The Focused International Business Program

In this format, the entire program is designed to educate students for business positions in other countries. Standard business courses are linked with international and comparative management courses and with state- or area-focused courses on the politics, history, and economics of specific countries. Usually, a commitment to "foreign" language competency is incorporated in the program. The mission of this type of program is clear and accepted by most of the faculty, staff, and student body who were attracted to these programs because of this well-defined focus.

While students are exposed to the language, culture, politics, and history of the countries studied, sometimes this is achieved at the expense of rigorous knowledge of the theory and practice of management. These programs often consciously develop an international atmosphere and culture that contribute to the socialization and learning of the students.

With such a clear mission, students self-select into this type of program. Students come from widely diverse backgrounds, but their objectives are probably more similar than would be found in most less focused M.B.A. programs. Since students want to live and work overseas, this allows the program to be more focused and provides opportunities for building international awareness and sensitivity. However, it does have the drawback of lack of exposure to peers with very different educational and career objectives. The specific focus may also result in a narrowness of sorts with students missing both the larger context of business and the extraordinary diversity of businesses and business people. Of particular concern is that students are so committed to the multidomestic approach that they fail to understand the overarching global context of business and the complexities of operation globally.

Faculty, too, self-select into teaching in this type of program. They are either comfortable with this kind of institution, and, if not, they do not seek employment at such a school. The focus and "fit" may enable a school to implement a more specific reward, incentive, and faculty development system, for the centrifugal force of the traditional disciplines may be ameliorated.

However, faculty in such programs may lose touch with or not be fully accepted by their discipline-based colleagues nationally and internationally. Their teaching and research may be less transferable than faculty at other kinds of institutions, and they often do not publish in the most highly regarded discipline-based journals. Therefore, personal mobility may be reduced, for there are relatively few schools adopting the focused international business approach.

Further, schools with this approach may find it difficult to recruit faculty from

the most prestigious Ph.D. programs since the mentors of these younger faculty often will urge them to make their professional mark in the discipline in which they have studied. The "best" new Ph.D.'s in traditional fields are likely to want to begin their careers where their efforts to expand knowledge of the discipline are appreciated, encouraged, and rewarded.

This selecting out of new faculty coupled with the reduced mobility of faculty already in the system may result in some intellectual self-absorption. The strong commitment to the school's mission and way of doing things may reduce the inflow of new ways of thinking and the encouragement of divergent views. The latest insights and approaches developed in the disciplines may be slow in reaching the students.

To some extent, a focused international business specialist program has many of the same attributes (positive and negative) as the geographically focused program.

The International Business Department

In the more traditional, broad-based business schools, a common organizational structure is to establish an international business department. Faculty in such departments usually view their functional discipline as international business, and they teach and research in that field primarily, though they may have linkages to one of the more traditional disciplines.

This structure provides a focus for international business within the school. International business faculty often become advocates for their field, urging the establishment of majors or concentrations at the B.B.A. and M.B.A. levels as well as in the Ph.D. program. Further, they can serve as an international "conscience" within the school and for their colleagues. They can lobby for required international courses, stress the importance of the AACSB international standards, urge the creation of study abroad or international exchange programs, and assist other colleagues to explore the international dimensions of their fields.

The existence of a faculty "home" encourages the hiring of additional international faculty. Faculty hiring, renewal, tenure, and promotion decisions, at least in the early stages of these processes, are in the hands of peers with international interests, people who understand and appreciate the field.

The international business department gathers together faculty with a strong commitment to the field. As a group they may be able to build a critical mass and collaborate with each other in teaching and research. The department becomes a focal point for in-depth and widespread knowledge about international business.

One of the difficulties caused by the creation of an international department is that noninternational faculty may therefore feel that they do not need to internationalize their courses. That task, they feel, can be left up to the international business department and the faculty experts/specialists. To the extent that the international dimension is not integrated into other disciplines and their courses, students are likely to get a skewed view of the importance of the global system. The curriculum itself is likely to convey the message that there are the traditional business disciplines and then there is the more specialized and perhaps "less important" international area. This may actually be reinforced by the international business faculty if they themselves take the multidomestic or the international business specialist approach to the field as opposed to an insistence that business and management are themselves global in nature.

If the international business faculty imply that their subject area is primarily a specialty in and of itself, the important connections with the other disciplines are likely not to be addressed. Students will not be exposed to how marketing, accounting, finance, human resources, information systems, and operations are all part of and subject to the global context. The narrowness of faculty is inflicted on their students.

A Matrix Approach

Some schools have attempted to avoid the problems of the departmental structure by establishing an international business affinity group composed of faculty who are formally members of other departments. This arrangement seeks to encourage collegial discourse on teaching and research issues on international business while at the same time maintaining the intellectual foundation of the more traditional disciplines. The affinity group faculty can provide direction and focus to the international curriculum and extra-curricular activities of the program or school. At the same time, it is expected/hoped that internationalization also takes place within the more traditional departments and disciplines. In this way, faculty and students may observe that the field of international business is an integral part of all elements of the curriculum.

For the matrix approach to work, there needs to be a widely shared commitment to and understanding of international business in its several dimensions. The affinity group needs to have institutional power in deliberations about the curriculum, allocation of research and teaching resources, and all aspects of the critical personnel decisions. A group without the ability to affect the outcome of these key issues will often become little more than a shell with few faculty participating. Faculty will correctly perceive that the traditional departments can generate the resources and influence the decisions that affect their careers. Therefore, it becomes sensible to respond to the values and standards of the department as opposed to the international affinity group.

This conflict occurs most dramatically in personnel decisions. Disagreements between the international affinity group faculty and the traditional departments over hiring, promotion, and tenure can be devastating to the cooperation necessary for the matrix system to work. Often the departmental faculty do not value as highly the internationally-oriented research and publication outlets of their colleague who is part of the international business interest group. Where the department has greater influence in these decisions, a clear message is sent to other faculty that the international business group is essentially irrelevant, and therefore time should not be "wasted" in participating in their efforts.

Strong and highly respected leadership of the affinity group can help make the matrix system work. A leader with substantial influence in the important councils of the school will be able to support the interests of the group. However, without a well-respected organizational base where the affinity group really counts, the effectiveness of the international faculty may be limited.

There is still the tendency on the part of faculty who are not part of the affinity group to do no more than pay lip service to the objective of internationalization. It is easier to leave the task to those who are designated "international" faculty. The result is that students perceive that international matters are not important to all subject areas since this approach is not reinforced by all faculty in all courses.

An additional problem may be that the specialist nature of the field becomes lost in the matrix. The international area is not treated in the same way as the more traditional disciplines, and this may cause students and faculty to think that it is less important or necessary as a specific function of management.

Where this system works well, students may gain a better understanding that international business is both a discipline in itself and one of the most critical environments within which business is conducted and therefore it affects all aspects of business. However, the mission of the school, a clear understanding of the multiple nature of the field, and effective implementation are essential to overcome some of the weaknesses of this type of organizational structure.

Schoolwide Commitment

A few schools are determined to thoroughly globalize the curriculum which requires the globalization of the faculty. Some schools in the United States have adopted this approach, and some of the leading European schools have made great progress in institutionalizing their approach. The ob-

jective is not to make international business teachers and researchers of each faculty member but rather to provide an education for students that prepares them for work in the global economic system. This means that the subject matter of each course must address and be addressed within the global nature of business. This is not the same thing as having a session or two on the international nature of a particular subject. Instead it requires the reworking of the course to teach the material as it is relevant to the broad issue of business operations in a global context. This does not necessarily mean that the faculty member must learn a new discipline, international business. It does mean that the existing discipline is recast in recognition of the international nature of business and its global context. It is an extension, sometimes substantial, of the basic knowledge of the discipline and not the necessity of developing a whole new area of international functional expertise.

To accomplish this, schools may seek faculty-wide commitment and involvement in the task. The departmental or interest group structure is not changed, and there may not even be an international business faculty in the traditional sense of the phrase. However, each faculty member, within the context of the existing organizational structure, is encouraged to globalize his/her part of the curriculum. Oversight committees can be established, but the basic principle of organization is not changed.

The problems with decentralizing the responsibility of globalization to the department or individual instructor levels are substantial. Achieving faculty acceptance and action may be difficult. The ability to impose globalization is limited. The package of incentives, disincentives, and rewards may not be enough to achieve globalization in the classroom of a particular professor. Schools are fairly impotent when it comes to resisting tenured faculty. Nevertheless, the resources at hand including the expectations of peers can lead to success. Indeed, an extensive and largely voluntary commitment by faculty can produce significant results. Further, faculty from many disciplines can stimulate one another in revising their courses to incorporate the global nature of business.

On the negative side, the lack of an institutional focal point may mean the absence of advocates for globalization or even the development of international business as a functional specialty. Since globalization is everyone's business, it becomes the responsibility of no one. This can be overcome by leadership from the school's administration and most importantly others on the faculty. Through incentives (resources), enthusiasm, and leadership, globalization can occur, but abject failure is probably as likely. However, a school- and faculty-wide commitment to globalization can bring about the changes desired.

The same issues of incentives, influence over key personnel and resource issues, and ownership/advocacy by specific faculty as noted above can be a problem with this approach. The incentives for globalization are either nonexistent or not likely to be implemented well.

CONCLUSION

The various archetypes discussed above are presented as if each existed in its pure form in schools of business. Obviously, this is not the case, and a mixture of organizational structures exists in most schools. One reason for this is that schools seek to get the advantages inherent in several organizational structures while reducing their disadvantages. Further, since there is no obviously "best" structure, schools experiment to achieve the "best" for their particular objectives and circumstances.

Actually, the diversity of approaches and structures is a positive development. Diversity brings richness, innovation, experimentation, success, and failure—all of which add to the existing knowledge of what can be done to prepare graduates for careers in the global business system. The diversity enables faculty and administrators to learn from the success and failures of others, plagiarize and adapt good ideas, and extend the efforts of others beyond where they have carried them. In this time-honored fashion, the profession of management education will slowly but surely improve its ability to educate students for success in a changing world of business.

An essential component of all these efforts to internationalize management education is that of leadership. Leadership by the administration and leadership by the faculty are critical. Certainly, incentives need to be provided and so too must there be the resources to help faculty enhance their knowledge and revise their courses. Further, the incentives and resources should not be available only once but rather should support continual change and improvement and assist faculty after they have taken the initial steps and now see how much more needs to be done to change their courses. Often, fairly systematic efforts to achieve globalization are important.

The leadership role of the administration in defining and articulating the international mission of the school is important. The administration should take advantage of opportunities to reinforce the approach and mission of the school as it relates to international business education. The role of the dean and her/his staff is crucial.

However, leadership by the faculty is equally as critical. Individual faculty or groups of committed faculty are able to urge their more reluctant colleagues to understand and respond to the challenges of a global business system. Peer leadership in a constructive manner can hasten the development of a broader commitment. Just as the dean can take advantage of many opportunities to emphasize the mission, faculty have many more opportunities to influence the thinking of their colleagues directly and indirectly in formal meetings and informal discussions. The encouragement and mobilization of faculty leadership are fundamental to the success of internationalizing management education.

NOTES

1. The global orientation may also be applied to various regions of the world such as the Pacific Rim, Middle East, or Central Europe.

2. This international business specialist approach for faculty was incorporated into the December 1991 survey conducted on behalf of the AACSB on internationalization efforts. Several questions asked respondents to indicate how important the availability or lack thereof of publication outlets for international research was in promoting or inhibiting internationalization. This kind of statement suggests that for faculty to internationalize they must research and publish in the international area. The view presented in this paper is that professors in each of the traditional disciplines must be encouraged to internationalize their traditional courses not necessarily become scholars of international business.

■ ■ ■

Institutional Barriers to Internationalization
William R. Folks, Jr.

INTRODUCTION

When the University of South Carolina first developed its Masters of International Business Studies degree program in 1974, we decided to incorporate a mandatory six-month offshore internship as a requirement for all U.S. nationals. When the time came to begin the arrangement of these internships, university faculty charged with that task ran afoul of a particularly binding state budgetary regulation. The amount of state appropriated funds allowed for foreign travel was restricted to $1,000 per trip, and any such travel required the advanced approval of the State Budget and Control Board, a group charged with the fiscal oversight of the entire budget of the state of South Carolina.

Foreign nationals admitted to the M.I.B.S. program, and foreign participants in our exchange programs, encountered an unusual insurance requirement. In addition to the usual and necessary health insurance, all foreign nationals were required to have insurance providing for the repatriation of their remains to their native land, should they be so unfortunate as to expire during their sojourn in Columbia, South Carolina. In addition to creating an unnecessary additional expense for the prospective student, it also transmitted a rather sinister message as to the climate of security (or just the climate) at the University of South Carolina. This nonstandard coverage resulted from the refusal of one foreign student's parents to pay for the local funeral expenses or repatriate the body of their deceased son. Although the incident leading to this requirement occurred, legend has it, in the 1950s, the insurance requirement remained in effect in the mid 1980s, when the university sought to expand foreign student enrollment in M.I.B.S. and the M.B.A. program through a number of exchange programs.

When the university's first Portuguese-track students sought their visas for their internship assignments in Brazil, they were asked by the Atlanta Consulate to provide, among many other items, a medical certificate, stating that they were free of certain diseases and in generally good health. In addition, the university was required to provide a certificate from the State Board of Medical Examiners that the doctor certifying the health of the individual student was indeed licensed to practice medicine. Substantial discussion occurred as to whether the State Board of Medical Examiners would require payment for issuing this certification, and whether it would be legal, according to state law, for the University of South Carolina to transfer funds to pay for this service.

Common sense, and the luster of contributing to a nationally recognized international business program, have resolved these and a host of other institutionally inflicted obstacles to the process of international business education. We generally contrive to get foreign nationals their visa documents prior to the time they need to be on campus for the start of their academic program, we are no longer restricted to using surface mail for sending out applications to prospective overseas students, and the State Budget and Control Board has changed the foreign travel limitation to $2,000 per trip. In general, we have learned that to be a world institution, the University of South Carolina must treat its foreign customers for education with the same quality of service as it does its domestic ones.

Nonetheless, in my experience, the obstacles to effective internationalization of the business school are far greater now than they were in 1974, when the M.I.B.S. program began operation. In my earlier paper in the program I defined internationalization of the business school as "the development of a global orientation to all aspects of a business school's endeavors, be it in teaching, research, or service, consistent with the school's overall mission." The development of such a global orientation is impeded institutionally by the increasing specialization of business school faculty, the organizational response which naturally follows from and facilitates such specialization, and the resulting dearth of incentives integrating interdisciplinary programmatic responses in research and teaching.

IDENTIFYING THE BARRIERS TO INTERNATIONALIZATION

It is the premise of this paper that in most major tier one and tier two research universities, it is becoming extraordinarily difficult to initiate the truly interdisciplinary international programmatic initiatives needed to respond to the globalization of business. First, the institutional barriers to the creation of truly responsive educational

programs are high. Further, institutionally driven organizational structures make it difficult or impossible to assemble the critical international business education expertise to administer and staff a truly multidisciplinary, comprehensive international business education academic program.

The purpose of this paper is to identify the institutional obstacles to internationalization of the business school. In this section of the paper I will briefly categorize the operational barriers which might restrict the ability to provide a meaningful strategic response to the globalization of business.

The first set of barriers might be characterized as barriers relating to availability and training of faculty to carry out the business school's internationalization strategy:

1. There are very few faculty available in the market who have sufficient international academic training to do research in or teach international topics.
2. Existing business doctoral programs provide little, if any, international capabilities for future scholars.
3. Existing business faculty without international capabilities are extremely resistant to acquiring these capabilities, and the alternatives for developing them are limited and expensive.
4. Faculty who have international capabilities generally tend to do less well in promotion and tenure decisions, and turnover is high.

Involvement by faculty in international activities is frequently viewed as damaging to the business school's research endeavors. In schools seeking to be known as top research institutions, barriers arise to protect the research dimension of the business school's mission:

1. Extensive internationalization activities take time and are seen as significantly hindering research productivity.

2. International research is viewed by faculty as not welcomed at the top scholarly journals.
3. International research is viewed by faculty as difficult for peers to understand, and consequently a relatively dangerous strategy to take if peer review is a major determinant of promotion and tenure.
4. Faculty view time spent in international travel is time spent away from productive scholarly research.

Further, there are significant organizational barriers to the development of an internationalized business school:

1. Initially, many deans have limited international experience, and do not have the expertise to operate in an international educational environment.
2. The expertise to manage international educational programs is absent on many campuses.
3. What international expertise there is available usually resides in faculty teaching in the international business discipline. To assign that faculty to administrative responsibilities precludes their research and teaching, and may hinder their eventual career progress.
4. Creation of a separate international business faculty group requires a substantial commitment of resources on the part of the administration of the business school, and develops a competitive response on the part of other faculty.
5. Injection of individual international faculty into various departments places them at risk from a tenure-promotion perspective and isolates them from other internationally active faculty.
6. Compartmentalization of faculty creates a narrower perspective which makes resource shifting to initiate international programs difficult.

Finally, the activities taken to internationalize a business school are relatively expensive, and mistakes can indeed be costly. Another set of possible barriers concern funding:

1. The necessary faculty acquisition and development activities to build international capacity in the faculty are expensive.
2. Typically, internationalization activities are supplementary to existing activities of the business school; additional resources must be found to fund them, or resources shifted from other activities.
3. Any type of special international activity, such as internships, foreign sabbaticals, or faculty exchange programs, are expensive to operate.
4. University and state regulations restrict foreign travel, and travel funds needed to carry out complicated research designs with an international component are simply not forthcoming.

In summary, barriers to internationalization of the business school include availability and current training of faculty, perceived conflict with the research objectives of the business school, existing organizational and incentive structures in business schools, and the incremental funding dilemma.

CONCEPTUAL BARRIERS TO INTERNATIONALIZATION

The primary conceptual barrier which leads to the operational barriers outlined in section 2 arises from a fundamental misunderstanding of the discipline of international business. Administrators and faculty fail to recognize that there is indeed a conceptual domain for a subject such as international business. While we may debate its boundaries and organizing principles, it is quite clear that inquiry into the impact of globalization of business on the individual subfields of business (for example, finance, marketing, or production) is not the same as inquiry into the impact of globalization on the firm as a whole.

It is my contention that this conceptual error gives rise to many of the institutional barriers which make a business school's efforts ineffectual. As my other paper in this program discussed the need for clear strategic thinking in developing international activities, in this paper I will focus primarily on obstacles which arise from the confusion which exists between education in the academic discipline of international business and the internationalization of the business education curriculum.

The establishment beyond question of the existence of a conceptual domain for international business is absolutely critical if meaningful progress in the elimination of institutional barriers is to be achieved. At one business school recently, in a discussion of the internationalization of the Ph.D. program in Business Administration, one faculty member suggested that the proper approach to internationalize the students in each discipline would be to dissolve the school's relatively successful International Business department and assign each faculty member to a functional field. The justification for the move was to provide each doctoral area of concentration with resident faculty who could be certain that the doctoral program provided exposure to the international dimension of the particular functional field. While the suggestion also ignores the synergy of multidisciplinary research which can take place within a group exclusively devoted to international business, its fundamental flaw is its complete exclusion of the possibility that there is a independently existent domain of inquiry for these international business faculty. Thinking along these lines would require that all business faculty (with the possible exception of accounting faculty) ought to be transferred to the economics department.

THE MISCONCEIVED EDUCATION: ERRORS IN DEFINING INTERNATIONAL BUSINESS

This conference has as its premise that there is a discipline known as international business, which has its own conceptual domain, theoretical foundation, and empirical tests of that foundation. Partially drawing from constructs developed in other academic disciplines, but developing a theoretical and empirical life of its own, the academic discipline known as international business now possesses a unique body of knowledge.

International business education can thus be viewed from two perspectives. In a narrow sense, international business education is the process by which the theory and related empirical research concerning international business as a whole, but without regard to the application of that theory to an existing functional field, is transmitted to those who need to learn it, whether at the undergraduate, professional graduate or doctoral level. In the broader sense the overall body of international business knowledge includes an additional component, the understanding of the impact of the existence of multiple nation states on managerial practice in specific disciplines. Traditionally, such fields have been known as international corporate finance or international marketing or international accounting. Thus our discipline of international business can be divided into (1) study of the core theory of international business and (2) study of the traditional managerial functions within the global firm.

The fundamental conceptual error which creates or contributes to the extensive barriers to internationalization of the business school is defining internationalization to be the development of expertise in the second of these divisions exclusively, and ignoring the existence of a core body of theoretical and empirical material in the separate discipline of international business. While the further subdivision of domains of inquiry is appropriate, current business school institutional arrangements focus only on the second aspect of our domain of inquiry. That loss of focus creates a fundamental barriers to successful internationalization.

The result of this first barrier is a misconceived definition of the educational task. Most administrators of business schools conceive of internationalization as a curricular matter (possibly one brought to their attention by the American Assembly of Collegiate Schools of Business) and not as a strategic response to a major environmental change. Internationalization activities are defined in terms of the educational experience of the student, with an emphasis on instruction only on the international aspects of particular functional disciplines.[1]

As my background is in international finance, my first example will be drawn from that discipline, which can briefly be defined as the study of the impact which the existence of multiple nation states have on the finance function. The primary organizing principle of most approaches to the field is the description of the differences or complicating factors in finance arising from the international environment. Because traditional finance theory assumes a one currency world, a significant effort is needed to explain the financial management significance of multiple currencies, an explanation that is necessary because of the existence of multiple nation states. A second significant effort, described under the heading of "political risk," is needed to explain the impact of multiple legal jurisdictions on the investment process. Other issues, such as capital budgeting, capital structure, and performance evaluation, are normally considered as extensions of "domestic" finance.

If one's objective is simply to provide this discipline-by-discipline perspective, at its most fundamental, the educational emphasis becomes one of providing "internationalized" functional field courses; by the term "internationalized" I mean a course in which some international material is infused into the traditional functional curriculum. At the conclusion of such a course, the student is aware that there is an inter-

national dimension to the practice of business, as it relates to that particular academic discipline.

The second order implication of the concentration on the international aspects of a particular discipline is that courses, texts and supporting materials are developed which try to provide a more complete treatment of that component of a specific managerial function. Most business schools have specialized courses in international business finance, or multinational financial management, or some similarly titled course. These courses provide, as we have seen, analyses of precisely how multiple nation states impact a particular business function.

An alternate approach is to assemble the international basics of each functional discipline into a particular course, nominally entitled an introduction to international business. Within that course are found modules which deal both with the theoretical basis of international business and the international dimension of a particular subject. Such an approach has the advantage of allowing for the treatment of the domain of international business with extensions as appropriate into its applications in the separate disciplines.

THE MISCONCEIVED ORGANIZATION: ASSEMBLING INTERNATIONAL BUSINESS EXPERTISE

There are essentially two fundamental approaches to assembling international business expertise. One is to concentrate on developing expertise regarding the core theories of international business. Individuals with this core expertise develop or may bring with them a specific functional orientation

Organizationally, the second and most utilized approach has been to develop expertise on the international dimension of a particular functional discipline, and assign or link faculty to that specific functional discipline. That expertise is deployed in both "internationalizing" the existing academic courses in the discipline and then, where there is sufficient complexity, developing "international expertise" courses in that discipline. The institutional benefits of such a strategy are clear. First, there are large numbers of faculty who are competent in many of the various academic disciplines. With some cost and effort these individuals can be "internationalized"; that is, given sufficient background in the specific international components of their discipline to allow them to teach both the basic international perspective functional course and the international expertise course. Secondly, the approach of "internationalizing" faculty allows them, at least initially, to retain their existing functional relationships. They still attend their functional disciplinary academic meetings, still remain current in the functional literature, still are attached to their existing functional department, still continue their research endeavors with the same functional methodology, and most importantly still relate to a functional peer group with regard to promotion and tenure.

Internationalizing functional faculty essentially forces internationalization of faculty research activities to be responsive to the editorial policy of the leading academic journals in the field, which may not rate research on international business theory or the international dimension of the particular field as useful. If faculty are granted promotion and tenure by a vote of functional field peers, they must respond to the criteria set for promotion and tenure by these peers, and the interpretation given to these criteria by these peers. These peers, in turn, have no incentive to give value to research of an international nature, as they have a possibly limited understanding of the issues which are of importance in the international dimensions of their field, and no interest in seeing issues on which they are not themselves adept take a more prominent role in the research agenda of the discipline.

Although this paper does not have as its objective the development of a position on the institutional barriers to international research, to the extent that the individual faculty member's research agenda controls that individual's willingness to participate in other international activities, and in particular the acquisition of knowledge which can then be incorporated into educational products, to that extent a fundamental institutional barrier to internationalization has been erected which accounts for the apparent opposition of otherwise brilliant individual faculty members to internationalization. If research in the international dimension of one's own discipline is not rewarded because it is not publishable, it will not be done, or only done peripherally. If teaching the international dimension of a functional field does not contribute to the research objectives of the individual, or to the research objectives of the discipline, it may be slighted.

My experience at a research active university would suggest that the entire model of functional internationalization is fatally flawed in a university environment which emphasizes research. Faculty are hired into the functional fields on the basis of their research potential from doctoral programs where little or no attention has been paid to the international dimension of the functional field, and none whatsoever to the theoretical analysis of global business. Teaching is rewarded from a stylistic perspective; limited content monitoring takes place, and limited feedback as to the appropriateness of content is provided.

The fundamental flaw in the internationalization approach is that the kernel of international business theory is never reached, either for research or instruction, because there is no one whose responsibility it is to reach it. Cross disciplinary research, which might lead one to that kernel, is neither encouraged or rewarded. No one is assigned to teach that fundamental theory, and many do not even believe it exists. Even if a functional faculty member thought that such a theory existed, there is no incentive to begin the intellectual quest to discover it.

Further, while programmatic efforts exist to support the classic internationalization effort, there is very little available other than doctoral programs in international business which would provide an individual with sufficient mastery of the core material in international business.

Thus the internationalization approach is an administratively convenient mechanism which minimizes the effort to satisfy external calls for a globally relevant business curriculum. For the functional field faculty who must master the international dimension of their subject so that they may teach it at a sufficient level of sophistication, the development of an international perspective in fact may provide substantial personal growth and intellectual rewards, but little career or institutional advancement results.

BREAKING THE BARRIERS

What is the alternative to the functional field recipe for failure? In my view business schools must put in place a critical mass of faculty trained in the theory of international business, whose primary organizational attachments are independent of existing functional units on campus. That group should be assigned the task of developing a curriculum which provides a comprehensive treatment of international business, both the core theory and the functional applications of that theory. The latter task may indeed be shared with functional specialists with an international interest, but the organizational structure should utilize the international business conceptual domain as the primary organizing principle rather than rely on functional disciplines alone.

Such an approach is the methodology of organization used at the University of South Carolina. As early as 1976 a separate International Business faculty group was created; two years later, when the college decentralized promotion and tenure recommendations to faculty units, it received separate status as a promotion and tenure unit. The International Business area was provided with organizational responsibility for all international courses in the curriculum except international accounting and international economics, and was indeed given responsibility for all aspects of the international business masters program.

In evaluating the success of the South Carolina program, it seems clear that the consolidation of faculty into an international group has been successful despite our failure to accomplish (except at the doctoral level) the full and comprehensive treatment of theoretical issues in international business. Rather, we have been able to provide well developed international dimensions courses, developed by faculty generally trained in the conceptual domain of international business. Partially because of the haziness of that domain, we have not yet completed our primary mission, which is to incorporate that conceptual domain completely into the master's and undergraduate curriculum. To that extent we will benefit greatly from the refinement of the conceptual domain accomplished in this conference.

In one sense, through the M.I.B.S. program, we have indeed incorporated at least the primary concepts of the conceptual domain of international business into the academic experience. Rather than treat that domain explicitly, however, we have "infused" it into our international dimensions courses. That infusion is possible because those who teach these courses are selected from an international business faculty well trained in those core concepts. Further, each student in M.I.B.S. takes all the basic international material, and international business area faculty within the program work to coordinate and relate these internationally oriented components.

The University of South Carolina experience has primary relevance for research active, relatively highly departmentalized business schools. The provision of a cadre of internationally oriented faculty, trained in the core discipline of international business and trained to apply that core discipline to various functional fields, overcomes the primary barriers to internationalization at such a school. It concentrates international academic expertise in such a way as to attract faculty from the limited pool of qualified international business faculty. Rather than drawing away from the research activities of these faculty, international activities support research agendas. Organizationally, faculty are placed where their expertise can be brought to bear on both programmatic and research initiatives, and the results of such activities valued for career advancement. The early managerial competence for our international master's level program and the initial leadership and idea generation for almost all of the major international initiatives of the college have all come from that group of experts, who have applied that expertise in the management of the international activities of the college. Finally, I have elsewhere demonstrated that a creative strategy of internationalization can generate substantial financial resources, overcoming in the long run the real barrier of expense (chapter 2, 62–79).

SUMMARY

To summarize my fundamental conclusion, the institutional obstacles to international business education are not operational but conceptual. Almost any clever administrator can discover ways to overcome the institutional barriers to the operation of an international business degree program. It is far more difficult to overcome a flawed conceptualization of international business as an academic discipline, and a flawed definition of the internationalization process. At the heart of the delay in providing the business community with managers prepared for a truly global operating environment is a fundamental misconception of the educational task which is set before the university.

Too long have we labored at merely adjusting what we do in managerial education to somehow accommodate the verity of business globalization. Enshrined under the heading "internationalizing the curriculum" is an administratively facile, resource saving game plan for failure.

Essentially, what is required is an alternate model of business education which incorporates a proper perspective of the domain of international business inquiry, and a corresponding revision of institutional structures congruent with that model. At research active business schools, the recommended prescription for success is to group a critical mass of international business experts and utilize them to develop programmatic and research international initiatives. While such a prescription may indeed be difficult to implement, it offers the only proven method of generating a radical shift in institutional direction.

When corporate executives and faculty from other business schools encounter the M.I.B.S. program and other international initiatives of the University of South Carolina, it typically does not take very long for the inevitable question to arise: why South Carolina? In part, the answer to that question is that the group who designed the M.I.B.S. program developed just the right technology for international business education. Just as important, however, is the answer that the University of South Carolina did not tie its future to the then conventional wisdom of gradual internationalization and matching organizational gradualism. Rather, to borrow a phrase, we successfully executed an international "great leap forward" by simultaneously matching to quality program design an organizational structure which gave free rein to the talents of my colleagues in international business. While no specific institutional structure can be recommended for all academic environments, for too long the concept of assembling international faculty as a group has been dismissed out of hand as either too expensive or too experimental. The experiment works.

NOTES

1. See Arpan (1991) for a treatment of the interactions of curriculum and administrative structure, including a classification of the objectives of curricular internationalization.

■ ■ ■

Institutional Structure for International Business Education

Lee H. Radebaugh

INTRODUCTION

The purpose of this paper is to discuss the institutional issues that influence international business education. In particular, I will examine different ways that universities include international business from an organizational perspective, as well as ways that universities include international business within the curriculum. The paper will be organized around the following key themes: the importance of a mission statement, educational philosophies, organizational issues, and organizational alternatives.

MISSION STATEMENT

Before dealing with educational philosophy and organizational structure, it is important to develop a mission statement that relates to international business. That allows the university to determine whether or not international education and research programs are consistent with the overall objectives of the university.

Three university groups that must be considered in terms of mission statements are the university at large, the business school, and the international business faculty. To illustrate the importance and impact of these mission statements, I will use BYU as an example.

On the entrance to the BYU campus is the phrase, "The World Is Our Campus." The David M. Kennedy Center for International Studies has the major responsibility for charting the international course of the university. The Kennedy Center is a cross-disciplinary center that involves faculty members from various departments at the university. The Kennedy Center does not

have any faculty slots so it relies on the various departments, such as Political Science, to provide it with key faculty members. It is responsible for establishing crossdisciplinary programs such as master's degrees in International Relations, Latin American Studies, Asian Studies, and so forth.

Any mission statement is going to reflect the characteristics of the institution as well as the leadership of that institution. BYU is unique because of the bilingual nature of its student body, which is 30% for all students, rising to 87% for the M.B.A. program. These students, most of whom are members of the Church of Jesus Christ of Latter-Day Saints, the sponsoring institution of BYU, have served two-year missions for the church; and many of those are foreign language missions. In 1991–92, for example, more than 1500 students were conversant in Japanese as a result of missions to Japan. In addition, although the largest population for the church resides in North America, the fastest-growing segment of the church population is outside of North America, especially in the developing countries. As a result, the student body at BYU is becoming increasingly international, a trend that will continue into the future based on the demographics of the church.

The mission statement of the Kennedy Center reflects this international background at BYU:

> The David M. Kennedy Center for International Studies has been established to support the global mission of Brigham Young University and its sponsoring institution, the Church of Jesus Christ of Latter-Day Saints. In supporting that mission, the Kennedy Center fosters love of God

through a genuine concern for our neighbor's welfare, encourages the pursuit of truth through a variety of disciplines, and strives to make international activity and experience one of BYU's pinnacles of excellence. Utilizing the university's unique international strengths, the center seeks to:

Prepare students for responsible and constructive careers in a world that increasingly requires international awareness and intercultural skills. Develop superior programs in teaching, research, and scholarly publications in international and area studies. Emphasize the fundamental importance of service as a means to promote peace, friendship, and dignity, and to combat poverty and degradation throughout the world. Facilitate collegial and productive exchanges between representatives of nations, the Church of Jesus Christ of Latter-Day Saints, and Brigham Young University.

Although it would seem obvious that the Kennedy Center has a unique international mission, its mission broadly covers education, research, and service.

The mission statement of the Maraud School of Management of BYU (MSM) is not as internationally comprehensive as that of the Kennedy Center, but international was included specifically. One of the three broad missions of the MSM is as follows: "To educate students with ethical values, management skills, and leadership abilities needed in organizations worldwide." The mission statement continues to identify how the MSM will accomplish that mission by identifying several major areas of excellence. One of those is as follows:

Educating principles for students who are dedicated to personal integrity and a strong work ethic; who have a mature view of their responsibilities to serve society in the work place as well as in the family, community, nation, and the world; and who will demonstrate: . . . a global perspective and an ability to work effectively in diverse environments and cultures.

Finally, the mission statement states that "faculty of the MSM will, through teaching, research, and service, improve the quality of management education in the world."

The importance of the mission statement is not just the words but the process in which they were developed. When Fred Skousen became dean of the MSM in 1989, he appointed a strategy task force to work on the mission statement. Several meetings were held with the faculty to discuss the mission statement at different steps along the way, so the entire faculty was well aware of what the dean and the strategy committee were thinking. By the time the mission statement was adopted, it was clear that one of the areas of focus of the MSM in the coming decade would be the international area. In addition, several task forces were established to look at different areas in more depth, and one of those was an international committee. Several faculty meetings were held to discuss the results of these task forces, so a clear picture began to develop on the international thrust.

Finally, a global consulting firm did a study for the Marriott School of Management on its competitive advantage. The firm concluded that international is an area of competitive advantage that needs to be enhanced. As a result of these experiences, the MSM decided to push ahead with international as one of its major areas of focus. The intensive level of discussion by the entire faculty was one of the major reasons why it has been possible to generate significant interest among a broad range of the faculty.

As noted by Blake (1992), however, "more often such a [mission] statement is lacking, or the school's implicit approach provides little guidance to faculty or students about what can be expected in the educational process." The mission statements described above are rather vague. However, that must be the case. Mission statements cannot be changed frequently. The value of the above mission statements is that they allow broad recognition of the principles of internationalization. They also allow for creativity and for program evolution.

A major concern, of course, is that we have not developed a mission statement for the international business program within the MSM. We have developed ad hoc mission statements for each new thrust within the international area, but we have not looked specifically at what we want to accomplish and developed a mission statement to capture that. It is important that the international business group at any university develop a sound mission statement as well.

BASIC EDUCATIONAL PHILOSOPHY

As noted by Blake (1992), there are three different approaches to educational philosophy: the multidomestic approach, the specific functional discipline approach, and the global approach. Blake points out the difficulty schools have in determining the correct approach: "Most schools have a haphazard blend of each of these objectives, more the result of accident than planning . . . there often is not a clear approach or philosophy that guides what the curriculum should achieve and what students will learn."

I believe that it is a mistake to look at these three approaches as discrete and mutually exclusive. Programs tend to evolve based on the mission of the institution, the nature of the students, the international experience and expertise of the faculty, and the market. I will now look at this evolution in three different phases.

Universities, like companies, evolve into the international area. The first phase is to ignore the international dimensions of business, with the exception of a few functional courses and a limited amount of international content in existing courses. However, the latter characteristic is more accidental than strategic.

This situation cannot last forever because neither students nor the external public will allow it. This became evident with the change of wording of the "common body of knowledge" by the Accreditation Council of the AACSB in 1974:

IV. CURRICULUM
The purpose of the curriculum shall be to provide for a broad education, preparing the student for imaginative and responsible

citizenship and leadership roles in business and society—domestic and worldwide.

Those last three words required business schools to move out of phase 1 and join the international community, or face losing or not gaining accreditation. However, there was not a clear interpretation of how schools were to accomplish that.

Phase 2 involves the recognition of more international business in the curriculum but without a clear strategic thrust. The easiest way to include the international is to provide a series of elective international "functional" courses and maybe limited inclusion of international issues in some of the core and noninternational elective courses. However, the absence of a clear strategy will doom phase 2 to mediocrity.

Phase 3 is both reactive and proactive. It is reactive in that it involves developing a program that is consistent with the university and college mission statements, the background and strength of the faculty, and the nature of the student body. It is proactive in that it might involve changing the mission statements to include a stronger international thrust, developing a faculty that is more internationally oriented, and attracting a student body with skills more suited to the international environment.

It is difficult to determine the optimum approach because university environments are very different from each other, and programs tend to evolve over time due to experience. Blake's three pedagogical views of international business are an interesting starting point for discussion. The multidomestic approach deals with differences in doing business in different countries. The functional approach involves educating students to become international business specialists, more than likely in a particular business functional area. The global approach educates students to be effective managers in a global business environment, whether in their own home country or working in a foreign environment. However, it is not country-specific as is the case with the multidomestic approach.

There are many different ways to deliver international business education, from programs that focus on international business to those that include international business in the context of the traditional business program. I will provide the latter perspective rather than the former. Clearly, programs develop within the constraints of the university while attempting to change those constraints to meet the needs of the future. At BYU, the current assumption is that we will not have a separate international program with a separate faculty. The program is developing within the context of the general undergraduate and graduate programs.

Another basic assumption that we have had for the past decade is that we do not want to have a specific international business major. At the undergraduate level, we are trying to minimize the number of courses that students take within the MSM, and we also feel that students should have strength in a specific functional area. For several years, we were much more like phase 1 or 2, with a few functional international business courses and a broad introduction to international business course for nonbusiness majors and business majors who might only want to take one international business course. However, it is also clear that we are looking at specific niches that might move away from this traditional approach.

In the early 1980s, the program began to take slightly more shape. At the undergraduate level, we established a coconcentration in international business. The coconcentration involved two different dimensions: foreign language fluency and required international business courses. The language requirement resulted from the high percentage of business students with foreign language fluency. At the undergraduate level, 40–60% of the business majors were fluent in a foreign language, usually having lived abroad for two years and taken three years of foreign language at the university level.

In terms of coursework, students were required to take a general international business course and a course in international finance or marketing. The students' transcripts actually stated that by fulfilling the requirements of the coconcentration, they would graduate in international finance or international marketing. However, the level of international coursework was rather limited in comparison with a major in international business.

At the graduate level, the coconcentration was also established. However, the level of foreign language expertise and interest was significantly higher: during the decade of the 1980s, between 80% and 90% of the students were bilingual. The coconcentration in international business required foreign language fluency and a series of three courses in international business. As was the case with the undergraduate level, the purpose of the coconcentration was to provide functional depth rather than managerial breadth. In addition, the absence of a requirement of international business for all students assumed that only a few would be prepared for careers in a global environment.

In addition to the coconcentration at the M.B.A. level, we also introduced a joint M.B.A.-Master of Arts in International Studies for students desiring to concentrate more specifically in international business. The M.A. degree is administered by the Kennedy Center and it provides a broad focus, such as an M.A. in International Relations, or a specific focus, such as an M.A. in Asian Studies. The M.A. degree also requires a master's thesis. The joint program usually involves only a few students a year.

A major change that took place in the mid-1980s was an expansion of the number of faculty more sensitive to international issues, a trend that needs to take place in all Schools of Management. This involved the addition of some key new faculty in the functional areas who were not international functional specialists, but who were very sensitive to global competitive issues and who also had the stature among the faculty to be listened to. Up to that point, a small core of faculty were carrying the international banner but with limited support from the rest of the faculty.

As we began to discuss curricular changes, it became obvious that international business was increasing in importance. The focus at that point was not on functional specialization but the recognition that all students need to be aware of global competitive issues so that they can function effectively in a global environment. As a result of these discussions, two important changes took place in the curriculum. At the undergraduate level, a new required international business course was added to the curriculum. Even though it is only a one-credit-hour course, it is an important introduction early in a student's academic program and helps set an international tone for the rest of the program.

At the graduate level, a required international business course was also added to the curriculum of the first year of the M.B.A. program. It is a broad course that focuses on global competitiveness, macroeconomics, and country analysis. The course focuses on key areas of the world rather than a specific country. The purpose of the course is to develop a global mind set.

The next level of development follows the multidomestic approach discussed by Blake. Although the functional and global approaches described above are the foundation of the international efforts at both the graduate and undergraduate levels, there is also a strong and relatively broad effort within the college to develop a series of country specializations. This interest has been sparked for three reasons: the development of strategic alliances, a perceived competitive advantage of students based on country and language expertise, and the internationalization of the faculty.

The first major strategic alliance involved Moscow State University. Historically, BYU has tended to establish facilities around the world for study abroad programs rather than enter into alliances with other universities. However, the relationship with Moscow State University, which was initiated by a member of our faculty

from Finland, caused us to think about what these strategic alliances should look like and how we could use them to further our international program. As a result of this experience with Moscow State University, we are in the process of developing strategic alliances with universities in Germany and Korea. The second reason listed above is the perceived competitive advantage of our students in terms of language and country expertise. Even though the largest number of students with bilingual fluency at BYU are in the more common languages of Spanish, French, and German, there is a surprisingly large number of students with fluency in Japanese, Korean, and Chinese. In addition, within five years BYU will have a large number of students with fluency in Russian and other former Soviet and East European languages backed up by two years of living experience abroad.

The first language that caused us to think of a multidomestic focus was Japanese. Due to the efforts of a new hire in business strategy with significant experience in Japan as well as Japanese language fluency, we are moving aggressively to establish a specialty in Japanese within the M.B.A. program. The program would require significant fluency before entering the program and will involve extra courses on doing business in Japan, some of which will be taught in Japanese, and an internship in Japan of six months to one year. The current planning of the program is also causing us to take a serious look at the structure of the M.B.A. program and how we might make it more flexible to allow the inclusion of additional language experience.

The third reason listed above is the increasing internationalization of our faculty. One of our recent faculty hires has had significant teaching and research experience in Germany, especially former East Germany. He was the director of the university's Vienna study abroad program during fall semester 1990 and followed that with a Fulbright award to conduct research at a German university during the first six months of 1991. When he returned to BYU, he was put in charge of the development of a German interest group that will be discussed in more detail in the organizational section.

These multidomestic efforts are not expected to be large programs. The goal is to design a program that involves language, country-specific courses to complement the broader global and functional courses, and internship and study abroad experiences. We hope to recruit a few outstanding students for each country focus in order to maintain a high level quality graduate. Although we are initially developing programs for Korea, Japan, Germany, and Russia, there are a number of other countries, such as Brazil and China, that are being considered for the next round.

It is evident from the above discussion that I do not favor one specific approach at the exclusion of all others. I agree with Blake that business schools must consider educating its students to compete in a global marketplace, whether or not the students are interested in working abroad. However, I also feel that students will get a job because they can do something well, not just because they speak a language and understand general international business. That is why any program must include international functional courses. In addition, I believe there is a place, especially in a graduate program, for specific country tracks, especially where the university has a competitive advantage due to unique faculty or student capabilities.

ORGANIZATIONAL ISSUES

As Blake points out, there are several ways to organize the international effort within the business school: the focused international business program, the international business department, and a matrix that brings together faculty from different departments. I would like to expand on these issues in the next section, but first it makes sense to discuss some key organizational issues.

At BYU, we have struggled with the decision on whether or not to recommend the establishment of a separate international business department, or a department that includes business strategy and international business. Whatever the final solution will be, there are clearly four areas which create organizational challenges:

1. Financial resources,
2. Faculty hiring and development,
3. The development and control of the international program; and
4. The determination and scheduling of classes.

The issue of financial resources is not trivial, because any area needs resources to survive. We have been able to establish a good international business group within the MSM because of a grant from the U.S. Department of Education. BYU, in partnership with the University of Utah, is one of the new Centers for International Business Education and Research (CIBER). With additional resources, we have been able to capture the attention of the faculty. Thus the allocation and management of financial resources is key.

Faculty hiring and development is the second major issue. When academic departments have control over their faculty slots, the international area has a difficult time hiring faculty with international expertise. In addition, the institution needs to determine how to evaluate performance, mentor new faculty members, promote and retain faculty members, and develop a sense of collegiality among the group.

The third major issue is the development and control over the international program. The discussion above of the development of a multidomestic strategy at BYU is a good example. It is important for the international group to have the power to develop ideas which it can bring to the faculty and for which it can get resources from the dean or from external sources.

Finally, it is important for the international group to have the power to determine what courses need to be taught for a strong international focus, when they will be taught, and who will staff them.

ORGANIZATIONAL STRATEGIES

Good plans still need implementation, and organizational structure is a difficult issue to resolve. The difficulties arise from a variety of sources. The first is the history and tradition of the business school. Departmental groupings vary from university to university based as much on historical accident as on any kind of strategic focus. For example, the international business group at Penn State University for many years was located in the Business Logistics Department, whereas it is located in the Management Department at Indiana University. There is no logical reason for preferring one department over another.

Another difficulty lies in the nature of the faculty themselves. In general, faculty tend to have a functional discipline first and an international background second. Of course that is not true in all cases, but it is generally true. Thus it is difficult for the faculty member to leave his or her functional department and move into an international business department.

Blake points out that schools are usually organized in one of three ways: as a focused international business group, as an international department, or in some form of matrix. An additional approach is to establish the international group completely inside a specific department, such as the management department. As noted in table 4.1, there are few universities that choose to have a separate international department. In most schools, international faculty tend to be housed in different departments, which implies some form of matrix organization must be taking place to coordinate efforts. As also noted in table 4.1, that was the case for the Academy of International Business Curriculum Surveys in 1974, 1980, and 1985.

Until recently at BYU, the international business focus was located in the Business Management department, one of six departments in the college, along with most of the faculty with a strong international interest. In terms of the leadership of the department, the international group was lumped together with the business policy group. There was no separate budget allocated to the international effort, and faculty slots remained in the functional disciplines of marketing, finance, policy, and operations. The only way to hire a person with an international background was to convince

Table 4.1 Handling of the International Business Programs

	1974 Daniels & Radebaugh	1980 Grosse & Perritt	1985 Thanopoulos & Leonard
In an International Business Department	4.1%	9%	5%
In another Department	24%	19%	25%
International Business material is incorporated into functional courses	**	29%	67%
Functional faculty teach specific International Business courses	**	40%	64%
Total responding schools	271	227	384(*)

(*) Multiple responses are possible for the 1985 survey.
Source: 1980 Survey (page 188) and present study.

(**) The same questions were not asked in the 1974 survey.
Sources: International Business Curriculum Surveys, 1974, 1980, 1986.

one of the functional groups that the candidate would help in both the functional and the international area. Although the policy/international group coordinator worked with the international group to staff classes, the major power was with the functional group. Functional classes were always staffed first, and international classes were staffed last. That led to high instability in course offerings and the frequent use of adjunct faculty rather than the full-time faculty to teach the international courses.

However, three things have caused the dynamics to change in the power of the international group. The first is the broadening of the international group to include more strong functional faculty from within the department as well faculty from other departments. Now, the international group includes at least two faculty members from each of the departments in the MSM. As the faculty base increased, so has the power of the group.

The second is the wide dissatisfaction of the existing international faculty with the status quo. The absence of power, autonomy, and control has cost the international group over the years. This has forced a recognition that international course offerings are just as important as functional course offerings. Faculty are allowed to decide how they want to allocate their teaching load. However, the system is far from perfect.

The third is the arrival of the CIBER grant from the U.S. Department of Education. That grant has forced the international group to look carefully at their activities and to focus their efforts more. The funding has also helped give the group a little more autonomy.

At this point, the MSM has opted for the matrix approach rather than a strong departmental approach. However, many of the issues discussed above are causing some problems. It is almost impossible to hire faculty in the international area unless they have a functional area that fits the hiring needs. Few resources have been forthcoming from the functional departments, so the

group has had to exist as a result of CIBER resources. In addition, we still have to fight with departments over the scheduling of courses and the assignment of faculty.

At first, the group was meeting as a committee of all faculty with an international interest. However, that has proven to be unwieldy due to the large numbers and diverse interests within the group. We have now decided to establish an executive committee and to break down the rest of the group into areas of interest. In addition to functional areas of interest, we have interest groups in crosscultural communications and in countries. As mentioned earlier, the initial country areas of interest are Japan, Korea, Germany and Russia.

The German group so far is the best organized. There is a group of twelve faculty members out of 115 full-time faculty with significant experience in Germany and German language fluency. In fact, many of the meetings are actually held in German to help strengthen the language fluency of the faculty. That group is responsible for designing a German focus within the M.B.A. program. Initially, they are concentrating on three areas: coursework, language instruction and experiences, and a strategic alliance with a German university. To help in the language area, a course on German business language for graduate students will be offered in 1992–93 for the first time. Students must already demonstrate significant language fluency before enrolling in the course. Similar courses are being offered in Spanish, Japanese, Korean, and Chinese. In 1993–94, courses in additional languages will be developed.

The matrix form of organization has many advantages. As noted by Bartlett and Ghoshal (1990) in their discussion of the matrix form of organization employed by multinational corporations,

> Its [the matrix form of organization] parallel reporting relationships acknowledged the diverse, conflicting needs of functional, product, and geographic management groups and provided a formal mechanism for resolving them. Its multiple

information channels allowed the organization to capture and analyze external complexity. And its overlapping responsibilities were designed to combat parochialism and build flexibility into the company's response to change.

In the university setting, since we draw on faculty members from several departments, we are able to keep at least one faculty member who has a strong international interest in each department. That faculty member can be conscience of the rest of the department and also serves as a liaison with that faculty group. That important source of communication in department meetings can help spread information about the activities of the international group and the CIBER. If the faculty members were outside of the department in a separate department, those natural lines of communication would be broken.

However, the matrix form also has some disadvantages as mentioned above. As Bartlett and Ghoshal point out, "the matrix proved all but unmanageable—especially in an international context. Dual reporting led to conflict and confusion; the proliferation of channels created informational logjams as a proliferation of committees and reports bogged down the organization; and overlapping responsibilities produced turf battles and a loss of accountability."

In the university context, the disadvantages include no separate resources except for what is raised from external sources, no faculty slots or power in hiring decisions, the need to bargain for faculty members from different departments to teach courses, no voice in the college executive committee, and marginal influence in the individual departments.

In order to move the internationalization process along faster, it is important to develop strategic alliances. These alliances need to be formed both inside and outside of the university. Those outside the university include strategic alliances in the United States and abroad.

At BYU, we are working with both the language departments and international studies programs inside the university. For example, we recently took the dean of the MSM to Japan and Korea to meet some of our best contacts and to help us establish new contacts. His presence added a lot of prestige to the visit and allowed us to meet with some people that would not have been possible without him and the status of his position. To assist in the Japan effort, we paid to have the chair of the Asian and Near Eastern Languages Department, who is a Japanese specialist, accompany us. We were able to share a number of good contacts with each other, especially for internship possibilities. In addition, we were able to spend significant time away from campus discussing with him a Japanese focus within the M.B.A. program. It was as excellent trip and generated a fruitful interchange of ideas.

The Kennedy Center, the home of our international studies programs, is also an important organizational unit within the university because it is charged with defining the international mission of the university. We are able to cooperate with each other in the joint M.B.A.-M.A. degree, the development of internship programs, and the development of study abroad programs. We have used our contacts at Moscow State University to establish a one-week travel study program for BYU students enrolled in the Vienna and London Study Abroad programs, none of which are business majors. We are also cooperating in the development of a broader alumni network involving graduates of the business school, the Kennedy Center, and the law school. These internal alliances have been very helpful to all parties involved.

There are two types of external strategic alliances that are part of our center thrust. One involves cooperation with the University of Utah, and the other involves the development of joint programs with foreign universities.

In the case of the University of Utah, we have a joint CIBER that is working together to develop courses, joint faculty research projects, and programs for the local business community. By pooling our resources, we are able to share faculty, discuss curriculum and research ideas, and pool our contacts in the business community.

Foreign strategic alliances are best exemplified by the German focus group, which is responsible for defining the nature of our relationship with the Technical University of Braunschweig. The Center Executive Committee has broadly defined the nature of relationships with foreign universities, but each country focus has significant leeway to determine what their specific relationship will include. We are concentrating initially on faculty and student exchanges, assisting in the establishment of internships for students from each university, and the development of joint research projects. Because the relationship with Technical University of Braunschweig is progressing so well, it will be a model for our work with other universities, along with our experiences with Moscow State University. We have begun preliminary talks with Seoul National University in Korea and with universities in other countries.

Another important dimension to our organizational structure is the use of business advisory boards. Unfortunately, we have too many business groups that we are working with and getting input from, but that does give us a diversity of viewpoints. We have an alumni board with an international committee, our MSM National Advisory Council with an international committee, the joint BYU–University of Utah Center Advisory Committee, and center representation on the local World Trade Association and District Export Council. These advisory boards are able to give us valuable insight into the attitudes of the business community toward our business programs. Sometimes they tend to be a little proactive, but they are still extremely valuable.

Although the matrix approach is useful at this point, we are still struggling with our organizational identity. There are two different and relatively strong opinions as to how we should organize. There are some faculty members who moved away from

their initial functional expertise and became much more interested in international business and/or business strategy. They would argue for a separate department in order to get access to resources, faculty slots, and greater cohesion and control. There are others who are comfortable with the matrix approach so that they can retain their allegiance to the functional departments but who would argue for greater power in making decisions. There is no easy answer.

The best way to internationalize the curriculum is to internationalize the faculty. We are proceeding slowly in that area, but we are making progress. For the past two years, we have sent four different faculty with little international experience to accompany our Executive M.B.A. students in their international trip. That has proven to be a valuable experience to those faculty in finding ways to include international issues in their traditional courses. We also fund faculty to conduct international research and to travel to foreign countries to present papers and attend conferences.

We are now working with the functional groups to determine how they want to internationalize themselves and their courses. The key is to present a menu of possibilities to the faculty groups and let them pick something that makes sense for them.

SUMMARY AND CONCLUSIONS

There is no clear and unique way to prepare students to be competitive in the global economy. We have chosen a very flexible approach that includes all three philosophies of international education: the multidomestic, the functional, and the global approaches. We believe, above all, that all students must have enough background to consider global issues in all business decisions. However, we realize that functional and country expertise is also valuable for a smaller subset of students.

Organizational structure is still problematic. Although we have opted for a matrix approach to bring together faculty members from various disciplines, we still have major issues that need to be resolved. However, the matrix approach has allowed us to include in the international effort faculty members that would not leave their disciplines to become part of an international department. A department would help a small core of international specialists, but we would still have to adopt a matrix approach for the rest of the faculty. There is clearly no easy answer.

■ ■ ■

INSTITUTIONAL STRUCTURE AND IB EDUCATION
Yair Aharoni

This session deals with the degree of functionality of the institutional structure to IB education. The authors of the papers are expected to examine institutional issues and organizational issues as well as ways that universities include international business in their curriculum. It is not expected to define missions and identify the conceptual domain of the IB field: These topics are discussed in other sessions in this conference. Yet since strategy and structure are interrelated, a major question to be dealt with must be the strategy of the IB education. The objectives of teaching international business must be clarified. Since these objectives are diverse, different schools will have different strategy and therefore different structure. Currently, most business students in the United States are taught to view business problems and opportunities through a narrow, U.S. lens instead of through a kaleidoscope of international issues. The reasons for this myopic focus must be understood and the means of changing this narrow attitude must be analyzed.

The three papers I was asked to discuss are very different in their implicit assumptions about objectives. Each presents a different philosophy, perhaps influenced by ideas prevailing in the school from which each comes. Professor Lee H. Radebaugh of

Brigham Young University quotes very explicit statements of mission. According to its mission, "the Marriott School of Management wants to provide students with management skills, ethical values, and leadership abilities needed in organizations worldwide." Professor William R. Folks of the University of South Carolina stresses that international business has a unique body of knowledge. This body of knowledge is related, in his view, to the impact the existence of multiple nation states has on managerial practices in a multinational firm. Therefore, when Folks illustrates his views by an example from the field of international finance, he excludes the possibility of cultural differences that have an impact on the ways the finance function is carried out in different countries. Clearly, there is a significant difference between teaching how organizations operate in different parts of the globe and how an organization is impacted by crossing national borders.

Dean David H. Blake emphasizes the richness of diversity. He proposes several reasons for the variation in organization of international business. First, he claims, the field is still emerging with different and changing perspectives about the field. In addition, the number of faculty of IB in most schools is small. They are, therefore, placed into an area based on convenience. (Folks agrees that there are very few available IB faculty, but feels a critical mass is essential). Further, the field is said to have a dual personality, being both a discipline and an all embracing context of global management. Blake also points out that different schools have different objectives: some attempt to produce technical experts on foreign countries and cultures. Graduates are expected to live and work in foreign countries. Others seek to create expertise in one or another IB function. Others want all their students to have a basic understanding of IB, stressing the global nature of business. Blake would lean toward an approach of globalizing both curriculum and faculty, stressing that professors in each and every traditional discipline must be encouraged to internationalize their traditional courses, not necessarily become scholars of international business. Blake leaves unanswered the question of the place in which the scholars of international business will flourish.

Given the described diversity in objectives and strategy, one would naturally expect a diversity in structure. Blake describes in detail three archetypal institutional structures. Although he does not do it, it is easy to connect the strategy and the structure. The foreign approach results in the focused IB program. Faculty of these programs self select and may lose their credibility and the respect of their discipline-based peers in other academic institutions. The experts are related to an international business department, often divorced from other departments in the business school but able to protect its turf and the tenure and promotion procedures. The global approach may be related to the matrix structure, in which each member of the IB faculty has also a discipline-based affinity group. Blake would rather see a schoolwide commitment, in which each faculty member is encouraged to globalize his/her part of the curriculum. Although Blake does not state who will encourage he implicitly suggests that this role is to be played by the dean as well as by the faculty.

The problems discussed by the three authors are of course of utmost importance and Blake's classification and topologies are very helpful. Each author makes some very thoughtful and extremely succinct points. Let me draw your attention to Folks's distinction between education in the academic discipline of international business and the internationalization of the business education curriculum. Another distinction must be drawn, I believe, between teaching international business and research in the field.

Folks's solution to the problems discussed by Blake is for schools to put in place a critical mass of faculty trained in the theory of international business, whose primary organizational attachments are independent of existing functional units in the campus. Having a separate status for a promotion and tenure unit indeed solves many problems. I am somewhat perplexed as to the feasibility of such a solution if applied to all business schools given the statement that a major barrier to internationalization is that there are few faculty available that have sufficient international training. Even if availability is no problem, is such solution possible for schools without M.I.B.S. type program and the amount of courses it entails? Radebaugh emphasizes IB as a crosscultural study. His students are expected to gain an ability to work effectively in diverse environments and cultures. He does not agree with Blake's cataloging of three distinct approaches. He proposes a three-stage model of programs' evolution. In the first phase, universities ignore the international dimension. In the second phase, schools do not have a strategy. In the third phase, schools are both reactive and proactive and attempt to develop an international business program. Radebaugh offers a statement of the progress and difficulties faced by his university—and discusses also the organizational implications.

My own bias on the question of curriculum and objectives is very near to that of Blake. I fully agree that for most business schools in the United States teaching and research in international business must be an integrative part of the total business school program. Indeed, my efforts at Duke University in the last five years were precisely in this direction. At Duke, as in other U.S. schools, we have been looking for methods for internationalizing the whole curriculum rather than developing a separate international curriculum. At Duke, we believe in an integrated international curriculum and a global viewpoint interwoven into the whole program. One main difficulty we had to face was to tackle with is the apathy of the existing faculty. Another has been to find ways to foster more research on international business without having an international business department. We believe that internationalization of business curriculum and faculty must address all aspects of crossborder business activity and therefore be taught by the various professions comprising a business school. Such curriculum modifications may easily be wiped out when the faculty is not continuously reminded of the need for awareness of the global dimensions of business. For this to happen, it is not enough to hire a few international business experts. Such a task necessitates a major change in the culture and a basic shift in norms within various professions, be it accounting, finance, or O.B.

Folks's interesting description of his experience at the University of South Carolina led him to conclude that "a group of critical mass of international business experts" is "the only proven method of generating a radical shift in institutional direction." I beg to disagree, based on my experience at Duke. I am sure there are also other possible methods. Indeed, having stated my bias, I would like to stress the right of different schools to diverse views and therefore organizational and institutional structure. I do agree that the problem is one of academic beliefs and norms, not of budget. Integrating the global dimension in each of the core courses in business requires no new faculty resources. It does require retraining of existing faculty so that they can teach the international dimension. Yet, unless there is a change in professional norms, the international dimension may be listed at the end of each course and may be left uncovered because of time pressures at the end of the term, or it may not be listed at all.

Now, in the 1990s, it must be obvious that the world has shrunk, economies are interdependent, and technological development in communications, transportation and information allow almost instantaneous movements of capital, people, technology, information and goods across borders, and business firms compete across national boundaries. Indeed, international business was added as a field of study since the late 1950s. In 1974 the American Assembly of Collegiate School of Business (AACSB) changed its accreditation standards to require that the curricula of busi-

ness schools reflect the worldwide as well as domestic aspects of business. By 1976, deans of 76% of the responding accredited schools reported they had begun to take steps to comply with the new AACSB standard. By the end of 1978, a National Advisory Committee background report noted: "It is our fond hope that business schools practice what they preach. We teach that the business firm should continually scan the environment so as to adapt the strategies of the firm to the environment in which they must operate."

If business schools had followed the same approach, they would have realized more than a decade ago that the business environment which their graduates would enter had changed dramatically. By the mid 1960s it was clear that the large international business firms (be they transnationals, multinationals, or simply international) were becoming a very important force in the international economy as well as in their individual national economies.

The larger companies employ, however, only a small portion of the total number of businessmen involved in the international aspects of business. One must also include all aspects of exporting and importing of goods and services, in addition to the various international licensing agreements, international leasing agreements, short-term investing or borrowing in other currencies, and a wide variety of other activities.

The need to understand the global environment is not restricted to the needs of those 35,000 U.S. businessmen who live overseas, nor just to those who work for the 30,000 U.S. business firms involved in exporting, nor to those who are directly involved in managing the international operations of the 6,000 large companies with direct foreign investments. International business and its indirect effects have become a pervasive factor in the business scene. It is difficult to point to a firm of any size which is not involved in or affected by some aspect of international business.

It seems strange, therefore, that relatively few business schools are responding to this need. Business textbooks continue to be written as though the world stops at the boundaries of the United States. The teaching of international business courses per se began to make inroads in business school curricula in the early 1960s, and by the early 1970s was widespread, but in most schools only a small percentage of students are taking any of these courses (AACSB 1979, 1).

By 1992, it is still true that relatively few business schools are responding to this need. There is a gap between declarations about the recognition of the globalization of product and financial markets, the interactions and interdependence of all open economies, and actual behavior in teaching and research. "Efficient markets" seem to be seen as a parochial U.S. phenomenon, based on its particular institutions. Marketing deals with U.S. marketing, not with global distribution and its peculiar problems and challenges. Organization design largely ignores the complexities of running a firm across national boundaries and across cultures. Further, the functional areas in a business school are not willing to divert a significant part of their existing research efforts to global problems and to generate more internationally relevant research. There is also a reluctance to reduce the U.S.-based course contents requirements, adding international elements or comparative analysis. Most business textbooks continue to be written as though the world stops at the boundaries of the U.S. (although more and more publishers come to the market with supplementary international texts).

Almost twenty years after the AACSB decision on internationalization, most faculty members in most U.S.-based business schools still do not believe theory is different in its international application. Multiplicity of foreign environments is perceived as an interesting aberration, that can be safely ignored. International business is very often equated with foreign business, and therefore is doing business within a foreign country rather than across national borders.

Clearly, almost all faculty members in the United States have not yet adopted a global outlook in their teaching and research and lack any incentives for such an adaptation. Thirty years after many studies have been published on the increasing importance of MNEs and 18 years after the majority of the deans told the AACSB they were working hard to internationalize their programs, there is yet very little progress.

Note that even the three papers presented in this session, despite their title, are restricted to a discussion of their respective topics only in the United States. To be sure, in his paper, Blake mentions European schools with his school and the one at SMU. Yet the organization of international business education in other countries is not discussed at all. The two other authors discuss mainly the experience in their specific schools. Such an omission is very typical of U.S. scholars who after all were born and raised in a country in which the "world" series in baseball consists of only U.S. teams. The omission, however, is very unfortunate. There is much one can learn from schools such as the University of Uppsalla and perhaps from our friends in Asia. In some European schools, for example, an attempt has been made to achieve a global view by purposely having teachers from different countries, teaching in several languages a diverse group of students from many countries. In at least one school in Europe, the international business program is combined with foreign languages departments. In many schools in small countries, international business is an integral part of all or almost all functional fields: teachers in a basic marketing course would deal as a mater of course with topics related to imports and exports, and teachers of finance with questions of international finance. In contrast, in the United States, international business does not carry much respect and most of the brightest young researchers deal solely with domestic issues. At most, international business may have been elevated in the eyes of some faculty from the status of nonexistent, nonscientific and unnecessary field to that of an acceptable nuisance, to be taught by professors with foreign accent, with very little rigorous research and relatively low status.

To achieve any modicum of the needed change in curriculum and research agenda, it is important to understand why internationalization has been less than successful in most U.S. academic institutions. What are the barriers to internationalization? Are they country specific? There are several possible reasons for the deplorable state of affairs regarding teaching and research in international business. My explanation above emphasized university's structure and the way the familiar territory of each discipline is nurtured and guarded. Before reflecting on the implications, other hypotheses may be considered.

One is that we underestimate the time it takes to make a major change. Indeed, Blake is not alone in stressing the newness of the IB field and its "still emerging" status. Yet Raymond Vernon's studies of the multinational enterprise are more than thirty years old (and so is my study of the foreign investment decision process as well as many other works). Mira Wilkins reminds us that Charlie Stewart and George Simmons compiled a 600–page bibliography of IB literature as early as 1964. IB emerging status may be compared to that of the infant industry protected firms in import substitution countries. These firms (and our field) attempt to continue be infant for quite a long time. I do hope our field, unlike the import substitution protected firms, is not perceived as infant because it is unable to face the competition of more efficient competitors (such as more prolific disciplines).

The "time it takes" explanation seems tenuous when the not very satisfactory progress achieved in the process of internationalization of the U. S. business school curriculum is compared to the changes achieved, say, in finance. In 1960, it may be remembered, finance was still largely corporate finance; and mainly institutional in scope. Since then, a stream of researchers shook up thinking by challenging folklore

with rigorous models. These pioneers were able to completely change the contents of the finance course, the culture and norms of the profession and the acceptable research requirements in the field as defined in almost all accredited programs. This change penetrated relatively rapidly to all U.S. and foreign business schools. Further, this change was academically initiated and has been a cause of a change in the financial industry. At least in this respect, the change in finance is very much different than another change since 1960—that of business policy to corporate strategy. The SBU concept, for example, started in General Electric and the portfolio concept was developed in Boston Consulting Group, not in academic business schools. Common to both is the much higher rate of change and of its much faster diffusion across all business schools than for the case of international business.

Data banks are also not an explanatory factor for the difference. Finance may have enjoyed the availability of huge data banks created to study securities prices (and the ease of data manipulation by computers). Business policy has had the same advantage when GE helped establish PIMS. However, international business had a similar fortune: Harvard Business School's data bank on the multinational enterprise developed in the 1960s. Based on this source a stream of Ph.D. theses, books and mimeographs have been published.

What, then, is the reason for the difference between ability to change in finance or business policy and the relative slow pace of change in international business? Certainly, as amply demonstrated in the three papers and in other papers, not lack of demand. AACSB publications (and Folks's paper) point out to the lack of faculty, but the greater demand should have convinced many Ph.D. candidates to move to the study of international business a long time ago. Why is it that such a move did not occur? Despite the efforts of many, there are still very few faculty members that are trained and motivated to teach and research international business issues.

My own explanation is that research has increasingly become discipline-oriented. International business has been perceived as interdisciplinary, and therefore as moving away from disciplines. This by itself has been its major disadvantage, since all professional schools and departments in universities have been unable to absorb interdisciplinary research. My own work on the foreign investment decision process has been fortunately widely quoted by international business teachers, but was rarely read nor influenced those dealing with the decision process.

The difficulty of organizing interdisciplinary teams stands in stark contradiction to the well known fact that many important scientific questions can neither be framed nor answered within the confines of traditional disciplines. The research must be interdisciplinary or at least collaborative. This is true in biochemistry or composite materials, not less in policy related sciences and is absolutely necessary in international business. Yet the structure of universities and the incentive mechanism for researchers impose constraints on the ability of researchers to move to international problems. Most universities would argue that quality interdisciplinary programs must be built on the base of excellence in the disciplinary departments. Those working in such fields do not have the incentives (or the opportunity) to learn from the common elements. Young scholars are very much confined to the language, rituals, style of research and the lingo of their profession. They perceive themselves as belonging to their disciplines, and attempt to identify with the norms of the discipline. Since most interdisciplinary research still lacks a body of agreed theory (or even concepts) most scholars, instead of attempting to achieve a new synthesis, still follow the values, habits, procedures, and restraints of their own disciplines. Otherwise, they are afraid, they would lose respect of their colleagues (and tenure). They certainly are not going to risk publishing on unacceptable topics in economic journals not recognized by the respective professions.

Unfortunately, the IB field is still confused about what is its core knowledge base and refuses to distinguish between "foreign" and "international." These problems are discussed by three eminent scholars (Wilkins, Boddewyn, and Toyne) in another session. Without such an agreed core, it might be difficult to reach an agreement on organization. Yet what Blake calls "multidomestic" is simply the study of foreign countries. Such a study certainly has a place under the sun, and clearly no one can be an expert on each and every one of the more than two hundreds countries in the world. I would suggest that experts on the USSR, Japan, Germany, the United States, or any other country or region are not experts in international business but scholars of a certain foreign culture, language, or geographical area.

Again, what has been achieved in international business has been done by a small number of schools in which international business has been alleviated to the status of a separate, often strong, department and in most of these cases with a Ph.D. program. Members of such departments enjoy the power of having a number of them concentrated in the same university. This allows crossfertilization of ideas, mutual reinforcement, and political strength in getting faculty members promoted.

Indeed, new knowledge on international business—that is, operations that cross national borders—may stem from a minority of schools specializing in the different areas of our discipline. I hope such new knowledge may also be initiated by scholars in a majority of the schools that cannot afford the luxury of a large international business department. London School of Business is one obvious example of such a school. Only a few schools can establish a department of international business since by definition this means hiring a large number of international business scholars—and therefore the offering of a large number of courses in international business to create an international business experts. These few schools have provided invaluable research on global issues and a great hotbed for young scholars of IB who found it easier to flourish in a school where tenure and promotion decisions were made by seasoned IB scholars. Yet by definition such schools have served a small number of students and business leaders of the total students' body. This is true not only in the United States but in other countries, too. The vast majority of business schools in the United States provided token attention to international topics, leaving a large majority of U.S. business students almost completely unaware of international business impact and lacking a minimum preparation to operate in a global environment. In contrast, as already noted, many schools in small countries integrated international business topics into each and every functional area. In some countries, the global element is further articulated by teaching each and every business course in several languages and interweaving examples from different cultures. In this sense the *Economist* evaluation of business programs is of great interest. It counts such things as number and diversity of countries from which the students body and the teachers came to the school. In both counts, schools such as IMD in Lausanne, Switzerland, or INSEAD in Fontainebleau, France, carved for themselves an important niche.

Many deans have been concerned lest IB would be less academically respectful and strong than other disciplines. They are concerned that having a separate IB department may be the first step to reduced academic standards. During the 1970s, several universities thought they would strengthen both the status of international business and the ability of the professors to internationalize by dismantling specialized international business departments and assigning or attaching the IB experts to the nearest functional field of their specialization. This organizational design was perceived as a way to enforce stringer quality standards on the different faculty members and attract scholars to the international business field. Faculty members understood they are expected to behave according to professional standards of their

own discipline and survive the scrutiny of their peers. The self-identification of the teachers and researchers was expected to be both in international business and in a profession of one kind or another. Unfortunately, to achieve and sustain a claim of professional status in both meant satisfying the well-established rules of the games of both. An economist, however, cannot gain respect among his or her peers unless he or she publishes in the acceptable refereed journals on well delineated subjects. Young economists take a gamble if they publish much of their work in journals based on other considerations in choosing topics of study. The same has been true in other professions.

Another attempt was the compulsory joint appointment, in which tenure or promotion had to be voted both by the IB department and by the functional field department. Both of these systems signaled to young researchers to avoid research in IB. Faculty members resented the joint appointment (and the need for double approval of tenure and promotion): they would rather remain in their respective departments. To be sure, a few could show the degree of virtuosity needed to survive the scrutiny of two groups of different peers. A few more sought personal satisfaction and, usually after they secured job tenure, escaped from the restraints of their professions, indulging in research that has been less acceptable by the well identified rules of their original disciplines. Most understand they would rather not deviate from the habits, norms and procedures of their designated professions by researching unconventional subjects, using unacceptable models and publishing in unacknowledged (by their professions) journals. Those who were attracted more to studying the implications of an interdependent world, and were willing to ignore the restraints of assumptions established by their disciplines, felt the need to survive outside the territory of their original discipline—and joined one of the schools in which relatively larger departments of international business continued to exist and in which a devoted core of academics attempted to achieve new synthesis. These departments attracted many of the top researchers, and allowed niches of teaching and research to flourish. These few, however, were unable yet to convince their peers that they have created a new professional field. They certainly had only a limited impact upon the values and norms of other institutions. Only few universities have been prepared to create a separate M.B.A. in international business. Certainly these few are not enough to satisfy the need for launching a major change in all M.B.A. programs. In most U.S. schools, the faculty find it risky to indulge in international business research and attempt to publish the type of work that will be acceptable to their peers in the domestic profession. It is much easier to be mainstream! In my view, the only solution is to make IB itself mainstream—not only in the specialized schools but everywhere. Any professor or scholar whose theories are country specific and his span of knowledge is restricted only to what is true for one country should feel his knowledge base is inadequate!

The major challenge of the U.S. academic business community is to interweave the international business topics into each and every part of the entire curriculum with global viewpoint interwoven into the whole program. Since most U.S. faculty are not international specialists, and the majority of them did not see foreign countries nor do they speak foreign languages, strong incentives are needed to support the internationalization of all courses. To achieve such objectives, IB faculty in most schools cannot be separated into its own department. One possible method is to institutionalize collaborative efforts and interdisciplinary work. A better means is to make sure that a minimum level of competence in the international dimension of each field would be made mandatory requirement for promotion, tenure and evaluation for everyone. The existing norms of the professions certainly do not encourage the faculty in U.S.

business schools, especially the nontenured one, to devote the additional time and effort needed to learn about international dimensions. The major U.S. journals will accept papers even if they do not refer at all to a situation outside the United States. Despite much lip service to the interdependent world, faculty members know very well the rules of the game—not only the declared but the espoused one. They have strong reasons to behave according to what they believe will be taken into account in a tenure decision. Even after tenure, faculty feel constantly evaluated by their peers, and are eager to gain the respect of these peers. Attempts to find some generalizations about complex problems such as the degree of cooperation among firms in worldwide research and development efforts, or the impact of Confucius on management thinking, are perceived as dangerous. Much safer is another small brick on a wall of established knowledge. These problems will not disappear by themselves. Because of their general nature, they cannot be efficiently dealt with by anyone dean in one business school. Rather, they are national in character and need national solutions.

The means of achieving such a state of affairs are only very partially controlled by deans. The major tool must be in a changing definition of the requirements of each and every functional field. I believe such a change will occur if and only if leaders of different fields would demonstrate the importance of the international topics by their own behavior and by their own choice of research topics. Michael Porter's publication of the *Competitive Advantage of Nations* is a case in point—making global strategy an integral part of the body of knowledge and legitimate if not an "in" field of research for strategy faculty.

Specifically, the major hope for globalization of the U.S. M.B.A. programs is for the various professions to make an explicit effort to recognize the international dimensions as an integral and necessary part of the professions. For this to happen, marketing professors should stop inquiring "What is different about international marketing" and start looking at U.S. marketing as a special case to be taught in an elective course, (and similarly for other disciplines). One means of achieving such a change is for editors of major management journals to reject papers narrowly confined to analysis of one nation and one culture (including the United States) and require international implications as a condition for acceptability. Another is for well-established, highly reputable tenured and senior professors to make the shift, bringing international business into the territory of their respective disciplines. If these changes will occur, moving to international research will cease to be too risky if made by young nontenured faculty. Only then will U.S. business schools be able to internationalize.

Once these changes will occur, international business experts will lose much of their uniqueness, but they will have much more impact. The international dimension of marketing, for example, will be integrated into each and every marketing topic. Faculty apathy and disinclination to depart from traditional (and familiar) course of study will disappear—as it did in finance. "International" will be fully integrated into the professions and acceptable referred journals will change their norms. One will be unable to survive in a profession without international knowledge within the confines of the disciplines.

Of course, the core studies will still be carried out by international business specialists. Thus, specialized international business courses and research will remain in topics such as international environment, International trade policy, or management issues in multinational environment. Professors will be willing to retool and teach courses if these courses will have a body of rigorous research behind them. Such a body will be made available by the shift of research into the fascinating problems of interdependent world. If global relevance will be a criteria of acceptance of a paper to a major journal, it will certainly be taken into account by the young scholars, too.

In this new world, most barriers discussed by the authors of the three papers will be dismantled. Without such a major shift, schools all over the world will continue to give a lip service to the need of internationalization, but will find it impossible to surpass the strong barriers to internationalization unless they will have the funds to assemble a core of enough tenured professors that might create a self-propelled and strong enough department of international business. For most schools, this is not a viable solution. For them, only the major shift in professional norms can offer a solution.

To recapitulate: structure is based on strategy. Strategy may be different in different schools. For the majority of universities all over the world, and particularly in the United States, the structure of a separate international business department is not compatible with the desired strategy. A solution must be found to encourage international business research even though the school would not have a separate international business department. Much of the research will be carried out in a small number of schools—such as the University of Reading in the United Kingdom or others in the United States—with a critical mass of international business experts. The teaching must be much more widespread and should cover all institutions teaching business administration.

■ ■ ■

INSTITUTIONAL STRUCTURE AND
INTERNATIONAL BUSINESS EDUCATION:
THE TRADEOFFS

Bernard M. Wolf

INTRODUCTION

The relationship between institutional structures and international business education is the theme of this session. The three authors deal with the theme by focusing on several key issues: the way the conception and status of international business as an academic field impacts on international business education; the objectives of international business education in conjunction with the structure of programs to deliver that education; and the institutional arrangements for organizing the faculty who teach international business. Clearly, in establishing institutional structures and programs there are tradeoffs to be made. In dealing with the issues, the authors of the three papers have referred to the experiences of their home universities in the United States, primarily at the graduate level. I will do likewise by discussing a Canadian case.

The debate over the role of international business education is at the very heart of the larger discussion taking place regarding the function and execution of management education in the 1990s and beyond. Certainly, international business education must play a leading role if management schools are to cope with the three "driving forces" identified by the (United States) Commission on Admission to Graduate Management Education, in their report *Leadership for a Changing World: The Future Role of Graduate Management Education* (1990, 4):

> accelerating rates of change and complexity in technology, globalization of markets, and increasing demographic diversity. By demographic diversity the Commission is referring chiefly to changes in the age, gender, racial and ethnic backgrounds, and national origins of the home workforce as well as the changing composition of the markets to which goods are supplied. However, one could add a further foreign dimension to demographic diversity. It is the increased interaction with foreign nationals as a result of foreign direct investment and international strategic alliances.

BARRIERS TO INTERNATIONALIZING BUSINESS SCHOOLS

All three authors discuss the institutional roadblocks to better and more extensive international business education. However, it is

William R. Folks, Jr., who gives the most systematic exploration by dividing the obstacles into operational and conceptual. The operational obstacles arise from four sources: lack of availability of trained faculty, conflict with the business school's research endeavors, organizational and incentive structures in business schools, and limited funding. The conceptual obstacle is failure to focus on the core body of theoretical and empirical material that comprises the conceptual domain of international business. Instead, attention is given only to the international aspects of particular functional disciplines. Perhaps this occurs because as viewed from the traditional business disciplines, international business is seen as being made up of contributions from each of these disciplines rather than as a coherent interdisciplinary field of study in itself. Moreover, as this conference illustrates, there is still some lack of agreement about what constitutes the field of study.

According to Folks, the most serious obstacle is the conceptual one. In my own mind, I am not sure whether operational or conceptual obstacles present the more formidable hurdle. Probably, they go hand in hand. Nevertheless, I certainly concur with Folks that in many places, international business as a separate and distinct field of inquiry is overlooked. Hopefully, the deliberations of this conference should help a little toward changing the situation.

DIVERSITY IN INTERNATIONAL BUSINESS EDUCATION

Dean David H. Blake has identified three philosophies of teaching international business. The first emphasizes the foreign geographic region in which business takes place. The second emphasizes international expertise with respect to a functional area of business. The third takes a more global approach attempting to internationalize the whole curriculum. Which of these philosophies is prevalent in a given school will help shape course content. As well, account must be taken of the pool of the students for admission,[1] the institution's resources and the job opportunities likely to be available to graduates. These obviously differ somewhat within the United States, but the distinctions are even more pronounced when one looks across countries. Given the differences, there is bound to be diversity in the delivery of international business education. Blake applauds the diversity for adding to our knowledge of what can be done to prepare students for the global business system.

Schools must decide how much international business all students take, and to what extent more specialized education in international business will be available for those desiring it. In both cases, decisions must be made on the degree to which all courses address business in a global context and whether there will be separate international business courses. The latter could emphasize the interdisciplinary nature of international business, deal with the international aspects of a functional field or concentrate on a geographical area.

Blake's school, the Cox School of Business at Southern Methodist University, has opted for thoroughly globalizing its curriculum. In so doing it is attempting to achieve what Luostarinen and Pulkkinen, in their survey of *International Business Education in European Universities in 1990*, call the mature model for organizing international business education, in which a management school, by no longer separating domestic and international business education, becomes an international business school. In Europe, in their sample of 197 schools providing international business education only two fully achieved this status: INSEAD in Fontainbleau, France, and IMD in Lausanne, Switzerland.[2] Although the INSEAD M.B.A. has a global focus, special attention is paid to European integration and its implications.

Professor Lee H. Radebaugh's school, Brigham Young University, has chosen a very flexible approach which embodies all three philosophies of international business education. All of BYU's students are expected to be able to consider global issues in all business decisions. For a smaller subset of students, courses developing functional and country expertise are available.

At our host school, the University of South Carolina, the Masters of International Business Studies (M.I.B.S.) degree was, as Folks describes in another paper he presented in session 10, "the first large scale commitment of resources to a separate graduate international business degree at a major business school" (Folks 1992, 5). The program took a global approach to the core material usually offered in M.B.A. studies and had a strong regional focus with a heavy dose of foreign language. Functional expertise was kept to minimum. However, recent redesigning of M.I.B.S. has put more emphasis on students creating a concentration in a functional field.

TARGETING INTERNATIONAL BUSINESS EDUCATION FOR A SUBSET OF STUDENTS

My institution, the Schulich School of Business at York University in Toronto, Canada, has adopted an international focus also using a flexible approach.[3] Students in the regular M.B.A. can take courses in international business and can develop a concentration in IB. For those master's level students wanting a more intensive exposure to international business with a language component, we created a separate International Master of Business (I.M.B.A.) degree in some respects similar to the M.I.B.S. program at South Carolina and to the Lauder Institute program at the Wharton School.

York's Schulich School of Business is predominantly a graduate management faculty. It has about 50 full-time master's degree students and 850 part-time master's students. The latter comprise about 255 full-time equivalents. It has the largest number of M.B.A./I.M.B.A. students in Canada. We also have about 900 undergraduate students and about 50 students in a growing Ph.D. program.

Let me describe how our international business offerings have developed. In the

1970s, we introduced into the M.B.A. program an international business course and international business electives in most functional specializations. These offerings were strengthened in the early 1980s. By taking a number of these courses an M.B.A. student could concentrate in international business. However, to enhance job prospects, as director of the international business program, I generally advised students instead to opt for a concentration in a functional specialization. This could then be rounded out with some of the IB electives offered. For example, a student could concentrate in finance with an emphasis on international finance.

By the mid-1980s, the faculty began to recognize the need for more international content in most core courses in order to provide all students with what Radebaugh calls the "global mind set." However, progress in internationalizing the core has been slow, as has been pointed out, since it requires the difficult task of internationalizing the faculty.

Also in the mid-1980s, the faculty began to see the need for a smaller master's level program for students seeking positions in international management at home or abroad. There were several reasons for the decision.

1. The Canadian economy was heavily dependent on exports. Yet, about 75% of exports were being shipped to the U.S. Most international business was conceived as business with the United States. The vast opportunities in the Pacific Rim and Europe heading for 1992 were largely not being exploited by Canadian firms.
2. Canada, and particularly Toronto, has a large immigrant population which has been encouraged to maintain its ethnic identity through a policy of multiculturalism. This provides potential students already familiar with foreign lands and foreign languages.
3. No other Canadian school had an extensive international business

program and certainly no international business master's degree, so there were first mover advantages that could benefit the school.
4. The existing nucleus of faculty specializing in international business could provide a critical mass to begin the program. In addition, our existing strategic alliances with a number of foreign universities could assist us in finding work internships abroad and provide opportunities for students to spend a semester abroad on an exchange program.

We decided not to apply the program to the whole school for the following reasons:

1. There would not have been a sufficient number of interested students with adequate language prerequisites. While it is true that even firms which are chiefly domestic must pay attention to the international dimension of business, most of York's graduates will spend little time in their employment dealing directly with international business. Hence, for them a functional specialization is more appropriate than an international business specialization.
2. The cost of the full program to encompass the whole school would have far exceeded available funds.
3. With existing faculty it would have been impossible in a short time to internationalize adequately all the necessary courses.

In 1987, a proposal for the new master's degree was made to the Ontario Provincial government, which provides the major portion of revenues to universities in the province. The proposal was part of a plan for an international business center to be run jointly with the University of Toronto.

In April 1988, the province agreed to provide funds to create the Ontario Centre for International Business. By the fall of

1989, York admitted its first students into the new twenty four month degree program, the I.M.B.A. The program has now reached its desired goal of admitting about fifty students per annum. These students take newly designed interdisciplinary courses in international business, core courses in functional areas which are expected to have a global perspective, and intensive courses dealing with a particular geographic area. In addition, they continue studies in one or more foreign languages. The languages currently offered are French, Spanish, German, Japanese, Mandarin, Russian, and English as a second language. In some cases the nonmainstream languages can be selected to fulfill the program requirements. While language courses are currently noncredit, students have to pass a proficiency examination before they graduate. Related to their geographic area and language studies, they have a compulsory three month work internship abroad and can undertake an optional one-semester study program abroad with one of our university exchange partners.

In designing a degree program for students to complete in twenty four months, it was apparent that there were tradeoffs. How many courses were to be devoted to interdisciplinary IB courses? How many were to have a geographical orientation? Should the geographical oriented courses provide more of an overview of all major regions or concentrate more on one region or country? What importance should be placed on the study of foreign languages? To what extent would there be functional specialization? There has always been ambivalence between taking a generalist and a specialist approach to M.B.A. education. However, when international management is the focus, the tradeoff is even more agonizing. We would like generalists knowing the basic elements of international business and a knowledge of all the key trading regions in the world. We would also like specialists having intensive exposure to a particular region and in-depth knowledge of the international application of their

major functional field. Given the opposing forces, a delicate balance must be struck.[4] Initially, geographically based courses won out over function-based specialization. Students were required to take one regionally based semester course and, if they did not chose to participate in a university exchange abroad, they would also take a country-based course. Drawing upon our experience in the first three years, the emphasis in the program has shifted somewhat toward functional specialization. We have become aware of the considerable importance both students and potential employers attach to graduates having some depth in a functional field.

Over time, the tradeoffs are likely to become somewhat less acute, as students utilize their undergraduate education to be better prepared for the I.M.B.A. For instance, students, knowing that the I.M.B.A. is an option, might improve their foreign language proficiency prior to entry.

The establishment of the I.M.B.A. has contributed to achieving the goal of internationalizing our faculty. For example, within the functional disciplines, more internationally oriented faculty have been recruited and library resources have been added. Core courses in the I.M.B.A. serve as prototypes for internationalizing core courses in the regular M.B.A. More strategic alliances with foreign universities have been instituted including new exchange programs open to all students, not just those in the I.M.B.A. program. With increased exchanges the number of foreign students has grown. Enhanced international activities also pave the way for further international activities. For example, as result of its international orientation, York obtained funds from the Canadian government to conduct a two-month summer program that trains managers from Eastern Europe and the former Soviet Union (East-West Enterprise Exchange).

Thus, the creation of the I.M.B.A. has proved to be a major stimulant for York. As the I.M.B.A. matures and gains strength, it will further help York to incorporate the international business dimension into all aspects of the faculty.

INSTITUTIONAL ARRANGEMENTS FOR INTERNATIONAL BUSINESS FACULTY

The effectiveness of international business education is related both to the institutional arrangements for organizing IB faculty and the leadership shown by the school's administration. In York's case, the dean, Dezsö J. Horváth, is a vigorous proponent of both the I.M.B.A. and internationalizing the regular M.B.A. program. He has obtained the strong endorsement of the faculty for having an international focus as one of its strategic goals. However, York has not gone as far as creating an international business department within the faculty, as Folks advocates. In fact, I think there will only be a very few schools that can amass sufficient IB faculty talent to create a separate IB department without considerably weakening departments in the functional fields. We continue to appoint international business faculty to functional field departments which initiate tenure and promotion decisions as well as hiring recommendations. This undoubtedly does skew loyalties in the direction of established disciplines.[5] But I would suggest that even with a separate IB department, faculty will generally anchor their research in a functional discipline rather than in international business because they cannot ignore the scholarly community outside their home university. Recognition there still stems largely from the functional field. As well, faculty members wishing to be mobile enough to consider alternative positions will have to follow the norms of their functional discipline. For this to change appreciably, international business as a field will need to obtain greater stature in the academic world. Then it will attract more new Ph.D.'s. The exchange of ideas, examination of issues and the discussion of problems taking place at this conference is a useful process for improving the field of IB and increasing its standing in the academic world.

CONCLUDING REMARKS

Blake, Radebaugh, and Folks have each discussed key issues that confront international business education today and each has presented solutions with which they have experimented in their home universities. In my comments I have tried to relate York's experience to theirs. Learning from other schools' experiences can provide valuable insights which can be used constructively in developing international business education.

In seeking these insights, international business education ought to be viewed from an international perspective that encompasses all major business schools around the globe, not just those in North America. This is beginning to happen, albeit slowly. Competition is intensifying between leading international business schools in various parts of the world. At the same time, cooperation in the form of strategic alliances is increasing among the competitors. If business can benefit from simultaneously competing and cooperating in a global web of coalitions, so should faculties engaged in international business education.

NOTES

1. With respect to potential students one must consider whether they are likely to be recruited locally, regionally, nationally or internationally. One must also consider their preparation including fluency in foreign languages. For example, some European schools can easily attract a large number of students fluent in one or more foreign languages.

2. The categories used to classify schools in Europe by type of IB education and the percentages achieving that status are:

IB as an extension in existing courses (16%)
Separate IB courses (39%)
Separate IB programme (44%)
International (1%).

In addition, 34 institutions in the sample provide no IB education at all. (Lustarinen and Pulkkinen 1991, 42.)

3. For a directory of international busi-

ness education in Canada, see Corporate Higher Education Foundation 1991.

4. The Commission on Admission to Graduate Management Education calls for managers that are "well versed in concepts and theory that can be used in coping with a wide range of general management problems, and they must be equally skilled in identifying and solving specific, functional problems" (13). At the same time the Commission calls for M.B.A. program to achieve an appropriate synthesis between academic rigor and managerial relevance (33).

5. The Commission on Admissions to Graduate Management Education points out that the strong grip of functional disciplines in business schools hinders the development of all types of broader multidisciplinary and interdisciplinary approaches (2).

REFERENCES

American Assembly of Collegiate Schools of Business. Internationalization of the Business School Curriculum, March 1979.

Arpan, Jeffrey S. 1991. Internationalization: curricular and administrative considerations—the Cheshire cat parable. Center for International Business Education and Research, University of South Carolina. Working Paper D-91-07.

Bartlett, Christopher A., and S. Ghoshal.1990. Matrix management: Not a structure, a frame of mind. *Harvard Business Review* 68, no. 4: 138–145.

Blake, David H. 1992. Organizing for education in International Business. Paper presented at a conference on Perspectives on International Business: Theory, Research, and Institutional Arrangements, 21–24 May, at the University of South Carolina, Columbia, South Carolina.

Commission on Admission to Graduate Management Education. 1990. *Leadership for a Changing World: The Future Role of Management Education*, Graduate Management Admissions Council, Los Angeles.

Corporate-Higher Education Forum. 1991. *Directory to International Business Education in Canada*, Montreal.

Daniels, John D., and L. H. Radebaugh. *International Business Curriculum Survey* (Academy of International Business, 1974).

Folks, William R., Jr. 1992. Strategic Dimensions of International Business Education at the University of South Carolina. Perspectives on International Business: Theory, Research and Institutional Arrangements. Center for International Business Education and Research, University of South Carolina, Working Paper.

Grosse, Robert, and G. W. Perritt. *International Business Curricula: A Global Survey* (Academy of International Business, 1980).

Luostarinen, Reijo, and T. Pulkkinen. 1991. International Business Education in European Universities in 1990, Helsinki, Helsinki School of Economics and Business Administration.

Thanopoulos, John. *International Business Curricula: A Global Survey* (Academy of International Business, 1986).

5

Institutional Implications for International Business Inquiry

PANEL CHAIR'S COMMENTS
(WILLIAM R. FOLKS, JR.)

The assignment of the Institutional Implications panel was to resolve the issue of the institutional structure necessary to support the internationalization activities of the business school. While the initial concern expressed in the call for the panel was the institutional location of international business endeavors in theory-building, research, and education, the papers prepared for the panel widen the scope of the discussion to incorporate many administrative aspects associated with the internationalization process. The institutional structure, as all panelists repeatedly note, depends on the mission of the institution. The primary conclusion of the panel, reached early and often, is that there is no one institutional structure which fits all situations. Rather, one can expect a rich diversity of institutional responses depending primarily on the assumptions which are made about the mission of the business school.

Emerging from the discussion, as well, is the importance to institutional structure of the answer to the fundamental question of this conference. If one accepts as a fundamental principle that all business in international business, the institutional response most likely will be fundamentally different form that made if one hold the view that there is in fact a body of international business theory in some way distinct from other fundamental bodies of theory forming the basis for the research and educational mission of the business school.

The purpose of this introduction is to provide a guide to the discussion contained in the papers prepared for the panel, with a summary framework to be used in assessing the conclusions of the panel. The three papers are prepared by individuals with substantial administrative responsibilities within their business school. David Bess is the dean of the College of Business Administration at the University of Hawaii. Both Edwin L. Miller of the University of Michigan and Duane Kujawa of the University of Miami have served as associate deans with responsibility for the international activities of their schools. In addition to these authors, Dean Robert Hawkins of the School of Management, Rensellaer Polytechnic Institute, also served as a panel participant. Professor Miller was unable to attend the panel session, and his place was taken by his colleague, Professor Gunter Dufey of the University of Michigan. In the narrative below, contributions from the papers and from the panel discussion will be used and notes as appropriate.

To stimulate panel discussion, each panelist was given a specific question for the panel discussion regarding a key aspect of their prepared papers. These questions will be noted in the framework below. As each question was directed to the appropriate panelist in the discussion session, there ensued a substantive discussion on each issue from all the panelists.

As one reviews the contributions of these distinguished and experienced panelists, there emerges a substantive framework of issues which describe the institutional aspect of the internationalization process. That framework is presented here as a structure for understanding the papers which follow, and is applied in the summary of the panel discussion which follows.

The mission of the particular business school is viewed as one of the primary determinants of institutional structure. Since the mission of the school ought to determine the strategy which the school follows, the process which a business school uses to determine its mission should be the starting point for our discussion. Hawkins was asked specifically to relate the mission of the business school to its institutional structure.

Critical to the process of defining the international mission of the business school is the perspective which is assumed regarding the status of international business as a discipline. Whether or not it is held to be discipline, the view which the mission developers and strategic planners of a particular business school take on this issue may be critical with regard to the resolution of the institutional issues which follow.

The second area of focus is the process of programmatic choice, or strategic planning for the business school. What is the international perspective of the strategic planners within the business school? How is the international dimension of the strategic plan implemented? Are there barriers to the proper inclusion of the international business perspective in the strategic decision-making of the institution? As is discussed below, many of the leading business schools have developed strong international programs because of the activities of one or more international champions. These international champions have provided the institutional impetus for the development of an international strategy. To what extent is the phenomenon of an international champion one which has replaced the orderly development of an international strategy for the business school? The question of the international champion was direction to Dufey as a representative of Miller.

The strategic choices of the business school fall within the general missions of all business schools: research, teaching, and service to the business community. As there are numerous environmental factors which bear on the menu of strategic responses available to business school planners, the panel sought to address those environmental aspects which have a particular international dimension. The context of strategic choice is determined both externally and internally, from the business community which the school serves, the environment of government support, the backgrounds, achievements and aspirations of current faculty and administrators, and a myriad of other factors. Part of the work of the panel is to explore the international dimensions of these limiting factors and sources of opportunity as they relate to organizational design.

The emphasis of the panel papers and discussions then turns to the task of strategic implementation. Because of the constraining nature of the context in which international business education has taken place, much of the discussion is devoted to the overcoming of external and internal barriers to internationalization.

However one seeks to characterize it, as the overcoming of barriers or the implementation of strategy, the particular set of activities required are common to most of the panelists. The papers focus on four major issues: the programmatic response of the business school, the organizational structure of the business school, the process within the business school in which resources are allocated, and the international development of the faculty of the business school.

The papers of the panelists generally assume that the programmatic response of the business school is to take place primarily within the redesign, or the internationalization, of existing programs rather than the introduction of new programs. Kujawa was asked to comment specifically on the organizational issues which arise in the development of the new M.I.B.S. program at the University of Miami was the most recent.

The primary discussion of the panel, as one might expect from its charge, concerned the organization of the business school. While recognizing that the organizational response of the school must fit its chosen strategy, the panelists did provide a rich analysis of the available forms of academic organizational structure and the conditions which favor the use of each. In general no agreement on the ideal structural format

was reached, although there was general agreement that structure was conditioned by strategic choice and that existing structures did indeed place limits on the ability to identify and implement strategic choices. There was uniform agreement that organizational structure must recognize the international dimension. Further, there was substantial agreement on the impact which various forms of structure would have on the accomplishment of institutional strategic objectives. In the panel discussion session, Bess was asked to comment on the organizational issues and the role of the business school dean in resolving them.

Resource allocation within the business school, and allocation to the business school form the university administration and the school's external constituencies, flows from and in turn impacts the programmatic initiatives and organizational structure of the business school. Perhaps the reliance on program redesign as the internationalization response of the business school is chosen because of the resource allocation mechanisms in place within business schools favor the continuation of past successful program endeavors. Clearly, whatever international strategy is chosen, the process by which resources are allocated within the university and business school affects the ability of the institution of higher education to implement that strategy. While resources should logically follow the strategy chosen, it is clear from the papers and discussion of the panelists that one of the major issues confronting institutional administrators is how the internationalization strategy of the business school is impacted by the method by which resources are allocated, and whether that allocation methodology needs to be altered fundamentally to secure international strategic objectives.

Faculty, present and future, are the distinguishing resource of the business school, and a major part of the discussion of the institutional issues raised by the panelists in their papers and discussion involve issues surrounding the development of faculty resources capable of carrying out the international activities of the business schools. Key faculty policies identified as part of the internationalization process include faculty hiring policies, evaluation, tenure and retention policies, policies toward development and change in faculty capabilities tin teaching and research, and issues regarding the international content of doctoral programs.

In summary, the papers and the discussion in the session form a reasonably complete survey of the issues which must be addressed when institutions of higher education seek to address the implications of the globalization of business. From the initial definition of the mission of the business school, the panelists turn to strategic development and implementation, with emphasis on the role of programmatic initiative, organizational structure, resource allocation and the key nexus of issues surrounding the faculty's role in the internationalization process.

■ ■ ■

Organizational Structure and the Nurturing of International Business
David Bess

Where should a business school's international business dimension be housed in order to simultaneously satisfy its theory-building, research and educational responsibilities? To answer this question, we must define "international business" (IB) and address the goals and priorities of the institution.

I have found no universally accepted definition of IB. Some people prefer "global business." Kenichi Ohmae speaks of interlinked economies, and there are some, myself probably included, who are unable to define "domestic business" (like an "American car") and conclude, if one cannot define domestic business, maybe all business (markets) are in fact international or global or interlinked.

In a parallel vein, I have found no one universally accepted set of goals and priorities for business schools. Some are private, some public. Some cater to the masses, others to "select" students. Some build reputations on doctoral programs, others on M.B.A.'s, and others on undergraduate programs. And some try to do all three. Each individual school has certain market segments it is trying to serve, and a unique focus on specific degree programs, and a mix of teaching, research and service which must be addressed in order to define the optimal organizational structure—and, hence, the best institutional home for IB.

To a considerable extent, the answer to the question of optimal structure hinges on institutional focus, heritage, and faculty interests and politics. There is no one best organizational arrangement. There are, however, some universal "truths" which each institution must face in arriving at its own organizational decision. I will address some of these "truths," and present organizational structures I believe best meet the nurturing needs of IB.

Before going forward, the reader must recognize that I am looking at this issue as a dean with oversight of an entire college offering undergraduate and graduate programs. Also, I have two sets of beliefs (biases?) which directly affect my personal opinions on this topic. The first set relating to IB directly accrues from being raised in a multicultural society where everyone was (and is) a "minority"; experiencing international travel and living, among them a year's residence in Sri Lanka at the age of sixteen, which dramatically affected my views on life in the United States and the United

States' role in world society; holding an interest and involvement in international shipping and shipping policy for over three decades; working for twenty-five years at a state university which has long held a strong Asia-Pacific focus; living in an economy which is immediately and directly affected by the actions of governments, corporations, and people of the Pacific, and, to an increasing extent, Europe. Growing from these experiences is the belief that the United States is but one country in our global society and that in order for a person to become "educated" one must understand why and how political, economic, business and social decisions in foreign countries directly and indirectly affect life in the United States and be sensitive to, appreciate, or— even better—feel, cultures which are different from our own. (This includes cultures both within and outside U.S. borders).

The second set of beliefs stems from the fact that all of my university education and post-Ph.D. employment has been with public institutions. Further, in the case of the University of Hawaii at Manoa, we are the only comprehensive university in the state offering the only AACSB-accredited programs available in Hawaii. Hence, while all of our faculty are actively engaged in research and we emphasize research activity, teaching sound B.B.A. and M.B.A. programs is at the very core of the college. (In this regard, I believe that public institutions must reflect the primacy of the education mission. Private institutions may substitute another focal mission, but even there most must place education first.) In fact, the preamble to the new AACSB accreditation standards states that "all member schools share a common purpose . . . the preparation of students to enter useful professional and societal lives."

Putting these two sets of beliefs together, I believe in the primacy of educating students and that all students must receive a strong foundation in IB if they are to be well-educated. These two beliefs are the foundation of that which follows, and, incidentally, my role as faculty member and executive at the University of Hawaii. In fact, if one analyzes our faculty hiring and retention policies, our curricula, our faculty development programs, promotion and tenure decisions, allocation of resources, and, yes, organizational structure, it is impossible to miss the fact that these two themes are prime (not exclusive) foundations of our college.

Before I delve into the specifics of international business issues, I think it appropriate to briefly consider the new AACSB accreditation standards. Two issues are of direct relevance to the organizational issues before us. The first is that the new standards hinge on a school's mission statement and the processes put in place to achieve the mission with overall high quality. This is important for now the needs of the clientele must be specifically addressed and clientele input to mission statements, processes, and evaluation of output is assumed.

Second, three themes are to cut across bachelor's and master's curricula: the management of technology; the management of diversity; and the management of a global context. Putting these together, there is an inference that these three themes must be of concern across the school and not the purview of one, or more, isolated units. These three overlapping areas must be integral and inextricable parts of the education process. Moreover, these must not just be subject areas for teaching. The faculty must "live" these truths and exhibit them in their activities.

In the question at hand, it would appear that an isolated IB unit under which "international/global" curricular issues are addressed might not meet the intent of the new standards. The AACSB has finally recognized that IB is a concern of the entire school—and this, in turn, has a direct impact on the organization of the school.

Now back to the issue of how to define "international business." Too often it reflects the concept of a corporate "international division"—in other words, doing business outside the United States. This, of course, is important. International trade has increased 24–fold in 30 years, but the U.S. has been losing market share since 1970, and the percentage of our GNP attributed to exports (7%) trails that of Japan (10%), Canada (25%), Germany (30%), and South Korea (35%). Moreover, U.S. firms have declined in relation to foreign competitors in the world (including U.S.) markets in almost every industry.

If the United States is to regain its leadership role, we must educate our people to increase U.S. competitiveness in foreign markets. This includes recognizing that governments increasingly include international business on their political agendas and play growing roles in business decisions—including the location of economic activity.

Business students must understand these realities. "International business" is far more than competition for foreign markets—it is also the competition for "domestic" markets. Over 70% of U.S.-produced goods face "foreign" competition within U.S. borders. Services, like tourism, face worldwide competition, and even restaurants face "foreign" competition in the United States. For all practical purposes, the day of a domestic U.S. market is behind us (if it ever truly existed).

There are no longer national products in the traditional sense—nor a national market. Article after article tries to tell us these facts. Robert Reichs's two *Harvard Business Review* articles "Who Is Them?" and "Who Is Us?," the November-December 1991 issue of *Business Horizons,* and the 17 June 1991 *Fortune* article entitled, "Do You Know Where Your Car Was Made?" are but a sampling. The education of our students, and research of our faculty must reflect the reality that "international business" is not a narrow restricting term or concept. Instead, we must recognize that it is an all-encompassing reflection of the reality that we live and operate in a global market. All business is international and the research we produce and education we provide must reflect this reality.

In the paragraphs above, I have discussed IB in the context of doing business across and within national borders. There is another aspect of IB, however, which is critical to addressing the whole of IB, and that is the issue of culture and of managing cultural diversity. Too often the management of cultural diversity refers to the management of diversity within the U.S. borders—a domestic issue.

This is too narrow a construct. The management of diversity and managing in a global context are two sides of the same coin. Read *Beyond Race and Gender* by R. Roosevelt Thomas, or his March-April 1991 *Harvard Business Review* article "From Affirmative Action to Affirming Diversity," and you cannot escape the conclusion that management in a multicultural environment (I prefer this term over "diversity") is not isolated to the domestic scene— it is an overreaching, critical issue of efficiency and competitiveness.

This is no longer a legal, social, or moral issue. It is an issue of longterm economic viability. Our students must, therefore, graduate with a deep understanding that they will no longer be managing in a basically monocultural (WASPish) society (Japan, incidentally, faces this same basic issue.) Instead, they will manage a workforce comprised of unassimilated diversity (not a melting pot as in the past), and it is only in effectively managing in such an environment worldwide that they will be able to become and remain players in the marketplace.

William B. Johnson, in his March-April 1991 *Harvard Business Review* article entitled "Global Workforce 2000: The New World's Labor Market," provides ample ammunition in this regard. To date, neither the U.S. government nor U.S. businesses have really grasped the full economic and competitive implications of the need to address culture as a competitive tool—nor, very frankly, have business schools.

Indeed, this is not a topic which should be introduced at the M.B.A. or even undergraduate level. It must start in K-12 education and continue in university work. This topic must not be limited to business school curricula—it must cut across the "liberal arts" core. Nor can it be relegated to marketing, or organizational behavior, or strategic management. It must cut across business schools in research, service and educational programs. It must become a way of life—not a specific subject researched or a course taught. For just like businesses, business schools are in a global market, and in order to remain competitive must take advantage of the full array of available human resources—both domestic and worldwide.

As an aside, it is useful to look at what is happening in Europe with regards to management education. First, the decade of the 1980s brought the number of business schools from about 50 to over 275—and growing. Second, at the five "leading" business schools, the number of home-country faculty ranges from 9% to 70%, and home-country students from 6% to 60% . . . and many have targets or caps on both to assure a multinational and multicultural education experience and, hence, business relevance. There is a message here to which U.S. business schools should listen.

One additional topic needs to be woven into this fabric, and that is the issue of educating for quality—as in Total Quality Management (TQM). This is a global issue—not just Japan- and U.S.-specific. It is also a concept which relates to technology, and managing diversity. Chapter 8 of *Beyond Race and Gender* is entitled, "Managing Diversity and Total Quality: An Integrated Strategy for Organizational Renewal." To overgeneralize, there is the message that both TQM and managing diversity stress improvement or involvement of employees, and both represent "way-of-life changes." Quality is an essential competitive tool—within and across national borders. Is it part of IB—or is IB a part of quality? The answer is not important. What is important is that they are linked and they both must cut across business school programs/activities.

To summarize: Business *is* international business. There are no national products or markets. The topic of managing diversity is essential, so is quality. Both are a matter of efficiency and competitiveness—and are not restricted to national borders. Our students are entitled to educations which reflect these truths. The question before us is how best to organize schools of business to achieve a setting where such educational programs can be offered.

First, let me identity what I see as the five major inhibiting factors to globalizing the curriculum and research of business schools—but in no order of priority.

1. The AACSB's inability or unwillingness to put teeth into the globalization-of-curricula requirement. For whatever reasons, the "old" AACSB accreditation process has not recognized the fact that there is no domestic business, nor are multicultural management and technology isolated from domestic business; hence, enforcement of the global education requirement has been given little real attention. (To illustrate: I was on more than one accreditation visit team which found the international coverage to be the last-chapter-in-the-book-covered-in-the-last-week-of-class pattern and, hence, often not covered. In one case, the compilation of such coverages [at least on paper] amounted to less than a semester of work—but that was enough to sail through with accreditation). The AACSB has the power to focus attention and to demand specific course coverages. It has not, to date, used this power to enforce an adequate coverage of global business.

2. Deans and others who set policy, control resources and oversee hiring and promotion and tenure have not been pro-active in encouraging the internationalization of teaching, research and service. In our college, "international" is one of

our major strategic thrusts. Our personnel policies dictate that faculty are not hired (or granted promotion or tenure) in departments like management and marketing unless they are "international." Resources for research, travel, and the like, put a premium on international involvement. We also set up a separate institute, PAMI (Pacific-Asian Management Institute), to help coordinate and encourage a wide range of international programs and faculty support, including the development of internationally-focused courses. Other schools have similar stories—but too often the proactive leadership has just not been there.

3. "Leading" journals have been slow to accept research on international topics. In our academic community where research is so often (and often incorrectly) judged by the name of the journal in which it is published, such policies have discouraged international research. In parallel with this are school and university promotion and tenure policies which reinforce this publishing syndrome. Change comes slowly in academe—and, in this case, it has hurt the internationalization of research and courses (as well as crossdepartmental research which international research fosters).

4. Doctoral programs which continue to see international as a separate field. This, of course, ties in with the points above. The AACSB has not looked at Ph.D. programs in the accreditation process, thus removing them from any AACSB encouragement to internationalize. Senior (and some not-so-senior) faculty who are involved with doctoral programs have not been visible proponents of interna-

tionalizing research. Often, they are editors of journals and leaders of academic associations and, hence, the "quality assurance" forces. They also want "their" students to undertake research which will be published in the "best" journals. No wonder that Lee Nehrt found that only 17% of the leading Ph.D. programs had any meaningful international coverage.

There is some hope in this regard, for the AACSB will now include doctorate programs in accreditation reviews. While the coverage of global issues is not mandated (as it is for undergraduate and M.B.A. programs), the standards do address the preparation for teaching responsibilities. I assume that in order for a new Ph.D. to teach a solid core course in management the course must reflect management in a global context and, hence, some IB competency expected.

5. In the same vein, and for many of the same reasons, our traditional fields divided into departments of kindred souls inhibit crossfield research, publishing in "nonleading" (for example, noninternational) journals, research support for nontraditional research, offering innovative new courses, and hiring of faculty members "not like them." In his 1991 USC Working Paper "Internationalizing the BA Faculty Is No Easy Task," Brian Toyne states, "Collectively, these trends suggest that the pool of current and future business educators will continue to consist of persons who are highly skilled, functionally parochial, narrowly focused, and internationally illiterate." In order to get around these departmental barriers, one sees special IB departments spring up (usually looked down upon by faculty in

the "traditional" fields). Such special IB departments too often relieve other faculty from internationalizing their courses or research.

Given the above, how should the international dimension be housed? There is no one solution, but any solution dictates that any structure:

1. Reflect the primacy of education and that all graduates must understand IB. Brian Toyne in "Internationalizing Business Education," *Business and Education Review,* January-March 1992, provides three approaches (I would call them stages) of internationalizing curricula: awareness, understanding, and competency. In this context, all curricula should aim for "understanding" and be moving towards competency. If not, graduates will be unable to "read" the *Wall Street Journal* and compete in the long run even if only employed within U.S. borders.

2. Recognize that in providing strong educational programs "IB" must cut across the core curriculum and be represented in elective offerings. Unless one is teaching calculus or statistics, this global context must be part of every department's agenda. There can be an IB department, but no department can be excused from this educational responsibility.

3. Permit and encourage the hiring, promotion and tenure of faculty who facilitate international education, research and service. This includes resource allocation which encourages such research and teaching. It implies certain "nonleading" journals which focus on international topics be given "leading" status for promotion and tenure. It also implies that the dean and other school leaders make sure university promotion and tenure decisions reflect these school internationalization policies.

4. Recognize that good global research requires increasing amounts of crossfield collaborative efforts. The structure must reflect this. Here, an international institute may be of help—but it must reflect the need for crossdepartmental research and teaching not only within the business school, but also the need for collaborative efforts with colleagues across campus (in anthropology, computer science, philosophy, political science, area studies, foreign languages, and other fields).

My personal bias is that I do not like separate IB departments unless they have a major role of bridging with traditional departments and are a vehicle to encourage international research, courses and service across the college. By definition, if they do a great job, either the whole school will become an "IB department" or the IB department will be phased out.

I believe it is easier to meet internationalization goals by encouraging each department to address global issues—to reflect the fact that all business is global business. There should be strong IB faculty in each department to serve as a resource and building block and bridge to other "international" faculty in departments across campus. Strong IB faculty in departments are essential to developing strong departments capable of meeting the essential research, teaching, and service agendas of business schools. An international institute may help to bridge department walls—but is not essential.

There are two critical ingredients to successfully address the theory-building, research and educational programs of a business school—and neither hinges on a particular institutional housing of IB.

The first is a faculty comprised of people who are committed to providing their students with an up-to-date, relevant education which prepares them for a management (or even faculty) career in the evolving business world they will enter. If these faculty members are truly knowledgeable of the managerial world, not just pieces of a specific field, they will demand the resources to provide courses which reflect the fact that the world is not flat—it is global—and so is the business world their students will enter.

Another part of this is to recognize that faculty reflect their research—their area of interest—in their teaching. Encouraging faculty to engage in international research almost by definition will be reflected in their courses and the Ph.D.'s they produce.

The second ingredient is the commitment of school leadership, from the dean through department and committee chairs, to point the school in the global direction and facilitate progress on the march to globalize. Often the dean is in a unique position to help. Deans are, by definition, responsible for understanding the external environment of the school (both in and out of the university) and conveying this to the faculty.

To contrast reading patterns, faculty must focus on narrower issues to remain at the cutting edge; deans must read widely to understand the broad contexts of where business and education as a whole are heading. If the dean does not get to the faculty the message in words and deeds that managing diversity, the management of technology, and managing in a global context are critical focal areas which demand attention, then no organizational arrangement will assure that the job will be done.

To conclude, organizational structures can help or hinder any globalization/IB efforts—but it is the people who must ultimately get together and get the job done. Focus and commitment are the keys. If business schools are to contribute to the effective management of organizations, there is a critical need to focus on strengthening our schools' commitment to producing graduates and research which promote competitiveness in the global marketplace. The optimal organizational structure of any business school which seeks to achieve such goals, therefore, hinges on the people at hand, and how best to crate that unique synergistic structure which will work in that distinct arena. The keys are the goals and the people working to meet these goals.

■ ■ ■

Suggestions for Internationalizing a Business School Faculty and Administration
Edwin L. Miller

International trade is skyrocketing, and no longer can any nation or company consider itself to be insulated from foreign competition. As business activity becomes increasingly international, U.S. firms have gradually come to understand that they will have to become global in their strategic perspective, their management decision-making processes and their organizational structures. U.S. corporations need managerial and technically grounded staff who understand the complexities of managing and performing in a globally-oriented environment. Responding to this need, the nation's business schools are coming to recognize that as part of their educational mission it is essential that (1) an international perspective be incorporated into the curriculum; (2) a development program will be implemented to guarantee that faculty members are prepared to present international dimension in their courses; and (3) scholars who are skilled in designing rigorous and important internationally oriented research projects. The result will be an institution whose structure, teaching and scholarship activities reflect an international orientation.

How does a business school faculty and administration organize itself in order to satisfy the demands for international business theory-building, research and educational responsibilities? Obviously there is no universally appropriate solution to the above question, and it would be naive to suggest that there is. It is my opinion that one must have a portfolio of options and use them wherever and whenever they are appropriate. Included in that portfolio will be certain personnel and organizational initiatives that can help to facilitate the institutionalization of an international business perspective throughout a business school's academic and research programs.

There are many more options for internationalization than those that I will discuss in this manuscript, but among the initiatives I would include the following:

1. gaining the commitment of the faculty and administration to the goal of internationalization of the academic and scholarship programs, utilization of the institution's personnel systems, the development of the faculty's international competence and the departmentalization of the faculty;
2. enlisting a respected senior faculty member to champion the institution's program for internationalization of the faculty and curriculum;
3. applying the reward system in such a way as to operationalize the faculty and administration's commitment to internationalization;
4. assisting the faculty in developing international business competence;
5. designing an organization structure that will enhance the internationalization process;
6. recruiting prospective faculty members who have an international business competence and encourag-

ing present members of the faculty to incorporate an international perspective in their respective courses and research projects.

Gaining faculty and administrative commitment to the internationalization of the curriculum, the faculty and the overall culture of the institution is a mandatory first step toward institutionalizing an international business dimension throughout the institution including its various programs and culture. Without this type of support and commitment, there is little likelihood of successfully achieving the restructuring and redirection of the institution, sharing of resources, impacting personnel decisions and influencing scholarship and curriculum development. Where does one begin?

I believe that the faculty and the administration must come to an understanding and agreement concerning the necessity for incorporating an international dimension throughout the academic and research programs and the institution's overall orientation and perspective. It has been my experience that most business school faculties accept the fact that global competition is an important challenge facing U.S. management, and business students should know something about international business. Unfortunately, there is little agreement about how a school should go about meeting this educational challenge. For example, faculty groups and administrators are confronted by a series of academic and administrative questions that naturally flow from the discipline orientation of the faculty. Who will be responsible for teaching the international dimensions of business? How does one go about incorporating internationally oriented issues into an already overstuffed array of courses? Should there be an international business faculty group, or should international oriented faculty members be located in the various basic discipline grounded departments? As an administrator, how does one prepare to cope with this nest of questions?

I believe that internationalization of the

faculty and the institution demands aggressive but skilled administrative leadership. The dean, as the administrative leader, is the critical variable in the equation for institutionalizing an international perspective throughout the school. The dean's public expression of commitment to international, the effective use of the reward system to help shape the desired behavior of the faculty in terms of teaching and scholarship, encouragement and support for faculty development and the possible restructure of the school are arenas in which the dean and his or her administration can exert leadership.

The dean must lead the faculty either by personal example or the selection of an individual who represents the dean and commands the respect of the faculty. The faculty member or administrator must be an influential leader within the faculty, committed to internationalization of the school, prepared to devote considerable time and energy to the internationalization process and border on becoming a zealot as he or she goes about his or her assignment. There are likely to be a wide variety of forces and individuals that may be pressuring the dean to quickly move to institutionalize an international perspective throughout the school. It may be advantageous to draw upon these as one seeks to obtain commitment and ultimately institutionalization of the organization. What are some of the pressures that may be impacting the dean? To name a few, I would include the following:

1. AACSB's accreditation requirements represent a motivating force for introducing an international dimension into the curriculum and the restructure of the organization. Questionably, accreditation requirements may be transformed into a hammer that can threaten an institution to adjust its academic programs or face such undesirable consequences as disaccreditation. One must recognize that there may be faculty members who are skep-

tical and unenthusiastic about the importance of internationalizing the institution, and they may be resistant to cooperating in any way with the internationalization process. However, the consequences of disaccreditation are such that the faculty or specific faculty members may be reluctant to stand in the way of the program to internationalize the institution.

2. As business schools across the country, and especially one's peer institutions, move forward with a program to internationalize themselves, there is little that one can do but fall into line and launch a similar course of action. Competition for students and funds as well as the possibility of an unfavorable comparison among peer institutions become powerful forces for change.

3. Critical comments from employers, alumni and members of external advisory boards frequently gain the attention of the administration and faculty. For example, employers who are actively engaged in global competition may demand that graduates be familiar with the implications of international competition and how they can impact the various functional areas of business. Given the current interest in the opinions of students, alumni and employers concerning the quality and comprehensiveness of an institution's educational programs, faculties and their administrators are likely to move quickly in an effort to respond positively to the constituents' comments.

4. Students are becoming much more vocal and critical about the quality of their education, and they are likely to express dissatisfaction if they believe that their program of study does not prepare them to

compete successfully in the job market.

The above pressures can enhance the forcefulness of a dean's desire to integrate an international dimension throughout the school. For the dean the assignment becomes that of making personnel and organizational decisions that will effectively and efficiently move toward the stated goal. In the next section, I would like to examine some of the initiatives that an administration may want to consider as it seeks to guarantee that an international perspective will permeate the institution.

Faculty development is the key to the institution's internationalization process because it is the faculty that must be competent in the international dimensions of their respective fields of expertise. The initiation of a wide variety of activities will be the result of a faculty and administration that is sensitive to international forces impacting business and the practice of management. The challenge becomes that of building competence in a faculty that is changing its membership because of additions, resignations and retirements, and whose teaching and research interests may not have prepared them for such a professional demand.

A successful faculty development program will rest upon a promotion and tenure process that recognizes the importance of international competence, a reward system that encourages faculty involvement in internationally oriented research and curriculum development and a supportive senior faculty and administration.

1. Participation in internationally oriented workshops and summer programs can be an excellent source of resource materials for the faculty member who is interested in redesigning a core or elective course. These workshops can be important for building professional competence in terms of teaching skills, perspective and self confidence.
2. Participation in corporate sponsored international internships is another alternative that can be used as a career development option. I have found some corporations willing to have a faculty member spend a period of up to six weeks in one of their overseas locations. The management of the overseas location jointly determines with the faculty member an assignment that will utilize the individual's expertise on a project or projects that are of interest to the management and professionally valuable to the faculty member. Such internships have been win-win situations for the corporation and the faculty member. Frequently, the faculty intern brings a different perspective and expertise to a management issue, engages in a series of activities that result in helpful management studies and becomes a potentially valuable corporate friend. When the individual completes the assignment, he or she has a better understanding of the company and what it is like to live and work abroad for a period of time. The result is that the faculty member becomes much more comfortable introducing internationally oriented topics in his or her courses, designing internationally related research projects and advocating an international business perspective throughout the institution.
3. Development of research projects that have an international dimension can quickly contribute to the development of a scholar's international competence. Joint research projects with a foreign scholar or a colleague represent a relatively efficient technique for stimulating the internationalization of the faculty. The faculty member learns something about the international aspects of his or her field of specialization, the results of the re-

search are discussed in the classroom, the addition of research data and knowledge to the existing body of theory or application and the expansion of his or her research network to include scholars from other countries.

As a corollary, the acquisition of data tapes that contain international data is another way to encourage the faculty member to become acquainted with the international dimensions of his or her field. Some forms of internationally oriented research can be accomplished by the use of these data tapes and the faculty member's research may be performed without leaving the country. Furthermore, these tapes may serve other faculty members interested in international business problems.

4. Participation in management or executive development programs offer faculty members an opportunity to meet and interact with management participants who have international experience. Such programs can be offered within the United States or abroad. Participation in a management or executive development program, especially those offered overseas, represents a relatively efficient method for gaining an international exposure in a short period of time. Going abroad to teach, having a class of international executives and interacting with them over several days can be an excellent international development experience.

If a dean wishes to introduce an international dimension in teaching and scholarship, the institution's reward system represents one of the most important tools available. The challenge confronting the dean becomes that of determining how to motivate individual faculty members to incorporate an international dimension into their courses and research programs.

The dean must focus on the challenge of creating conditions such that faculty members can achieve their goals best by directing his or her efforts toward expanding their professional competence resulting in a better understanding of the international dimensions of their fields. Goals of the faculty are multiple and varied but assuredly promotion, tenure, recognition and financial rewards are among some of the more apparent ones.

Monetary incentives are some of the most commonly available rewards or incentives that can be used in order to influence a faculty member's behavior. For example, rewarding faculty members who publish internationally oriented articles and incorporate an international dimension into their respective courses are among some of the more tangible means to impact the individual's compensation package and the type of behavior that will reach the desired goal.

Rewards should not be limited to the compensation package. School support for internationally oriented research projects, international travel and release time are other means for encouraging internationally oriented scholarship. Not to be overlooked, promotion and tenure decisions are among the most tangible examples for recognizing and rewarding desired behavior.

From the institution's perspective, criteria included in promotion and tenure decisions can include an emphasis on the amount of internationally oriented activities the candidate has engaged in. The administration can establish conditions such that an internationally oriented perspective will be integrated into one's research. Teaching and professional development will be judged as necessary but not sufficient requirements in promotion and tenure decisions. Such criteria will have an important impact on the professional development of junior faculty. However, among the senior faculty, promotion and tenure decisions will have lost their personal appeal and ability to influence behavior, and consequently, the administrator will have to rely upon the application of monetary rewards as a means to achieve the desired behavior of the faculty.

Recruitment of new faculty members represents still another technique that can be used to accelerate the pace of internationalization. If the dean and the faculty have committed themselves "to ramping up" the international perspective of the institution, the recruitment of new faculty members becomes an important tool. Most likely, the criteria for joining the faculty would require that each candidate possess a strong basic business discipline and a demonstrated international orientation to one's teaching and research. Faculty recruitment represents a relatively painless way to intensify the international perspective and competence of a faculty.

Should there be an international business department or should international business expertise be distributed throughout the basic disciplines of the institution? Any discussion about a proposed organizational structure that will most adequately guarantee that resources will be directed toward international business theorizing, scholarship and education will immediately generate a heated discussion.

In the long term one should expect structure will become a nonissue and an international business perspective will be an integral part of the institution's fabric. Unfortunately that is not the situation in most institutions. For me, the overriding concern is intellectual excellence of the faculty— their scholarship and the quality of the academic program. Without that as the criterion, issues of organization structure are literally meaningless.

I wonder if a school's understanding and commitment to internationalization can be translated into its budget commitments, personnel and organization design decisions? When an institution offers nothing more than a single international business course, it is likely that promotion and tenure decisions will not be influenced by any consideration of the instructor's internationally oriented scholarship or teaching. The traditional departmental structure will suffice and the international business course will be housed in one of the more traditional functional areas. International business is interpreted as an add on, and not much thought is given to the relevance of the subject or its impact on the institution and the instructor's career.

As the institution moves toward the introduction of an internationally oriented course in each of the basic fields, there is likely to be the dawning of an awareness of international business and its relevance to the business curriculum. The introduction of "international" generally takes two paths. Among the departments, there may be an attempt to introduce international material into the introductory course in the discipline or field and perhaps other courses where it is appropriate. Unfortunately, experience suggests that international issues are not discussed or the term ends before "international" is considered. The other path is to design a discipline specific international course. Frequently, the development of an international competence has little impact on the faculty member's career. Academic merit will be judged on the value of scholarly contributions to the discipline. Involvement in an internationally oriented research project adds little to one's academic stature, and personnel decisions continue to be on the basis of publication in the discipline's most prestigious journals. Sadly, much internationally oriented research never survives the referee process because it is judged to be inappropriate for a discipline's main line journal.

Establishment of an international business department is a familiar organization design, and it can provide a degree of energy for internationally oriented faculty to produce high quality and high output research. We have many examples of that among the various U.S. schools of business. However, one must be cautious about isolation, faculty inbreeding and reduction in the intellectual quality of the research produced by that particular faculty group.

If there is to be an international faculty, it is essential that its boundaries be easily penetrated by faculty members from different disciplinary orientations. Functionally oriented international research should be encouraged and bridges and opportunities should be made available to faculty members who are in the basic business fields.

Promotion, tenure and merit increase decisions are somewhat more clearly defined, but there is frequently some resistance flowing from the faculty who remain skeptical of international business research. Frequently, the criteria for judging intellectual excellence of the faculty member will be reduced to those standards which are applied to the more well established basic disciplines.

In the final analysis, the pace of internationalizing a business school depends on the academic and administration leadership's ability to create a vision that influences the institution's mindset. An institution that has successfully integrated an international dimension into its academic and research programs thinks about business and the management processes in a different way.

No longer will there be a distinction between global and domestic priorities, and the smokestack mentality evidenced by the way disciplines insulate themselves from their colleagues in other disciplines will no longer be applicable.

CONCLUSION

Throughout this manuscript I have emphasized the importance of the role played by the administration. However, the administration cannot succeed without active support and involvement by the faculty. The administration may strive to exert leadership and communicate vision, but it will be the faculty that must decide if it will accept the leadership of the dean and administration. Without that acceptance there will be little that an administration can do organizationally to institutionalize an international or global dimension in the fabric of the school.

■ ■ ■

On the Academic Institutional Setting and the Nature of a Supportive Environment for International Business
Duane Kujawa

The question to be addressed by this panel appears to be both straightforward and simple: How should the international business dimension be institutionally housed in order to simultaneously satisfy its theory-building, research, and educational responsibilities? In fact, the question is not so simple. It is normative, suggesting that there is an optimal "institutional" setting for international business in academe. This, in turn, requires some conceptual development as to the criteria for determining an "optimal setting" and the structural or organizational elements involved in implementing or meeting these criteria in the real world of academe.

Related to this, the question is also assertive as it relates to "theory-building." Theory-building can only result if there is a theory base upon which to build. Moreover, the extent to which this core theory is generally recognized by academic professionals influences the optimal institutional setting for international business. Whether or not there is a generally recognized core theory of international business is itself then a core issue as to the institutional validity of international business as a social science discipline. Certainly the academic institutional environment for international business research and education will be significantly

198

affected by the status of a core theory of international business. The issue of the institutional housing of international business is then a "derivative" or secondary issue.

Having a core theory of international business would favor an institutional arrangement characterized by an international business (IB) department. If IB did not have a distinctive, recognized theory base, then research and education in international business would best be served institutionally via the functional structure only, that is, research and education on the international dimensions of, for example, production, marketing, or financial management, carried out only in functionally identified departments. At best, the institutional setting in this situation would identify an "IB champion," that is, someone charged with encouraging the internationalization of faculty, courses, and programs. The "power" of such a champion would be contingent upon the amount of budget support and programmatic control the administration and faculty would be willing to assign to that person. Institutional mission would appear to be a determining factor in this scenario.

These considerations then lead to a need for further discussion of what I consider to be a fundamental issue: Is there a recognized, core theory of international business? This question needs to be addressed first. Then the normative framework regarding institutional arrangements and the research and educational aspects of international business will be addressed.

IS THERE A RECOGNIZED, CORE THEORY OF INTERNATIONAL BUSINESS?

Most scholars in international business would certainly agree there is a core theory of international business. Moreover, given the dynamic nature of the subject matter—the multinational enterprise (MNE)—one would expect the theory to be advancing over time as it seeks to explain changing behaviors. We have seen, for example, the advent of the international product cycle theory to help explain the emergence of the

MNE in the 1960s, and the development of internalization theory and of the eclectic model in the 1970s and 1980s to help us in our understanding of the growth and the nature of the competitive advantages associated with MNEs.

Other illustrations of an international business theory core could be noted. The pertinent point here though is that this core exists, is relevant, and is generally accepted by those concerned with the need for such a theory base.

But this does not mean the institutional environment outside of international business (that is, for our purposes, faculty in functional departments) universally recognizes this theory core as academically acceptable. The reasons here are varied. Many "functionalists" just consistently discredit all interdisciplinary areas. In addition, their behavior in this regard may well become increasingly evident in the face of calls for more frequent and more comprehensive interdisciplinary research on international business topics. Other "functionalists" seem to be just unaware and unappreciative of the theory core in international business. This may well be affected by the apparent disinterest, or inability, of IB scholars to extend the theory core to substantially explain functional behaviors distinctive to the MNE.

This brief discussion, of course, simply illustrates one observer's assessment of the "institutional state" of the core theory of international business. And the assessment is that, while the theory is timely, rigorous, and relevant, it is not recognized in any universal or extensive sense. This impacts the institutional location of international business within academe, as well as research and education in international business.

The discussion now turns to an alternative perspective—that is, the institutional setting of international business and its impact on IB research and education.

THE INSTITUTIONAL SETTING

The lack of any substantial status of international business theory in the perspectives and values of the "functionalists" suggests the institutional setting of international business will generally be dominated by the "functionalists." (Of course, the historical evolution of power within the academic setting supports this view.) This in turn suggests "no room" for dedicated international business research or education. It also suggests there be no international business department (with control over the tenure-granting process). (In this case, the identification of a "champion" of internationalization may well be the best that can be expected.) In any event, and perhaps with but a few exceptions, the institutional environment will be characterized by a fragmented presence of international business. Is this "good" or "bad"? It depends.

THE CHANGING NATURE OF THE "FUNCTIONALIST" ENVIRONMENT

A traditional, cynical view on the nature of an academic environment dominated by the "functionalists" is that international business theory, research, and education is not relevant and is not to be tolerated, or certainly not encouraged. This view—the sinister view—discredits publishing in anything but the top functionally oriented journals, likely favors functionally focused research and educational programs—especially at the doctoral level where depth of functional competency can be emphasized, is likely more theory-based than applications-oriented, and is generally intolerant of interdisciplinary research and theory-building because of the need to protect the intellectual integrity and acceptance of the functional area itself.

While some may argue the pervasiveness of this sinister view across and within different academic institutions is varied and debatable, it is my sense key environmental changes are making this view increasingly less tenable and less evident. These changes are being market-driven by employers and by students themselves. Their roots, ultimately, are to be found in the increasingly globally competitive nature of business.

Evidence of the market pressures favoring, in not forcing, the internationalization of business schools abounds. Corporate recruiters are looking more often for employment candidates with international business majors or minors. Job offers to applicants with foreign language capabilities are often more forthcoming. More business schools are developing specialized majors and even specialized degree programs in international business. Student enrollments in such programs are growing. More students are participating in study abroad programs than ever before.

As if market pressures were not enough, powerful influences by the two lead management education associations are also having an effect. The American Assembly of Collegiate Schools of Business (AACSB) has long advocated worldwide business dimensions be evident in the curricula. AACSB-sponsored research on trends in business and business education has clearly identified the globalization of markets and of businesses as the dominant trend which needs to be addressed by business schools as the 21st century unfolds. Similarly, the Graduate Management Admissions Council (GMAC) supports expanded internationalization of academic programs and faculty competencies.

THE CHANGING NATURE OF THE "FUNCTIONALIST"

In response to market pressures and institutional interests, the business school "functionalists" have become increasingly interested in the international aspects of their individual specialties.

The sinister view, concerned with its own preservation, would contend this change is the result of administratively led changes in the reward systems affecting academics. In other words, deans and others concerned with meeting budgets and being customer-oriented will influence financial reward, recruitment, tenure-granting, and promotion decisions in favor of faculty accomplishments supportive of the institutional shift towards programs and competencies supportive of a stronger institutional presence in international business. An alternative view would be that faculty, serious about maintaining competencies that are "mainstream" to their disciplines, would want to become increasingly internationalized because that is the way their disciplines are going.

In any event, the barriers between the "functionalists" and those interested in interdisciplinary, international business studies appear to be coming down. The "functionalists" are becoming more international. So much so, in fact, and so quickly, that I have heard traditional internationalists voice concerns. These concerns are that the internationalists' (heretofore) "leadership" positions—weak as they were, or aspirations regarding internationalization, were being preempted by the expanding international interests of the "functionalists"—people who traditionally are often the centers of academic power—making decisions on programs, budgets, and personnel advancement.

THE CHANGING NATURE OF ACCREDITATION STANDARDS

The institutional setting has also been seriously affected by the recent changes in the accreditation standards of the AACSB. In the new environment, AACSB accreditation is driven by the suitability of a business school's mission and what the school does to accomplish its mission. Different standards are applied depending on the mission itself. Regarding our area of interest, schools that wish to do so may emphasize international business in a variety of programmatic ways; those that see their interest otherwise may elect to focus on more traditional disciplines and strengths, or on other interdisciplinary areas, such as quality management, telecommunications management, or health administration.

I see the changing accreditation standards as having a substantial, positive impact on international business research and educational programs. This is because one key feature of the new standards is to make the standards more market-based—as the markets are "read" and understood by different business schools. Our earlier discussion concluded business schools are becoming increasingly aware of the globalization of markets and companies, and that students, aware of these trends, were wanting to major in International business, to become more competent professionally in the international dimensions of their functional specialties, and/or to develop skills, such as language skills, supportive of what they perceive as necessary for success in international business careers.

We can summarize then that the changing environment of the "functionalists," characterized by changes in the employment interests of corporate recruiters and the programmatic interests of students, is putting pressures on "functionalists" to become more international and that, indeed and in response, "functionalists" are becoming more international. The changes in AACSB accreditation standards appear to present an opportunity for a generally more tolerant, if not more supportive environment at the business school level.

NORMATIVE IMPLICATIONS

Having identified some key shifts in the academic environment affecting international business research and education, the discussion now returns to the mandated question of "how should international business be institutionally housed to satisfy its theory-building, research, and educational responsibilities?."

This paper presents powerful trends affecting the answer to this question. One of these trends presents a "pull through" situation regarding international business programs, that is, our customers—employers of our graduates and our students themselves—want expanded and more comprehensive international business educational offerings. Another is working on the "push" side, as we find faculty increasingly interested in, if not dedicated to, expanding their own professional international business competencies.

From the institutional perspective, these trends result in a proliferation of international business interests across academic departments (that is, functional specialties). I would contend these newly directed interests are deeply felt, are growing, and will be durable over the long term.

Our question of interest then becomes "how can highly diversified faculty interests in international business be institutionally focused and supported to encourage international business theory-building, research, and educational responsibilities?." I would expect the answer to this question would vary depending on the mission of the academic institution. Let's look then at what one institution did—the University of Miami—where during the past few years a transition process has unfolded which has profoundly changed the business school.

The transition process began with a faculty-led redefinition of the mission of the school which featured the concept that our graduates would be knowledgeable about and competitive in the global marketplace. This objective, along with the results of a "base line" study on the school's then current international business course offerings and programs, faculty competencies and interests in international business (broadly defined to include foreign language capabilities and experiences in teaching and conducting research abroad), the international profile of the student body, and the international involvements of the South Florida commercial and financial communities, in turn led to decisions on the budgets and structures required to achieve this goal. The faculty and administration proceeded along several directions simultaneously.

Regarding curriculum, a faculty committee developed a proposal for a new Masters in International Business Studies program. Graduates from this dual degree program will earn a new "internationalized" M.B.A. (where every course will feature either international or comparative content) and a Master of Science in International Business degree featuring coursework in area studies, crosscultural communications and management, leadership and other skill development topics, and an internship. Foreign language competency is also a requirement of the program, as are one-on-one student-faculty and student–local businessperson mentoring activities.

The program, it should be noted was developed with massive inputs from the faculty, current students, and the local (international) business community. It was developed over a period of but four months and overwhelmingly (82% in favor) approved by the faculty. It should also be noted the strategic intent of the new program is to provide the business school with a program of distinction, and a basis for developing expanded course content and faculty experience which will be useful in internationalizing all the other academic programs in the school. Faculty will rotate through the new program and take what they have developed and apply it to the other programs in which they teach.

Developing broad-based faculty expertise in international business means investing in faculty development. Business school funds, supplemented with support from the school's Center for International Business Education and Research (CIBER) and other external sources, were allocated to pay for faculty research (for example, foreign travel, purchases of data bases, and summertime compensation) and foreign language training. A faculty committee allocated these funds in response to specific, faculty-generated proposals.

In addition, the business school established linkages with business schools outside the United States to provide our faculty with an opportunity to teach abroad and to conduct international business research in a foreign country in collaboration with a foreign faculty colleague. We have, for example, established a joint faculty research and executive seminars program with INCAE in San Jose, Costa Rica, and an extensive relationship with the Universidad Gabriela Mistral in Santiago, Chile, to assist in the development and delivery of that institution's own new international M.B.A. program, and to develop faculty expertise in both our institutions through a collaborative faculty research program. We are in the process of finalizing yet a third cooperative relationship in this case with the Instituto de Empresas in Madrid, Spain.

This then is how one business school has been moving to direct highly diversified faculty interests towards international business institution and faculty "building." The Miami experience, thus far, has certainly had its accomplishments in institution and program structuring and in the redirection (and generation) of selected financial resources, but much remains to be done.

The issue of having a more prominent international business component in tenure and promotion decisions has yet to be structurally addressed. To some extent, this will be accomplished because of the shifting faculty research and publishing interests towards international business. A model we developed in our health management program may be useful in this regard. In the health program, a "program faculty," that is, those teaching and doing research on health management issues, formally offer advisement to the functional departments as candidates for promotion and tenure are being reviewed.

How will institutional restructuring, such as that evident at the University of Miami and elsewhere, affect theory-building in international business? My expectation is that it will favor expanded faculty interests in interdisciplinary studies. Increased faculty exposure to the more complex and diversified environmental and competitive challenges in international business should drive this trend. Of course, this is all longterm.

CONCLUSIONS

Returning to the original question to be addressed by those on this panel, I believe the institutional housing of international business has to follow the broader interests and mission of the academic institution itself. Traditionally, this has meant interna-

tional business theory-building, research, and educational responsibilities have "suffered" because of either the benign or deliberate neglect of the "functionalists."

But times are changing. Students' and companies' needs are changing. Internationalism is more relevant today. The "functionalists" are undergoing a serious redirection to expand their international business competencies. This bodes well for expanded research in international business and more academic programs with international content.

The institutional restructuring accommodating these trends will likely be varied, but quite deliberate. Massive faculty involvement in the restructuring process is one key, as is a careful analysis of faculty and institutional strengths and weaknesses and the development of a viable and appropriate institutional strategy.

■ ■ ■

SUMMARY OF THE PANEL CONCLUSIONS
William R. Folks, Jr.

INTRODUCTION: MAJOR ISSUES IN INTERNATIONALIZATION

This section of the paper provides a summary of the conclusions reached in the paper and panel discussion, with additional analysis and comments by the panel chair. Given the framework developed in the introduction to the panel paper, the panel discussion summary contains four main parts: (1) mission setting; (2) strategic planning, external analysis, and programmatic development; (3) implementation (organizational structure and resource allocation); and (4) faculty issues.

MISSION SETTING

The paper by Miller provides a clear statement of the importance of internationalizing the mission of the business school.

203

He argues that it is essential that the business schools in general incorporate an international perspective into the curriculum, and that scholars become skilled in the conduct of rigorous internationally oriented research projects. He further includes in the mission the development of faculty who can incorporate the international dimension into the curriculum, which perhaps is more of a derived that a primary mission.

Bess asserts the primacy of educating students as the mission of the business school and that all students must receive a strong foundation in IB if they are to be well educated. However, Bess rightly distinguishes the wide range of possible missions which might be adopted by a given business school, arguing that each school has a particular set of market segments which it tries to serve and a strategic mix of teaching, research and service which responds to that mission. In his panel comments, Hawkins reports the interaction of the defined mission of Rensellaer with that school's decision to concentrate on the inclusion of international business in the school's curricula, with, by way of implication, at least, limited attention to the development of theory and the conduct of empirical research. One can clearly see the defined mission of the University of Miami reflected in its strategic decision to launch a specialized master's degree in IB.

In general, Miller provides the broadest possible definition of the mission of the business school relative to international activity. In the comments of other panelists, the mission of the business school is not explicitly related to international business as a field of inquiry, but rather is derived from the historical mission of the business school to educate in whatever concerns business. Responding to the traditional mission rather than reshaping that mission leads one into programmatic responses which in turn drive the organizational structures created to serve them.

STRATEGIC PLANNING

No doubt because the strategic planning issue is not directly addressed in the ini-

tial structuring of this session, it is not possible to extract from these papers a detailed analysis of the particular international aspects of this process. There are numerous comments about gaining administrative and faculty commitment to the strategy selected, and from those comments and discussion the following themes do emerge:

1. The influence of leadership from the top appears to be a major factor in the development of an international strategy. Bess, in the wider context of a discussion of ingredients for a successful strategy, sees leadership in internationalization flowing from the dean through department and committee chairs, with responsibilities for understanding the external environment of the school, and conveying that understanding to the faculty.

2. The process of building an international strategy appears historically to have been usefully supported by the emergence of a champion for the international area in the strategic process. Indeed, although not specifically using the term "international champion," Miller argues that if the dean chooses not to lead by personal example, he should select an individual to provide that leadership who must be an influential leader within the faculty, committed to the internationalization of the school, prepared to devote considerable time and energy to the internationalization process and border on becoming a zealot in accomplishing this task.

3. If, therefore, it is necessary either for the dean or for the international champion to provide significant leadership in the process of internationalization, by implication there are barriers in most organizational structures which preclude faculty from providing the strategic initiative which would

lead to a proper recognition of the international dimension of a business school's mission. As it is widely recognized that barriers to internationalization of curriculum and research do exist, and as most major discussions assume that faculty must be internationalized in some manner, it can be concluded that a strategic planning process which is predominantly faculty driven would be inclined to miss certain key international issues.

Thus, one major inference which can be drawn from the papers and discussions is that the strategic planning process of the business school must provide special support to provide for the inclusion of the international dimension in the business school's strategic plan. Possible means for providing this support are special leadership efforts by the dean, the utilization of a well-respected international champion, or the exertion of substantial external influence for internationalization on faculty involved in the planning process.

Further, if the dean is to provide the special support for internationalization in the strategic planning process, the underlying attitude toward internationalization of the dean becomes a major determinant of the direction which the college takes. Perhaps the best illustration of the impact of the underlying assumptions of the planner on the strategic choice are the comments by Bess regarding his personal beliefs about the nature of international business and the nature of the education task in a public education institution, when he states that those beliefs are prime but not exclusive foundations of his college.

Kujawa addresses the problems of the interaction of the strategic planning process with the recognition of the functional existence of international business. He clearly recognizes that "functionalists," his term for faculty whose primary academic loyalty resides in an existing academic discipline, may not recognize international business as

a discipline, even after they have completed the process of mastering the international dimensions of their particular functional discipline. Kujawa has an optimistic perspective on the impact of market pressures for internationalization of the functionalists. He argues that external pressures on the business school will lead to an internal incentive program which will prod functionalists in the direction of greater international activities. Further, the academic integrity of these functional specialists will lead them to an effort at personal development to remain in the mainstream of their specific disciplines.

Kujawa then notes the proper concerns of traditional internationalists, those who view international business as a separate domain of study, with the possible preemption of leadership positions in the international development of the university by internationalizing functionalists, because of their entrenched position in the academic hierarchy of the institution. Should international business indeed be a separate domain of academic inquiry, a corollary to Kujawa's concern is the likelihood that this domain would be excluded from business school activities in curriculum design and research agenda definition and resource allocation decisions.

Thus, whether or not international business is considered a separate discipline becomes an important factor in defining the structure which a school adopts for its strategic planning process. Failure to consider international business as a separate domain of study ensures that the constituency on campus which is involved in theory-building and empirical research will not be represented in the strategic planning process. As it is generally accepted by the panel that organization follows strategic choice, the conclusions of this conference regarding the existence of international business as a separate discipline in a sense become a conditioning factor in the conclusions of this session. If a business school's strategy is formulated on the basis that "all business is global," it is apt to be quite different from a strategy formulated on the basis that there is a separate discipline such as international business. Even if the school's curricular response is to expose all students to the global business environment, the appropriate strategy for meeting that mission is conditioned on how the content to which the students are to be exposed is organized from a conceptual perspective.

Thus, one major conclusion that can be drawn from the panel discussion is the need for each academic administrator charged with designing an internationalization strategy for a particular business school to provide an institutional response to the issue of whether international business is a separate discipline. If it is not, the globalization of the school can proceed with no fear of omission from either the curriculum structure or the research program of the business school.

External Issues in Strategic Planning

In their papers and discussion, the panelist seek to characterize the issue of the external factors which are providing the impetus to the internationalization process. In both the research and educational dimension, the driving force behind the internationalization of the curriculum has been the globalization of the enterprise. Hawkins, in a separate paper prepared for this conference, dates the emergence of the research interest in the global corporation to the 1960s, following a historical process of descriptive, explanatory, evaluative and normative research which has parallels in other disciplines of academics. In his opinion the barriers to research have proven more tractable, perhaps, than the barriers to education, possibly in part because research is certainly more personal than institutional in nature, or at least consumptive of institutional resources. Scholars, fired by the emergence of a major phenomenon, the multinational corporation, have been able to overcome institutional barriers to the extent that at least the question of whether the discipline should be organized separately can be addressed.

However, academic institutions appear to need more driving from external sources

to integrate that understanding of the global firm into the curriculum. Miller argues that the issue is one of disagreement about the means rather than the ends. However, Dufey, in his comments describing the process by which the deans of the University of Michigan Business School became sensitized to the need of integrating international business into the curriculum of the school, stresses the need for demand from students and visiting committees of corporate constituents to globalize the curriculum.

The stakeholders of the business school who are most active in pressuring for increasing global involvement are the students and the businesses who hire them, as both Miller and Kujawa note. The programmatic emphasis of the internationalization efforts at most business schools is not surprising, as pressure from these stakeholders provides the impetus for deans to address the internationalization of educational programs first, rather than research. Kujawa characterized the process of internationalization as a "pull through" situation driven by the customers of the business school. In effecting change, this pressure is likely to be stronger than the "push through" approach resulting from faculty internationalization initiatives.

Miller, Bess, and Kujawa all see the accreditation requirements of the AACSB as a fundamental environmental force which has shaped the institutional response to incorporate global materials in the curriculum. Miller views disaccreditation primarily as a potential but questionable institutional hammer used to obtain the cooperation of resistant faculty. Bess views the original globalization of curricula requirements as lacking enforcement, which confirms the lack of usefulness of the threat of disaccreditation in enforcing meaningful change. As accreditation standards have changed, to bring standards more in line with the mission of the business school, Kujawa forecast a substantial and positive impact of the accreditation standards on the strategy of the institution, as these standards have become more market-based. In the panel discussion, in response to a questioning of the relevance of accreditation standards based almost exclusively on the institutional perspective of the United States, Kujawa argues that past accreditation standards had been confining, failing to allow for enough professional creativity or individual institutional responsibility, and refers to the rich diversity of business programs outside the United States as evidence of the hindrances by these standards. In the ensuing discussion, Bess concludes that membership in the AACSB is unlikely to be granted to non-U.S. business schools.

Strategic and Programmatic Initiatives

Because of the identified external pressures for globalization of education programs, the process and content of change in education programs are primary concerns of the panelists in considering the impact of institutional arrangements on the internationalization of business. Miller poses as two of his three fundamental questions: Who will be responsible for teaching the international dimensions of business? How does one go about incorporating internationally oriented issues into an already overstuffed array of course? In view of his assertion that all business is global business, it is not surprising that Bess further argues that management of business in a global context must permeate all programmatic initiatives of the business school. The globalization of curriculum is more central to the mission of most business schools than the development of a major international business research capability.

As Hawkins also observes, basing internationalization on curriculum developments rather than research endeavors provides the opportunity for smaller schools to make progress without the need for developing a significant international business research capacity. He calls for significant innovation in the approach to internationalization, basing curriculum content on a detailed analysis of the needs of program's graduates. Such an approach requires a redefinition of course ownership, and Hawkins argues that

it may be necessary to secure a separation of the educational program structure from the organizational structure of the business school. In that manner faculty members from different functional disciplines have the opportunity to work together in team-taught or sequential case method to inject international material into the core curriculum. Hawkins call for the educational experience to be a seamless educational stream rather than a bunch of discipline-based three credit hour courses. At Rensellaer this approach has been incorporated into an executive M.B.A. program, with the intent to apply that model to the M.B.A. program and then to the undergraduate program.

In both the paper and panel discussion, Kujawa addresses the issues which led to the development of the new international business degree initiative at the University of Miami. After conducting a baseline study of current programs, faculty, student body and constituents, Miami identified internationalization as the institution's number one mission objective. This conclusion was based on the opportunity which was offered by the international area for the development of a program which would be able to sustain itself on a zero-based budgeting basis. The decision was undertaken in the context of a private university in which the business school represents more than twenty per cent of the university enrollment.

Miller notes that the internationalization of peer institutions also provides a powerful motivation for business schools to internationalize their own programs. In response to the perceived market demand for greater international content, business school are seeking to globalize their curricula. However, as Dufey points out in the panel discussion, that globalization can occur only to the extent that the business school devotes resources to the development of faculty and meaningful curricula. Dufey indicates his concern about the possible development of what he refers to as a "globalization race." He is concerned about "amateur hour," in which the globalization effort is placed in the hands of faculty who do not have sufficient understanding of the institutional framework of international business. Ill-judged globalization efforts create a negative opinion toward international business if these efforts fail.

In the panel discussion the tension which exists between the career objectives of internationally oriented faculty and the goals of administrators to develop programmatic initiatives with an appropriate and accreditable global content clearly emerged. The identified domain of international business as a field of inquiry does not span the material needed for programmatic initiatives. Hawkins argues that the assembly of a group of international business scholars active in research is neither a sufficient nor necessary condition for the development of an educational process which provides the appropriate level of understanding of the global dimension of business for the majority of students in the business school's academic programs. However, the development of an organizational structure based purely on programmatic initiatives fails to provide any incentive for the theory-building activities needed for the discipline, and only limited incentives for the empirical research needed to sustain that theory-building effort.

STRATEGY IMPLEMENTATION—THE DESIGN OF ORGANIZATIONAL STRUCTURE

All authors and panel members consider the development of the proper organizational structure for the international activities of the business school. In the panel discussion particularly, the organizational issue is usually restricted to the proper organization for the delivery of instruction relating to the international business discipline. In his paper Miller rightly adds the need to consider organizational issues concerned with the academic tasks of international business theory-building and empirical research. Kujawa links issue of the

appropriate organizational structure to the issue of whether or not there exists a distinctive theory base for the discipline, arguing that the existence of a core theory of international business would require the formation of an international business department to institutionalize the study of that core theory. Kujawa argues persuasively that such a core theory exists. However, as noted above, he further recognized that the institutional environment outside of international business fails to accept the academic validity of this core theory.

The strength of the departmental structure in the business school seems to play little role in the form of organization selected. Michigan's small International Business core group is embedded in a departmental structure which is classified as "weak." The University of Hawaii's approach of requiring an international background of all newly hired faculty has taken place in a situation where the departmental structure is viewed as not strong relative to the power of the dean. The University of South Carolina's separate international business department structure has been formed in an environment of relatively weak department (actually called program areas), although it has survived and indeed hardened as the departments have become more separated. In contrast to the relatively weak departmental structure prevailing at these schools, the University of Miami's experience in developing their M.I.B.S. program took place in what Kujawa characterizes as a strongly departmentalized organizational structure. As noted above, Kujawa argues that the organizational barriers to internationalization arising from the current departmentalized structure of business schools, are falling. The concerns of leadership capture, which Kujawa expressed in this discussion of the strategic planning process, also are present in the implementation process, as well.

Kujawa's rather optimistic attitude may be influenced by the reported success of the University of Miami in overcoming possible departmental barriers in the development of their M.I.B.S. program. With the initial impetus for program development coming form the dean's response to the increasingly international environment of the Miami community, the task was accomplished through the creation of a twenty person task force incorporating members from all eight departments. The final design of the curriculum structure itself reflects the representation of all departmental interest, including the opportunity for students to select from a wide variety of functional majors. The experience of the University of Miami demonstrates the importance of developing faculty consensus in sustaining international programmatic initiatives. Although the initial direction for an international business program came from the business school's dean, the program survived a major administrative shakeup which took place at the same time as the program's introduction.

Dufey also stresses the importance of informal contacts between the core IB group and faculty with international interests organizationally located in other academic areas. Techniques used include involvement in international research projects, inclusion in seminar invitations and interest groups. He refers to this process as "creating converts," faculty from other areas who develop an international capability, and further recommends the judicious use of sabbatical assignments to achieve this objective. In his comments regarding their organizational structure, Dufey indicates the value of cooperation and coalition building in securing adequate resources from the administration of the business school. The strategy selected a the University of Michigan was to create a group of five or six international faculty members fully committed to IB, with four of them tied to functional departments as well through dual appointments. Dufey argues that placing the faculty exclusively in functional departments has the effect of submerging their international interests.

In his comments for the panel, Hawkins

argues that the organizational structure is subordinate to the educational mission of the school. Miller relates the approach taken in instruction to the institutional housing of international business. Schools that offer a single international business course tend to house that course in one functional department. In a later stage of development, schools utilize existing departmental structure to infuse international issues into the functional area core course or create specific international course for the specific disciplines. This, Miller's view agrees with that of Hawkins in relating the organizational structure adopted to the approach adopted for teaching intentional business.

However, the panel did not address in depth the interactions of that mission with the research and service mission of the business school. Miller does note that housing international business as a component within one or more other disciplines provides disincentives to international research projects and fails to introduce an international dimension into the performance evaluation process. Further, Miller does indicate that the creation of a separate international business department does provide a degree of energy for internationally oriented faculty to produce high quality and high output research. He cautions, however, that such a department may become isolated, inbred and produce research of a reduced intellectual quality, and suggests that an openness to faculty members from different disciplines be practiced to overcome these dangers.

However, much remains to be said about how the business school develops an institutional structure which contributes to the development of a strong international business research capability. In almost every discussion of the issue, faculty research is viewed as either idiosyncratic to the individual faculty member, who needs to be bribed to internationalize, or as an opportunity to develop international capabilities which contribute to the quality of the internationalization of teaching.

Strategic Implantation and Resource Allocation

Dufey, in his review of the development of international programs at the University of Michigan, stresses the importance of providing the appropriate incentives to faculty to engage in international activities. Faculty released time from teaching responsibilities has prove to be the most effective in the Michigan experience, with incremental funding for travel second in order of attractiveness to faculty. Bess argues as well for the efficacy of travel funding. Kujawa includes released time and summer or additional compensation for faculty research and foreign travel support as significant in the University of Miami effort, and makes the further point that the method by which these resources may be allocated can help to gain acceptance for their utilization in international activities. Miami uses a faculty committee to allocate these funds in response to faculty generated proposals, perhaps in response to the strong departmental structure at the school.

However, Hawkins makes the major point that international activities must be clearly related to the mission of the given business school. Those activities in turn compete with other activities, such as the sensitization of all students to the management of technology and total quality management, for the limited resources of the business school. Even though Hawkins identifies areas of potential complementarity between international business and other dimensions (particularly cultural diversity and ethical issues), he argues that there is a tradeoff which must be addressed between these competing demands. Factors which need to be considered include the size of the business school, whether it is publicly funded or private, and whether it has an undergraduate mission only or a commitment to excellence at the graduate level. Hawkins argues that international business education is accomplished through diversity, and that the Japanese accomplish their international business education through a number of departments, institutes and centers not connected with a Western-style management education program.

Although the panel may have assumed that the potential for resource generation is axiomatic in the review of programmatic initiatives, perhaps it should be emphasized, in line with Dufey's remarks about the "globalization race," that properly designed international business educational endeavors have a large potential for generating higher levels of resources for those business schools who run well in the race. While it is not appropriate at public academic institutions to require each academic program to be self sufficient, the experience of the University of South Carolina demonstrates that substantial financial advantages accrue to the institution from the establishment of a strong market position in international business education. The issue rather becomes one of insuring that the academic programs thus developed continue to provide adequate support to the international endeavors of the business school.

FACULTY ISSUES

Hiring Policies

The panel devoted a considerable portion of its time to the administrative issues regarding faculty, in the area of hiring, evaluation, promotion and tenure and personal development issues.

The first method of developing an internationally oriented faculty is through control of the hiring decision. Miller suggests that the criteria for joining the faculty would require that each candidate possess a strong basic business discipline and a demonstrated international orientation to one's teaching and research. Reviewing the experience of the University of Hawaii, Bess notes that no one is hired unless they come from an international background, regardless of the academic discipline to which that position is assigned. This hiring policy is related to the policy at Hawaii of internationalizing the entire college. As Bess indicates in the panel discussion, the ability to utilize this method for rapid results depends on the maturity of existing faculty when the process starts. As Bess rightly points out, a faculty dominated by mid-career faculty required a different approach.

The policy of hiring faculty internationally competent in their particular disciplines is an ideal, and even if reached does not provide expertise in the core discipline of international business. An international faculty whose primary conceptualization of research problems is functional will, most likely, fail to provide the cross disciplinary perspective necessary for theory-building in international business.

Dufey reports major success at the University of Michigan in the use of adjunct, non–tenure track faculty in international business education, and notes the necessity of limiting their teaching loads and involvement in other academic duties. The use of adjunct faculty in this manner is to be applauded, provided they are drawn from sufficiently high levels of management involved in global business to provide an integrated treatment of international business issue in the classroom. However, adjunct faculty do not usually add to the research capability of the business school.

Evaluation, Compensation, Promotion, and Tenure

Faculty who commit to international activities, whether teaching or research, confront a number of issues regarding performance evaluation, which in turn impacts their incremental compensation, promotion and tenure. Miller warns that they may encounter resistance from faculty who are skeptical of the value of international business research, and may be evaluated on standards which are applied to the more well established basic disciplines. In his discussion of experience at New York University, Hawkins notes the difficulty of providing appropriate positions for international business specialists in the matrix organizational structure used at that school, perhaps caused by an absence of rigorous

analytic training. Even given the much stronger methodological and theoretical training of current doctoral graduates in international business, Hawkins considers that they would probably not survive in a school with a rigid departmental structure.

Yet international business faculty bring major strengths to the universities where they are located, in the sense that they have the ability to interact with faculty from two or three other disciplines. The interdisciplinary nature of their capabilities, which normally is advantageous to the academic institutions, may prove to be a disadvantage if research collaboration is viewed negatively in the promotion-tenure process. Bess argues that for this reason promotion-tenure requirements for internationally directed faculty allow credit for publication in nonleading academic journals which encourage interdisciplinary research.

Administrative support for the evaluation of international activities in the tenure and promotion decision is one of the major methods which can be used to set the international direction of an institution, as evidenced by the strong support provided by Bess in the process at the University of Hawaii. Kujawa notes that Miami has not yet addressed the issue of incorporating international activities in the tenure and promotion evaluations of its faculty. In reviewing the comments of panel participants, it appears that the approach of providing specific international criteria for promotion and tenure is currently not very common. The inclusion of international activities as a factor in this decision is being accomplished more by leadership and commitment of various deans, rather than through institutional efforts. Such an approach may be dangerous in an era where the average dean's appointment is approximately four years, significantly shorter than the usual time to tenure.

In assessing the impact of the tenure decision, many of the panelists argued that the tenure system is a detriment to the internationalization process. Miller notes that faculty who have received tenure are less likely to be able to be motivated by that system to increase their international commitment, but require financial compensation for the effort. However, Dufey argues that faculty who have received tenure are less constrained in their research agenda, and have greater freedom to engage in the types of activities which develop their international skills. Dufey agrees that the tenure incentive in later career is replaced by the compensation incentive, with the market working effectively to allow faculty mobility. Bess adds that appropriate use of the tenure incentive in a business school with substantial administrative input into the tenure decision provides a strong incentive to junior faculty to engage in international activities.

Faculty Development

The demand for international educational programs requires a substantial commitment to the development of appropriate teaching and research competencies on the part of the existing faculty, and the continued future development of those who are appointed with international capabilities. Miller refers to faculty development as the key to the institution's internationalization process, and devotes a major portion of his paper to this process. Once again, the primary thrust of the proposed faculty development activities is directed towards those which have as their primary function the development of teaching competence.

Miller recommends participation by faculty international workshops and summer programs, and strongly recommends the use of faculty overseas internships. Another strategy which Miller recommends is the assignment of faculty to participate in management or executive education programs in which the participants have international experience. Kujawa notes the importance of the establishment of linkages with institutions outside the United States, which can be used to provide vigorous assignments to faculty to develop their international expertise. Faculty can be used in teaching in academic an executive programs at the partner institution, and in providing technical assistance in the design of curriculum.

Involvement in international research also develops the international competence of faculty. Miller views such an activity as promising a quick contribution to faculty development, and recommends joint research projects with foreign scholars or with internationally competent colleagues. However, given the negatives associated on some campus with joint research and the logistical and data acquisition difficulties of conducting research on international issues, the opportunity cost of such research endeavors may be higher and the timing of the impact on the individual later that might initially be thought. Kujawa argues that such research may be rewarding if conducted with faculty from linked overseas institutions, where other points of contact between U.S. and foreign faculty are in place.

It is now clear that the technology for the development of international teaching and research competencies for functional specialists exists, and that the primary stumbling blocks to developing such capabilities are high cost and faculty reluctance. As indicated earlier, Kujawa views these barriers as breaking down, and foresees the emergence of international "functionalists" as a response to the educational change in direction of the business school.

The Role of Doctoral Programs in Faculty Development

If business schools are to be successful in hiring internationally oriented faculty, they must have a source for such faculty, and that source primarily is the doctoral degree granting business schools. Thus, one of the major issues addressed in the panel is the proper approach for doctoral education. Two separate tasks were identified in the panel discussion. The first is the preparation of specialists in the international business domain, and in the international dimensions of particular academic disciplines. The second task is to provide functional field specialists. Hawkins argues strongly for a combination of academic rigor and inter-disciplinary creativity in the training of doc-toral students in international business, but warns of the dangers of subsequent affilia-tion with institutions organized on a strong disciplinary basis. Bess notes that one result of the international development of current faculty is the transmission of that orienta-tion to doctoral students with whom they interact. Nonetheless, the absence of inter-national training in the doctoral programs of currently recognized business disciplines, and the existence of a core body of theory in the international business discipline argue for a major strengthening of the in-ternationalization efforts at the doctoral level. Regrettably, these programs will not be internationalized until their customers (the academic departments who hire new faculty) demand it, and given the functional field structure of most business schools, those departments are not yet ready to com-mit to hiring requirements like those of the University of Hawaii. As Bess clearly identifies, the absence of AACSB accredita-tion of doctoral programs removes them from any external incentive to internation-alize, although Bess does see a change in the approach of the AACSB to doctoral pro-gram accreditation forthcoming, through the requirement that doctoral students be adequately prepared for their teaching re-sponsibilities, which clearly include incor-poration of the international dimension of the functional field in basic courses.

SHOULD THERE BE AN INTERNATIONAL BUSINESS DEPARTMENT?

This summary is concluded by a review of the authors' answer to this basic ques-tion. The issue itself is a derived rather than fundamental issue, in the sense that a sepa-rate international business department is an organizational decision derived from the re-search and program strategy decisions of the given business school, which are in turn derived from the school's mission. Thus, as all panelists are clear in concluding, there is no single answer to this question.

However, from the authors' papers and comments one can identify those situations where an intentional business department would be useful. Clearly, where there is a recognition that international business is a separate field of inquiry, and where the busi-ness school chooses to make a major re-search commitment to that field of inquiry, an international business department is use-ful for developing a critical mass of similarly inclined faculty. Where the theory devel-oped by the scholars of the discipline is ignored by functionalists, even international-ized functionalists, in the design of curricu-lum, an international business department would provide the expertise to preclude such an omission. Such a department might serve as a bridge or temporary organization until the internationalization of the functionalists is completed, as Bess discusses.

An international business department may prove useful as a device to concentrate resources on the international endeavors of the university. As Miller argues, a separate international business department, success-fully implemented, does lead to the produc-tion of high quality and high output research. It provides a useful mechanism for providing some degree of protection with regard to the promotion and tenure process, allowing a more liberal evaluation of what constitutes peer acceptance in refereed jour-nals. Finally, an international business de-partment is the logical place for the granting of an international business doctorate, if such a degree is to be concerned with the discipline's core theory.

However, it is clear that the creation of a separate international business faculty is inappropriate if the institution's only inter-est is in international business education, or if the resource base of the business school would be strained by the acquisition of this specialized expertise. Even if an interna-tional business department is created, Miller observes that special precautions must be taken to protect it against isolation, inbreeding of the faculty, and reduction in the intellectual quality of research pro-duced. What emerges from both Bess's and

Miller's comments is that the international business department be a very open one; in Miller's apt phrase, its boundaries must be easily penetrated by faculty members from different disciplinary orientations.

A FINAL WORD

The vision which the panelists brought to the issue of the institutional setting of the international business research and education task of the business school has provided significant insights into a myriad of issues internal to the business school. Probably because of the design of the charge, what has not been addressed in any systematic fashion is the relationship of the international efforts of the business school itself to the other international research and programmatic efforts of the university where that school is located.

Little has been said and much needs to be about the international strategy of the university as a whole and the relationship of the business school to it. At the University of South Carolina, the emergence of the faculty and programs in International Business have had a major campuswide demonstration effect, providing an incentive to other areas of the university to seek to develop international competence. Although the role of the College of Business Administration has been to draw along other partners in international endeavor, and to provide a platform of achievement and credibility on which the rest of the university can build, it is of necessity a two-way process, with the international achievements of other disciplines providing an impetus and support for the further development of International Business and the College of Business Administration. There is no reason precluding the development of crosscampus alliances and structures on some campuses, in which the business school combines with other units in the development of programs of re-search and education. The possibilities for such combinations represent potentially fertile terrain for business schools to explore. If alliances among departments may secure additional international resources, carefully crafted alliances among larger university units may be even more rewarding.

REFERENCES

Blum, Debra E. 1992. Harvard's Kennedy School of Government, facing criticism, examines its mission. *The Chronicle of Higher Education* 26 February, A20.

Edelstein, Richard. 1992. International Affairs. Speech read at the AACSB to CIBER Conference, 7 February, at the University of Washington, Seattle, Washington.

Erskine, James A., M. R. Leenders and L. A. Mauffette-Leenders, 1981 *Teaching with Cases*, London: University of Western Ontario.

Hawkins, Robert G. 1984. International business in academia: The state of the field. *Journal of International Business Studies* 15(3): 15.

Leenders, M. R. and James A. Erskine. 1989. *Case Research: The Case Writing Process*. 3d ed. London, Ont.: University of Western Ontario.

Menacker, Julius, and S. Tanabe. 1991. Education law and policy: An American-Japanese comparison. *The Educational Forum* 55, no. 3 (Spring): 215–31.

Perry, Richard J. 1992. Why do multiculturalists ignore anthropologists? *The Chronicle of Higher Education*, 4 March, A52.

Selz, Michael. 1992. For many small firms, going abroad is no vacation. *The Wall Street Journal*, 27 February.

Selz, Michael. 1991. Foreign language studies still rate at U.S. B-schools. *M.B.A. Newsletter* (November/December).

Stabler, E. A., and Michael O. Suilleabhain, eds. 1990. *Research on the Case Method in a Non-Case Environment*. Cologne: Bohlau.

Yin, Robert K. 1984. *Case Study Research: Design and Method*. Beverly Hills: Sage.

6

Educational Implications for International Business Inquiry

EDITORS' COMMENTS

We believe—as we are sure most do—that to be of value, the knowledge developed by scholars must be eventually disseminated. Thus, for any discussion of a field's development and its output to be complete, it must include discourse concerning teaching and learning. But, as the authors in chapter 2 collectively point out, teaching and learning take a variety of forms. Also, these forms need to be responsive to shifts in a field's conceptual domain and perspectives, and the new findings these shifts bring about. Accordingly, this panel was charged with addressing the following question: What are the most effective and efficient teaching and learning forms for insuring the IB dimension is adequately and properly covered in formal degree programs, in-house executive training programs, and so on?

Four leading IB scholars were asked to draft papers and participate in a panel discussion that addressed this central question. The panel included J. Frederick Truitt, Willamette University; Franklin R. Root, the University of Pennsylvania; Paul W. Beamish, University of Western Ontario; and John D. Daniels, University of Richmond. David A. Ricks (Thunderbird) served as the panel chair.

While Paul Beamish addresses the pedagogical value of the case method for IB education and the barriers faculty must overcome in order to write cases, the other three authors take a more comprehensive view and address such varied topics as the field of IB, methods for internationalizing the curriculum, the need to incorporate the study of a foreign language, the intended audience of IB education, and whether business schools need to become truly international in order to remain viable in the years ahead.

Because of their more inclusive discussions, it is not surprising to find that Truitt, Root, and Daniels highlight some similar issues and problems, and identified some common concerns. These include a need for foreign language instruction, the multidisciplinary nature of IB education, and the problem of repetition that arises because of specialization. At the same time, there is some disagreement. Whereas Daniels and Root accept the inevitability of knowledge and skills overlap, and too some extent believe that repetition is conducive to learning, Truitt tends to disagree and suggests that IB "transfusion" rather than IB "infusion" would reduce the unneeded duplication that might occur if IB were taught as a core course, and material repeated in functional courses.

In the first panel paper, Truitt states that IB teaching is faced with extraordinary opportunity. In order to grasp this opportunity, however, four initiatives are required. First, business schools need to be internationalized and integrated (that is, the current functional and crossdisciplinary fragmentation emphasis needs to be abandoned). Additionally, the poor achievement of U.S. students must be addressed directly and openly, and existing M.B.A. programs reformed accordingly.

Second, business schools need to distinguish between two different missions: the internationalization of all business students, and the education and training of international business specialists. Moreover, the appropriate level of internationalization can be judged using three key elements: cognitive, behavioral/experimental, and foreign language. That is, the successful internationalization of all students requires the introduction of new teaching materials and

the use of cases to change their intellectual understanding of the way the global political economy actually works. Furthermore, it is desirable that they should have the opportunity to interact effectively and affectively with the international dimensions of the global political economy. Lastly, students need experiences that provide insight and appreciation of the role of language and culture in shaping approaches to economic problems. In contrast, the cognitive (knowledge), behavioral (experience), and foreign language elements are more or less essential to the education of IB specialist.

Third, of the four models of internationalization followed by business schools— IB core course, international functional electives, infusion, and transfusion—Truitt favors the transfusion model which he describes as an IB specialist teaching modules in the various functionally-oriented classes. Supportive of our liberated business education recommendation explored more fully in chapter 7, he also favors a joint liberal arts undergraduate degree (usually foreign languages) combined with a business graduate degree. This particular recommendation is in response to what he, and many others at the conference, sees as the multidisciplinary nature of IB education.

In the second paper, Root summarizes the evolution of the field of international business from the 1920s to the present, and suggests that IB is not just multidisciplinary, it is a metafield, permeating as it does all fields. He observes that IB, as a field of study, depends on political economy, culture, sociology, and comparative management. As a consequence of its metafield status and multidisciplinary nature, he further believes that IB as a major will wither away as U.S. business schools internationalize their programs.

He continues by suggesting that since U.S. schools are in a global marketplace for both students and faculty, they will have to become International Schools of Business. That is, to survive, they will need to design and implement global strategies in business education.

In the third paper, Paul Beamish presents an excellent discussion of the "ins and outs" of using the case method in IB education, and how to improve both the students' and teacher's experience. At the same time, however, he observes that it is difficult to learn of the existence of some cases, and suggests that perhaps we need a clearinghouse for international business cases. Also, partly because of the view taken toward the writing of cases, he explains why many cases need coauthors. He concludes by reminding us of the need to match cases with the audience.

In the fourth and final paper, John Daniels recommends that business schools need to expand on their missions since IB education is of potential benefit for several audiences, including such untapped markets such as nonbusiness majors. He further recommends that the teaching of IB needs to be varied according to the audience's level of proficiency and need (for example, informed citizens, knowledgeable managers, and IB specialists). Although he recommends an interdisciplinary approach to the teaching of the subject, he recognizes that course overlap will occur and should be accepted.

In a vein somewhat similar to ours in chapter 1, he further notes that the environment of international business is changing rapidly and that it is therefore more difficult for IB scholars to remain current. This is made even more challenging because of the crossfunctional and crossdisciplinary nature of the IB field. Finally, he makes several additional recommendations that bear on pedagogy and the need to interact more directly and more frequently with non-IB faculty.

PANEL CHAIR'S COMMENTS
(DAVID A. RICKS)

As noted in the introduction, the panel's discussion was based on papers by Frederick Truitt, Franklin Root, Paul Beamish, and John Daniels. Their four papers immediately follow these opening comments.

Fred Truitt, in his paper, distinguishes between two different missions of a business school: the internationalization of all business students, and the education and training of international business specialists. He then suggests that we approach these two missions differently while considering three key elements: cognitive, behavioral/experimental, and language. His conclusions are summarized in table 6.1.

Franklin Root summarizes the evolution of the field of international business in his paper. He then suggests that international business is not a single discipline, but it is a multidisciplinary field of study. Moreover, international business education should focus on the nexus of national governments and international firms. He observes that the real core of international business is made up of three areas: the political economy, national culture/society, and international (global) management.

Perhaps to generate discussion, Franklin Root also predicts that international business as a major will disappear in the next

Table 6.1

	All business students	International students business specialists
Cognitive	Essential	Essential
Behavioral/ Experiential	Desirable	Essential
Language	Optional - "Appreciation" level	Desirable - Essential

decade. He believes that the challenge for the 1990s is to transform the business school into an international institution; to make the business school a member of a global educational network. He concludes that for business schools to survive, they will need to have global strategies in business education.

Paul Beamish discusses international business education via the case method. He observes that it is difficult to learn of the existence of some cases and that perhaps we need a clearinghouse for international business cases. He also explains why many cases need coauthors and the value of forcing students to take different perspectives. He concludes by reminding us of the need to match cases with the audience.

The major point John Daniels makes in his paper is that we need to expand our missions. He suggests that we teach a greater variety of people (for example, more non-business students). Daniels also concludes that he is not concerned with the fact that researchers' efforts overlap or that our domain is not clearly focused. In fact, he considers this good. Furthermore, he does not believe it wise to limit ourselves with narrowly defined domains.

Daniels also suggests that an abstract service for research and a clearinghouse for data on visual aids and cases be created. He closes by suggesting that we have workshops on teaching international business.

I recommend these four thought-provoking and informative papers to you.

. . .

Perspectives on International Business: Theory, Research and Institutional Arrangements: "Teaching Implications for IB"
J. Frederick Truitt

INTRODUCTION

The world of business is internationalized to an extraordinary degree. National economies are intertwined and interdependent as never before and competition is increasingly global. It takes a team of government cost accountants to decide if Canadian made Japanese Hondas assembled in Ontario for the U.S. market are or are not North American. Ford Crown Victorias are converted from domestic to foreign with the flip of a few parts from U.S.-made to Mexican-and European-made. Boeing worries about competition from a crossnational European consortium that did not even exist a generation ago. McDonnell Douglas responds to international competition by selling 40% of its commercial aircraft business to a Taiwan government owned corporation that never built a plane and did not even exist last year. Caterpillar worries about Komatsu, while GE and Westinghouse keep their eyes on Rolls Royce and Siemens and ASEA-Brown-Boveri. Foreign markets and foreign suppliers are increasingly important to medium and small firms as the international challenge and international response come earlier in the life cycle and push down into the ranks of smaller firms. "Small companies that once would not have dreamed of operating on a global basis are moving very quickly to establish foreign offices" (*Wall Street Journal* 1992). The teaching of International business is at the crest of a great wave of extraordinary opportunity. The question is how to take advantage of this opportunity.

INTERNATIONALIZE AND INTEGRATE

The twin need to internationalize and integrate U.S. higher education—especially higher education in business and management—is obvious now even at the "street level." U.S. higher education for business and management, in spite of calls to internationalize which were first made twenty years ago, has remained remarkably domestic and insulated not only from the developments outside the U.S. but insulated also from the rich offerings of other departments and disciplines within the larger university setting, and even in too many cases insulated from the developments and insights offered by other departments in the business school. Students' behavior in class too often reflects this insulation. How often have we heard students say something to the effect of, "But this is a course in marketing; I took economics last year. Don't expect me to incorporate what was covered in that course into this course." That was that and this is this, and never the twain shall mix, especially not in teaching or learning. Hence the concern of, for example, FIPSE (Fund for Improvement of Post Secondary Education) to encourage innovation in post secondary education which emphasizes genuine and effective cross disciplinary exchange and reform that "works" in the classroom.

Exhortations to internationalize have been a steady, if largely unheeded, feature of teaching international business for a generation. As teachers of international business we are more aware of these exhortations

than our purely domestic colleagues in accounting and marketing, hence many of us have slightly ambivalent feelings toward the recent flood of calls for internationalization, and invitations to endless workshops and conferences that offer to internationalize and initiate the domestic and uninitiated. In fact, some of us might even twitch quietly with the same chagrin felt by anthropologists who now must witness colleagues in the social sciences and humanities "discover" multiculturalism, and in the process of discovery, experience profound (and politically correct) shifts in their personal worldviews. English professors and other nonanthropologist Johnny-come-latelys behave as though no one had ever before thought to study cultures or societies beyond the Euro-American sphere (Perry 1992, A52).

But recently the call to internationalize has become more urgent. Two strands of societal criticism of education lie behind this new, urgent call. First, there is the growing realization of a mounting dissatisfaction with the poor achievement of U.S. students. While the U.S. spends more on education as a percent of GDP than most developed countries, the achievement scores of U.S. students—especially in science and math—rate down low at a Third World level. While numerous studies point out our weaknesses, the National Commission on Excellence in Education's *Nation At Risk* (April 1983) is probably the key to start understanding contemporary dissatisfaction with U.S. education—dissatisfaction which has most recently led to the White House's "America 2000 Plan" at the national level and a plethora of plans and reform initiatives at state and local levels.

A second strand of criticism focuses more directly on business education—in particular M.B.A. education. (Upon receipt of 1990 Nobel Prize in Economics, Chicago professor Merton Miller was asked why U.S. competitiveness lagged Japan. He said we should not worry too much about this lag because given the increasing number of Japanese studying in U.S. M.B.A. programs Japanese business performance would soon be dragged down to the U.S. level.) But *Business Week*, the *Economist*, and *Fortune* have all raised questions about the M.B.A. course of education recently and encouraged reform. For better or worse, international business is at the confluence of both the dissatisfaction with U.S. K-12 education and criticism of M.B.A. education. The precipitating catalyst of concern for education and internationalization is the palpable loss of industrial preeminence to Japan, Germany and the NICs—especially the Asian NICs. The belief (one could argue this is an irrational and misplaced belief) that sustaining U.S. competitiveness in the new global economy by teaching new skills and attitudes—especially international skills and crosscultural attitudes—and teaching them better, grows from a core U.S. belief in the value of education in societal change and reform. But this belief is countered by another, conflicting strand in U.S. culture that has created predispositions that make educational accomplishment less attractive and harder to attain (Menacker and Tanabe 1991, 215–31).

Hence we see for example: (1) the emphasis in FIPSE's call for proposals on internationalization of business school programs which reach out to integrate other areas in the university; (2) AACSB and AAC (Association of American Colleges) team up to administer a grant from KPMG Peat Marwick Foundation to enhance internationalization through increased cooperation between business schools and liberal arts programs; and (3) U.S. Department of Education continues to fund Centers for International Business Education and Research at $5 million per year and favors proposals that join business schools and other parts of the university.

If current exhortations to internationalize and integrate are to be more successful than past exhortations, we need to understand why previous calls to internationalize were so often unheeded. In the past, U.S. institutions were reluctant to act on this need to internationalize because natural barriers between disciplines were reinforced by narrowly defined academic reward systems. Furthermore, the isolation of U.S. higher ed-

ucation was reinforced by its large scale and location (distant from challenging examples in Asia and Europe). Ethnocentric, provincial customs prevented people in the United States from adopting a "What can we learn from them?" attitude even when we did engage in international "exchange." Finally, we were insulated by superior wealth, income and economic performance. Only the cumulative and mounting losses of U.S. business and economic preeminence underscored by the palpable success of Japanese economic performance (both in Japan and abroad in such places as Marysville, Ohio, Flatrock, Michigan, and Smyrna, Tennessee), the credible promise of invigorated European economic performance after the completion of the common market in 1992, and the multiplying success stories of the Asian Newly Industrialized Countries have forced some in the United States to see the need for change toward internationalization. The question remains, "Will the reward structure in higher education open and shift enough to encourage and sustain difficult and risky moves to internationalize and integrate?" The answer to this question is still clearly unclear. But it is clear that international business is uniquely able to offer both internationalization and integration to the reform movement in higher education.

THREE ELEMENTS OF INTERNATIONALIZATION

Ideally, international competence has three key elements: cognitive, behavioral/experiential, and foreign language. Consequently, successful internationalization will work at three levels of understanding and appreciation—even though not all graduates will be functionally fluent in a foreign language. At the cognitive level the introduction of new teaching materials and use of cases is needed to change students' intellectual understanding of the way the global political economy and political society actually work. At the behavioral/experiential level students need the opportunity to interact effectively (and affectively) with international dimensions of the global political economy. With respect to foreign language and culture, students need experiences that provide insight and appreciation of the role of language and culture in shaping approaches to economic—and especially to management—problems (Edelstein 1992). The second two components are both more controversial and more difficult and more expensive to provide to students. But they are more important to the new mission of the reformed business school. However, at this point we must distinguish between two different missions of a business school—the internationalization of the education of all students and the education and training of international business specialists. (See table 6.1.)

Certainly the view that business schools should include foreign language instruction is quite controversial. Clearly schools outside the U.S. take foreign language competence as a given prerequisite (for example, INSEAD) for either generalists or specialists. We have much to learn from several Scandinavian schools. Such institutions as Norwegian School of Economics and Business Administration's MIB (Master of International Business) program, the Norwegian School of Marketing, Copenhagen Business School (where students study five to six years to master two foreign languages through intensive language programs overseas and foreign internships), and Denmark's FUHU (Danish Society for the Advancement of Business Education) use interesting and innovative approaches to combining language and business.

THREE MAIN MODELS OF INTERNATIONALIZATION: "ANOTHER COURSE," "SMORGASBORD," OR "INFUSION"

Schools which have heeded the call to internationalize have usually followed one of three models from the menu of internationalization: add a new core course; let students choose from a smorgasbord of international prefix courses; or "infuse" international into existing core courses—usually by sending core faculty off to Hawaii or Co-

lumbia, South Carolina, to hear the gospel of internationalization and get ideas about exactly what to do and how to do it.

The pros and cons of these three basic approaches have been thoroughly and widely discussed in international business circles and there is no need here for more than the briefest summary.

Only a few schools have adopted the "additional core course" approach because of the controversy surrounding the net expansion of the core course requirements, the difficulty of replacing an incumbent course, and the shortage of faculty to teach such a course.

Other schools offer a "smorgasbord" of "international" courses—international accounting, for example, or international marketing. The disadvantage of this approach is that the "international" component is not systematically covered, and students taking more than one international course face haphazard, repetitive, elementary introductions to "international."

Most schools remain content with the more expedient "infusion" approach. That is, faculty—often with no training or inclination in "international"—are exhorted to "Add an international dimension or element to your core course," and "Be sure to cover the sixteenth chapter in your text." (The sixteenth chapter—for a fifteen-week course—is the international chapter added by the author in the third edition to respond to exhortations to internationalize.) Certainly, some faculty in some schools do make a serious attempt at internationalization by infusion. But our belief is that such infusion is pretty weak tea on the average, and successful infusion is the exception rather than the rule.

THE CORE OF THE INTERNATIONAL CORE

This year, in response to an invitation to internationalize the core of a master's program, I found myself invited to make a list of essential topics and issues which international had to offer all students in the core. Table 6.2 is the short version of that list.

Table 6.2 LINE: Learning Objectives for International

LINE's international segment commits Atkinson Graduate School of Management to provide students with an awareness of and understanding of and appreciation for the global environment within which business and government organizations operate. This commitment covers not only the world economy and international competition but also differences between business and economic systems and differences between the cultures which create and manage national businesses and political economies.

Specific components and objectives:

I. International political economy—competition between the philosophies of central planning and market economies and the movement toward free markets
   A. Bretton Woods institutions: GATT, IMF, IBRD
   B. Three political economic models
      1. Sovereignty at Bay—markets
      2. Dependencia—central planning
      3. Neomercantilism—government participation and direction
   C. Patterns of trade and investment in post-WWII economy
      1. Main trading and investing countries and blocs
      2. Main product flows
      3. Relative impact of trade and investment on national economies
II. International trade environment, fundamental trade theories and practices
   A. Trade theories
      1. Comparative advantage as explained by Ricardo and Heckscher Ohlin
      2. "New" theories of trade: intraindustry, product life cycle, strategic trade theory, LDC-dependencia-primary product issues
   B. Trade restrictions
      1. Theory of protection: tariff, non tariff barriers (NTBs)
      2. Practice of protection—how tariffs and NTBs are used by industry and governments
   C. Institutional framework and reduction of trade restrictions
      1. GATT and multilateral trade negotiations in Uruguay Round
      2. Bloc trade
         a. EC, EC 1992
         b. FTA (U.S.–Canada)
         c. North American Free Trade Area (proposed)
      3. What do U.S. government institutions do to promote and regulate trade
         a. Main institutions: Commerce, Treasury, ITC, etc.
         b. Adjustment assistance
         c. VER
         d. Dumping and subsidy remedies
III. The international financial environment
   A. How are international transactions paid for
      1. Foreign exchange and foreign exchange markets
      2. Institutions
      3. Balance of Payments
   B. Introduction to theory
      1. PPP (purchasing power parity)
      2. IFE (international Fisher effects)
   C. Institutions: IMF
IV. Direct foreign investment (FDI)
   A. Portfolio vs direct
      1. Diversification
      2. Market imperfections
      3. Product differentiation
      4. EOS
   B. Political economy of FDI
      1. Host–MNC–Home views of
      2. Cost-benefit analysis of FDI
V. Understanding the effect of different cultures on management in the global economy requires a familiarity with cultural diversity, how different values and customs affect management practices and government regulation, and negotiation.

TRANSFUSION, NOT INFUSION— SIMPLE MODEL

In contrast to infusion approaches which rely on several core course instructors to infuse international into their courses, I propose a "transfusion" program. The first essential step in this program is for core faculty to agree that internationalization should take place. (In fact, in our situation faculty agreed that in addition to international, law, negotiation and ethics should also be infused into core courses over a two- or three-year time period. See table 6.3 for other LINE learning objectives.) The second essential step is for core faculty to agree to "give up" about three or four class sessions from their teaching and allow me to teach the international part of their internationalized core course. (See figure 6.1.) Where possible, this internationalization uses the international competition case study introduced in the Compass Week orientation to introduce or facilitate the study of international business topics and issues. For example, the Caterpillar-Komatsu case series encourages us to use the yen-dollar exchange rate in Second Semester Finance Core to focus our discussion of foreign exchange and foreign exchange markets on how foreign exchange rate changes affect international competitiveness. Caterpillar's Code of Conduct, FCPA, 1977, and ethics in the international context are discussed in Fall Semester Organizational Theory and Behavior, while Japan's industrial policies and its impact on Komatsu is discussed in Spring Semester Political Economy-Government.

The implementation of this kind of transfusion relies heavily on:

1. prior agreement—strong, committed agreement—that the international learning objectives are important and need to be met;
2. a matrix-based plan for allocating the topics among core courses;
3. a flexible enough schedule so that the international instructor can accommodate as well as be accommodated at many different class meeting times during the year; and
4. a dean committed enough to internationalization to allow a full course teaching credit to the international instructor.

TRANSFUSION—COMPLEX MODEL

In the simple transfusion model only one inhouse faculty member—the international business instructor—takes all responsibility for transfusing international content and perspective into the core. A more complex model addresses the need for more genuine cross disciplinary teaching by having the international components in the core introduced by a team of faculty drawn from across campus, but most especially from economics, political science, religion and ethics, Japanese language and Japan studies, and Spanish and Latin American studies. (This more complex version of the transfusion model is illustrated in figure 6.2.)

This case-based, crossdisciplinary integrative learning experience benefits both students in business and students in the liberal arts. Business students benefit from hearing the authentic and credible voice of political science, economics, religion, Japanese, and Spanish faculty. Liberal arts students benefit from teaching materials developed by liberal arts faculty, and the clearer picture of "business" which liberal arts faculty bring back from their sojourn among the Philistines.

Of course, this more complex form of transfusion will be even more interesting and valuable if the faculty from liberal arts can address issues of gender and non-Western views with greater authenticity and credibility than the typical WASP business core faculty.

TWO FOR THE PRICE OF ONE AND A HALF

One of the most interesting features of the growth and evolution of teaching and international business programs in the U.S. is the retarded and reluctant recognition that foreign language training and requirements should be part of our programs. But we were not alone—Harvard University's John F. Kennedy School of Government, the national leader which purports to have expertise in in-

Figure 6.1 Internationalizing the Masters Core

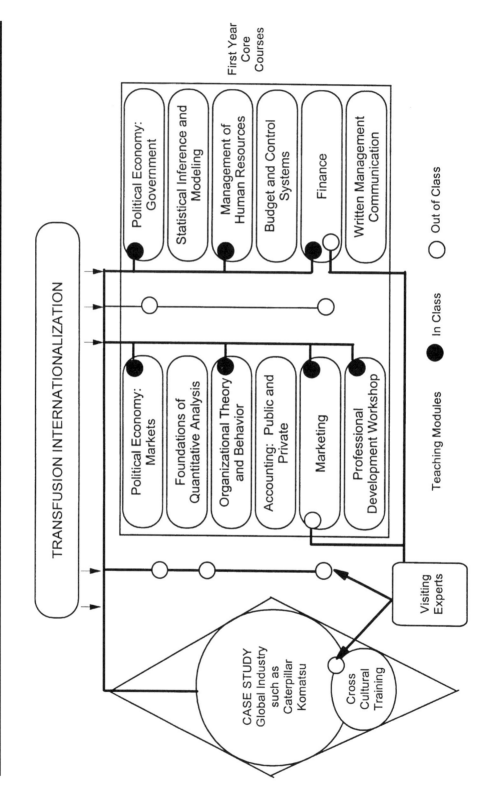

Figure 6.2 Team Teaching the Internationalization of the Masters Core

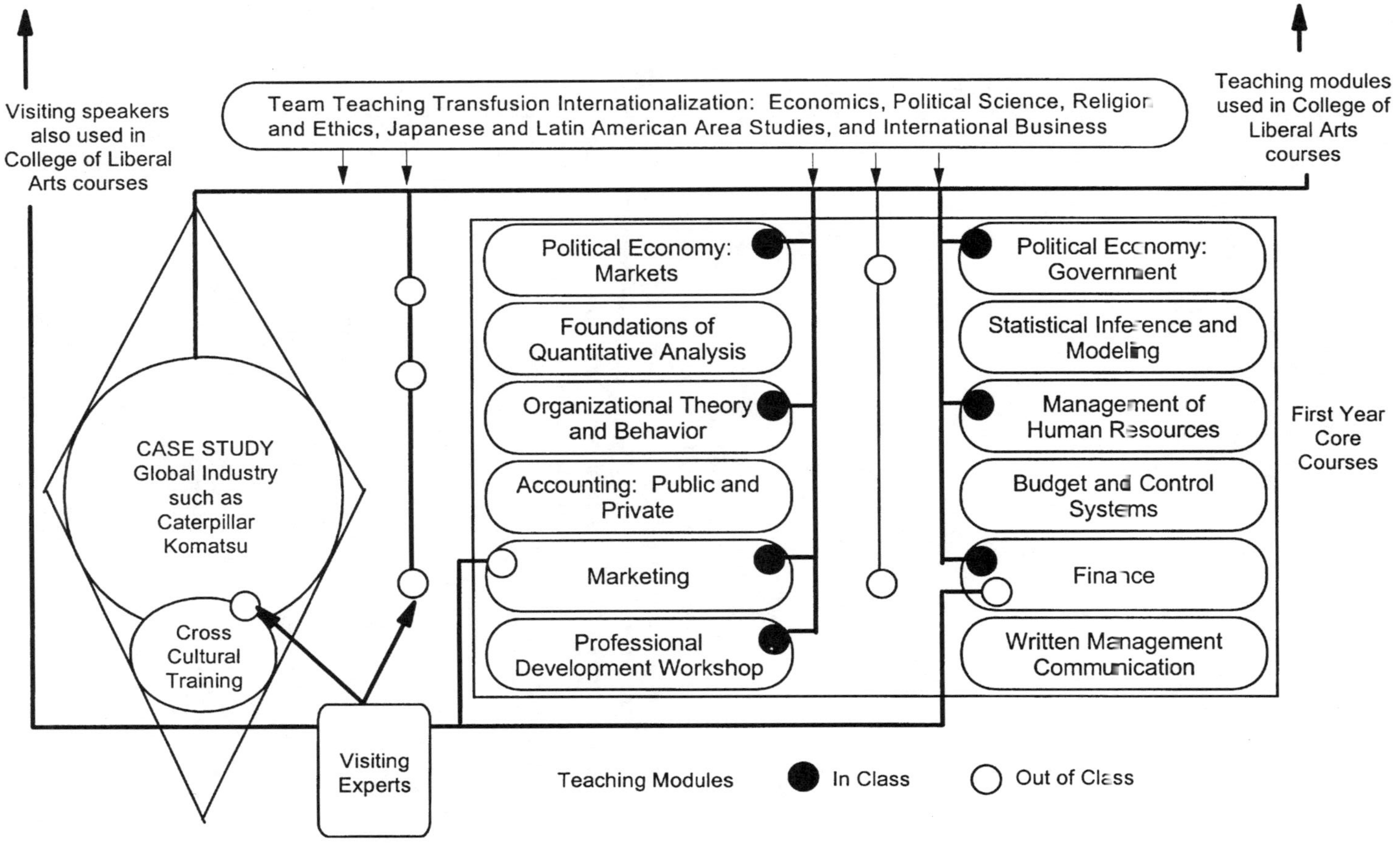

Visiting speakers also used in College of Liberal Arts courses

Teaching modules used in College of Liberal Arts courses

Team Teaching Transfusion Internationalization: Economics, Political Science, Religion and Ethics, Japanese and Latin American Area Studies, and International Business

First Year Core Courses

Political Economy: Government

Statistical Inference and Modeling

Management of Human Resources

Budget and Control Systems

Finance

Written Management Communication

Political Economy: Markets

Foundations of Quantitative Analysis

Organizational Theory and Behavior

Accounting: Public and Private

Marketing

Professional Development Workshop

CASE STUDY Global Industry such as Caterpillar Komatsu

Cross Cultural Training

Visiting Experts

Teaching Modules   ● In Class   ○ Out of Class

ternational relations and development studies, has had no foreign language requirement (Blum 1992, A20). Certainly there were always individual proponents of language instruction in international business programs, and the American Graduate School of International Management always required language, and the M.I.B.S. program at South Carolina wisely required language from its start and achieved such good results one would think it would have been copied and cloned more frequently. Nevertheless, it is remarkable how few business schools require foreign language and how suddenly the interest in language training for business has blossomed (*M.B.A. Newsletter* 1991).

But this change in emphasis means that many schools will be presented with an extraordinary opportunity to take advantage of this newfound interest in "language for business," and at the same time address other needs for reform in higher education—increased learning across disciplines and increased attention to cost effectiveness. Let me explain how this might work in an institution with good undergraduate programs in language and area studies and a graduate program in business. While I draw on the specific experience and plans of Willamette University, the conditions which permit this kind of program are widespread. Willamette University is fortunate in having a Combined Three-Two Liberal Arts and Management program structure in place (see table 6.4 for details). The cost effective essence of this program is that it allows a student to obtain two degrees—an undergraduate B.A. or B.S. degree and a master's degree in management—in five years instead of the usual six. We propose to develop a special version of the Three-Two Combined Program emphasizing foreign language and area study as preparation for a graduate degree with a specialization in international management.

Several factors lead us to develop the Spanish language and Latin American area studies track in the International Three-Two Combined Program first, followed by Japanese second. Enrollments in Spanish, interest in intensive Spanish as a spoken language (as opposed to a vehicle to analyze literature),

and Latin American area studies are at an all time high and are rising. The World Views interdisciplinary freshman seminar introduced in fall 1991 (students read Carrasco, *Religions of Mesoamerica*; Allende, *House of Spirits*; Menchu, *I, Rigoberta Mendez*; Hecht and Cockburn, *Fate of the Forest*; Fuentes, "Chac Mool" and "The Mandarin"; Vargas Llosa, *The Storyteller*; Amado, *Tent of Miracles*; and Neruda, *Heights of Macchu Picchu*) has increased student interest in Spanish and Latin American area studies. Two new faculty positions in Spanish have been filled with faculty having a primary interest in teaching intensive Spanish for use in applied situations, and the College of Liberal Arts dean is committed to adding intensive Spanish language instruction and a Latin American area studies major. Finally, interest in Latin American area studies is sustained by an energetic and capable group of faculty cutting across several disciplines, sustained by a growing collection of Latin American journals in the library. The development of a Japanese language and Japanese studies major will follow the implementation of the Spanish/Latin American track. Schematically, the ideas presented above are represented for the Spanish/Latin American area studies track in figure 6.3.

Creation of a clear, viable track in the Three-Two Combined Liberal Arts and Management Degree Program with the capacity to train well-educated specialists in international management and marketing, who are fluent in Spanish and ready to move into the real world of global business and the world economy, will be an important contribution to education in international business.

CONCLUSION

This is an exciting time to be teaching international business in the United States. We are presented with extraordinary opportunities to refine and extend the learning experiences of students as business programs redefine themselves and the challenge of preparing the next generation to compete in the increasingly global economy takes on new urgency and meaning.

Figure 6.3 Language Area Study and the Three-Two Combined Degree Program

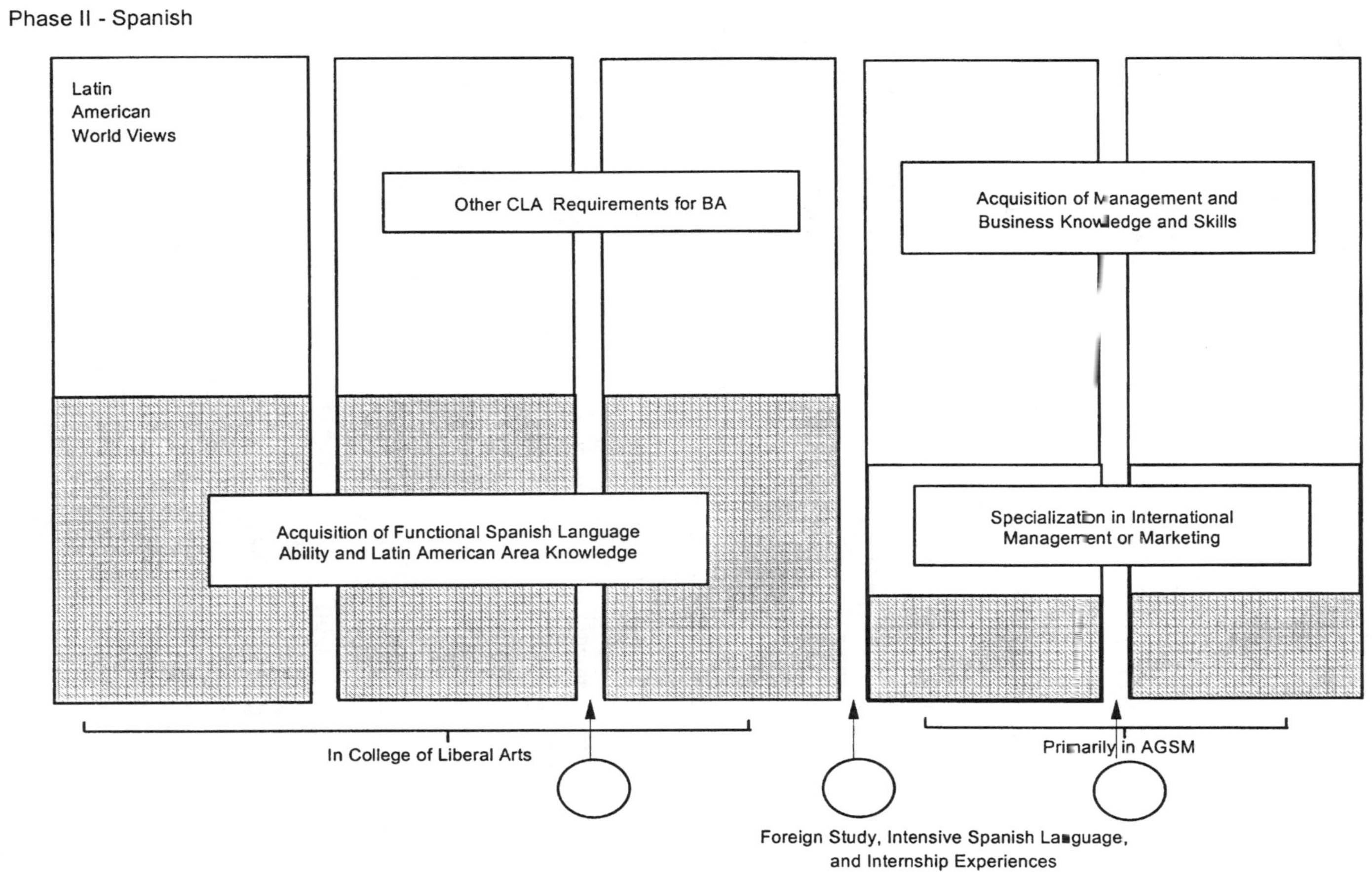

Phase II - Spanish

Latin American World Views

Other CLA Requirements for BA

Acquisition of Functional Spanish Language Ability and Latin American Area Knowledge

Acquisition of Management and Business Knowledge and Skills

Specialization in International Management or Marketing

In College of Liberal Arts

Foreign Study, Intensive Spanish Language, and Internship Experiences

Primarily in AGSM

Table 6.3 Line Learning Objectives

LINE Management Program

LEARNING OBJECTIVES FOR LAW

The objectives of the legal learning portion of the curriculum are to instruct students in how to comply with the law and in how to use the law to protect their interests, including knowing when to seek legal advice. About half of our efforts should address substantive areas of the law while the other half should focus on legal process and administrative procedures and the effects of the law upon the success of the firm and its processes.

1. Substantive areas of the law; the scope and content of the law.
   a. Contracts and torts: explore the sources of legal liability (product, service and stakeholder), the damages involved, and how to avoid them.
   b. Antitrust and trade regulation: which trade practices, including restraints of trade, are prohibited and what are their associated penalties.
   c. Business organization and SEC regulation: SEC required and prohibited behaviors and the penalties for violation and the personal liabilities that may arise as the result of organization form and role.
   d. Environmental, health, safety, and employment regulation and their effects on Business: which behaviors are required and the penalties for misconduct.
2. Legal processes and administrative procedure.
   a. Law making: factors which provoke rule-making by legislative bodies, administrative agencies, and common-law judges and the methods used to influence rule-making outcomes.
   b. Execution: how law is privately and publicly enforced and the ways to influence the enforcement process outcomes.
   c. Adjudication: attention on court procedures, how cases or controversies come to the attention of judges, and ways to influence the outcomes of adjudication.
3. Effects of the law upon the success of the firm and its processes.
   a. Threats to survival: what actions lead to bankruptcy and how to avoid them.
   b. Threats to success: how the law can positively and negatively influence competitive positioning.
   c. Effects upon internal action: legal implications for the design and operation of reporting systems, records, and administrative controls.

LEARNING OBJECTIVES FOR NEGOTIATION

1. Analyzing inputs:
   a. conflict analysis: diagnose the underlying problem facing **all** parties (e.g., allocating resources versus assuring performance on commitments)
   b. historical analysis: identify the economic, political, and interpersonal imperatives for **every** party involved
   c. resource analysis: evaluate the value of *every* participant's (e.g., subordinates, superiors, peers, suppliers, clients, regulators) resources, including alternatives to a negotiated agreement.
   d. structural analysis: treat *all* interactions at work as negotiations, taking into account constraints such as existing policies, procedures, and formal authority.
2. Analyzing outcomes:
   a. estimate the limits to joint gains (common interests; exploit differences in valuations, forecasts, risk aversion and time preferences to maximize joint gains)
   b. specify objective criteria for agreements (efficient, best available information, stable, fair, etc.)
   c. recognize the limits of "winning" (zero-sum mentality)

Table 6.3 Line Learning Objectives *(continued)*

3. Managing negotiations:
   a. understand power tactics in competitive bargaining: concessions, appeals to favorable norms, threats, coalitions, effective persuasion, agenda control
   b. understand collaborative tactics that overcome tension about disclosing information in variable sum bargaining: conditional commitments, joint fact-finding, third-party intervention
   c. tie the elements of negotiation (issues, interests, and parties) to effective processes (persuasion, learning)
4. Selecting appropriate roles for managers (facilitators, mediators, arbitrators, enforcers, auditors, problem-solvers)
5. Incorporating the interpersonal dynamics of negotiation (personal style; tools for managing psychological factors; ethical dilemmas)

LEARNING OBJECTIVES FOR ETHICS (12/5/91)

The intent of the Ethics component is to teach a process of principled reasoning and decision-making that is likely to produce ethical behavior. Ten percent of our effort should go into theory; 90% into applications. Student skills we want to help develop include:

1. Awareness:
   a. integrating formal, professional codes of conduct governing confidentiality, competence, integrity, objectivity, etc., into our considerations, identify the ethical implications of conduct at various levels of management decision-making:
      (1) personal, e.g., career path choices, professional integrity
      (2) organizational policy, e.g., product safety requirements, guidelines on deception, performance evaluation programs, internal controls
      (3) system, e.g., fair competition, free speech
      (4) strategic implementation, e.g., top management commitment to social responsibility, allocation of resources
   b. discuss the ethical implications of the theories and models we use in our disciplines, and the impact of ethical constraints on our functional decisions
   c. understand the influence of corporate climate and culture on ethical perspectives
2. Competent Reasoning:
   a. identify the intellectual basis for moral values applied to the workplace (e.g., deontology, utilitarianism, etc., from Kant, Mill, others)
   b. highlight some fundamental aspects of logic such as identifying fallacies, avoiding circular arguments, ad hominem reasoning, and logical errors, and maintaining argumentative consistency
   c. clarify reasons to respect the rights of others, while explaining the range of professional and personal ethical views, and the influence of societal values on the resolution of ethical dilemmas
   d. discuss the role of judgement and bias in the evaluation and prediction of the likely consequences of conduct, especially where facts are incomplete or ambiguous
3. Managerial Problem-Solving:
   a. explore the ethical dimensions of managerial decision-making via the posing of practical ethical problems, and the exposing of conflicts of interest
   b. motivate pragmatic and competitive behavior that creates value in relationships (including trade and exchange relationships) by identifying its benefits
   c. apply corporate policies and programs, and professional codes of conduct as practical means to implement and control ethical decisions

Table 6.4 Combined Three-Two Degree Program in Management

The College of Liberal Arts cooperates with the Atkinson Graduate School of Management in offering a 3–2 Program, through which a student can earn, in five years, a Bachelor's degree in an undergraduate major and a Master of Management degree.

Under the 3–2 Program, students study for three years in the College of Liberal Arts, earning at least 24 credits and completing most (in some cases, all) of their required courses for their major during that time. The actual number of required courses a given student needs to complete during the first three years will be determined by that student and his/her major department or program. In order to receive the baccalaureate degree upon completion of the first year of study in the Graduate School of Management, students in this program must have satisfied all of the College of Liberal Arts graduation requirements. The students are then ready to complete the second, and final, year of the Atkinson program and receive the Master of Management degree.

Application for admission to the Master's segment of the 3–2 Program is made during the first semester of the junior year. Applications are considered on an individual basis. Generally, a student will be considered qualified for admission if he or she:

1. has the written consent of the undergraduate department or program concerned;
2. has maintained normal progress during the first four semesters of undergraduate residence and accumulated a "B" average or better;
3. has scored 550 or above on the Graduate Management Admissions Test (which should be taken by February of the junior year);
4. has demonstrated effective communication abilities in writing and speaking;
5. has, or will obtain, sufficient background in mathematics to succeed in courses required for the M.M. degree (normally up to, and often including, the first course in calculus);
6. has completed introductory courses in economics, political science, and either psychology or sociology.

■ ■ ■

Some Reflections on the Evolution of International Business as a Field of Study: From Periphery to Center
Franklin R. Root

INTRODUCTION

This paper addresses three questions relating to the evolution of international business as a field of study. How has the field evolved in this century? Is international business a discipline? Will U.S. schools of business become international schools of business?

My reflection on these questions come out of nearly forty years of experience in developing and teaching courses in international business in the United States. The longevity of this experience has nothing to do, of course, with the "truth" of my reflections, but it does mean that I have been involved in international business education since the 1950s when it began to evolve out of international commerce focused on export management.

INTERNATIONAL BUSINESS AS A FIELD OF STUDY

As a field of study, international business has passed through several phases responding to developments in the real world of international enterprise. These phases cannot be neatly dated; nor can they be regarded as watertight compartments. Rather each phase is identified by a dominant perspective on international business although other perspectives may accompany it.

The first phase, which started in the 1920s and ran until about 1960, may be designated International Trade. The field of study was a mix of international economic theory (comparative advantage, foreign exchange, and the balance of payments), national trade policy, and export/import trade. The latter subjects constituted the "business" content of the field. Emphasis was placed on intermediaries (such as, export-import houses, export agents, and foreign distributors) with little attention paid to manufacturers. The approach was descriptive and institutional rather than analytical or managerial. A leading text during this phase was *Foreign Trade: Principles and Practices* by Huebner and Kramer (1930).

The inadequacy of the international trade perspective became evident about the mid-1950s with the growing prominence of direct investment in foreign operations by large U.S. firms. The start of a second phase in the evolution of international business as a field of study was marked by the appearance of Fayerweather's *Management of International Operations* (1960). The dominant perspective now became that of the manufacturing firm (usually U.S.) with foreign operations in production. This firm was called a multinational corporation, and its study was called international business. During the 1960s decade, scholars worked to demarcate this new field of study, and in doing so they became generalists who regarded all aspects of international business (including international applications of the traditional business disciplines) as their domain of inquiry. However, they tended to focus on the international investment decision, foreign operations and expatriate staffing of those operations. It is appropriate, therefore, to designate this phase as Foreign Operations.

The Foreign Operations phase was an exciting time for international business scholars. They saw themselves as missionaries who were bringing enlightenment to ethnocentric, domestic-centered business school faculties. It is not coincidental that the Academy of International Business was founded in 1959 at the start of this phase. In addition to Fayerweather, other scholars, such as Robinson (1967), Farmer and Richman (1966), and Aharoni (1966), published texts that set the boundaries of field. A fitting marker for the end of this phase was the publication of Vernon's *Sovereignty at Bay* (1971). The multinational enterprise, ignored by scholars a decade or so earlier, was now portrayed as an institution wielding immense economic power but not accountable to any single national or international political body. In this phase, the international business field drew primarily on economics and general management and secondly on political science and cultural anthropology.

Internationally-oriented texts in the functional disciplines began to appear in the 1960s, particularly in marketing Fayerweather (1964) and Cateora (1966). But it was not until the 1970s that they began to proliferate, marking the beginning of a third phase in the evolution of international business as a field of study. Three notable texts introduced in the early 1970s were by Dymsza (1972), Terpstra (1972), and Eiteman and Stonehill (1973). During this phase, most young scholars in international business became specialists in international management, international marketing or international finance. Also, a small but growing number of economists started to develop theories to explain the behavior of multinational firms, including Dunning (1974), Buckley and Casson (1976), Caves (1982), Hennart (1982), and Kindleberger (1983). International business became progressively defined as an aggregation of the international dimensions of the traditional business disciplines and of economics. And so, we call this third phase, International Functional. It should be noted, however, that general international business texts following the broad approach of the second phase continued to appear, such as those by Robock and Simmonds (1973) and Daniels and Radebaugh (1976).

About the mid-1980s, the dominant perspective on international business began to shift away from business functions toward global competition. The stimulus of this reorientation was the dramatic intensification of worldwide competition in several key industries coupled with a perceived decline in the international competitiveness of U.S. firms. The publication of Porter's *Competition of Global Industries* (1986) set the tone for this phase. Another sign of change was the publication of texts on international strategic management, such as Bartlett and Ghoshal's *Transnational Management* (1992). Even general texts in international business began moving toward a global strategy perspective (Grosse and Kujawa 1988). Accordingly, we name this fourth phase, Global Strategy. Because it focuses on the strategic behavior of firms competing in global markets in a world of nation-states, this emerging field of interna-

tional business draws on economics, political science, cultural anthropology, strategic management, organizational design and behavior, human resource management and the other business disciplines (see table 6.6).

IS INTERNATIONAL BUSINESS A DISCIPLINE?

Does the study of international business have features that distinguish it from the traditional functional fields and economics? In 1969, I offered this definition of international business as a field of study:

As a field of study, international business is primarily directed towards the description, analysis, explanation and prediction of the actual and normative behavior of the private business enterprise as it strives to achieve its strategic and operational goals in a multinational environment. In particular, the study of international business centers on the crossnational interactions that create dynamic linkages among the elements comprising the international

Table 6.6 Evolution of International Business as a Field of Study

Phase	Period	Dominant Perspective	Disciplinary Bases
I. International Trade	1920s to early 1960s	Export/Import Operations	Economics/ Marketing
II. Foreign Operations	Early 1960s to early 1970s	MNC Operations Abroad	Economics/ Political Science/ Culture Anthropology/ Management
III. International Functional	Early 1970s to mid-1980s	Functional/ Economic performance of MNC	Economics/ Business Functions
IV. Global Strategy	Mid-1980s to ?	Global Competition	Economics/ Political Science/ Cultural Anthropology/ Strategic Management/ Other Business Functions

enterprise (intra-enterprise interactions) and between the enterprise as a whole and its multinational environmental systems (extra-enterprise interactions). (Root 1969, 18)

I went on to elaborate this conception in a model that regards the business firm as an international enterprise interacting with three environmental systems which, in turn, are subsystems of a nation-state system. The study of crossnational interactions may draw on political science, economics, cultural anthropology, the functional business fields, and other disciplines. I continue to regard this conception of the field of international business as useful because it identifies the distinctive nature of international business, namely interactions between the firm and two or more nation-states. These interactions lie outside the domain of domestic business studies. The evolution of international business as a field of study may be regarded as a study of crossnational interactions with respect to export/import operations, foreign operations, business functions, and global strategy.

I regard interactions between the international firm and national governments as the most distinctive of the international business field. In a world of laissez-faire with no interference by national governments in domestic economies or their external relations, international business as we know it would not exist. There would remain intercultural business, but this would also occur within multicultural nations and hence would not be uniquely international. It follows, then that the fundamental distinctiveness of international business lies in the fact that it is transacted across different national political systems whose governments follow policies intended to promote the national interest. Some of these policies seek to influence international flows of products, people, technology and enterprise generated by international firms. Not content with laissez-faire outcomes, national governments try to enhance the benefits and lower the costs of their participation in the international economy.

Because markets ignore equity, their viability depends on political systems that can redistribute economic outcomes so as to achieve the balance demanded by society. It is a peculiarly U.S. misconception that markets and governments are always mutual enemies. But they become antithetical only when either governments smother markets through central planning or excessive regulation or when markets alone determine the distribution of economic outcomes under laissez-faire conditions. The misconception of the role of government as only a source of market imperfections has led to a dismissal of political forces in economics and the functional business fields.

Clearly, the distinctive core of international business does not fall within any traditional field of study: it is multidisciplinary. And for that reason, it has been in a weak power position vis-à-vis the unitary fields of the functional business disciplines and economics. The educational establishment of business schools is reluctant to reward multidisciplinary research. Furthermore, international business research is more time-consuming and costly than domestic business research. Consequently, it has been downgraded by young professors seeking tenure.

This institutional domination of the functional business fields has thwarted the development of multidisciplinary programs and research in international business. It follows that international business is unlikely to achieve recognition as a separate discipline with its own conceptual framework and received body of theory. I believe, however, that this is for the good. As a field of study—not as a discipline—international business is now poised to transform the entire business school curriculum.

In my judgment, the focus of international business education should be the nexus between national governments and international firms as each attempts to maximize its net benefits (economic and otherwise) from crossnational flows of products, technology, factors of production, and enterprise. The core of international business as a field of study, then, rests on three pillars: political economy (drawing on economics and political science); national culture/society (drawing on cultural anthropology, sociology, and comparative management); and international (global) management. Because this core is common to an international approach to all business functions, there is a strong argument for covering it in one or more courses required of all students.

THE COMING INTERNATIONAL SCHOOLS OF BUSINESS

The evolution of international business as a field of study has been accompanied in schools of business by a shift from a peripheral field pursued by a few misguided professors to a central field that promises to encompass the entire business school, much to the bewilderment of the many ethnocentric professors. International business has become a metafield—a field of fields. Hence, the probable disappearance over the next decade of international business as a major or concentration represents a triumph—not a defeat—for those persons who have tried to "internationalize" business schools over the last quarter-century.

The challenge of the 1960s and 1970s was to introduce international business as a course or program into schools of business. This challenge has been met in all the leading schools. But the challenge of the 1990s is more formidable. It is to transform the business school into an international institution that matches a global economy.

What would an international (or global) business school be like? Its primary mission would be the education of persons to manage international firms pursuing global strategies. It would draw is faculty from several countries and each faculty person would have experience as a researcher or teacher in at least two countries. The faculty would approach their individual fields from a global orientation. For example, professors of marketing would regard domestic marketing as a subset of global marketing. Similarly, professors of management would teach from a global viewpoint, no longer assuming that management theories that "work" in one country will necessarily work in another.

The international business school would draw half or more of its students from countries other than the country where the school is located. Each student would be required to read, write, and speak a language other than his or her own. Moreover, instruction in some courses would be in languages other than the school's official language. There would be cooperative arrangements with schools in several countries. In effect, the business school would become a member of a global educational network.

Clearly, U.S. business schools have a long way to go to become international schools. True, everyone talks of "globalization" today. And perhaps business school deans know the direction in which they want to move their schools. But almost certainly they underestimate the changes needed to become an international business school in philosophy, faculty, curriculum and students. And yet, if U.S. business schools fail to internationalize, they will become increasingly irrelevant to the needs of business. Leadership in business education will pass from the United States to other countries, notably to Europe where business schools are already ahead in multinational faculty and students. In the future, U.S. business schools will be competing worldwide for faculty and students. Willing or not, business schools will enter a global marketplace. To survive, they will need to design and carry out global strategies in business education.

■ ■ ■

International Business Education via the Case Method
Paul W. Beamish

INTRODUCTION

Firsthand experience represents the most effective way of learning about business. Yet experience is an inefficient teacher. Because most managerial positions have a large repetitive component and are narrow in scope—particularly in the early years—the opportunity to confront new and varied learning situations on a frequent basis is limited.

Classroom instruction is designed to be both a preparation for, and an aid to, actual experience. While lacking the richness and intensity of the actual experience, the classroom can serve to condense examination of a range of issues into a short time frame.

Classroom instruction can take place via a variety of forms—lectures, games, exercises, the case method, and so forth. All have their place. The focus of this paper is on one form of instruction—the case method. This paper will not review the general merits and limitations of using the case method,[1] nor attempt to compare and contrast the relative merits of the case method with other forms of instruction.

The position taken herein is that the case method is among the most frequently used teaching and learning forms for insuring that the International Business dimension is adequately and properly covered in formal degree and executive programs. Is it the most effective and efficient form? In my opinion,[2] it can be the most efficient form, yet sometimes falls short. To that end, the purpose of this paper will be to examine ways of making more effective and efficient use of the case method in international business education. Of specific interest is the international case study. Here a case study refers to a comprehensive, field-based, decision-oriented description of an administrative issue, and does not include case histories or minicases or vignettes.

Every business professor wants to finish each class with a sense that the objectives for the session have been achieved. Broadly speaking, there are three reasons why class objectives may not be met: the students, the material used, and the professor. While professors have limited control over who their students are, they have almost total control over what they teach, and how they do it.

Here what is taught, and how it is taught corresponds in part to case selection and case teaching. The first section of this paper looks at the international case-writing process, and ways it can be improved. The underlying assumption is that instructors will always be hampered in their selection of the most appropriate case as long as they have an incomplete understanding of how particular cases are written. The ways in which international cases are prepared can, and do, vary widely. They vary with respect to such things as the quantity and quality of underlying research, the care taken in preparation, and the willingness of authors to subsequently revise their material, and edit it to a length (typically under 20 pages) which students will actually read.

The second section of this paper looks at the in-class experience. Teaching with cases is not formula teaching. Numerous innovations are available for improving the overall experience for both student and professor.

The third section of this paper concludes with some comments on using cases in degree versus executive programs. The essential message here is the need to match the complexity of the material with the international experience of student.

Many of the points which follow will be familiar to experienced international casewriters/teachers. To them, this paper is intended to serve as a reminder of the Kaizen principle—the need for continuous improvement. The larger intended audience however is those international business professors with limited familiarity with the case method, and those who may have underestimated its potential.

INTERNATIONAL CASE-WRITING AND SELECTION

Current Material

Keeping the international course current is a unique challenge. International cases are among the most expensive (at Ivey, costs are estimated at $10–12,000 each) and complex ones to write (due to travel, language and cultural issues), they are often used in elective rather than required courses (with more limited opportunity to amortize the costs over larger student usage) and can become dated more quickly. No where does the environment in which cases are set change as quickly as with international cases. Cases set ten years ago—even when written more recently—are often viewed as irrelevant by many students. No simple solution to this dilemma exists. For the instructor wishing to keep a course current, there are partial solutions.

The first of these is to ensure the home school maintains a complete set of case directories or has access to the COLIS system, and alerts faculty to their availability in a timely fashion. The largest institutional producers of business case studies (Harvard, Ivey, and Darden) each maintain a complete and easily accessible infrastructure for case-writing registration and distribution. (Ivey regularly produces directories devoted exclusively to international cases.) Other schools which produce a more limited number of international cases, publish case directories on an irregular basis. Despite the fact they could advertise in the AIB newsletter or other low cost sources, they unfortunately seem less attuned to making the academic public aware of their existence. A computerized on-lime system (COLIS) has been designed by the European Case Clearing House (ECCH) to allow interested parties to search for specific case materials. COLIS contains bibliographic data on the major case collections and can be accessed in Europe (through ECCH) and in North America (through Babson College in Boston).

For faculty who write international cases at other schools, awareness can be created via word-of-mouth registration of the case in general systems such as Intercollegiate Case Clearing House in Boston, the Case Clearing House of Great Britain and Ireland, or publication of the case in the *Case Research Journal* or a casebook.

A second solution is to personally engage in case-writing to fill desired gaps. For a variety of reasons, individuals may not wish to engage in case-writing to fill course gaps. In addition to those already noted, they may be inexperienced (although some institutions offer one-week training programs), or at a university which neither funds nor values cases as a legitimate academic activity. Notwithstanding these problems, personally engaging in the activity can be beneficial since it:

1. will fill a tangible need,
2. can be publishable (albeit not in *JIBS* or most journals),
3. provides insights which build the intellectual capital of the writer, which in turn is useful with any form of subsequent research,
4. will improve in-class credibility with students—especially executives—and
5. makes one a better judge when assessing other available cases.

Source of Case Content

Whether international or not, cases will typically be either interview-based, or public sources–based. While the most managerially oriented will be interview-based, a relevant issue to the instructor trying to select an international case will be to try and determine from the case and teaching note who was interviewed. Most times it is not made explicit (a) the number of managers interviewed, (b) their seniority in the organization, and (c) whether interviews (and visits) were made with people from the host country. Just as joint venture research which involves only a single perspective has limited value, so too does an international case which does not entail at least one visit to the host country. Without such a visit, the case runs the risk of being too superficial and missing key contextual issues.

For the casewriter, the problem of obtaining in-country observations can also be alleviated by the use of a local coauthor. This inevitably adds a richness to the case which might otherwise be absent.

Case-writing in culturally dissimilar environments—especially when one does not possess the local language skills—almost 'requires' a coauthor. Here the coauthor can serve to contribute not just to the intellectual content of the case, but as interpreter during the interview(s), translator of background material, and can make the itinerary arrangements. In my experience in Japan and China, for example, use of a local coauthor unquestionably also improved the quality of information which was obtained from the host country managers interviewed.

Reducing Cultural Misunderstandings with International Cases

The vast majority of cases ask the student to adopt the perspective of the home country decision maker. A low-cost alternative is to select a case which requires the home country student to adopt a host country (foreign) perspective. This forces the home country student to more rigorously consider another national perspective, hence reducing ethnocentric tendencies.

A more complex and expensive case-writing methodology we have been experimenting with is to write two cases about the same issue—one from the home country perspective and one from the host country perspective. Some information in each case differs, and each student is assigned one or the other case. This methodology requires excellent access—sometimes in two companies—and potentially two case release forms to be signed. The increases the risk of noncompletion.

On a variety of occasions we have worked with foreign colleagues who wish to translate our cases and teaching notes into their language. This experience has improved the quality of the original English-language versions by forcing us to (a) recognize and remove undesirable colloquialisms; (b) clarify ambiguous words or phrases; (c) eliminate any ethnocentric biases which may have unconsciously crept in.

This interactive process between authors and foreign colleagues to critically review international cases prior to translation is time consuming, even for the original author. However, just as questionnaires designed for use in other languages are often back-translated as a quality control check, so too is it worthwhile to ensure accurate case translations.

CASE TEACHING

Using Case Supplements

There are many prerequisites for effective case teaching; preparation for class, classroom management, postclass evaluation of the experience. In addition to the case itself, numerous aids to the classroom experience are available. These include multipart cases, video, computer analysis, and current readings.

One of the general criticisms of the case method is its historic, static nature. To overcome this, many instructors will use a case series—for example, Kentucky Fried Chicken in China (A) (B) (C)—which can be taught over three (not necessarily consecutive) sessions. In addition, short supplementary cases designed for in-class reading and discussion can be an effective means of putting a sense of current reality to the case.

Video supplements are increasingly being used as part of the class experience. The supplements shown are typically 5–30 minutes of an 80–90 minute class and range in content from corporate promotional material, to "talking heads," to "what happened" explanations from the relevant executive. Video supplements vary greatly in terms of both quality of production and how well they are integrated into the suggested class plan. An excellent example of a video tightly integrated into the class plan would be the Panda Furniture case,[3] where four short segments of 2–7 minutes each are shown.

Another innovation available is the potential to use computers for scenario and spread sheet analysis. While this most frequently occurs as part of a homework assignment it can be used with good results in-class, for example to conduct cost-benefit analysis from the perspectives of a multinational enterprise and host government.[4]

With the rapidly changing international business environment, ample opportunity exists to link cases to current events via current readings or developments. In the light of new political boundaries, trade agreements, regulations and so forth, the class discussion—even for cases studied months earlier—can be given a currency which is difficult to achieve for noninternational cases.

The single most frequently used aid to the case teacher is a teaching note. When well written, they represent one fully developed perspective on how the case can be taught, which the reader can follow or not. Yet problems abound here. Teaching notes are often never written by the case author, not made available or registered (despite solid evidence that it improves frequency of use), written prior to class testing of the case, never revised once written (despite new insights from the case experience), and/or not explicitly linked to the underlying theory or research.

This link between theory and practice is important. As the 1984 Academy of International Business Presidential address noted, teachers are among the group who are called upon to "interpret and pragmatize" the results of scholarly research (Hawkins 1984). Given that much of this interpretation may ultimately occur via the case method, the anchoring of class objectives to International Business research—in the teaching note—would be a positive step.

Class Structure

Most case teachers utilize a variety of approaches to enhance the traditional in-class case experience. These include frequent use of role-playing to engage the students, combining minilectures with case discussions, team-teaching the case from several functional perspectives, linking case classes to crosscultural exercises (that is, Bafa-Bafa), running classes back-to-back, combining the case discussion with comments from a guest executive speaker, requiring students to work in groups prior to class in different dimensions of the issue, and so forth. While these approaches are not specific to international business, some can be enhanced by the unique student composition and cultural diversity of many IB classes.

In most universities, international business is still taught as an elective rather than required course. Through self-selection, many of the students who gravitate towards IB courses have lived, worked, and/or traveled abroad, speak another language(s), and so forth. As well, with the increase in the number of international student exchange programs, there are many more foreign students studying abroad for a term or longer. In our experience, these students frequently also enroll in international business courses. As a result, the international business case teacher has a uniquely qualified pool of students with whom to confront International Business issues. Role-playing is especially effective if nationals of the countries in question are present. To add further realism, in some role plays it is possible to have students conduct their part of the discussion in a foreign language, with one of the groups serving as interpreter.

CASES IN DEGREE VERSUS EXECUTIVE PROGRAM

The case method can be effectively used with both the twenty-year-old undergraduate degree student and the forty-five-year-old executive program participant. As well, some individual case studies are robust enough—permitting analysis at several levels of sophistication and/or on different issues—to be used with both audiences.

Given, however, the enormous breadth of the international business area, many case studies are not suitable for all audiences. A twenty-year-old undergraduate who has never been inside a business—let alone a large diversified multinational enterprise—cannot truly appreciate the intricacies of managing within a large MNE. Consequently teaching a course on transnational management to such a student—as opposed to prefacing it with one on internationalization—is ineffective. As figure 6.4 suggests, the teaching material selected (including cases) must be tailored to the target market. In practical terms, there is a need to match the complexity of the material to the international experience of the audience. Case studies for executive are designed less to impart fact or base knowledge (due to the executive's extensive work experience) and more to provide a vehicle to develop judgment through consideration of various (often more subtle) contingencies.

NOTES

1. A substantial literature exists on the use of the case method in business education and elsewhere. See, for example, Yin 1984, Leenders and Erskine 1989, and Erskine, Leenders, and Mauffette-Leenders 1981. WACRA, the World Association for Case Method Research and Case

Figure 6.4 Teaching International Business

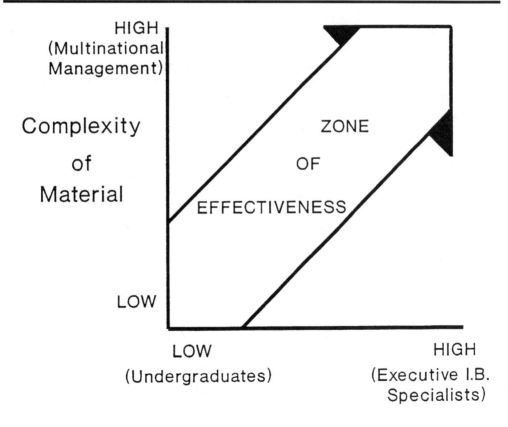

Method Application, held its eighth international conference in 1991, and the Ivey Business School annually offers a one-week Teaching With Cases Workshop. For additional perspectives, see Stabler and Suilleabhain 1990.

2. These observations are based on 15 years of experience teaching with cases on degree and executive programs, and authorship or coauthorship of 60 cases in the international strategy area. These cases have been published in

40 books or journals, and variously translated into French, Japanese, Chinese and Russian. Notwithstanding this experience, I readily acknowledge that some of the suggestions provided for making more effective use of the case method apply to my own work.

3. Developed by Peter Killing while at IMD and polished further upon his return to Ivey.

4. Lou Wells spoke on this subject at the 1990 AIB meetings in Toronto.

■ ■ ■

International Business Teaching Implications
John D. Daniels

George Bernard Shaw quipped, "He who can does. He who cannot, teaches," a quip that has relevance to me, my presentation, and this session. This point can be illustrated by a brief incident that occurred during my first day at Indiana University nearly five years ago. A small newspaper item accompanied my joining the faculty, which mentioned my move from Lima, Peru. That led a local farmer to phone me about how he might import llamas. I responded in a true academic manner, that is to say I evaded direct answers but blurted out everything I could recall that had any possible connection with llama trade. For instance, I recounted a Lima newspaper blurb on Peruvian export controls to maintain high llama prices and another that reported on nonacceptance of a llama shipment by health authorities, but I could not recall what country it was. I also suggested that he contact the U.S. Department of Commerce and the nearest Peruvian consular office, a suggestion that is usually an effective cop out for "real world" questions. But my elusive tactic did not satisfy him. He told me in no uncertain terms that neither I nor my institutional colleagues knew much (or anything) about our purported areas of expertise. Thus, this experience establishes why I am teaching international business rather than "doing" it. However, my conference assignment to talk about teaching must, by an extension of Shaw's logic, mean that I can neither do nor teach international business.

The topic of the session, "Teaching Implications for IB," is itself perplexing primarily because we know so little about the optimum content, structure, and evaluation of teaching in general, much less the specifics of teaching international business. For example, those of us who have been involved on promotion and tenure committees have been frustrated by having to rely almost entirely on student evaluations when considering teaching records, in spite of all the shortcomings that are readily apparent. For this session, we do not even have that level of data but the five panelists collectively have over 100 years of teaching experience. Personally, I would feel much more comfortable if I could present an analysis of substantial relevant data. But I cannot. In the absence thereof, I must perforce make a largely impressionistic and anecdotal presentation. But hopefully, my nearly 25 years of IB teaching will allow me to make some meaningful insights in much the way that anthropologists sometimes correctly describe a "national character" from a strictly observational rather than a measurable basis. As an aside, let me mention that I received copies of the other panelists' papers before I started preparing mine and have taken the opportunity to emphasize niches not covered elsewhere. My comments will address the issues of IB teaching in terms of (1) the audience, (2) the content, and (3) the method of delivery.

THE AUDIENCE

The term "Think Globally, Act Locally" has recently been popularized. As teachers of international business we are so-called global thinkers who I believe have focused too locally for too long, at least in terms of the market for international business training. We have concentrated on the internationalization of business students and have measured progress in terms of the number of business students served. This was perhaps appropriate in what Franklin R. Root has designated as the international trade, foreign operations, international functional, and global strategy phases of international business. But, if we accept that international business is becoming a metafield within business, have we then solved the problems of international business training? I do not believe so. Even if we were to become successful in training all business students, we would shortchange the needs of society and business organizations. According to 1989 figures for the United States, the 35–44 age group had the highest incidence of college completion, about 28%. At the same time, roughly 24% of university degrees were in business and management. Therefore, a rough estimation is that, at best, we have been setting our goals at serving about 7% of the adult population. What does this mean?

Jeffrey Arpan has outlined three levels of students' proficiency: awareness, understanding, and competency. He contends that all business students should reach the first two levels. I prefer a slight modification of the terms, so that they are citizenship requirements, knowledgeable worker requirements, and specialist requirements. I also take a broader view on who should be trained in the requirements. The informed

citizen requirement includes a realization that nations are irreversibly interdependent, that competition in almost any industry involves firms headquartered in different nations, that national interests may differ from the best interests of companies headquartered therein, that there are gains and losses from any national policy that regulates international business, and that new opportunities and risks exist from international operations. I argue that nearly all adults, not just business majors, should reach this level so that they can make more informed decisions when they vote or otherwise influence legislative directives.

The knowledgeable manager requirements should be reached by any employee, or at least any manager, whose job and company may be affected by international competition—even though they may never have direct international responsibilities. The knowledgeable worker requirements include a realization of whether and how operations must be adjusted to succeed against foreign competition and how foreign operations affect the overall corporate well-being. The third category, the specialist requirements, are needed by only a select group of students who need to be able to handle direct international responsibilities. Depending on their functions and on their geographic work location, these specialists may need, in addition to the informed citizens and knowledgeable workers requirements, more technical expertise within their functional majors, an understanding of global strategies, specific knowledge about foreign environments, and the tools to improve their international learning and adaptability. But most employees in business organizations, including managers, lack formal business training and the business-specific requirements of categories two and three. As international business specialists, we may be failing overall if we do not succeed in alleviating the shortfall for nonbusiness students through credit and noncredit programs. Another possibility is that we push for more international business content within nonbusiness courses in much the way that we have pushed for the inclusion of information about foreign environments within domestically oriented business courses.

Aside from the need to service nonbusiness majors because of their future roles as employees and citizens, there are other practical reasons to reach out. By almost all indicators, the bloom is off the growth in demand for majors in business relative to majors in other university programs. Furthermore, as we move within the internationalization process to require business students to take more nonbusiness course, such as in foreign languages, there is a smaller percentage of their courses that they can take in business. For business schools to maintain or improve on their market share within the universities, it may be possible or even necessary to service more nonbusiness students—perhaps in newly-designed courses that more closely fit their needs. Furthermore, as we call on liberal arts faculty to help internationalize business students, we are in a much stronger bargaining position if we can show some quid pro quo to help their students.

In summary, as international business educators it behooves us to seek out new and expanded means to reach a much broader clientele than the one for which our sights now seem set. As we reach out to a broader clientele, we should be mindful of the lessons we have learned in trying to internationalize business programs, such as the tradeoffs between specialized courses versus infusion of content into existing courses.

THE CONTENT

Domains

There is a growing consensus that no formula is suitable for all students. Teaching needs differ, for example, among undergraduate, masters, and doctoral students, between business and nonbusiness majors, among functions, by whether students seek direct versus indirect international responsibilities, and by the types of organizations that are their eventual employers. Given

these differences I am concerned about international business domain statements that implicitly attempt to delineate what is and is not appropriate content for the teaching of international business. If the domains are set too broadly, such as by saying that all business studies are a subset of international business, they tend to be meaningless. If the domains are set too narrowly, they may not allow for sufficient continuation of the field's content evolution, for example, a domain statement in the early 1960s would have necessitated either new statements in subsequent years or a limitation of the field to teach only about import-export mechanics. Domain statements also smack of protectionism now that some members of our ranks predict the possible demise of specialized international courses. The setting of territorial boundaries will not prevent intrusions, nor should they. If they did, history shows a likelihood of stifling the field's development, such as occurred in thirteenth century Italian universities when they attempted to monopolize knowledge by preventing faculty movements or what has occurred in many industries protected from import competition.

Next, regardless of what we attempt in domain statements or definitions, there are bound to be areas (1) on which we individually make some arbitrary decisions that fall into different camps and (2) that are so gray they must be treated as exceptions. For example, on the former point, one camp excludes unilateral transfers within its definition of international business, but another includes them. Each has made an arbitrary definition akin to the debate on whether life begins before conception, at conception, somewhere between conception and birth, at birth, or when the children move out of the house. No camp is likely to abandon its position, regardless of domain statements or evidence presented in debates. An example of the latter point are international transactions of state enterprises that lack a profit motive and do not quite fit some of our definitions of business. Yet, I cannot think of anyone in the IB field who would exclude these transactions just because of a definition or domain statement. This does not bother me. Biologists, for example, have defined reptiles and mammals but have conceded that some animals must be arbitrarily classified as one or the other because they fit only part of either definition. This does not seem to hinder the field of biology, nor should similar exceptions to definitions hinder the field of international business. It is also interesting to draw an analogy on the debate of who is male and who is female. The Olympics continue with no demonstrated significant differences in results, regardless of whether visual observation or saliva tests are used to make a determination. The content of what is taught in international business is probably not apt to change a great deal regardless of whether we treat it as a separate discipline or not.

More fundamentally, what difference does it make if international business overlaps with other fields of business? The only drawback seems to be that content is repeated in more than one course. Even this is a drawback only if we assume that students remember everything from the first coverage, a very tenuous assumption at best. Furthermore, repetition is rife

1. within functional business majors, such as reemphasizing in advanced courses what was taught in introductory course;
2. between business functions, for example, some finance and accounting courses overlap considerably and some work in finance and in business economics is increasingly converging;
3. between tool and applied areas, such as the teaching of tools in a quantitative analysis department and then the review of them in the operations department; and
4. between business and nonbusiness courses, such as some in psychology with some in organization behavior or some in business law with those in the law school.

This repetition does not seem to bother anyone; and we should make this point strongly if we feel that an overlap exists between IB and other areas *and* is justified from a learning perspective.

When we look beyond business schools, overlap is even more pervasive. I think it is safe to say that promotion and tenure in some of the physical and behavioral science fields is more dependent on the securement of outside research funds than has been the case for business, and that the current "name of the game" for securement of these funds is crossdisciplinary collaboration. As professors have collaborated more across those disciplines, there are many examples of very similar courses that have emerged in more than one. For example, there are now many collaborative animal behavior research projects and courses reporting the same output in both psychology and biology. In essence, although the crossfunctional or even crossdisciplinary content of international business may be at odds with the current popularity of functional isolation within business schools, it is certainly consistent with the overall movement within universities. Instead of asking the question of what should be in a particular domain, we should concentrate on identifying important issues and then taking as broad a view as necessary to teach and research effectively about them.

The Evolving Content

The last few years should have taught us how rapidly the international environment can evolve and how difficult it is to predict how environments will change. This should indicate a need from a content standpoint to cover as broad a spectrum of international theories and issues as possible, regardless of whether coverage is within internationally specialized courses that cut across functions, is within specialized functional courses, or is within all courses through infusion.

This rapid evolution poses problems for staying current in teaching international business. This problem is probably less prevalent in most other fields. A second problem is the crossfunctional and crossdisciplinary nature of most issues that are discussed, which means that relevant research appears in a wider number and broader spectrum of publications. These two problems create difficulties in keeping current in what we teach. I rely heavily on our library service which regularly sends me tables of contents from about 30 journals; however, even by supplementing these with retrieval searches, such as INFOTRAC, I find the process less satisfactory than when *International Executive* published short abstracts from several hundred publications. IB probably has a greater need for an abstract service than most other fields that have such a service; therefore, I wish that somehow we could resurrect what we once had.

Tools versus Environmental Specifics

In spite of the inherent problems of keeping current, we have probably done a better job in presenting IB content (that is, the covering of knowledge specific to foreign environments and operational nuances) than in training on interpersonal awareness and adaptability (that is, tools to make students more sensitive and adaptive to other cultures and environments, such as through sensitivity training). For example, the former may tend to remove some of the fear and aggression that are aroused when dealing with the unknown; however, the understanding of the different culture does not necessarily imply a willingness to adapt to that culture. There are also other dichotomies that relate to these two approaches. Given time constraints within a standard curriculum, it is reasonable to assume that more emphasis on one approach may necessitate the tradeoff of emphasis on the other. I am not at all sure that any of us has really analyzed the optimum combination of approaches in terms of student needs. Instead, we may be like the surgeon who seeks a cutting solution because a knife is the only tool he has.

There are other tradeoffs as well. For example, since the 1988 joint Gallup/National Geographic Society publication showed that U.S. respondents were less able to locate global places on maps by memory than were many other nationalities, there has been some increased emphasis on map memorization—even at the university level. But U.S. respondents scored better than other nationalities on map interpretation, such as calculating the distance between pairs of hypothetical cities. Should map reading tools be emphasized rather than locational literacy?

Foreign language training is once again in vogue, and I have no intention of rehashing all the arguments, pro and con, that we have all heard for requiring a language. However, I do wish to raise the question of tradeoffs. At the undergraduate level in my own institution, either foreign language training or area studies courses are acceptable substitutes for taking courses with an international business content, thus raising the issue of whether communications tools fulfill or should suffice to replace the informed citizen and knowledgeable worker requirements outlined earlier. If so, should some other communications tools be considered as well, such as how to work effectively with people whose second language is English and how to perceive better through nonverbal communication regardless of what language is used. As long as I am on the subject of foreign language I cannot resist responding to two statements that I'm sure we have all heard. First, the statement that "one can never understand a foreign culture without learning a foreign language" seems to assume that (1) there are no quality translations and (2) there are no foreign cultures that share one's home language. Second, the statement that "one learns a foreign culture by learning a foreign language" presupposes a substantial cultural component in foreign-language courses and ignores a current debate splitting the Modern Language Association of America (MLA) on whether cultural content is appropriate within language courses.

Universalities

We recently interviewed over a hundred of our business faculty members to determine the international content of their courses. We wanted to tabulate the extent of international infusion but were stymied on how to treat courses or parts thereof where principles seem to have universal applicability. Realistically, there are such courses, and the universal application of materials is probably growing rather than diminishing. After considerable discussion, we concluded that if we emphasize only the nuances in the internationalization process, we may be doing a disservice to students by developing overly polycentric attitudes. At the same time, we cannot necessarily assume how students will interpret the applicability of concepts to foreign operations when no explicit linkage is presented. On the one hand, they may accept that there is universality when, in fact, the professor is teaching from a domestic perspective. On the other hand, they may reject universality when it is possible. We must encourage faculty to make explicit delineations.

Curriculum Structure

The infusion and separation models have been well-discussed as have their subsets, such as the use of transfusion to achieve infusion and the use of crossfunctional versus functional alternatives within internationally separate courses. I will not attempt to add to the pros and cons that have been presented. My one observation is that the debate has gone on for many years and is not likely to abate in the near future. However, the continuing debate has implications for doctoral programs. Given the different needs of business schools that are taking different approaches to internationalize, the programs should probably be turning out some Ph.D.'s who can infuse international within their functional specialties and others who will fill the demand for teaching in crossfunctional courses.

METHOD OF DELIVERY

Visual Materials

The lack of international experience by most students, particularly at the undergraduate level, imposes more constraints on learning "by words alone" than in other areas of business. One approach suggested for alleviating this problem is the use of relevant films or videotapes. Frederick J. Truitt wrote an early piece on the validity of incorporating film materials into IB courses, and I did a follow-up piece several years later. Of course, these visual materials present problems, such as cost, length, obsolescence, and bias. Because these problems have been well-publicized, I shall not elaborate on them. I do, however, wish to make a case for easier access to relevant titles and descriptions. At one time there was a dearth of good materials, but this is no longer the case. One must depend either on being on distributors' mailing lists, a haphazard approach at best, or access the computerized retrieval systems for information on videotapes. The latter approach is difficult and time consuming because the systems do not index in a way that relevant international business materials can be easily identified. In essence, one needs to examine entries item-by-item. The same is true of the catalogues for the Vanderbilt News Archives that lend news segments from ABC, CBS, and NBC. Either a change in indexing systems or a clearinghouse of information would greatly facilitate the introduction of up-to-date visual materials in IB classes.

Cases

Paul W. Beamish's paper makes a strong point of using cases in IB courses, a position that I endorse. Cases share the same accessibility problem that is common for films and videotapes in that there is no central clearing house to list, describe, and disseminate materials prepared in multiple locations. I perceive another problem as well. Whereas there has been an increased availability of good visual materials, my impression is that international business case-writing has been waning. In addition to the cost and time problems that Beamish has outlined, I believe another factor has been the downplay of case-writing in many so-called research institutions, which once authored many cases. Cases are no longer considered as legitimate research within their promotion and tenure reviews. Instead, they are considered as part of the teaching component, and few people successfully pass through the promotion and tenure process in these schools on the basis of a teaching record. It is unlikely that these schools will change their review processes. However, now that AACSB is recognizing institutions with a predominately teaching mission, it may be possible to develop incentives whereby faculty therein become more active in case-writing.

Games and Exercises

Many computerized and noncomputerized games, exercise, and simulations are potentially useful as well for teaching such aspects of international business as global strategy, MNE-government negotiations, cultural sensitivity, export mechanics, and foreign exchange trading. These include Bafá-Bafá, Core II, Export to Win, Foreign Exchange Simulator, INTOPIA, Foreign Investment Negotiations Simulation, PC Globe, Meeting the Challenge of Foreign Expansion, Multinational Management Game, Upside Down, Where in the World, and World Games. There are, of course, equivalent games for teaching business in a domestic context; however, the international ones have perhaps a higher learning potential because more students lack any experience of operating in a global business environment. Precise information on the extent to which these are used is sketchy at best; however, my impression is that they are used very sparsely. Why is this?

First, there are faculty perceptions of pedagogical limitations, for example, that students spend an excessive amount of time learning the rules and that they concentrate on beating the games rather than making intelligent decisions. Although, these are valid concerns, I suspect that there is some faculty overreaction. Second, there are institutional impediments which tend neither to reward teaching efforts highly nor to permit resources to flow easily to the acquisition of tools needed for pedagogical experimentation and improvement. For example, my recent experience to acquire a license to put the Foreign Exchange Simulator onto the mainframe for use in a graduate course is probably typical for faculty at many institutions. The paperwork to justify the expenditure was substantial; and there was high uncertainty that a proposal would be approved. Overall, the process of application, approval, acquisition, installation, and testing took about two years. In turn, I'll use the exercise for only one day in class per semester. Therefore, one may question whether the time justified either the learning experience or the rewards to the instructor. Third, there is a perceived high risk of failure, particularly when one must learn the games and exercises from manuals that do not allow for clarification through two-way communication. Because instructors are often risk-aversive, they forego the use of tools in class that have not been personally demonstrated to them. In summary, we need to find means to minimize impediments, such as through easier funding for the acquisition of games, as well as workshops to learn how to use them efficiently.

CONCLUSION

Something that worries me about this conference and others I've attended on IB through the years is that we continue to be a group of internationalists talking to the already converted and rehashing much of what has been said for at least the last twenty years. The one theme that has had more emphasis in this conference, at least within the teaching dimension, has been the growing consensus of need for more language and area studies training for business students—a plea undoubtedly influenced by requirements from recent U.S. funding sources. If we are indeed to move toward more dependence on liberal arts faculty, why are there no tracks at this or other international business meetings to elicit their opinions, such as from language and area studies? For that matter, why are there no tracks for faculty who believe international infusion is unnecessary because of the universality of principles they teach? Why do we talk almost only of the resources liberal arts students can bring to our students rather than what we can offer them? Until we face up to these questions, I'm not sure how far we shall progress in implementing necessary changes.

Panel Discussion

The discussion that followed a brief summary of the four papers by Ricks and presentations by the four authors centered on five questions raised by members of the audience (shown in italics to distinguish them from the panel's answers). The following abbreviated report is presented primarily to provide the reader with a sense of the discussion that occurred during the panel session. It is not meant to be a verbatim report. It is made possible partly because of the taping of the session, and partly because of the excellent work done by three doctoral students at the University of South Carolina: Nick Athanassiou (now at Northeastern University), Jim Ondracek (now at Minot State University), and Debbie Francis (now at the University of Alabama). They were willing to undertake the demanding task of being the panel's reporters.

MARK COLLINS

Some countries do not distinguish between domestic and international business, they just teach international—they essentially teach with a global orientation. Which approach do the panel members feel is more appropriate—infusion or transfusion of international content into a domestic curriculum?

FRANKLIN R. ROOT

I think the infusion model is working. It is working fitfully, slowly, and agonizingly slowly for many of us. As I argue in my paper, IB is a metafield with a core that does not exclude other disciplines. Since it includes political economy, national culture, and international global management, there is a strong argument for covering it in one or more courses required of all students. Business schools are in a global competition for students and faculty and this will become increasingly evident. Unless they adapt to the world as it is rather than the world they think it is, they are going to lose out. But I also believe that schools and universities are buffered against the real world in all sorts of ways, tenure being one of them incidentally, that the process is going to be slow, fitful, and some schools are going to take the lead.

FREDERICK TRUITT

Infusion is slow and not effective. [He continued by then explaining the value of the transfusion model as presented in his paper.]

PAUL BEAMISH

The University of Western Ontario (UWO) has eliminated the international strategy elective and has transformed the core strategy course into an international course. UWO is doing this on an area by area basis.

JOHN DANIELS

At Indiana University, such classes as international marketing and international finance have been downplayed, since the core marketing and finance courses are sufficiently internationalized.

WILLIAM R. FOLKS

The University of South Carolina's approach is similar to Truitt's approach. I think that if there is a separate body of knowledge that cannot be accessed from the functional fields, and it ought to be taught, then you must have someone who is familiar with this body of knowledge introduce it in an appropriate fashion.

HUBERT GATIGON

Wharton is moving to the extreme and is converting courses to a global approach so that they have global orientations. Wharton even offers one marketing course in French. [*Editors' comment:* The University of Utah has also experimented with teaching business courses in other languages.]

DAVID A. RICKS

One effective way to internationalize a business course is to first assign the introductory chapter of whatever textbook is being used and then assign the international chapter of the book (a chapter which is usually near the back of the book). The international chapter is really also an introductory chapter—the introduction to the international dimensions. Then, when the other chapters of the book are assigned, the instructor is free to add international dimensions that are relevant to the chapter. The student is already introduced and prepared for whatever international dimensions are added.

JEAN BODDEWYN

I am concerned with the delivery of internationalized concepts in regular business courses. Does the functional area professor have the sophistication to teach the international dimension of his or her specialty effectively?

STEPHEN KOBRIN

Why should we be surprised at this development? Several of us have been working for twenty years for this to happen. Its been a goal to have the functional faculty become interested in the international dimensions and to introduce them into their classes.

FRED TRUITT

I agree with Steve's remarks, and personally pleased to see this happening.

JOHN DANIELS

I also agree with Steve, and further ask if any of us really understand all of the nuances of any topic. Some of us may do well in an international course, but may be a bit weak in its functional aspects. No course or instructor is perfect.

DAVID RICKS

It might be interesting to have an entire conference on the implications of successfully getting all business school faculty involved in the international dimensions of business. We could consider the implications in teaching, research, careers, publication opportunities, etc. The conference would have to address the positive and the potentially threatening or negative aspects.

JOHN MACKLIN

What is the panel's position on the use of courses outside the traditional business curriculum for IB enhancement? For example, language courses and courses in the humanities—particularly in social science disciplines.

JOHN DANIELS

I believe that these disciplines are valuable and we should encourage greater use of them. We should listen to what these people have to say, and even use them in our business courses.

[The prediction that we will use other disciplines more as we internationalize our business schools was supported by the panel's members, who stated that the more we use international concepts, the more we will find relevance across campus.]

RHY-SONG YEH

What emphasis should institutions put on geographic knowledge?

FREDERICK TRUITT

Schools cannot do everything, they must lead from their strengths.

DAVID A. RICKS

The easy answer to the question is that it depends on the institution. For example, the University of Hawaii puts emphasis on Asia and it make sense—that it fits into its strengths.

JOHN DANIELS

We do not really know the answer. Education has its tradeoffs. Do students need more knowledge about functional areas of business or about geographic areas of the world? Perhaps what we should concentrate on is providing them with the tools to find the geographic information on their own.

BERNIE WOLFE

I suggest that schools with particular interests might form strategic alliances. These groupings might then lead to new educational approaches.

ATUL PARVATIYAR

What should be taught in international business, what in the functional areas, and what in the international functional areas? As more and more students get international dimensions in their functional courses, for example, marketing, and take an introductory course in international business, then it is harder to know what is left to teach in such courses as international marketing. What is left? Who coordinates? Are all these courses needed?

JOHN DANIELS

I agree with your concern and acknowledge that it's a real issue.

DAVID A. RICKS

I think you make a good point. It is another example of the problems of successfully internationalizing business schools. But this kind of problem is not really totally new. We have always had the challenge—in such courses as international marketing—of trying to teach two different kinds of students. The international marketing course, for example, has usually been taken by some marketing majors and some international business majors. The marketing majors do well in the micro topics but struggle with the macro ones. The international majors are just the opposite. Inevitably, we end up duplicating some material for each group. Coordination helps reduce the problem, but it never fully goes away.

ATUL PARVATIYOR

One possible solution is to use more cases.

DAVID A. RICKS

I agree that cases might help, but even the use of cases requires coordination and information. At South Carolina, for example, regular marketing professors are now sometimes using international marketing cases in introductory marketing courses. It is no longer enough to simply canvas the international business faculty to discover which cases have been used. The entire business school needs to be contacted. Otherwise, you might end up trying to use a case your students have already discussed. This again is another example of the problems of successfully internationalizing a business school.

LEE RADEBAUGH

One way we are trying to solve the problem of repetition, is to modularize our program rather than seeking to create discrete courses to cut down on repetition. That is, we need to be more creative in curricula design.

DAVID A. RICKS

I think that is an excellent point and it has come out in several of the conference's sessions. We need to start thinking more flexibly, more creatively. We also need to start teaching more flexibly. For example, by components, by changing times, and by changing course length. We have got to recognize that the world has changed, and the old teaching models no longer may work.

REFERENCES

Aharoni, Yair. 1966. *The Foreign Investment Decision Process*. Boston, Mass.: Harvard Business School.

Bartlett, Christopher A. and S. Ghoshal. 1992. *Transnational Management*. Homewood, Ill.: Richard D. Irwin, Inc.

Buckley, Peter J. and M. Casson. 1976. *The Future of the Multinational Enterprise*. London: MacMillan.

Cateora, Philip R. 1966. *International Marketing*. Homewood, Ill.: Richard D. Irwin, Inc.

Caves, Richard E. 1982. *Multinational Enterprise and Economic Analysis*. Cambridge: Cambridge University Press.

Daniels, John D. and L. H. Radebaugh. 1976. *International Business Environments and Operations*. Reading, Mass.: Addison-Wesley Publishing Company.

Dunning, John H., ed. 1974. *Economic Analysis and the Multinational Enterprise*. New York: Praeger Publishers.

Dymsza, William A. 1972. *Multinational Business Strategy*. New York: McGraw-Hill, Inc.

Eiteman, David and A. I. Stonehill. 1973. *Multinational Business Finance*. Reading, Mass.: Addison-Wesley Publishing Company/

Farmer, Richard N. and B. M. Richman. 1966. *International Business: An Operational Theory*. Homewood, Ill.: Richard D. Irwin, Inc.

Fayerweather, John. 1960. *Management of International Operations*. New York: McGraw-Hill, Inc.

———. 1964. *International Marketing*. Englewood Cliffs, N.J.: Prentice-Hall, Inc.

Grosse, Robert and D. Kujawa. 1988. *International Business, Theory and Managerial Applications*. Homewood, Ill.: Richard D. Irwin, Inc.

Hennart, Jean-Francois. 1982. *A Theory of Multinational Enterprise*. Ann Arbor: University of Michigan Press.

Huebner, Grover G. and R. L. Kramer. 1930. *Foreign Trade: Principles and Practices*. New York: D. Appleton and Company.

Kindleberger, Charles P. 1983. *The Multinational Corporation in the 1980s*. Cambridge: the MIT Press.

Porter, Michael E., ed. 1986. *Competition in Global Industries*. Boston, Mass.: Harvard Business School Press.

Robinson, Richard D. 1967. *International Management*. New York: Holt, Rinehart and Winston, Inc.

Robock, Stefan and K. Simmonds. 1973. *International Business and Multinational Enterprises*. Homewood, Ill.: Richard D. Irwin, Inc.

Root, Franklin R. 1969. A Conceptual Approach to International Business, *Journal of Business Administration*. Summer: 18–28. Reprinted in John C. Baker, et al. 1988. *International Business Classics*. Lexington, Mass.: Lexington Books.

Terpstra, Vern. 1972. *International Marketing*. Hinsdale, Ill.: the Dryden Press.

Vernon, Raymond. 1971. *Sovereignty at Bay*. New York: Basic Books, Inc.

7

Responding to the Educational Challenges of International Business
Brian Toyne and Douglas Nigh

INTRODUCTION

Should all business graduates—Ph.D.'s, M.B.A.'s, B.B.A.'s, and their equivalents—receive IB training, or just a select few? Should IB's contributions to the body of business knowledge be viewed as augmenting the traditional business curriculum, or as necessitating a complete transformation and refocusing of the traditional business curriculum? More fundamental, perhaps, should the emphasis on business and IB training be replaced with an emphasis on education?[1]

A review of the literature on the "internationalization" of U.S. business education leaves the reader with the impression that IB is merely an extension of, or an enlargement of, the traditional business curriculum. In fact, most deans and their business discipline-oriented faculties view the internationalization task as simply the modification and augmentation of the business curriculum's content. They are not alone in thinking this way. Many IB scholars also approach the internationalization task making the same assumption, since they cling to what we described in chapter 1 as the extension or crossborder management paradigms of international business. That is, international business has traditionally been viewed merely as the adjustment or adaptation of domestic practice, and the management of environmental diversity from a parochial perspective (for example, Nehrt, Truitt, and

Wright 1970). However, from the perspective of the interaction paradigm, also described in chapter 1, we argue that this assumption is no longer valid. To understand and to respond correctly to the outcomes arising from the interaction of two or more business processes involves more than the mechanical application of a number of local business concepts and principles within a different context. It involves a deeper, yet more inclusive, understanding of the social embeddedness of business processes and their outcomes than is normally provided in the traditional business program. It also requires that the IB student receive an internationalized education that may or may not include elements of training.

When viewed within the context of the interaction paradigm, IB education involves a multidisciplinary, perhaps even an interdisciplinary, approach that places increased emphasis on the contributions made by the humanities and social sciences. Thus, the "internationalization" of the business curriculum, particularly the undergraduate curriculum, cannot be achieved simply by "infusing" existing business courses with an "international dimension," or by adding an assortment of IB courses designed to provide students with "international business competency" (see, for example, Cavusgil 1993). At its most fundamental level, "internationalization" involves "liberating" the business student from his or her parochial roots; a consequence of a parochial K–12 education coupled with and thus reinforced by local socialization. Thus, as shown in figure 7.1, the internationalized undergraduate and graduate business curricula need to be viewed as a formal educational process that starts in the freshman year and is designed to introduce the student to the richness and diversity of human activities and expression.

In increasing numbers, institutions and their business faculties are recognizing that business training must give way to business education.[2] Given the rapidity of economic, political, social, and technological change this is occurring locally, regionally, and globally. Thus, future business graduates need to be broadly educated, not narrowly trained. They must also be instilled with a desire to learn, adapt, and grow as the world about them changes, to appreciate the knowledge and understanding provided by the humanities and social sciences, and

to be capable of recognizing when additional education and training are needed. This is to say that they need to be intellectually equipped to appreciate, enter into, and enjoy lifetime learning that contributes to their professional growth in a world that is increasingly international. They also need to be taught the fundamentals of critical thinking, and to enjoy the intellectual virtues of disinterested curiosity, patience, intellectual honesty, exactness, industry, concentration, and doubt. These, as depicted in figure 7.1, need to permeate the entire educational experience from the undergraduate to the doctoral degree.

These claims flow from the conclusions presented in our companion book, *International Business Inquiry: An Emerging Vision* (1997). The first conclusion is that IB researchers and educators address two fundamental challenges that places them at the

Figure 7.1 Liberated Business Education: The Education/Training Process of Higher Education

Life-time learning (continue to learn, adapt, and grow)				
Self-learning (recognize when additional education/training is needed)				
Global awareness (become comfortable with a variety of cultures, and aware of global trends and their implications, etc.)				
Critical thinking (learn to enjoy the intellectual virtues—disinterested curiosity, patience, intellectual honesty, exactness, industry, concentration, and doubt)				
Ethical/moral thinking (develop an integrated value system which guides everything each of us does)				
Technological literacy (develop the ability to be comfortable with rapid technological change, and appreciate the implications that this change may have for society and for business)				
"Globalized" University Core and Electives Education	"Globalized" Business Core Education	"Globalized or Regionalized" Functional Education and Training	"Globalized or Regionalized" Graduate (MBA) Education and Training	"Globalized or Regionalized" Doctorate (PhD) Education and Training
Undergraduate Education and Training			Graduate Education and Training	

very core of business inquiry and thus at the very core of business education. The first challenge is to provide students with an understanding of why and how learned knowledge and behavior vary across cultures, and thus an understanding of the essential richness of our dissimilarities, our diversity. The second challenge is to provide students with (1) an understanding of why and how people with dissimilar learned knowledge, information, and experiences can purposefully and successfully interact, and (2) an understanding of what is likely to be the involved parties' outcomes (for example, how people and the organizations they create are changed as a consequence of international interaction, involvement, and reconciliation).

As a consequence of these two challenges, we further argued that a more emergent, a more inclusive view of international business practice must be central to our efforts as IB scholars and educators. If one truth became abundantly clear at the conference at which the papers in this book were first presented, it was that we need to actively enter into meaningful dialogue with our colleagues in the business disciplines, in other fields of human inquiry, and, importantly, in other parts of the world. In particular, we need to listen to what anthropologists, historians, political scientists, sociologists, psychologists, and many others have to say about why and how we interact as individuals, groups, and organizations, and how we may be changed as an outcome of these interactions. Unlike many of our parochial business colleagues, we need to deliberately and purposefully reach out and forge both research and educational linkages with our colleagues in the humanities and social sciences.

In this book, we built on these conclusions by asking well-known, respected IB scholars to comment on the issues and challenges they see confronting business school administrators and faculty as they go about fulfilling their responsibilities as educators. The papers and panel discussions included in this book, along with our observations in

chapter 1, are focused on two topics: the institutional attempts that have been made and are being made to encourage IB research and education, and the various approaches currently used to disseminate the body of knowledge that is the output of IB inquiry. We conclude this book by addressing a number of questions that center on the implications that our claims have for the future structuring of business schools and their educational responsibilities. More specifically, we focus on the future development of IB education assuming the centrality of IB inquiry in the study of business, and the need for constructive dialogue with members of the humanities and social sciences.[3] Who benefits from IB education and why? Is a new alliance needed between the business disciplines and IB? Is a new symbiosis to be sought between the humanities, the social sciences and business? And, as a finally question, is an interdisciplinary, multidisciplinary, or unidisciplinary pedagogy to be endorsed or emphasized? In addressing these specific questions it will also become clear that the way to foster a constructive, serviceable relationship among business scholars and IB scholars is to show how their interdependent scholarly activities jointly contribute to a richer, albeit more complex, understanding of the role and functioning of business in society and to the education and training of future practitioners and future business scholars.

PERSPECTIVES, NEW CHALLENGES: IB'S FUTURE LEADERSHIP ROLE IN BUSINESS EDUCATION

Leadership Role of IB Inquiry

The relative importance and significance of a field's contributions to the body of knowledge depends on the paradigms guiding its inquiry, and whether they are contractionary or expansionary in terms of the (restricted) body of knowledge upon which they depend for their insights.[4] The difference, of course, is extremely important when setting out to understand a "new" or ill-defined phenomenon, such as international business. For example, the extension and crossborder paradigms described earlier in chapter 1 are contractionary in that the contemporary knowledge used to speculate about and to explain the examined phenomena is parochial and increasingly specialized and thus increasingly separated from other bodies of knowledge; as noted earlier, the body of knowledge dealing with business is the result of separating the examined phenomena from the processes that created them, and then examining them from the researcher's blinkered point of view (for example, disciplinary, cultural, economic, political, and/or national perspectives). The emerging interaction paradigm, on the other hand, is expansionary since it speaks directly to the processes that create, and subsequently maintain the phenomena under examination (see, for example, Toyne and Nigh 1997). In this case, to understand or explain the phenomena under examination the researcher must have some understanding of the pool of contemporary knowledge, information, and experiences that are informing these processes and their outcomes (for example, firms, trade organizations). This, of course, is an investigative activity that is intrinsically expansionary. It is also multidisciplinary and international in scope.

The reason that the contributions of IB inquiry can successfully challenge the fragmented and increasingly specialized and narrowing developments occurring in the business disciplines is that its research agenda is focused on a major and singular source of business innovation.[5] While the research undertaken by IB scholars is mostly a result of the questions raised by the extension and crossborder management paradigms, and utilizes the theories, concepts, and methodologies supported by the introspective, parochially-inclined business disciplines, the phenomena studied are the product of the changes brought about as a result of the interaction and involvement of two or more socioculturally distinct business processes that are learning and changing in

dissimilar ways. That is, IB inquiry and education are at the frontier of business process change, and are focused on changes in organizational activities and practices that are generally ahead of those being examined by parochial business scholars. As stated in our companion book, it is not in the relatively quiet backwaters of national business processes that fundamental change is likely to occur. It is more likely to occur at the intersection of dissimilar business processes. Moreover, the reason that IB scholars are able to recognize the innovations that are a consequence of business process change is that they generally are not as specialized (blinkered) as their parochial counterparts, and thus relatively more open to the use of more comprehensive, multidisciplinary approaches and methodologies when examining business-related phenomena.[6]

The Educational Role of IB Inquiry

Business educators can select from at least three paradigms when making decisions about the role to be played by IB inquiry in the education and training of future business leaders. As shown in table 7.1, these paradigms draw on different, even additional bodies of knowledge when developing business programs, and invoke different needs in the way of pedagogy, crosscampus linkages, and external linkages.

IB education predicated on the extension paradigm can be viewed as the student's introduction to the conceptual limitations and boundaries of domestic business practices and activities as expressed in the metaphors, theories, concepts, and principles of the various business disciplines. At the core of this paradigm are four assumptions that have a pervasive, even dominant influence over what is taught and how it is taught:

1. business is business wherever practiced, only the contextual circumstances change;
2. business environments are converging as economic development occurs;

Table 7.1 Example of Internationalized Undergraduate Business Education: Paradigms, the Body of Knowledge, Pedagogical Approaches, and Linkages

Elements of the business education process	Paradigms Guiding the Educational Process		
	Extension	Cross-Border	Interaction
The body of knowledge	*Non-IB Majors:* Draws on domestic (U.S.) generated body of business knowledge. IB input based on a multi-disciplinary, albeit domestic (U.S.) interpretation of the international and foreign environments, and the operational issues of firms involved in international business. *IB Majors:* Assumes domestic business lore has universal value (validity), so draws on the domestic (U.S.) body of business knowledge. IB input based on a multi-disciplinary, domestic (U.S.) interpretation of the international and foreign environments, and on the body of knowledge dealing with the management of firms involved in foreign business activities. Recently, increased emphasis has been placed on a foreign language skill requirement.	*Non-IB Majors:* Draws on domestic (U.S.) generated body of business knowledge. IB input based on a multi-disciplinary, albeit domestic (U.S.) interpretation of the international and foreign environments, and on the operational issues of firms involved in international business. *IB Majors:* Assumes domestic business lore has universal value (validity), so draws on the domestic (U.S.) body of business knowledge. IB input based on a multi-disciplinary, domestic (U.S.) interpretation of the international and foreign environments, and on the body of knowledge dealing with the management of firms involved in foreign business activities and with the management of environmental diversity (i.e., within- and across-country). Recently, increased emphasis has been placed on a foreign language skill requirement.	*Non-IB Majors:* Assumes local business processes vary for socio-cultural, economic, political, historical, and learning differences. Students are provided an internationalized education, inclusive of university and business school's cores. Draws on a multi-country, multi-disciplinary body of business knowledge that presents multiple interpretations of the international and domestic environments, and on the operational issues of firms from different countries involved in foreign business activities. *IB Majors:* Builds on the "internationalized" body of business knowledge provided all business students by focusing attention on the body of knowledge with a focus on the management of firms involved in foreign business activities and with the management of environmental diversity (i.e., within- and across-country). Students are required to have a foreign language competency, and to complete at least one semester at a foreign institution.

Table 7.1 (*continued*)

Elements of the business education process	Paradigms Guiding the Educational Process		
	Extension	Cross-Border	Interaction
Pedagogical approaches used for IB topics	*Non-IB Majors:* A single "survey" course that introduces the student to the international business environment, and the operations of firms involved in international business. Various other business courses "internationalized," usually through modules or "infusion."	*Non-IB Majors:* A single "survey" course that introduces the student to the international business environment, and the operations of firms involved in international business. Various other business courses "internationalized," usually through modules or "infusion."	*Non-IB Majors:* All appropriate courses (general and business cores) are internationalized to insure students have a basic understanding of the roles that culture, socialization, history, etc. have on (1) world views and their tenuous influence on societies, individual behavior, etc.,(2) business processes and their outcomes (organizations), and (3) relations that cross national and cultural boundaries.
	Emphasis is on adding to the traditional content utilizing a reductionistic educational approach (i.e., by functional specialization).	Emphasis is on adding to the traditional content utilizing a reductionistic educational approach (i.e., by functional specialization).	Greater emphasis is placed on education, less on training, and utilizing a more integrated, holistic educational approach (i.e., multi-disciplinary, interdisciplinary approaches).
	IB Majors: Same as for non-IB majors	*IB Majors:* Same as for non-IB majors	*IB Majors:* Same as for non-IB majors
	Increasingly require students to participate in exchange programs, study-abroad programs, international business internships (domestic and foreign). Foreign language competency recommended.	Increasingly require students to participate in exchange programs, study-abroad programs, international business internships (domestic or foreign). Foreign language competency strongly recommended.	All IB students required to complete some of their education at a foreign institution, and take an internship at a foreign location. Foreign language competency required.

Required cross-campus linkages	*Non-IB Majors:* Satisfaction of the general educational core. Relations with other institutional departments tend to be passive. *IB Majors:* Same as for non-IB majors except for the development of formal relations with Foreign Languages, and possibly Political Science (e.g., International Relations) that seek predetermined programmatic inputs.	*Non-IB Majors:* Satisfaction of the general educational core. Relations with other institutional departments tend to be passive. *IB Majors:* Same as for non-IB majors except for the development of formal relations with Foreign Languages, and possibly Political Science (e.g., International Relations) that seek predetermined programmatic inputs.	*Non-IB Majors:* Business faculty highly involved in cross-campus partnerships to develop appropriate general education internationalized courses. Also, involved in cross-campus teaching coordination teams and the development of multi-disciplinary and interdisciplinary business courses. *IB Majors:* Same as for non-IB majors, plus highly involved in formal partnerships with Foreign Languages faculty, Political Science faculty (e.g., International Relations), Internationally-oriented Sociology faculty, and International Economics faculty in the development of multi-disciplinary and interdisciplinary IB courses.
Required external linkages	*Non-IB Majors:* Except for internships, practicums, and field studies, no external foreign linkages sought or required. *IB Majors:* Except for internships, practicums, and field studies that essentially focus on parochial firms' adjustments to foreign environments, no external foreign linkages sought or required.	*Non-IB Majors:* Except for internships, practicums, and field studies, no foreign external linkages sought or required. *IB Majors:* Except for internships, practicums, and field studies that essentially focus on parochial firms' adjustments to foreign environments, no external foreign linkages sought or required.	*Non-IB Majors:* To provide all business students with an internationalized education, students need to be encouraged to take semesters abroad. Thus, foreign external linkages need to be developed that includes all schools and departments of the university or college. *IB Majors:* Same as for non-IB majors except greater emphasis placed on the students acquiring cross-cultural skills needed for actually working in a culturally-diverse environment. This is probably best accomplished through internships at foreign-owned, foreign-based, and foreign-operated companies.

3. there is a propensity for business practices to converge (as an effect), and

4. since, to many, the U.S. is the most economically advanced country in the world, U.S. practice leads the world in new styles and new techniques.

Thus, the educational goal is to provide the student with an appreciation of environmental differences, and the adaptations and adjustments that (U.S.) enterprises need to make to be successful in other countries.

IB education predicated on the crossborder management paradigm builds on extension-based education by focusing additional attention on the problems and challenges associated with the movement of goods and capital across national boundaries and the simultaneous monitoring, control, coordination and integration of operations existing in two or more countries.

Vernon (1964, 9–10) suggests that the crossborder management paradigm highlights and addresses environmental peculiarities not addressed by parochial business education. That is, IB inquiry contributes in a distinctive way to the business training process by exploring such things as the strategic and operational issues and challenges associated with the management of environmental diversity. Notwithstanding these training-enriching contributions, this paradigm adds nothing in the way of uniqueness to the educational experience (for example, pedagogy, crosscampus linkages). It is a firm-level paradigm that does not challenge the core assumptions of the extension paradigm (that is, the universality of business practice, business practice convergence, and U.S. leadership in new business styles and techniques). Nor does it challenge the U.S.-bound traditions of business training that build upon a deterministic, reductionistic approach to the study of human behavior and experience.

Finally, IB education predicated on the emerging interaction paradigm, requires that the business student gain the knowl-

edge needed to appreciate and understand business as an integrated, multilevel, societal process. That is, it requires the student to see and understand business as a social as well as an economic process that varies from one setting to another because of the cumulative and interactive effects of culture-influenced learning at multiple levels (for example, supranational, national, industry, firm, individual). It also requires that the student be aware that when two or more business processes interact, each is affected, sometimes radically, sometimes subtly, but always differently. This requires the student to have some understanding of the sociocultural mechanisms at work in bringing about these different outcomes.

A Short Commentary on Business Education and Business Training

The U.S. business community's growing concerns regarding higher education are well known and have been widely discussed in the media, and among administrators of business schools.[7] The fundamental changes that are being sought at U.S. institutions in responses to both the concerns of business and also to the changing economic, political and social fabric of our society are also well known and widely discussed.

Organizations such as the AACSB–International Association for Management have changed or are changing their accreditation requirements to reflect a new reality (for example, managerial relevance, globalization, and cultural diversity of the workplace) that is already reflected in the education and training offered by foreign institutions. Business schools across the United States are responding in increasing numbers to the call to internationalize their missions, programs, and faculties. They are also seeking to develop new partnerships with business and government in order to respond to and shape developments in higher education (Anon 1994). While many of these responses are laudable, original, and daring, unfortunately, many lack vision, and are short-term, shallow solutions to longterm, fundamental problems, or are mere copies of what others have done, or are doing.

Students, regardless of their lifelong and career aspirations, will have to be prepared for a future that has become highly uncertain and geographically expansive. It should go without challenge that to prosper in such a future, academic excellence must be demanded of all students. At the same time, however, they need to be freed from their parochial constraints. That is, a parochial business education, whether defined in geographic, knowledge, or experiential terms, can no longer be assumed adequate. Nor can business students just be well *trained* in the skills and tools currently in use. They must be *educated* in ways that will enable them to appreciate, understand, and respond correctly to changing situations that are both fundamental, rapid, and to some extent, indeterminate. That is, they must enter into, and enjoy the benefits of lifelong learning.

At the same time, the business community must stop confusing training and education, and better understand the role played by colleges and universities in these activities. Restating Oakeshott (Fuller 1989, 95), an organization's activity, whether parochial or international, is ultimately bounded by the collective knowledge its people have of how to engage in this or that activity. Moreover, an organization's activity is not generally undertaken in pursuit of a single, predetermined goal. Essentially, an organization is an ongoing, evolving set of decisions and actions that are collectively indeterminate. What holds an organization together and gives it impetus and direction is not a known goal, but the knowledge its people (from, for example, CEO on down) have of how to successfully interact at any point in time with a changing environment. The decisions and actions of its people are not superimposed upon the knowledge they have, but emerge as a result of their current and future knowledge (that is, it is not knowing how to draw a cat ten different ways that is important, it is knowing how to see the cat).

The business educator's education and training roles should not be confused. As trainer, the objective is to provide the student with a time-bound bag of tricks of a trade that adds value for the employer. As educator, particularly within a university or college setting, the objective is to insure that the student acquires the discipline of study, a grasp of consequences, and a body of knowledge sufficiently broad as to provide meaning within situations that are in flux.

THE FUNDAMENTAL QUESTIONS TO BE ASKED OF IB EDUCATION AS LIBERATED BUSINESS EDUCATION

Who Benefits from IB Education?

In the mid-1950s, when the field of IB was in its embryonic stage, the purpose was to educate a select group of men and women destined for international managerial careers (Fayerweather 1986). In many ways, the task was a simple one since it was guided by the extension and crossborder paradigms that essentially suggested that all that was needed was an understanding of when and how to adapt or adjust domestic strategies and operations because of environmental differences.

Since then, however, several events have occurred that have transformed both the role of business education and the environment in which business takes place. Today, it is increasingly apparent that all business graduates, irrespective of their future occupations, require some understanding of how the world beyond their national borders may impact them professionally, economically, politically, and socially. For example, many of the accreditation changes implemented by the U.S.-based AACSB since 1991 are in response to the emergence of an interdependent world and the regionalization of markets (for example, EU or NAFTA).

The serious effects that the EU, NAFTA, MERCOSUR, and other similar regional agreements will have on the business processes of Asian, European, and Western Hemisphere countries cannot be ignored by their business schools. For example, in the very near future, most, if not all, graduates of U.S. business programs will need to have at least some understanding of foreign currency exchange and the business approaches and competitive practices of their Canadian and Mexican counterparts. In this regard, representatives of Canada, Mexico, and the United States have acknowledged that reform is necessary, and trilateral collaboration required (Anon 1994). Thus, in a very short span of thirty years, IB education has moved from being a peripheral activity undertaken for a small, select group of students to a central activity that should be undertaken for all business students.

Is a New Alliance To Be Sought between the Business Disciplines and IB?

As a result of the rapid and extensive globalization of business, and the shift of IB inquiry to a more central position within the business academy, the relationship between the various business disciplines and IB has also changed. Yet, while the business disciplines and IB recognize that "something" needs to be done to their business programs in terms of content, they have not recognized that their relationships have also changed, and changed in such fundamental ways that a realignment of formal and informal ties are needed.

Initially, IB inquiry was viewed as augmenting the parochial inquiry of the business disciplines. Many IB scholars viewed their investigatory roles as describing environmental differences using parochially inspired concepts, metaphors, and principles, and applying the paradigms and methodologies of their domestic-focused colleagues to test theories and concepts developed to explain local business practices. But, as noted earlier, the role of IB inquiry is moving from one of peripheral value to one of central value in the pursuit of generalized business knowledge. Simultaneously, the need for new theories and new methodologies has arisen because of the multidisciplinary, even interdisciplinary nature of IB inquiry (Toyne and Nigh 1997). It is this fundamental shift in the growing impor-

tance of IB inquiry, and IB's growing need for new theories that has not been recognized by parochial business scholars, many IB scholars, and the institutions with which they are associated.

Thus, we argue that a new alliance between the business disciplines and IB is mandated by the very nature, and the growing maturity of business inquiry. The problem, however, is in defining and structuring an alliance that must in some way accommodate the pronounced ontological (for example, given versus unfolding, local versus universal), epistemological (for example, deterministic versus holistic), and educational (unidisciplinary versus interdisciplinary) differences exhibited by the business disciplines on the one hand, and IB on the other hand.

In chapter 1, we recommended that a three-dimensional framework should be use by business schools when exploring the options they have for internationalizing their knowledge generation and knowledge dissemination activities. For example, we noted that their internationalization options were somewhat constrained by their faculties' research and teaching interests and the relative emphasis placed on teaching and research by the institution. That is, the school's environment limiting and molding its international options could be characterized using three dimensions: (1) the relative emphasis placed on teaching versus research dominance, (2) the type of teaching approach used (integrative versus unidisciplinary), and (3) types of research undertaken (discovery, integration, application). We further suggested that a school's internationalization options would be influenced by the IB paradigm adopted to focus and frame the questions ask of itself.

These recommendations were not exhaustive as the authors in later chapters demonstrate. When addressing the knowledge generation and dissemination responsibilities of the business school, an additional structural factor needs to be considered. As Folks, Bess, Miller, and Kujawa, just to name a few, suggest in their papers the internationalization option selected by a school is also dependent on the structure that has historically been used to house its IB-oriented faculty. For example, the IB faculty can be placed in an IB department, as done at the University of South Carolina, or dispersed throughout the school's various departments, as done at the University of Hawaii. As Daniels intimated in his paper, these extremes have different implications regarding the "culture" of the school, and the socialization and receptivity of its faculty regarding the treatment of the "international dimension." Thus, it needs to be recognized that these internal idiosyncrasies also impact a school's programmatic reforms, and the initiatives and incentives needed to "internationalize" the faculty.

Finally, the external loyalties of the faculty need to be addressed. Given the proclivity for faculty members to be loyal to their disciplines as defined by their academies, their reorientation toward, and their receptivity of, the growing importance of IB inquiry and its methodological needs may be extremely difficult. While the extension IB paradigm, and to a lesser extent, the crossborder management IB paradigm, do not pose serious threats to the entrenched disciplines, the multidisciplinary and interdisciplinary attributes of the interaction IB paradigm do pose serious threats to these disciplines. Thus, for a new alliance to be formed between the business disciplines and IB, the business school is confronted with two challenges. One challenge is the development of a structure and a set of mechanisms designed to encourage and support scholarly activities predicated on differing, even conflicting ontologies and epistemologies. The other challenge is the development of educational initiatives that while multidisciplinary and interdisciplinary do not threaten narrowly trained, highly specialized unidisciplinary faculty.

Is a New Symbiosis Needed between the Humanities, the Social Sciences, and Business?

Should business schools forge crosscampus linkages with the other schools and departments that make up their universities, and, if so, what should be the nature of these linkages? It is apparent that many of the authors whose papers appear in this book support the development of linkages designed to provide business students with predetermined skills and bodies of knowledge. For example, they recommend developing relations with such departments as economics, foreign languages, and political science for the purpose of providing specific inputs for their "internationalized" business programs (for example, oral and writing skills in a foreign language, political area studies). However, anthropology, history, psychology, and even sociology are less often mentioned, if at all, because of the domination of the economic-based extension and crossborder management paradigms.

Should these linkages be departmentally specific? Should they be program-specific (for example, for IB students)? Should they be content specific (that is, business faculty determine what is and is not important in the knowledge contributions to be made by the humanities and social sciences)? The answer to such questions, of course, depends only partially on the business and IB paradigms framing the relationship.

It also needs to be recognized that in recent years U.S. business education has reached an impasse. On the one hand, U.S. business education stands accused by the business community, by accrediting organizations, and by some business scholars of not producing graduates who can write clearly, speak eloquently, think critically, and who are inclined to seek lifelong learning.[8] On the other hand, it also stands accused of failing to provide graduates with skills that provide immediate "value-added" for their employers. The problem, of course, is a culture-bound tension between education and training; culture-bound in the sense that the U.S. business community and the U.S. business education community tend to see the problem as an "either-or" dilemma. As noted by Locke and by Grosse (chapter 2), the Germans and

Japanese do not face such a dilemma since they have resolved the education-training dilemma; their business communities are responsible for training, and their business education communities are responsible for education. They also appear to assume that education precedes training. Figure 7.1, of course, is predicated on the assumption that education is needed before training can occur, and that education continues after training has been initiated.

Unfortunately, the dilemma is more than just one of differing cultures. Oakshott (Fuller 1989, 104) has articulated the dilemma that now confronts business education, particularly a university-based educational delivery system such as in the United States as follows:

> A university will have ceased to exist when its learning has degenerated into what is now called research, when its teaching has become mere instruction and occupies the whole of an undergraduate's time, and when those who came to be taught come, not in search of the intellectual fortune but with a vitality so unaroused or so exhausted that they wish only to be provided with a serviceable moral and intellectual outfit; when they come with no understanding of the manners of conversation but desire only a qualification for earning a living or a certificate to let them in on the exploitation of the world.

Thus, the question of creating a new symbiosis between the humanities, the social sciences and business is an important one. The question suggests that the present arrangement of constituencies, goals, and structures are being challenged. Resolution of the education-training debate requires that we either bring the humanities and social sciences into the business school in order to provide the students with the specific skills and knowledge we deem important, or we develop a constructive, mutually rewarding dialogue with these schools. It should be noted, however, that simple dialogue that focuses exclusively on business education reform is insufficient. It is doomed to failure because of the artificial

barriers that have been erected, and the need to provide future business graduates with a multidisciplinary, explicitly integrated education.

Reductionism or Holism?

The question of whether a new set of relations are required between IB, the business disciplines, and the humanities and social sciences is partly the result of the traditional reductionistic approaches used (particularly in Anglo-American universities) to examine human activity, human creation, and human interaction. Following in the steps of the "hard sciences," the relentless drive by most business scholars, including IB scholars, has been to understand business as a collection of essentially mutually exclusive activities (for example, finance, marketing, personnel). This, of course, has had repercussions for business education, since the body of knowledge generated by the respective disciplines has been highly specialized and segregated. For example, when examining the need for a major revision to the Business Policy capstone course, Schendel and Hofer (1979) noted that students were expected to synthesize a number of departmentalized courses defining business without any substantive content and pedagogical support.

Put differently, is an interdisciplinary, multidisciplinary, or unidisciplinary pedagogy to be endorsed, or emphasized? The interaction paradigm described in chapter 1 suggests that at a minimum, a multidisciplinary approach is required. Essentially, IB inquiry has demonstrated that business cannot, and should not, be separated artificially from its social and political roots through their exogenization.

The need for multidisciplinary and interdisciplinary research is most strongly felt in the international business arena. While firms recognize the merits of specialization, they also recognize the compelling need for integration (Lawrence and Lorsch 1967). This dual requirement for differentiation and integration has been recognized by some IB scholars, primarily those who emphasize a more complex understanding of business (for example, Bartlett, Doz, and Hedlund 1990). Most IB scholarship that can be classified as integration has involved the study of firm-level activity that crosses borders and needs to handle environmental diversity. This, of course, is to be expected since this form of management is the most advanced and most complex, and demands an integrated (holistic) appreciation of a firm's totality as an organization (that is, the firm is more than an economic entity).[9]

We believe that a more integrative educational approach can be achieved even in institutions that stress unidisciplinary teaching and research. For example, in a tightly organized, unidisciplinary institution, a comprehensive, overarching educational framework can still be developed that insures that the business student receives systematic inputs at particular points in his or her program. The well-known M.I.B.S. program at the University of South Carolina was an innovative step in this direction.

CLOSING REMARKS

A major purpose of the conference and this book was to report on the ongoing discussion among IB educators concerning the objectives to be achieved by internationalizing business programs, and approaches to be used to achieve these objectives. However, in order to widen this discussion, we felt it was necessary to first identify and make clear the educational implications inherent in the various paradigms that focus and guide IB inquiry. Thus, we felt that it was important to pointed out in chapter 1 that the structure, content, and pedagogy employed to educate business students to the current and future implications of international business was based on an unvoiced (articulation) IB paradigm.

On the basis of our particular view of international business, we have made the following three assertions. First, IB inquiry is at the leading edge of organizational change in the global business community. As such, we believe that as IB educators, we have the responsibility to insure that all business students receive a liberated business education. This, in our opinion, requires that they receive an "internationalized" education that starts with their freshman year.

The second assertion we make is based on the unfolding globalization of business, and is that the primary role of the business school is to provide their students with a business education. Thus, we also believe that as IB educators, we have a responsibility to assume a leadership role in providing input, even direction, for this liberated education. That is, IB educators must take the initiative in reestablishing constructive, education-oriented linkages with the rest of the university. Moreover, the purpose of these linkages, we believe, should be the development of programs that have as their focus and purpose, the education of students to the richness and diversity of human expression, regardless of where it may occur.

The third and final assertion we make in this book is that the faculty governance of business institutions needs to be fundamentally changed. That is, we strongly recommend that a more accommodating research environment needs to be created, and a more interdisciplinary approach adopted for the education of future business students.

NOTES

1. As noted in chapter 1, we view training as preparing someone for a known "present" or anticipated future, thus emphasis is placed on acquiring skills and the regurgitation of "temporal facts." On the other hand, educating a person is to prepare them for an unknown, uncertain future by providing them with a fundamental, liberal, broadly-based understanding of human endeavor, the ability to think critically, and the ability to modify their learned behavior when circumstances change.

2. Notably, these assertions support the recent position adopted by the AACSB (Urban 1996).

3. The implications for IB research were explored fully in our companion book, *International Business: An Emerging Vision*.

4. This section draws heavily from Toyne 1997.

5. We elaborate on this argument in our companion book (Toyne and Nigh 1997).

6. This statement refers primarily to U.S.-based business scholars, thus the situation may be changing as the U.S. specialized functional disciplines become more interested in the study of international phenomena.

7. See Fuller 1989 and Adler 1982 for more developed discussions related to training and education.

8. Also see recent pronouncements by the AACSB (Urban 1996) and others such as the GMAC report (1990), Porter and McKibbin 1988, and Hayes and Abernathy 1980.

9. This principle draws on Toyne's "The Conceptual Frontiers of International Business" in Islam and Shepard (1996), 54.

REFERENCES

Adler, Mortimer J. 1982. *The Paideia Proposal: An Educational Manifesto.* New York: Collier Books.

Bartlett, Christopher A., Y. Doz, and G. Hedlund. 1990. *Managing the Global Firm.* New York: Routledge.

Cavusgil, S. Tamer, ed. 1993. *Internationalizing Business Education: Meeting the Challenge.* East Lansing: Michigan State University Press.

Fayerweather, John. 1986. A history of the Academy of International Business from infancy to maturity: The first 25 years. *Essays in International Business.* Columbia: The University of South Carolina Press, no. 6 (November):1–63.

Fuller, Timothy. 1989. *The Voice of Liberal Learning: Michael Oakshott on Education.* New Haven: Yale University Press.

Graduate Management Admission Council (GMAC). 1990. *Leadership for a Changing World: The Future Role of Graduate Management Education.* Los Angeles: Graduate Management Admission Council.

Hayes, R. H., and W. J. Abernathy. 1980. Managing our way to economic decline. *Harvard Business Review* 58 (July/August): 67–77.

Lawrence, Peter R., and J. W. Lorsch. 1967. *Organization and Environment.* Cambridge, Mass.: Harvard University Press.

Nehrt, Lee, Truitt, and R. Wright. 1970. *International Business: Past, Present, and Future.* Bloomington: Indiana University Press.

Report on the International Symposium on Higher Education and Strategic Partnerships. 1994. Vancouver: Inter-American Organization for Higher Education (Quebec) and the Open Learning Agency (Vancouver).

Schendel, Dan E., and C. W. Hofer. 1979. *Strategic Management: A New View of Business Policy and Planning.* Boston: Little, Brown.

Toyne, Brian. 1997. The conceptual frontiers of International Business. *Current Issues in International Business.* Ed. Islam, Iyanatul, and Shepherd. Aldershot, U.K.: Edward Elgar.

Toyne, Brian, and D. Nigh, ed. 1997. *International Business: An Emerging Vision.* Columbia: University of South Carolina Press.

Urban, Glen L. 1996. *A Report of the AACSB Faculty Leadership Task Force.* American Assembly of Collegiate Schools of Business, St. Louis, Mo.: March.

Vernon, Raymond. 1964. Comments. *Education in International Business.* Ed. Stefan H. Robock and L. C. Nehrt., Bloomington: Indiana University Press.

Contributing Authors

Yair Aharoni is rector of the College of Management in Tel Aviv, Israel. He was the J. Paul Sticht Visiting Professor of International Business at Duke University's Fuqua School of Business, and the Issachar Haimovic Professor of Business Policy at Tel-Aviv University. His visiting appointments include Boston University, the City University of New York, Columbia University, Harvard University, New York University, Stanford University, the University of California at Berkeley, and the IMEDE Management Development Institute, Switzerland. In addition to 23 books and manuscripts, he has authored over 70 articles and chapters, and more than 100 cases in Hebrew and English. Dr. Aharoni serves on the editorial boards of several journals, including *International Studies in Management* and *Management Science*. He is a Fellow of the Academy of International Business and the International Academy of Management, and is a member of the Academy of Management and the Strategic Management Society.

Jeffrey S. Arpan is the chairman and James F. Kane Professor of International Business at the University of South Carolina. He has also served as the initial director of USC's federally designated National Center for International Business Education and Research, USC's Center for Industry Policy and Strategy, and from 1982–1988 as director of USC's Master of International Business Studies (M.I.B.S.) program. After receiving his doctorate from Indiana University in 1971, he taught at Georgia State University for nearly a decade before moving to USC. He is currently the President of the Academy of International Business and member of the National Academic Advisory Panel to the AACSB Subcommittee on Internationalization. For more than two decades he has assisted business schools throughout the world to internationalize their curriculum and faculty.

Paul Beamish is the Royal Bank Professor of International Business at the Richard Ivey School of Business, University of Western Ontario, London, Canada. He served as editor-in-chief of the *Journal of International Business Studies (JIBS)* for the 1993–97 period. He is the author or coauthor of 20 books, and over 100 articles, contributed chapters, or published case studies. His books are in the areas of strategic management, international management, and especially joint ventures and alliances. His articles have appeared in *Academy of Management Review, Strategic Management Journal, JIBS, Journal of World Business,* and *Academy of Management Executive.* He has received Best Research awards from the Academy of Management, the Academy of International Business, the European Foundation for Management Development, and Administrative Sciences Association Canada. He worked for the Proctor & Gamble and Wilfred Laurier University before joining Ivey's faculty in 1987. At Ivey, he teaches on a variety of school programs and is founding director of Ivey's Asian Management Institute.

David Bess is currently professor of management and transportation and was dean of the College of Business Administration at the University of Hawaii at Manoa. He is the author of articles and two books in the area of maritime transportation. He has served as a speaker on the topic of the internationalization of management education at numerous academic and professional meetings. He chaired the Graduate Management Admission Council's International Management Education Task Force, and has served on the American Assembly of Collegiate Schools of Business International Affairs Committee for four years, the last two as Chair. In 1988, the Fellows of the Academy of International Business awarded him the International Dean of the Year Award.

David H. Blake is dean of the Graduate School of Management at the University of California, Irvine. He has specialized in the areas of business strategy and corporate planning, particularly the development of global business strategies for corporations. He also focuses on strategic leadership and leadership development. Blake is author or coauthor of seven books and monographs including "The Politics of Global Economic Relations," and "Managing the External Relations of Multinational Corporations." Blake has served on a number of State Department and United Nations advisory boards, and sits on the Boards of several companies. For many years he was a member of the Board of Directors of the American Assembly of Collegiate Schools of Business, and was president during 1996.

John D. Daniels is the E. Claiborne Robins Distinguished Chair at the University of Richmond. He was formerly professor of international business and director of the Center for International Business Education and Research (CIBER) at Indiana University. He is a past president of the Academy of International Business and is a dean of that organization's Fellows. He is also past chairman of the International Division of the Academy of Management and has served on the editorial boards of a dozen journals including the *Academy of Management Journal, Journal of International Business Studies,* and *Management International Review.* His coauthored text, *International Business : Environments and Operations*, is now in its eight edition.

Leonid Evenko is dean of the Graduate School of International Business at the Academy of National Economy for the Russian Federation. His distinguished career in education, government and business includes positions with the Institute of the USA and Canada of the Russian Academy of Science, with the Russian consulting company, INFORCOM, with the joint venture AMSCORT, and with several universities in the United States as visiting professor. He is on the Board of the Liberal Economic Society of Russia, and is editor-in-chief of the *Business and Management Journal.*

William R. Folks, Jr. is professor of international business and the Buck Mickel Distinguished Foundation Fellow in the College of Business Administration, University of South Carolina, Columbia. He also serves as director of the Center for International Business Education and Research, and is currently president of the National Association of University Centers for International Business Education and Research. His primary area of teaching and research is international finance, and he is the author of over forty research papers, published in journals such as *Financial Management, Management Science*, the *Journal of Financial and Quantitative Analysis*, and the *Journal of International Business Studies*, and is the coauthor of the text *International Dimensions of Managerial Finance.*

Robert Grosse is director of research at Thunderbird, the American Graduate School of International Management. He also directs the Thunderbird Center for International Business Education and Research, CIBER. He has a B.A. degree from Princeton University and a doctorate from the University of North Carolina, both in international economics. He has taught international finance in the M.B.A. programs at the University of Miami, the University of Michigan, and at the Instituto de Empresa (Madrid, Spain), as well as in various universities in Latin America. His research focuses on theory in international business, financial and managerial strategy of international firms, and business in Latin America. Recent studies include "Technology Transfer in Services to Latin America" (*Journal of International Business Studies,* 1996) and "The Future of the Global Financial Services Sector" (*The International Executive,* 1997).

Robert G. Hawkins is dean of the Ivan Allen College of Management, International Affairs and Policy at Georgia Institute of Technology. He was previously dean of the School of Management at Rensselaer Polytechnic Institute, and vice dean of the faculty of business administration at New York University. Earlier, he was chair of the International Business Area at NYU, and he served as president and dean of the Fellows of the Academy of International Business, and member and chair of the International Affairs Committee of the AACSB. Professor Hawkins' research interests are international finance and the economics of multinational firms.

Duane Kujawa is professor of management and international business at the University of Miami's School of Business Administration. He is also a founding faculty member of the university's Graduate School of International Studies, where he directs the International Business Program. Dr. Kujawa has served as associate dean for Graduate Programs and for International Programs and as interim dean at the School of Business Administration, and as co-Director of the university's Center for International Business Education and Research (CIBER). He is coauthor of *International Business: Theory and Managerial Applications* (1995) and is currently researching and writing on Chile's multinational companies. Dr. Kujawa has served as president of the Academy of International Business and is a Fellow of the Academy of International Business and of the International Academy of Management.

Robert R. Locke has been professor of economic and social history at the University of Hawaii at Manoa since 1974. He received his Ph.D. in history at UCLA in 1965, and has received numerous grants and awards during his career. He has had a summer fellowship at Meiji University in Tokyo in 1991, and has been visiting professor in the Department of Economics at Reading University, England, since 1989. Additionally, he has been Keynote Speaker at the 1991 and 1996 conference of the British Academy of Management. He has published several books: *The End of the Practical Man, Entrepreneurship and Higher Education in Germany, France, and Great Britain, 1880–1940* (Greenwich, Conn., 1984), *Management and Higher Education Since 1940: The Influence of America and Japan on West Germany, France, and Great Britain* (Cambridge, England, 1989), *Collapse of the American Management Mystique* (Oxford, 1998), and *History of Management Education* (Aldershot, England, 1998). Published articles include: "Business Education in Germany: Past Systems and Current Practice," *Business History Review* (1985), "Education and Entrepreneurship: An Historian's View," in M.B. Rose, ed., *Entrepreneurship, Networks and Modern Business* (Manchester, 1992), and "Higher Education and Management: Their Relational Changes in the 20th Century," in Nobui Kawabe and Eisuke Daito, eds., *Education and Training in the Development of Modern Corporations* (Tokyo, 1993).

Edwin L. Miller is associate dean for research and professor at the Michigan Business School, The University of Michigan. He received his Ph.D. degree from the University of California, Berkeley, and he joined the Michigan faculty in 1964. Professor Miller is affiliated with a variety of professional societies, and he has held leadership positions in many of them. In 1984, he was elected a Fellow in the Academy of Management. Over his career, he has continuously contributed to the body of knowledge on international human resource management. More than 40 articles on comparative and international management and personnel/human resource management have appeared in refereed journals. Professor Miller's research interests are currently focusing on management development and succession planning among multinational corporations.

Richard W. Moxon is associate professor of international business at the University of Washington. From 1990 to 1997, he was the director of the Center of International Business Education and Research (CIBER). He has been a visiting professor at Stanford University, INSEAD, the Instituto de Empresa in Spain, the Helsinki School of Economics, and the International University of Japan. Professor Moxon's research has focused on international competition and the management of multinational corporations. He has published articles on international strategic alliances and on foreign investment in developing countries. His work has appeared in journals and books, including the *Journal of International Business Studies*, and the *Handbook of International Management*. Professor Moxon received B.S. and M.S. degrees in Industrial Engineering from Stanford University, and the D.B.A. from Harvard University.

Douglas Nigh is associate professor of international business at the University of South Carolina (USC) and research director of USC's Center for International Business Education and Research. He received his Ph.D. in international management from UCLA and his M.B.A. and B.A. in economics Indiana University. He was recently elected program chair of the International Management Division of the Academy of Management. Also a co-founder of the International Association for Business and Society, he is past president of this association. His latest book is *Foreign Ownership and Foreign Direct Investment in the United Sates: Beyond Us and Them,* coedited with Douglas Woodward.

Lee H. Radebaugh is the KPMG Peat Marwick Professor of Accounting and International Business at Brigham Young University, director of the School of Accountancy and Information Systems, and codirector of the BYU-University of Utah Center for International Business. He received his D.B.A. from Indiana University in 1973. He previously taught at the Pennsylvania State University, and has been a visiting professor at New York University and at Escuela de Administracion de Negocios para Graduados (ESAN) in Lima, Peru. In 1985, he was the James Cusator Wards Visiting Professor at Glasgow University, Scotland. He was associate dean of the Marriott School of Management from 1984 to 1991. Professor Radebaugh's teaching interests are international business and international accounting. He is a member of the American Accounting Association, the European Accounting Association, an the Academy of International Business. He has written several books, including *International Business Environments and Operations* (8th Edition); *International Accounting and Multinational Enterprises* (4th Edition); and *Introduction to Business: International Dimensions*. He has also published several other books, monographs, and articles on international business and international accounting.

David A. Ricks is Distinguished Professor of International Business at Thunderbird, the American Graduate School of International Management. Prior to this appointment, he was professor of international business at the University of South Carolina, where he was the Wilbur S. Smith Distinguished Faculty Fellow, director of the Ph.D. program in international business, director of the faculty development in International Business Program, and acting director of the nationally recognized Masters of International Business Studies (M.I.B.S.) Program. Dr. Ricks has also served as editor-in-chief of the *Journal of International Business Studies*.

Franklin R. Root is Emeritus Professor of International Management at the Wharton School. A graduate of Trinity College, he has an M.B.A. from the Wharton School and a Ph.D. from the University of Pennsylvania. Professor Root has lectured in several countries in the fields of international business and economics. He has served on the faculties of the University of Maryland, the Copenhagen School of Economics and Business Administration, and the Naval War College. During the summer of 1970, he was regional advisor on export promotion for the Economic Commission for Latin America in Santiago, Chile. Professor Root has engaged in extensive consulting with business and government agencies. He has led several workshops, sponsored by the American Association of Collegiate Schools of Business (AACSB), to "internationalize" business schools in the United States. His latest books are *International Trade and Investment* (7th edition, 1994), and *Entry Strategies for International Markets* (1994). Professor Root is a past president of the Academy of International Business, and a past dean of the Fellows of the Academy of International Business.

Brian Toyne is the Emil C.E. Jurica Professor of International Business at St. Mary's University, San Antonio, Texas, Prior to 1993, he was professor of international business and Business Partnership Fellow at the University of South Carolina. He served as the acting director of the International Business Area, director of the Ph.D. international business program, and the series editor for *Critical Issues Facing Multinational Enterprises* for the University of South Carolina Press. His research is multidisciplinary, and has recently focused on the interface between human resource management and corporate business strategy within the international context. He is the author or coauthor of numerous books and articles dealing with international management and marketing issues. He has also served as the associate editor of the Journal of International Business Studies, and as vice president of the Academy of International Business. He has also served as chair of the International Management Division, Academy of Management.

J. Frederick Truitt joined the faculty at the Atkinson Graduate School of Management at Williamette University, Oregon, in Fall of 1991. His previous experience included working in the International Division of Cummins Engine Company and in the economics department of Sun Oil Company. He joined the faculty of the Schools of Business Administration at the University of Washington in 1969 and served as visiting lecturer or visiting professor at the Manchester Business School in England, Institute for International Studies and Training in Fujinomiya, Japan and the Helsinki School of Economics and Business in Finland. Dr. Truitt holds M.B.A. and D.B.A. degrees from Indiana University.

Bernard M. Wolf is professor of economics and international business at the Schulich School of Business, York University, Toronto, Canada, where he played a key role in developing the International M.B.A. Program. His research has focused on the role of MNEs in economic integration. Recent work deals with changes in the global automotive industry, a "Keynesian" approach to currency and exchange rate issues, and intellectual property legislation. He is the author of chapters in such books as *Multinationals and World Trade: Vertical Integration and the Division of Labour in World Industries*, edited by Mark Casson (1986); *Foreign Investment, Technology and Economic Growth*, edited by D. McFetridge (1991); and *Research in Global Strategic Management: Corporate Responses to Global Change*, edited by Alan Rugman and Alain Verbeke (1992). One of his articles, "Industrial diversification and internalization: some empirical evidence" has been reprinted in *International Investment*, edited by Peter J. Buckley (1990). He is on the editorial board of the *North American Journal of Economics and Finance*.

List of Conference Participants and Affiliation at the Time of the Conference

Nancy J. Adler
McGill University, Canada

Sanjeev Agarwal
Iowa State University

Yair Aharoni
Duke University and
Tel Aviv University

Li Aimin
World University Service of Canada
Canada

Michele Akoorie
University of Waikato, New Zealand

Harvey Arbelaez
Penn State at Harrisburg

Jeffrey S. Arpan
University of South Carolina

W. Graham Astley
University of Colorado at Denver

Catherine N. Axinn
Ohio University

Alan D. Bauerschmidt
University of South Carolina

Paul W. Beamish
University of Western Ontario, Canada

M. D. Beckman
University of Victoria, Canada

Jack N. Behrman
The University of North Carolina

H. David Bess
University of Hawaii at Manoa

Rabi S. Bhagat
Memphis State University

David S. Bigelow
Rensselaer Polytechnic Institute

Linda Bleicken
Georgia Southern University

Mark S. Blodgett
Georgia Southern University

Jean J. Boddewyn
City University of New York

Nakiye Boyacigiller
San Jose State University

Thomas L. Brewer
Georgetown University

T. J. Byrnes
University College-Dublin
Ireland

Claudio Carpano
University of North Carolina at Charlotte

Mark C. Casson
Reading University, United Kingdom

S. Tamer Cavusgil
Michigan State University

Joseph Cheng
Virginia Polytechnical Institute

Kang R. Cho
University of Colorado at Denver

Philip L. Cochran
The Pennsylvania State University

J. Markham Collins
The University of Tulsa

Kerry Cooper
Texas A&M University

Charles T. Crespy
Miami University

John D. Daniels
Indiana University

Sayeste Daser
Wake Forest University

Candace Deans
Wake Forest University

Jose de la Torre
University of California at Los Angeles

Yves L. Doz
INSEAD, France

Gunter Dufey
University of Michigan

John H. Dunning
Rutgers University and
Reading University, U.K.

John Dutton
North Carolina State University

Carl L. Dyer
University of Tennessee at Knoxville

Richard Edelstein
AACSB

William G. Egelhoff
Fordham University

Kamal Elsheshai
Georgia State University

Peter Enderwick
University of Waikato, New Zealand

Vihang R. Errunza
McGill University, Canada

Leonid I. Evenko
Graduate School of International Business
Academy of National Economy, Russia

Daniel C. Feldman
University of South Carolina

Karin Fladmoe-Lindquist
University of Utah

William R. Folks, Jr.
University of South Carolina

Robert Edward Freeman
University of Virginia

J. Stanley Fryer
University of South Carolina

Paul Garner
University of Alabama

Hubert Gatignon
University of Pennsylvania

Sumantra Ghoshal
INSEAD, France

Edward M. Graham
Institute for International Economics

Robert Grosse
University of Miami

Robert G. Hawkins
Georgia Institute of Technology

Xiaohong He
Quinnipiac College

Gunnar Hedlund
Stockholm School of Economics, Sweden

Jean-François Hennart
University of Illinois at Urbana-Champaign

Peter Herne
California State University at
Dominguez Hills

Michael A. Hitt
Texas A&M University

Robert H. Hogner
Florida International University

Hartmut H. Holzmuller
Wirtschaftsuniversitat Wien, Austria

Giorgio Inzerilli
Erasmus University Rotterdam,
Netherlands

James F. Kane
University of South Carolina

Sara L. Keck
Texas A&M University

Ben L. Kedia
Memphis State University

Thomas I. Kindel
The Citadel

Stephen J. Kobrin
University of Pennsylvania

Bruce M. Kogut
University of Pennsylvania

Christopher M. Korth
University of South Carolina

Charles O. Kroncke
University of Texas at Dallas

James A. Kuhlman
University of South Carolina

Duane Kujawa
University of Miami

Chuck Chun-Yau Kwok
University of South Carolina

Michael Landeck
Laredo State University

Paul Latortue
University of Puerto Rico, Puerto Rico

Donald R. Lessard
M.I.T.

Peter Li
University of Dubuque

Neng Liang
Loyola College

Robert R. Locke
University of Hawaii at Manoa

Thomas J. Madden
University of South Carolina

Zaida L. Martinez
South Carolina State University

Briance Mascarenhas
Rutgers University

Ann B. Matasar
Roosevelt University

Lars-Gunnar Mattsson
Stockholm School of Economics
Sweden

Mike McCormick
Jacksonville State University

Mark Mendenhall
University of Tennessee at Chattanooga

Edwin L. Miller
University of Michigan

Joseph M. Moricz
Robert Morris College

Allen Morrison
Thunderbird

Ken Morse
State University of New York at Geneseo

Murayama Motofusa
Chiba University, Japan

Richard W. Moxon
University of Washington

Douglas Nigh
University of South Carolina

Van N. Oliphant
Memphis State University

Lars Oxelhelm
School of Economics, Sweden

Terutomo Ozawa
Colorado State University

Jong H. Park
Kennesaw State College

Young-Ryeol Park
University of Illinois at Urbana-Champaign

Atul Parvatiyar
Emory University

Jean Pasquero
University of Quebec at Montreal, Canada

Karen Paul
Florida International University

Clotilde Pérez
University of Puerto Rico, Puerto Rico

William H. Phillips
University of South Carolina

Zhou Ping
World University Service of Canada
Canada

Juan Antonio Poblete
Universidad Gabriela Mistral, Chile

Rebecca Porterfield
University of North Carolina at Wilmington

Thomas A. Poynter
The Transitions Group, Inc.

Michael W. Pustay
Texas A&M University

Lee Radebaugh
Brigham Young University

Chandra Rajam
University of Colorado at Denver

Amitabh Raturi
University of Cincinnati

S. Gordon Redding
University of Hong Kong, Hong Kong

David A. Ricks
Thunderbird

Jonas Ridderstrale
Stockholm School of Economics, Sweden

Franklin R. Root
University of Pennsylvania

Kendall J. Roth
University of South Carolina

Stephen Salter
Texas A&M University

Rakesh B. Sambharya
Rutgers University

Saeed Samiee
University of South Carolina

Ravi Sarathy
Northeastern University

Karl Sauvant
The United Nations

Hans Schollhammer
University of California

Robert Scott
University of Maryland

Manuel G. Serapio, Jr.
University of Colorado at Denver

Michael Sibley
Loyola University

Harvey Starr
University of South Carolina

Donald L. Stevens
University of Colorado at Denver

Arthur Stonehill
Oregon State University

Jeremiah Sullivan
University of Washington

Michael A. Taku
Tarleton State University

Jorge Talavera
ESAN/CLADEA, Peru

Stephen Tallman
University of Utah

Hans B. Thorelli
Indiana University

Paz Estrella Tolentino
United Nations

Richard Torrisi
University of Hartford

Brian Toyne
University of South Carolina

J. Frederick Truitt
Willamette University

Gerardo R. Ungson
University of Oregon

M. Reza Vaghefi
University of North Florida

Kanoknart Visudtibhan
George Washington University

Hans-Gerhard Wachsmuth
University of South Carolina

Ingo Walter
New York University

Steven Wartick
University of Missouri at St. Louis

Nikolai Wasilewski
Pace University

D. Eleanor Westney
M.I.T.

Mira Wilkins
Florida International University

Bernard M. Wolf
York University
Canada

John Wong
Iowa State University

Donna J. Wood
University of Pittsburgh

Ryh-song Yeh
Pennsylvania State University

John Z. Yang
Fordham University

Srilata Zaheer
University of Minnesota

University of South Carolina Ph.D. Students in Attendance:

Name	Department (Major)
Allen Amason	Management
Nicholas Athanassiou	International Business
Steven Barnett	International Business
Ernie Csiszar	Management
Dorothee Feils	Finance
Debbie Francis	Management
Tom Hench	Management
Insik Jeong	International Business
James Johnson	International Business
Frances Katrishen	Management
Tomasz Lenartowicz	International Business
David McArthur	International Business
Martin Meznar	International Business
Andrew Morris	Management
Carolyn Mueller	Management
James Ondracek	Management
Emmanuel Onifade	Accounting
Joy Pahl	International Business
Chul-whi Park	International Business
Randolph Piper	Management
Russell Teasley	Management
David Thomas	Management
Bijoy Sahoo	Finance
Cheryl van Deusen	Management
Alan Wallace	International Business
Carolyn White	International Business
Yangjin Yoo	International Business

ABOUT THE EDITORS

Brian Toyne is Emil C. E. Jurica Professor International Business at St. Mary's University in San Antonio, Texas. He has served as a member of the editorial boards of several journals, including *Journal of Business Research* and *Journal of International Marketing*. The author or coauthor of many books, Toyne has also published more than 60 articles in such journals as *Academy of Management Review, Columbia Journal of World Business, Journal of International Marketing, Industrial Marketing Management,* and *Journal of International Business Studies*.

Douglas Nigh is associate professor of international business at the University of South Carolina and research director of USC's Center for International Business Education and Research. His latest book is *Foreign Ownership and the Consequences of Direct Investment in the United States: Beyond Us and Them,* coedited with Douglas Woodward. Professor Nigh is past president of the International Association for Business and Society and chair-elect of the International Management Division of the Academy of Management.